Environmental Technologies to Treat Sulfur Pollution

Integrated Environmental Technology Series

The *Integrated Environmental Technology Series* addresses key themes and issues in the field of environmental technology from a multidisciplinary and integrated perspective.

An integrated approach is potentially the most viable solution to the major pollution issues that face the globe in the 21st Century.

World experts are brought together to contribute to each volume, presenting a comprehensive blend of fundamental principles and applied technologies for each topic. Current practices and the state-of-the-art are reviewed, new developments in analytics, science and biotechnology are presented and, crucially, the theme of each volume is presented in relation to adjacent scientific, social and economic fields to provide solutions from a truly integrated perspective.

The *Integrated Environmental Technology Series* will form an invaluable and definitive resource in this rapidly evolving discipline.

Series Editor

Dr Ir Piet Lens, Sub-department of Environmental Technology, The University of Wageningen, P.O. Box 8129, 6700 EV Wageningen, The Netherlands. (piet.lens@algemeen.mt.wag-ur.nl)

Published Titles

Biofilms in Medicine, Industry and Environmental Biotechnology: *Characteristics, Analysis and Control* (ISBN: 1 84339 019 1)
Decentralised Sanitation and Reuse: *Concepts, Systems and Implementation* (ISBN: 1 900222 47 7)
Environmental Technologies to Treat Sulfur Pollution: *Principles and Engineering* (ISBN: 1 900222 09 4)
Phosphorus in Environmental Technology: *Principles and Applications* (ISBN: 1 84339 001 9)
Water Recycling and Resource Recovery in Industries: *Analysis, technologies and implementation* (ISBN: 1 84339 005 1)

Forthcoming titles

Pond Treatment Technology (ISBN: 1 84339 020 5)
Resource Recovery and Reuse in Organic Solid Waste Management (ISBN: 1 84339 054 X)

www.iwapublishing.com

Environmental Technologies to Treat Sulfur Pollution

Principles and Engineering

EDITED BY P. LENS AND L. HULSHOFF POL

IWA Publishing

Published by IWA Publishing, Alliance House, 12 Caxton Street, London SW1H 0QS, UK
Telephone: +44 (0) 20 7654 5500; Fax: +44 (0) 20 7654 5555; Email: publications@iwap.co.uk
Web: www.iwapublishing.com

First published 2000
Reprinted 2004

Final page layout by Bruce Goatly, Perfect Page, London, UK
Cover design by Hilary Basing, Cato Street Design, London, UK

British Library Cataloguing in Publication Data
A CIP catalogue record for this book is available from the British Library

Library of Congress Cataloging- in-Publication Data
A catalog record for this book is available from the Library of Congress

ISBN 1 900222 09 4

CONTENTS

PAGE

PREFACE [xi]
LIST OF CONTRIBUTORS [xiii]

PART I. THE SULFUR CYCLE

1	**The chemical sulfur cycle**	**1**
1.1	Introduction	1
1.2	Elemental sulfur and hydrophobic sulfur sols	4
1.3	Sulfide and polysulfides	10
1.4	Sulfites, thiosulfates, dithionites and dithionates	16
1.5	Polythionates and hydrophilic sulfur sols	20
1.6	Sulfuric acid and sulfates	23
1.7	Disproportionation of elemental sulfur in hot water	23
1.8	Organic derivatives of type R-S_n-R (organopolysulfanes)	25
1.9	References	28
2	**The geochemical sulfur cycle**	**33**
2.1	Introduction	33
2.2	Global sulfur reservoirs and fluxes	35
2.3	The sedimentary sulfur cycle	38
2.4	Sulfur isotopes	43
2.5	References	45
3	**The biological sulfur cycle**	**47**
3.1	Introduction	47
3.2	The inorganic sulfur compounds of the biological sulfur cycle	49
3.3	Dissimilatory oxidation of reduced sulfur compounds	51
3.4	Dissimilatory reduction of oxidised sulfur compounds	68
3.5	Sulfur assimilation	74
3.6	Acknowledgements	76
3.7	References	76

PART II. TECHNOLOGIES TO DESULFURISE RESOURCES

4	**Catalytic removal of sulfur from diesel oil by hydrotreating**	**87**
4.1	Introduction	87
4.2	Conventional HDS of diesel fuel	89
4.3	Deep HDS of diesel fuel	97
4.4	References	103
5	**Bioleaching of sulfide minerals**	**105**
5.1	Introduction	105
5.2	Rate-determining steps in bacterial oxidation of sulfide minerals	107
5.3	Chemical oxidation of sulfide minerals	112
5.4	Bacterial oxidation of sulfide minerals	115
5.5	Chemical versus bacterial oxidation: practical considerations	122
5.6	References	123

PART III. TREATMENT OF WATERS POLLUTED BY SULFUROUS COMPOUNDS

6	**Sulfur transformations during sewage transport**	**131**
6.1	Introduction	131
6.2	A general overview of sulfide formation and exchange in sewers	132
6.3	Factors affecting the formation of sulfide	133
6.4	Prediction of sulfide formation in pressure sewers	135
6.5	Prediction of sulfide formation in gravity sewers	137
6.6	An integrated aerobic/anaerobic approach for prediction of sulfide formation in sewers	139
6.7	Effects of sulfide formation in sewers	144
6.8	Methods for control of sulfide formation	148
6.9	References	149
7	**Biological treatment of organic sulfate-rich wastewaters**	**153**
7.1	Introduction	153
7.2	Anaerobic treatment of sulfate rich wastewaters	154
7.3	Biological sulfate removal	161
7.4	Novel developments and perspectives	164
7.5	References	169

8	**Biological removal of sulfurous compounds from inorganic wastewaters**	**175**
8.1	Introduction	175
8.2	Sulfur-rich wastewaters associated with mining activities	176
8.3	Prevention, containment and treatment of AMD	183
8.4	Biological remediation of inorganic wastewaters: constraints and opportunities	196
8.5	References	202
9	**Anaerobic of sulfate rich wastewaters: process modeling and control**	**207**
9.1	Introduction	207
9.2	Models	208
9.3	Sensors	213
9.4	Control	216
9.5	Fault detection and isolation	224
9.6	References	228
PART IV. TREATMENT OF GASES POLLUTED BY SULFUROUS COMPOUNDS		
10	**Survey of H_2S and SO_2 removal processes**	**237**
10.1	Introduction	237
10.2	Impurities in gases	240
10.3	Product gas requirements	242
10.4	Application of various gas-treating processes	243
10.5	Survey of H_2S removal processes	246
10.6	Survey of SO_2 removal processes	260
10.7	References	264
11	**Novel biological processes for the removal of H_2S and SO_2 from gas streams**	**265**
11.1	Introduction	265
11.2	High pressure natural gas desulfurisation by the Shell-Thiopaq process	266
11.3	Biological SO_2 removal from flue gas	273
11.4	References	279
12	**Biological treatment of gases polluted by volatile sulfur compounds**	**281**
12.1	Introduction	281

12.2 Biotechnological waste gas treatment methods 282
12.3 Practical aspects of reactor selection for VSC treatment 289
12.3 Biotechnological removal of H_2S 290
12.5 Biotechnological removal of organic sulfur compounds 294
12.6 References 300

13 Methods of odour measurement and assessment 305
13.1 Introduction 305
13.2 Methods of odour measurement 306
13.3 Odour assessment and prediction techniques 312
13.4 References 326

PART V. TREATMENT OF SOILS AND SEDIMENTS POLLUTED BY SULFUROUS COMPOUNDS

14 Treatment of solid materials containing inorganic sulfur compounds 329
14.1 Inorganic sulfur transformations in water / solid phase systems 329
14.2 Sulfur-containing materials with environmental concern 336
14.3 Treatment strategies 339
14.4 Sulfur in solid materials: novel aspects of environmental technology 348
14.5 References 350

15 Agricultural aspects of sulfur 355
15.1 Introduction 355
15.2 Sulfur demands of and release by agricultural crops 356
15.3 Sulfur supply to avoid deficiency 362
15.4 Phytoremediation of sulfur-enriched soils 366
15.5 References 370

PART VI. OTHER APPLICATIONS OF SULFUR CYCLE BIOCONVERSIONS IN ENVIRONMENTAL ENGINEERING

16 Biodegradation of sulfonated aromatic compounds 377
16.1 Introduction 377
16.2 Biodegradation 380
16.3 Biotechnological removal of aromatic sulfonates from wastewater 386
16.4 References 389

17	**Metal effects on sulfur cycle bacteria and metal removal by sulfate reducing bacteria**	**393**
17.1	Introduction	393
17.2	Metal inhibition and interaction	397
17.3	Heavy metal removal	401
17.4	Metalloid oxyanion reduction	406
17.5	References	409
18	**Interactions of the sulfur and nitrogen cycles: microbiology and process technology**	**415**
18.1	Introduction	415
18.2	Microbial aspects of oxidative processes	420
18.3	Technological aspects of sulfide oxidation	425
18.4	Microbial and technological aspects of reductive reactions	439
18.5	References	441
PART IV.	**PROBLEMS RELATED TO SULFUR CYCLE BIOCONVERSIONS**	
19	**Sulfur-storing bacteria and bulking of activated sludge**	**449**
19.1	Introduction	449
19.2	Filamentous organisms in activated sludge	452
19.3	Selection/growth of filamentous bacteria in activated sludge	456
19.4	Control of filamentous bulking by favouring the floc-forming population	458
19.5	References	464
20	**Sulfur problems in anaerobic digestion**	**467**
20.1	Introduction	467
20.2	Inhibition phenomena during anaerobic digestion of wastestreams with high levels of oxidised sulfur compounds	468
20.3	Techniques for quantification of sulfide toxicity on microbial populations involved in anaerobic digestion	472
20.4	Sulphide toxicity in anaerobic digestion	476
20.5	Effect of sulfide on process operation	482
20.6	References	485
21	**Corrosion and sulfur bacteria**	**491**
21.1	Introduction	491
21.2	Reduction reactions	492
21.3	Oxidation reactions	505

21.4 Acknowledgements 509
21.5 References 509

22 Recent developments in research on biogenic sulfuric acid attack of concrete 515
22.1 Introduction 515
22.2 Mechanism of biogenic sulfuric acid attack 516
22.3 Sulfide formation in sewers 519
22.4 Sulfuric acid production in sewers 528
22.5 Acknowledgements 536
22.6 References 536

Index 543

PREFACE

In recent years, natural as well as man-made ecosystems have been increasingly affected by sulfur pollution. Adverse effects of sulfur pollution are well known: acid rain; odour nuisance from polluted rivers, landfills or treatment systems; corrosion of steel and concrete; heavy metal and acidity release from oxygen-exposed sediments and mineral ores. In contrast to the enormous efforts dedicated to the abatement of pollution by organic compounds, nitrogen and phosphorus, treatment of sulfur-induced pollution has received rather limited attention. Nevertheless, a whole spectrum of environmental technologies has recently been developed. This book presents these technologies and informs the reader about their function and operation.

Sulfur is an essential element for life on Earth. Its redox conversions are of significance in the biogeochemical sulfur cycle. These sulfur conversions involve the metabolism of several different groups of bacteria (i.e. sulfate-reducing bacteria, phototrophic sulfur bacteria, colourless sulfur bacteria), specialised to use sulfur compounds in their different redox states. In Part I of this book, the reader is introduced to the (geo)chemistry of sulfur compounds and the organisms involved in their biotransformations.

In addition to its basic scientific importance, the sulfur cycle also plays a significant role in environmental technology and bioremediation. The subsequent parts of the book focus on a whole set of biological technologies developed to minimise the impact of, or completely overcome pollution by, sulfur compounds. In this respect, prevention of pollution is a first approach. Part II of the book describes possibilities for avoiding pollution by desulfurisation of resources, i.e. oil, coal and minerals. Subsequently, environmental engineering aspects of techniques to treat wastewaters (Part III), off gases (Part IV), and solid wastes and soils (Part V) are presented. The chapters of these sections present the chemistry, microbiology and process technology of various treatment options. Special attention is given to novel biotechnological processes. Some of these technologies allow a complete removal of sulfur from waste streams by its conversion into insoluble elemental sulfur. Thus sulfur can be reused, which fits well in the concept of sustainable environmental protection, which is based on remediation techniques focusing on resource recovery.

The conversions occurring in the sulfur cycle are not restricted to conversions involving the element sulfur. The sulfur cycle can also be utilised for the degradation of (recalcitrant) organic matter and the removal of heavy metals or nitrogen from wastewaters, soils and sediments. Part VI of this book covers practical aspects of these novel process applications.

Potential adverse effects of sulfurous compounds on the operation of established environmental technologies, i.e. sulfide inhibition, activated sludge bulking and corrosion, are addressed in the final part of the book. The chapters included in Part VII describe the mechanisms of how sulfur pollution can lead to process failure. Then, remediation techniques to overcome the various adverse effects are given.

Our main aim in assembling the chapters has been to present major, up-to-date reviews of the different technological applications. Each contributed chapter is presented on a stand-alone basis, so that the reader will find it helpful to consider only the theme of each chapter. There are, nevertheless, many connections between what may at first seem to be quite different fields of application. It was one of our purposes to draw out and emphasize these interdisciplinary linkages. For this reason, a comprehensive index is included to facilitate cross-reference. We hope that the work described in this book will inspire those working in the field, and will encourage those who are beginning to investigate the subject.

We thank all the contributors of this book for the enthusiastic and prompt compilation of their contributions. We also want to express our appreciation for the front cover design by Jasper Hulshoff Pol. Finally, we are grateful to Alan Click and Chris Purdon of IWA Publishing for their help and support in realising this book.

Wageningen, October 1999
Piet Lens
Look Hulshoff Pol

LIST OF CONTRIBUTORS

Anne Beeldens
Laboratory Reyntjens for Concrete Research, Department of Structural Engineering, University of Leuven, Kardinaal Mercierlaan 92, 3001 Leuven, Belgium

Nicolas Bernet
Laboratoire de Biotechnologie de l'Environnement, INRA, Avenue des Etangs, 11100 Narbonne, France

Mieke Boon
Delft University of Technology, Kluyver institute for Biotechnology, Julianalaan 67, 2628 BC Delft, The Netherlands

Thomas Brüser
Institut für Mikrobiologie und Biotechnologie, Universität Bonn, Meckenheimer Allee 168, D-53115 Bonn, Germany

Philippe M. Chazal
Laboratoire des Sciences de l'Eau et de l'Environnement, Faculté des Sciences, 123 Avenue A. Thomas, 87060 Limoges Cedex, France

Emer Colleran
Department of Microbiology, University College Galway, National University of Ireland, Galway, Ireland

Nele De Belie
Department of Agro-technology and Economy, University of Leuven, Kardinaal Mercierlaan 92, 3001 Leuven, Belgium

Henk Dijkman
Paques Bio Systems B.V., P.O. Box 52, 8560 AB Balk, The Netherlands

Dick H. Eikelboom
TNO Institute of Environmental Sciences, Energy Research and Process Innovation, P.O.Box 342, 7300 AH Apeldoorn, The Netherlands

Wilfried H.O. Ernst
Department of Ecology and Ecotoxicology of Plants, Faculty of Biology, Vrije Universiteit, De Boelelaan 1087, 1081 HV Amsterdam, The Netherlands

Richard A. Fenner
Water Engineering Research Group, University of Hertfordshire, Hatfield, AL10 9AB, UK

Jim A. Field
Wageningen University, Department of Agricultural, Environmental and Systems Technology, Sub-department of Environmental Technology. P.O. Box 8129, 6700 EV, Wageningen, The Netherlands

Per Halkjær Nielsen
Environmental Engineering Laboratory, Aalborg University, Sohngaardsholmsvej 57, 9000 Aalborg, Denmark

Oliver J. Hao
Department of Civil and Environmental Engineering, University of Maryland, College Park, MD 20742, USA

Veerle Herrygers
Department of Organic Chemistry, Faculty of Agricultural and Applied Biological Sciences, University of Gent, Coupure Links 653, B-9000 Gent, Belgium

Look W. Hulshoff Pol
Sub-Department of Environmental Technology, Agricultural University of Wageningen, 'Biotechnion', Bomenweg 2, P.O. Box 8129, 6700 EV Wageningen, The Netherlands

Thorkild Hvitved-Jacobsen
Environmental Engineering Laboratory, Aalborg University, Sohngaardsholmsvej 57, 9000 Aalborg, Denmark

Albert J.H. Janssen
Paques Bio Systems B.V., P.O. Box 52, 8560 AB Balk, The Netherlands

Guido Janssen
Paques Bio Systems B.V., P.O. Box 52, 8560 AB Balk, The Netherlands

Barrie Johnson
School of Biological Sciences, University of Wales, Bangor, LL57 2UW, UK

Jan A. Lagas
Stork Engineers and Contractors B.V., Radarweg 60, P.O. Box 58026, 1040 HA, Amsterdam, The Netherlands

Juan M. Lema
Department of Chemical Engineering, University of Santiago de Compostela, Avda. das Ciencias s/n, 15706 Santiago de Compostela, Spain

Piet N.L. Lens
Sub-Department of Environmental Technology, Agricultural University of Wageningen, 'Biotechnion', Bomenweg 2, P.O. Box 8129, 6700 EV Wageningen, The Netherlands

Brenda J. Little
Naval Research Laboratory, Stennis Space Center, Mississippi 39529, USA

Jack J. Middelburg
Netherlands Institute of Ecology, P.O. Box 140, 4400 AC Yerseke, The Netherlands

Rene Moletta
Laboratoire de Biotechnologie de l'Environnement, INRA, Avenue des Etangs, 11100 Narbonne, France

Joke Monteny
Lab. Magnel for Concrete Research, Department of Structural Engineering, University of Gent, Technologiepark Zwijnaarde 9, 9052 Gent, Belgium

Jacob A. Moulijn
Faculty of Chemical Technology and Material Science, Delft University of Technology, P.O. Box 5045, 2600 GA Delft, The Netherlands

Vincent O'Flaherty
Department of Microbiology, University College Galway, National University of Ireland, Galway, Ireland

Francisco Omil
Department of Chemical Engineering, University of Santiago de Compostela, Avda. das Ciencias s/n, 15706 Santiago de Compostela, Spain

Robert K. Pope
Naval Research Laboratory, Stennis Space Center, Mississippi 39529, USA

Richard I. Ray
Naval Research Laboratory, Stennis Space Center, Mississippi 39529, USA

Hank R. Reinhoudt
Faculty of Chemical Technology and Material Science, Delft University of Technology, P.O. Box 5045, 2600 GA Delft, The Netherlands

Erik Smet
Trevi, Sint-Denijslaan 317, B-9000 Gent, Belgium

Ralf Steudel
Institut für Anorganische und Analytische Chemie, Technische Universität Berlin, D-10623 Berlin, Germany

Jean-Philippe Steyer
Laboratoire de Biotechnologie de l'Environnement, INRA, Avenue des Étangs, 11100 Narbonne, France

Richard M. Stuetz
School of Water Sciences, Cranfield University, Cranfield MK43 0AL, UK

Luc Taerwe
Laboratory Magnel for Concrete Research, Department of Structural Engineering, University of Gent, Technologiepark Zwijnaarde 9, 9052 Gent, Belgium

Nico C.G. Tan
Wageningen University, Department of Agricultural, Environmental and Systems Technology, Sub-department of Environmental Technology, P.O. Box 8129, 6700 EV, Wageningen, The Netherlands

Richard Tichý
Institute of Landscape Ecology and University of South Bohemia, Faculty of Biological Sciences, Na Sádkách 7, 370 05 České Budějovice, Czech Republic

Hans G. Trüper
Institut für Mikrobiologie und Biotechnologie, Universität Bonn, Meckenheimer Allee 168, D-53115 Bonn, Germany

Dionys Van Gemert
Laboratory Reyntjens for Concrete Research, Department of Structural Engineering, University of Leuven, Kardinaal Mercierlaan 92, 3001 Leuven, Belgium

Herman Van Langenhove
Department of Organic Chemistry, Faculty of Agricultural and Applied Biological Sciences, University of Gent, Coupure Links 653, B-9000 Gent, Belgium

Willy H. Verstraete
Laboratory of Microbial Ecology and Technology, Faculty of Agricultural and Applied Biological Sciences, University of Gent, Coupure Links 653, 9000 Gent, Belgium

Elke Vincke
Laboratory of Microbial Ecology and Technology, Faculty of Agricultural and Applied Biological Sciences, University of Gent, Coupure Links 653, 9000 Gent, Belgium

1

The chemical sulfur cycle

Ralf Steudel

1.1 INTRODUCTION

1.1.1 Oxidation states and redox potentials

Chemically, sulfur is one of the most interesting but also one of the most difficult elements; geochemically it is abundant, and biochemically it is most important. The complexity of sulfur chemistry originates from the many oxidation states sulfur can assume, as well as from the tendency of sulfur in the zero oxidation state to catenate, forming chains and rings of an astonishing variety. In addition, more than 200 years of scientific research on sulfur and its compounds has resulted in a vast body of literature which cannot easily be searched for reliable information. Moreover, this literature is full of errors and contradictions because earlier workers, not having the methods available that are standard today, often made claims that have not always been confirmed subsequently. However, reliable reviews are available: above all, the many volumes of the *Gmelin Handbook of Inorganic Chemistry* in which the chemical literature is critically and exhaustively

Table 1.1. Oxidation states of sulfur in common compounds

Oxidation state	Compounds
–2	dihydrogen sulfide H_2S, hydrogen sulfide ion HS^-, sulfide ion S^{2-} as in FeS
–1	disulfane H_2S_2, disulfide S_2^{2-} as in pyrite FeS_2
0	elemental sulfur S_n, organic polysulfanes R-S_n-R
+1	dichlorodisulfane Cl-S-S-Cl
+2	sulfur dichloride SCl_2, sulfoxylate SO_2^{2-}
+3	dithionite $S_2O_4^{2-}$
+4	sulfur dioxide SO_2, sulfite SO_3^{2-}
+5	dithionate $S_2O_6^{2-}$, sulfonate RSO_3^-
+6	sulfur trioxide SO_3, sulfate SO_4^{2-}, peroxosulfate SO_5^{2-}

evaluated. On sulfur and its compounds 22 volumes have appeared, dating from 1939 to 1996 and covering literature up to 1991! Unfortunately, no further volumes will be produced. Other reliable reviews have been published by Holleman-Wiberg (1985), Karchmer (1970), Müller and Krebs (1984), Nickless (1968), Schmidt and Siebert (1973), Steudel (1982; 1998) and Szekeres (1974).

In Table 1.1 the nine *oxidation states* of sulfur are illustrated by typical examples. Most of them play a role in aqueous systems in which redox reactions occur either as a result of microbiological activity or simply following the thermodynamics of the system in non-enzymatic reactions. However, chemical systems are not always composed according to the requirements of thermodynamics. High activation enthalpies may keep certain reactions from proceeding at ambient temperatures, resulting in a chemical composition sometimes far from equilibrium.

The equilibrium composition of an aqueous system containing sulfur and oxygen is shown in the Pourbaix diagram in Figure 1.1. Depending on the redox potential, the pH value, the temperature and the overall concentration of sulfur, the relative stability areas of sulfide (HS^-), elemental sulfur (S_8), and sulfate (SO_4^{2-} or HSO_4^-) are shown (Garrels and Naeser 1958; Williamson and Rimstidt 1992). The different areas of this diagram indicate which species will predominate at a given potential and pH value. As lower the overall sulfur concentration as smaller the existence area of elemental sulfur becomes. Sulfite, thiosulfate and other sulfur oxyanions (e.g., polythionates) never predominate, regardless of pH and potential. In other words, these species exist in water only under non-equilibrium conditions or as minority species.

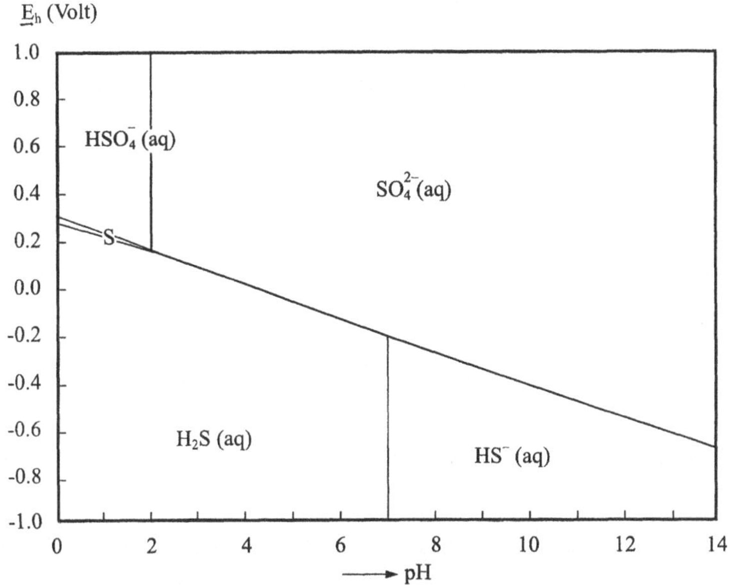

Figure 1.1. Pourbaix diagram for the binary system sulfur/oxygen in water at 25 °C and 1.013 bar with the sum of the activities of all sulfur-containing ions equal to 0.1 mM.

1.1.2 Catenation of sulfur atoms

As in hydrogen sulfide H–S–H and dimethyl sulfide CH_3–S–CH_3 sulfur atoms can form two covalent bonds with other atoms or with itself to form chains like –S–S–S– of practically unlimited length ("catenation"). These chains may be terminated by single atoms like H or Cl, by groups like CH_3 or SO_3H, or may "bite their own tail" forming rings of various sizes. Such species are listed in Table 1.2. A special case are the polysulfide anions in which the chains are terminated by negatively charged sulfur atoms (S^-) which are iso-electronic with Cl atoms. Therefore, polysulfide anions ^-S–S_x–S^- are iso-electronic with dichlorosulfanes Cl–S_x–Cl.

Table 1.2 gives those values of *n* which have been determined in compounds isolated in pure form (column 2) or which have been detected in mixtures by high-performance liquid chromatography (HPLC) or by nuclear magnetic resonance (NMR) spectroscopy (column 3). From these data it is obvious that there is no limitation to the values of *n*. It is just that the preparation of the high-molecular species becomes more and more difficult because the solubility decreases with increasing values of *n*.

Table 1.2. Ring sizes and chain-lengths n in sulfur-rich compounds

Species	Formula	Isolated as pure compounds	Larger species detected in mixtures only (method)
Homocycles	S_n	$n = 6...20$	$n = 21...25$ (HPLC)
Polysulfanes	$H-S_n-H$	$n = 1...8$	$n = 9...35$ (^1H NMR)
Polysulfides	S_n^{2-}	$n = 1...8$	–
Organic polysulfanes	$R-S_n-R$	$n = 1...13$	$n = 14...35$ (HPLC)
Polythionates	$^-O_3S-S_n-SO_3^-$	$n = 1...4$	$n = 5...22$ (ion chromatography)
Polysulfur	S_n ("S_μ" or "S_∞")	$n > 10^5$	–

Polymeric sulfur S_μ is insoluble in all solvents (except liquid sulfur) and therefore is considered to consist of very long chains and/or very large rings.

All compounds containing S–S bonds are light-sensitive because on irradiation with wavelengths shorter than 420 nm, homolytic cleavage of these bonds occurs and radicals are formed which trigger chain-reactions resulting in mixtures of compounds (Steudel *et al.* 1989c). Often daylight is sufficient to induce photochemical decomposition of sulfur-rich compounds.

1.2 ELEMENTAL SULFUR AND HYDROPHOBIC SULFUR SOLS

1.2.1 Sulfur allotropes

Isolated sulfur atoms (S_1) have a very high enthalpy of formation (279 kJ mol^{-1}) and therefore cannot exist at ambient temperatures. However, sulfur atoms are known as transient species from high-temperature vapours, from photolysis experiments and from electrical discharges. Therefore, equations like $H_2S + O \rightarrow H_2O + S$ are unreasonable since neither oxygen nor sulfur exist as atoms under standard conditions. At ambient pressure and temperature elemental sulfur exists as rings of different sizes (S_n) or as polymeric chains of high molecular mass (S_μ) (Steudel 1982). Of all these allotropes, the orthorhombic α-S_8 is most stable thermodynamically (enthalpy of formation is zero by definition). It is the main constituent of commercial sulfurs and can be prepared in a pure form by recrystallization of commercial sulfur from organic solvents (Steudel and Holz 1988). Pure α-S_8 is of greenish-yellow colour and easily soluble in carbon disulfide (CS_2) (33 weight-% at 25°C). S_8 is less soluble in methylene chloride CH_2Cl_2,

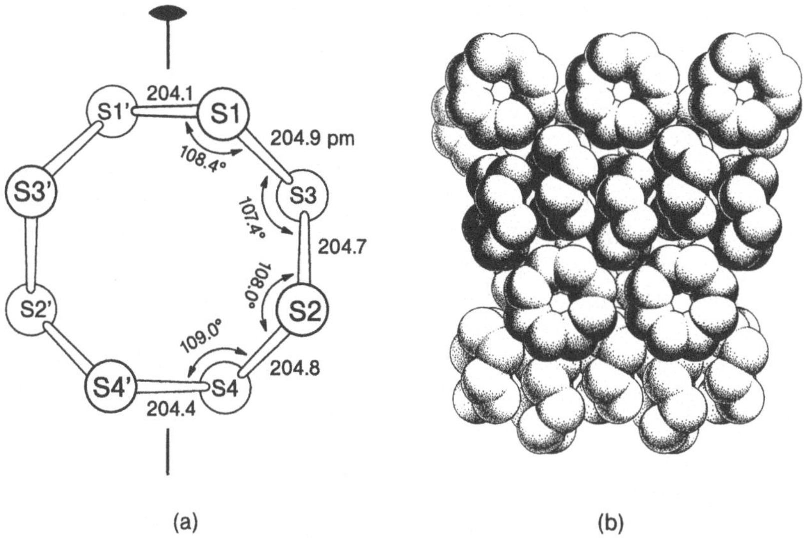

Figure 1.2. Crystal and molecular structure of orthorhombic α-S_8, the stable allotrope of sulfur at 20 °C/1.013 bar. (a) Bond lengths and bond angles. The two-fold rotation axis is indicated. (b) Molecular packing.

chloroform $CHCl_3$ and toluene $C_6H_5CH_3$ from which solvents it can nevertheless be recrystallized. The other components of commercial sulfur are polymeric S_μ and traces of S_7 which both originate from liquid sulfur from which all commercial sulfurs are produced. It is the small S_7 content that causes the "bright yellow" color of commercial sulfur (Steudel and Holz 1988).

The molecular structure of S_8 is shown in Figure 1.2. This species crystallizes as α-S_8 at temperatures below 96 °C, as monoclinic β-S_8 in the region 96-120 °C, and as monoclinic metastable γ-S_8 (obtained from certain solvents). The crystal structures of all these allotropes are quite complex as can be seen from the structure of α-S_8 in Figure 1.2.

By suitable chemical syntheses sulfur rings with between 6 and 20 atoms in the ring can be prepared and the following species have been obtained as pure substances (Steudel 1987; Steudel *et al.* 1998):

S_6 S_7 S_8 S_9 S_{10} S_{11} S_{12} S_{13} S_{14} S_{15} S_{18} S_{20}

These molecules form yellow to orange-yellow solids which are soluble in carbon disulfide and, to a lesser degree, in certain other organic solvents.

The solubility decreases with increasing ring size and with increasing symmetry of the molecule. The structures of some of these homocycles are shown in Figure 1.3. The melting points are found in the range 39-148 °C, with S_7 having the lowest and S_{12} the highest value. α-S_8 melts at 115 °C and β-S_8 at 120 °C. All species convert rapidly to the more stable molecule S_8 on heating to temperatures close to the melting points. In the case of S_6, S_7, S_9, S_{11}, S_{13} and S_{15} this decomposition occurs slowly at room temperature. All sulfur allotropes are hydrophobic. The densities of the crystalline materials are in the range of 1.9-2.2 g cm^{-3}.

All forms of elemental sulfur are practically insoluble in water at 20 °C, but neutral surfactants like Triton X-100, Tergitol 7 or sodium dodecylsulfate added to the mixture increase the solubility by several orders of magnitude provided the surfactant concentration exceeds the critical value needed for micelle formation. It is believed that the sulfur molecules are dissolved inside the micelles (Steudel and Holdt 1988).

If S_8 is dissolved in highly polar solvents like methanol or acetonitrile at 20 °C an equilibrium between S_8, S_7 and S_6 is established within 24 h. At a total sulfur concentration of 0.15 mg L^{-1} the three species are present in a molar ratio of 1000:8:3 as determined by HPLC analysis (Tebbe *et al.* 1982).

1.2.2 Liquid sulfur

Liquid elemental sulfur is an industrial product of outstanding importance, being produced on a huge scale (ca. 35 million tonnes annually). It is used mainly for the manufacture of sulfuric acid but also for rubber vulcanization. When α-S_8 is heated as a powder it transforms at 96 °C into β-S_8 which melts at 120 °C. In contrast, single crystals of α-S_8 do not transform easily to β-S_8 and therefore melt at 115 °C. Only in the very beginning do these melts consist entirely of S_8 molecules, because a decomposition reaction takes place which finally results in a complex mixture of molecules:

$$S_8 \rightleftharpoons 8/n\ S_n\ (n = 6, 7, 9, \ldots S_\mu) \tag{1.1}$$

Depending on the purity of the system it takes up to 12 h to establish the equilibrium composition (Mäusle and Steudel 1981; Steudel and Mäusle 1981). The composition has been determined by HPLC as well as by Raman spectroscopy; some of the HPLC data are given in Table 1.3 (Steudel *et al.* 1985). Apart from the main constituent S_8, the liquid also contains the more reactive rings S_7, S_6, S_9, etc. as well as polymeric S_μ which can be recovered

Table 1.3. Molecular composition of liquid sulfur after equilibration at various temperatures (values in mass-%)

Species	Temperature			
	116 °C	122 °C	159 °C	220 °C
S_8	93.6	93.1	83.4	54.3
S_7	3.1	3.3	5.2	4.6
S_6	0.5	0.6	0.9	0.9
S_9	0.3	0.4	0.6	0.6
S_{10}	0.1	0.1	0.2	0.2
S_{12}	0.4	0.4	0.5	0.4
S_x	1.8	1.9	6.2	4.8
S_∞	0.2	0.2	3.0	34.2

by quenching the melt in liquid nitrogen, dissolving the yellow powder-like product in carbon disulfide at 20 °C and filtering off the insoluble S_μ. The clear solution was then analysed by HPLC to produce the data in Table 1.3.

The additional species S_n ($n \neq 8$) lower the freezing point (triple point) of liquid sulfur to 115 °C (from 120 °C) corresponding to a molar concentration of non-S_8 molecules of 5%. The density of this liquid at 115 °C is 1.80 g cm^{-3}.

Even when the melt is *slowly* cooled to 20 °C some of the reactive rings survive and are built into the crystal structure of S_8 as "lattice defects" (Steudel and Holz 1988).

The concentration of rings larger than S_{12} (termed as S_x) was calculated as difference to 100%. S_∞ is used as a symbol for polymeric sulfur which is unsoluble in carbondisulfide.

1.2.3 Gaseous sulfur

Sulfur vapour exists either in equilibrium with solid and/or liquid sulfur (saturated vapour) or as a single phase under a pressure lower than the saturation pressure (unsaturated vapour). In a vacuum sulfur sublimes at temperatures below its melting point, a process used for purification. The molecular composition of sulfur vapours can be determined conveniently by mass spectrometry. At temperatures of up to 1000 °C sulfur vapours consist of all molecules S_n with n ranging from 2 to 8 which are in equilibrium with each other (Rau *et al.* 1973):

$$S_8 \rightleftharpoons 8/n\, S_n \qquad (1.2)$$

The lower the pressure and the higher the temperature, so the smaller the average molecular size becomes. At temperatures below 500 °C the

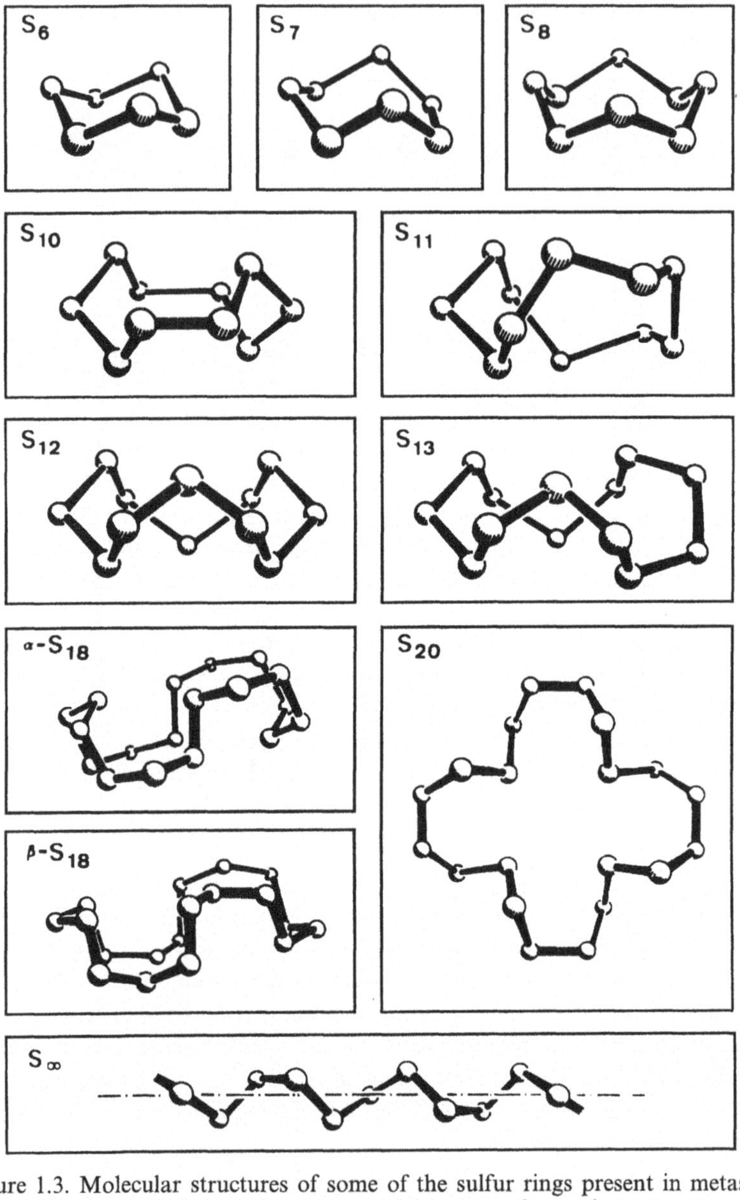

Figure 1.3. Molecular structures of some of the sulfur rings present in metastable sulfur allotropes which have been prepared in a pure form; these structures have been determined by X-ray crystallography. In addition the structure of the molecules in crystalline polymeric sulfur (S_∞) is shown.

molecules S_8, S_7 and S_6 are the main constituents of the saturated vapour. Above 800 °C the S_2 molecule is the predominating species in both saturated and unsaturated sulfur vapors. However, S_2 molecules can also be prepared at moderate temperatures in solution by decomposition of suitable organic precursor molecules (Steliou 1991; Tardif *et al.* 1997). Unless a trapping reagent (e.g., an unsaturated organic compound) is present to react with S_2 it will polymerize to S_6, S_7 and S_8.

1.2.4 Sulfur sols from elemental sulfur (WEIMARN sols)

All allotropes of sulfur are practically insoluble in water at 20 °C. The solubility of α-S_8 has been determined as 5 μg kg^{-1} (Boulègue 1978). Therefore, if a solution of S_8 in an organic solvent like ethanol or acetone is either diluted by water or poured into an excess of water, the sulfur must precipitate. In this way "sulfur sols" are prepared (Steudel 1999). Since sulfur is a hydrophobic material these sols are termed hydrophobic sulfur sols. They are of milky-white appearance and when freshly prepared, they consist of tiny droplets of liquid sulfur dispersed in the aqueous phase ("colloidal solution" or "emulsion").

It is highly interesting and informative to analyse the molecular self-organization that takes place when such a sulfur sol is prepared. In an organic solvent the S_8 molecules are dispersed molecularly, i.e. as single molecules surrounded by solvent molecules. If these S_8 molecules are now transferred into an aqueous phase they start forming clusters (dimers, trimers, ...) owing to their hydrophobic nature. This effect, termed "hydrophobic interaction", reduces the surface area of the hydrohopbic substance exposed to the hydrophilic environment because a cluster of n S_8 molecules has a smaller outer surface than n single S_8 molecules combined. When the cluster of S_8 molecules grows it will eventually form a tiny liquid-like droplet in which other hydrophobic molecules may also dissolve, e.g. hydrophobic solvent molecules or the hydrophobic parts of proteins which might be present in the mixture. Very small droplets of liquid sulfur have a tendency to supercooling, i.e. they stay liquid at 20 °C for quite some time. The hydrophobic sulfur sol prepared in this way is stable for a few days but eventually the sulfur crystallizes and settles to the bottom of the vessel.

Hydrophobic sulfur sols can be stabilized by the addition of surfactants like Triton X-100 or sodium dodecylsulfate to the aqueous phase whereas di- or trivalent cations destabilize the sol and accelerate precipitation.

The reactions of elemental sulfur with hot water are given in Section 1.7.

1.3 SULFIDE AND POLYSULFIDES

1.3.1 Hydrogen sulfide and sulfide ions

Hydrogen sulfide (H_2S) is a constituent of certain types of natural gas ("sour gas") from which it is recovered and oxidized to elemental (liquid) sulfur by the CLAUS process. H_2S is also formed in huge quantities on desulfurization of crude oil by the HDS process (hydrodesulfurization) carried out in refineries. Furthermore, H_2S is a frequent component of certain wastewaters and waste gases. It is also present in volcanic exhalations and abundant in the water exiting from hydrothermal vents on ocean floors. In anaerobic aqueous environments H_2S may be formed by sulfate reducing bacteria (Barton 1995).

The solubility of H_2S in water at 20 °C is only 0.40g/100g H_2O (0.12 M) at a total pressure (H_2S+H_2O) of 1.013 bar. As expected, the solubility decreases with increasing temperature. In aqueous solutions the very weak acid H_2S will be more or less deprotonated depending on the pH value:

$$H_2S + H_2O \rightleftharpoons H_3O^+ + HS^- \qquad K_1 = 1.0 \times 10^{-7}\ (20\ °C) \qquad (1.3)$$
$$HS^- + H_2O \rightleftharpoons H_3O^+ + S^{2-} \qquad K_2 = 0.8 \times 10^{-17}\ (20\ °C) \qquad (1.4)$$

The extremely low value of K_2 excludes the formation of sulfide ions S^{2-} in almost all solutions except at pH values near 14 (in the older literature lower values for K_2 have been used). Therefore, the reactions of sulfide in water at pH values of up to about 12 are practically the reactions of either H_2S or of the hydrogen sulfide ion HS^-. In a solution of pH = 7 with 1 mM total sulfide, 50% of the sulfide will be present as HS^- and the remainder as H_2S; at pH = 10 practically all sulfide is present as HS^-. If one talks about "sulfide solutions" one usually means solutions containing both H_2S and HS^- ions rather than S^{2-} ions.

The two acid dissociation constants of H_2S depend on the temperature. Only K_1 will be discussed here. According to various investigations K_1 has its maximum value of 2.0×10^{-7} near 100 °C while $K_1 = 0.50 \times 10^{-7}$ has been measured at 0 °C. At 200°C K_1 is about as small as at 0 °C (0.65×10^{-7}) (McCampbell Hamilton 1991).

Hydrogen sulfide ions react with many metal cations to form insoluble sulfides, e.g. the black iron(II) sulfide:

$$Fe^{2+} + HS^- \rightleftharpoons FeS\downarrow + H^+ \qquad (1.5)$$

In hydrochloric acid the reverse reaction takes place. Depending on their solubility products, metal sulfides are soluble in hydrochloric acid (ZnS, CdS, FeS, CoS, NiS, MnS) or not (HgS, As_2S_3, Sb_2S_3, Bi_2S_3, Cu_2S, Ag_2S). Using tabulated solublity products one has to make sure that these have been calculated using the most recent pK_a values of H_2S (see above). The sulfides of the alkali and alkaline earth metals are soluble even in neutral water but these solutions are subject to hydrolysis and therefore strongly alkaline, e.g.:

$$Na_2S + H_2O \rightleftharpoons NaHS + NaOH \quad (1.6)$$

Alkaline sulfide solutions react with SO_2 almost quantitatively to thiosulfate (Gmelin 1960):

$$2\ Na_2S + 4\ SO_2 \rightarrow 3\ Na_2S_2O_3 + H_2O \quad (1.7a)$$

or

$$2\ HS^- + 4\ HSO_3^- \rightarrow 3\ S_2O_3^{2-} + 3\ H_2O \quad (1.7b)$$

Even mild oxidants turn aqueous H_2S and in particular hydrogen sulfide ions into elemental sulfur which precipitates from the solution. The rate of autoxidation of such solutions depends on the concentration, the pH value and the temperature (Chen and Morris 1972; O'Brien and Birkner 1977; Chiu and Meerhan 1977). Elemental sulfur is only formed in near neutral solutions. Alkaline sulfide solutions and solid $Na_2S{\cdot}xH_2O$ are oxidized by air to polysulfide, thiosulfate and eventually sulfate (x has values between 7 and 9 in commercial sodium sulfide). Therefore commercial sodium sulfide always contains these products, which can only be removed by careful recrystallization from deoxygenated water in an inert atmosphere (Steudel *et al.* 1989a). The initial reactions of the autoxidation are as follows (Degrand and Lund 1979):

$$HS^- + O_2 \rightarrow HS^\bullet + O_2^{\bullet -} \quad (1.8)$$

$$2\ HS^\bullet \rightarrow H_2S_2 \rightleftharpoons S_2^{2-} + 2\ H^+ \quad (1.9)$$

The further oxidation of disulfide yields thiosulfate as explained in Section 1.3.2. The two radicals formed by electron transfer in reaction (1.8) may combine to give the sulfinate anion which is highly reactive and eventually is oxidized to sulfate:

$$HS^\bullet + O_2^{\bullet -} \rightarrow HSO_2^- \quad (1.10)$$

$$HSO_2^- + O_2 \rightarrow HSO_4^- \quad (1.11)$$

In addition, $HS^{\bullet}$ reacts with O_2 to the radical anion $SO_2^{\bullet -}$ (Zhu *et al.* 1991). The catalytic oxidation of hydrogen sulfide ions by metal ions will be discussed in Section 1.3.4.

1.3.2 Polysulfides and polysulfanes

Alkaline sulfide solutions dissolve elemental sulfur with the formation of polysulfide anions which form an equilibrium mixture of several species:

$$HS^- + S_8 \rightleftharpoons H\text{–}S\text{–}S_7\text{–}S^- \rightleftharpoons H^+ + S_9^{2-} \tag{1.12}$$
$$S_9^{2-} + HS^- \rightleftharpoons H^+ + 2\,S_5^{2-} \tag{1.13}$$
$$2\,S_5^{2-} \rightleftharpoons S_4^{2-} + S_6^{2-}, \text{ etc.} \tag{1.14}$$

These reversible reactions strongly depend on the pH value and on the total sulfur concentration. Low pH values result in the formation of H_2S and in precipitation of elemental sulfur, whereas very high pH values favour the shorter chains above the longer polysulfide anions. Polysulfide anions can therefore exist in water only at pH > 6. It is generally agreed that the *maximum chain-length* in water is 9 although there is no *a priori* reason why longer anions should not exist in small concentrations. If an aqueous solution of Na_2S is saturated with sulfur the *average chain-length* is 5.5, and tetra-, penta- and hexasulfide anions are believed to predominate in such mixtures (Teder 1971; McCampbell Hamilton 1991). The only available method to detect single polysulfide species in solution is absorption spectroscopy in the ultraviolet and visible region (UV-VIS) (Giggenbach 1974).

The above reactions are typical *nucleophilic displacement reactions*, which are quite common in sulfur chemistry. Both HS^- and S_x^{2-} anions are strong nucleophiles (Lewis bases) which are capable of opening S_8 rings and which tend to cleave also the S–S bonds of chain-like species. Other strong nucleophiles are the anions CN^- and SO_3^{2-} which react with sulfur-rich compounds to thiocyanate SCN^- and thiosulfate $S_2O_3^{2-}$, respectively. Reactions of this type are very fast if all species involved are present in a homogeneous solution.

Aqueous polysulfide solutions are subject to rapid autoxidation when exposed to air (Steudel *et al.* 1986). The major product is thiosulfate which forms according to the equation:

$$S_x^{2-} + 3/2\,O_2 \rightarrow S_2O_3^{2-} + (x-2)/8\,S_8\downarrow \tag{1.15}$$

Elemental sulfur is observed only if the average sulfur content x of the polysulfide solution is larger than 2. Because polysulfide solutions are yellow to orange whereas thiosulfate is colourless, the discoloration of the original solution and the formation of a precipitate (for $x > 2$) indicate that reaction (15) takes place. For this reason, thiosulfate is usually a component of aqueous polysulfide solutions.

Hot aqueous polysulfide solutions contain additional species originating from the homolytic dissociation of the dianions, e.g.:

$$S_6^{2-} \rightleftharpoons 2\,S_3^{\bullet -} \tag{1.16}$$

Such radical anions can be detected by electron spin resonance (ESR), UV-VIS and resonance Raman spectroscopy. For example, $S_3^{\bullet -}$ is blue due to an absorption at 610 nm (Chivers and Drummond 1972). It has also been prepared in non-aqueous solutions, and it is the blue pigment in the silicate mineral *lapis lazuli*. It is obvious that radical anions are more reactive than the dianions, and it has been suspected that they are responsible for the rapid autoxidation of polysulfide solutions.

If a sodium polysulfide solution is poured into ice-cold concentrated hydrochloric acid a yellow oil is formed in addition to crystalline NaCl:

$$Na_2S_x + 2\,HCl \rightarrow 2\,NaCl + H_2S_x \tag{1.17}$$

The polysulfane mixture H_2S_x is almost insoluble in water; it may contain all species from H_2S up to H_2S_{30}, especially after ageing, as has been demonstrated by ^{1}H-NMR spectroscopy (Hahn 1985). The spectra and other properties of the polysulfanes H–S_x–H show that these molecules are of chain-like structure. The two acidity constants of the polysulfanes are not known accurately. An older determination (Schwarzenbach and Fischer 1960) has been shown to be in error since too simple a composition of the polysulfide mixture had been assumed (McCampbell Hamilton 1991). However, the two pK_a values must be smaller by orders of magnitude than the corresponding values of H_2S. This follows from the chemical behaviour of polysulfides in aqueous solution: It is known that these anions are completely deprotonated in aqueous solutions and no hydrogen polysulfide ions HS have ever been observed. In the gas phase the hydrogen polysulfides are strong proton donors (Brønsted acids) (Otto and Steudel 1999).

As pure materials, hydrogen polysulfanes H_2S_x are unstable even at 20 °C, and slowly decompose to H_2S and S_8, especially in the presence of alkaline substances or rough surfaces of glass or quartz which act as catalysts (Fehér 1975):

$$H_2S_x \rightleftharpoons H_2S + (x\text{-}1)/8\ S_8 \qquad (1.18)$$

This reaction is also reversible: if H_2S gas is bubbled into *liquid* sulfur (at 120 °C or higher) the formation of hydrogen polysulfanes can be demonstrated spectroscopically. This reaction is slightly endothermic and responsible for the fact that the solubility of H_2S in liquid sulfur *increases* with temperature.

Organic derivatives of the polysulfanes are discussed in Section 1.8.

1.3.3 Polysulfido complexes of transition metals

Certain transition metal sulfides like Cu_2S and Ag_2S dissolve in aqueous polysulfide solutions forming chelate complexes with bi- or even tridentate polysulfido ligands (forming two or three bonds to a metal ion). Often such complexes contain heterocyclic MS_x units with ring sizes from 3 to 10. In the case of copper the following anions have been isolated in corresponding salts which have been structurally characterized: $[Cu_2S_{20}]^{4-}$, $[Cu_3S_{12}]^{3-}$, $[Cu_3S_{18}]^{3-}$, $[Cu_4S_{12}]^{2-}$, $[Cu_4S_{15}]^{2-}$ and $[Cu_6S_{17}]^{2-}$ (Figure 1.4).

Other metal cations like Zn^{2+}, Ni^{2+}, Mn^{2+}, Hg^{2+}, and Ag^{+} form also mononuclear and/or polynuclear complexes (Müller and Diemann 1987). These anionic species are soluble in water and in polar organic solvents. Complexes with S-donor ligands (e.g., the thiolate anion of cysteine) play an important role as redox catalysts in biochemical reactions, e.g., in rubredoxins and ferredoxins (Ueyama and Nakamura 1992).

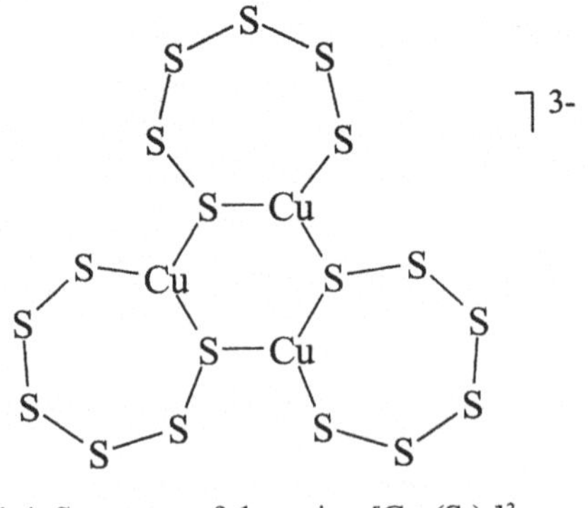

Figure 1.4. Structure of the anion $[Cu_3(S_6)_3]^{3-}$ containing tridentate tetrasulfide ligands (schematic).

1.3.4 Oxidation of sulfide and polysulfide ions by metal ions

The removal of sulfide ions from aqueous solutions by oxidation to elemental sulfur is one of the most important desulfurization processes (Steudel 1996). This process can be achieved either chemically or microbiologically (i.e., enzymatically). In both cases transition metal complexes are involved. The metal ion must be able to exist in at least two different oxidation states between which it can shuttle back and forth. Suitable metal ions are for instance iron(III/II), copper(II/I), manganese(IV/III/II), and vanadium(V/IV). The complexes of these metal ions have to be soluble at the pH of the sulfide solution, and the following type of reaction is expected to occur:

$$HS^- + 2\,M^{n+} \rightarrow 1/8\,S_8 + H^+ + 2\,M^{(n-1)+} \qquad (1.19)$$

The reduced metal ion is then oxidized by dioxygen and consequently serves as a catalyst only:

$$4\,M^{(n-1)+} + O_2 + 2\,H^+ \rightarrow 4\,M^{n+} + 2\,OH^- \qquad (1.20)$$

The overall reaction can therefore be represented by the equation:

$$2\,HS^- + O_2 \rightarrow 1/4\,S_8 + 2\,OH^- \qquad (1.21)$$

As O_2 is a paramagnetic molecule whereas all other species are diamagnetic, reaction (21) is forbidden by the *law of spin conservation* and therefore requires a paramagnetic catalyst as described by reactions (1.19) and (1.20).

Equation (1.19) just summarizes the chemical changes in the system but does not represent the reaction mechanism. If hydrogen sulfide ions HS^- are oxidized by a one-electron oxidant the initial product will be the radical $HS^\bullet$ which is expected to dimerize to the disulfide ion:

$$2\,HS^\bullet \rightarrow S_2^{2-} + 2\,H^+ \qquad (1.22)$$

The disulfide ion will then be further oxidized either by a metal ion or by a $HS^\bullet$ radical:

$$S_2^{2-} + M^{n+} \rightarrow S_2^{\bullet-} + M^{(n-1)+} \qquad (1.23)$$

$$S_2^{2-} + HS^\bullet \rightarrow S_2^{\bullet-} + HS^- \qquad (1.24)$$

The disulfide radical anion $S_2^{\bullet -}$ is expected to dimerize rapidly to the tetrasulfide ion S_4^{2-} which will be further oxidized and so forth, eventually yielding a mixture of polysulfide anions of different chain-lengths. It has in fact been observed that oxidation of hydrogen sulfide ions by dioxygen in the presence of heme or by iron(III) or ruthenium(III) complexes in the absence of dioxygen yields polysulfides from which elemental sulfur may then be formed according to the reverse reactions in the above Equations (1.12)-(1.14).

All these reactions will take place in desulfurization plants working by the Stretford, Sulfolin, Lo-Cat, SulFerox or Bio-SR process (Steudel 1996). The electrochemical oxidation of aqueous sulfide at pH = 9.3 also yields polysulfide anions first and from which elemental sulfur is then formed (Szynkarczuk *et al.* 1994).

Metal cations are known to catalyse also the oxidation of organic thiols (RSH) to disulfanes (RSSR) and water by dioxygen. Thiyl radicals ($RS^{\bullet}$) are likely intermediates in these reactions.

1.4 SULFITES, THIOSULFATES, DITHIONITES AND DITHIONATES

1.4.1 Sulfur dioxide, sulfite ions and sulfurous acid

Sulfur dioxide dissolves in water with the formation of what is called sulfurous acid H_2SO_3. Since the equilibrium

$$SO_2 + H_2O \rightleftharpoons H_2SO_3 \tag{1.25}$$

is largely on the left side the concentration of H_2SO_3 is very small. In fact, the molecule H_2SO_3 has never been detected in condensed phases because this acid exists mainly in the deprotonated form:

$$H_2SO_3 + H_2O \rightleftharpoons H_3O^+ + HSO_3^- \tag{1.26}$$

The hydrogensulfite anion occurs in two tautomeric forms which have been detected by ^{17}O-NMR spectroscopy (Horner and Connick 1986):

$$H\text{–}SO_3^- \rightleftharpoons HO\text{–}SO_2^- \tag{1.27}$$

The first species containing an H–S bond dominates at 25 °C and an ion strength of 1 M but the other tautomer is present also. In concentrated aqueous solutions of SO_2 another equilibrium is established:

$$2\ HSO \rightleftharpoons S_2O^- + H_2O \qquad (1.28)$$

Therefore, the determination of reliable pK_a values for sulfurous acid is difficult but an acid constant K_a of 1.5×10^{-2} mol L^{-1} has been reported for reaction (1.26). The disulfite anion has the connectivity $[O_2S–SO_3]^{2-}$. Sodium disulfite is produced by saturating aqueous NaOH with SO_2. Sulfite solutions are strong reducing agents which are oxidized by dioxygen, iron(III) salts or iodine to give sulfate. In alkaline solution, the autoxidation is particularly fast although this reaction is formally forbidden by the law of spin conservation. However, in practice the reaction is strongly catalysed by even traces of metal cations like Cu^{2+}, Mn^{2+}, Fe^{3+} and Co^{2+} which are present in most aqueous systems (Connick *et al.* 1995; Connick and Zhang 1996). On acidification of sulfites, SO_2 is evolved (reverse reactions 1.26 and 1.25).

1.4.2 Thiosulfates and thiosulfuric acid

Sodium thiosulfate is produced commercially by refluxing concentrated aqueous Na_2SO_3 with elemental sulfur:

$$Na_2SO_3 + 1/8\ S_8 \rightleftharpoons Na_2S_2O_3 \qquad (1.29)$$

This reaction is another example for a nucleopilic displacement. The nucleophile SO^- attacks the S_8 molecule and opens the ring with the initial formation of a chain-like dianion $^-S–S_7–SO$ which is rapidly attacked by other sulfite ions with formation of thiosulfate until the sulfur chain has been reduced stepwise to one remaining S–S bond.

The reverse process takes place when aqueous thiosulfate is decomposed by acid. The first product is the hydrogen thiosulfate anion which has been isolated as ammonium salt from methanolic solutions (Steudel and Prenzel 1989):

$$(NH_4)_2S_2O_3 + H_2SO_4 \rightarrow (NH_4)HS_2O_3 + (NH_4)HSO_4 \qquad (1.30)$$

The connectivity in $HS_2O_3^-$ is $H–S–SO_3^-$. The free thiosulfuric acid $H_2S_2O_3$ is certainly present in such solutions also but due to its instability it has never been observed directly nor has it been isolated.

In water the ions $HS_2O_3^-$ and $S_2O_3^{2-}$ react with each other with exchange of sulfur atoms resulting in the growth of a sulfur chain in the sulfane sulfonate anions formed:

$$HS_2O_3^- + S_2O_3^{2-} \rightleftharpoons HS\text{–}S\text{–}SO_3^- + SO_3^{2-} \quad (1.31)$$
$$HS_3O_3^- + S_2O_3^{2-} \rightleftharpoons HS\text{–}S\text{–}S\text{–}SO_3^- + SO_3^{2-} \text{ etc.} \quad (1.32)$$

After the sulfur chain has grown to a sufficient length the cyclic molecules S_6, S_7 and S_8 can be split off; all of them have been detected in such mixtures by HPLC analysis:

$$H\text{–}S_n\text{–}SO_3^- \rightleftharpoons cyclo\text{-}S_n + HSO_3^- \quad (n > 5) \quad (1.33)$$

Because HSO_3^- will decompose to H_2O and SO_2 in acidic solution and since the SO_2 formed will escape as a gas, the reaction goes to completion. The precipitating sulfur also shifts equilibrium (1.33) to the right. Additional reactions yield polythionates as will be shown in Section 1.5.

A highly characteristic reaction of thiosulfate anions is their facile oxidation ("oxidative coupling") to tetrathionate which when carried out with elemental iodine is used in quantitative analysis. This reaction is another example of a nucleophilic displacement (actually two such reactions):

$$I\text{–}I + {}^-S\text{–}SO_3^- \rightarrow I^- + I\text{–}S\text{–}SO_3^- \quad (1.34)$$
$$I\text{–}S\text{–}SO_3^- + {}^-S\text{–}SO_3^- \rightarrow I^- + {}^-O_3S\text{–}S\text{–}S\text{–}SO_3^- \quad (1.35)$$

The progress of this reaction can be followed by the decolorization of the formerly brown iodine solution. The tetrathionate is a stable anion in neutral and acidic solutions.

Metal ions (Fe^{3+}, Cu^{2+}, Au^{3+}) oxidize thiosulfate ions by a 1-electron oxidation to the $S_2O_3^{\bullet-}$ radical anion which then dimerizes to the tetrathionate ion. Cytochrome *c* and hexacyanoferrate(III) $[Fe(CN)_6]^{3-}$ are also suitable oxidants to transform thiosulfate into tetrathionate.

Reduction of thiosulfate by dithionite results in a cleavage of the S–S bond and therefore yields sulfide and sulfite:

$$S_2O_3^{2-} + S_2O_4^{2-} \rightarrow HS^- + 3\ HSO_3^- \quad (1.36)$$

In slightly acidic concentrated solutions, thiosulfate reacts with SO_2 at 20 °C to trithionate (Fehér 1975):

$$3\ SO_2 + 2\ S_2O_3^{2-} \rightarrow 3\ S_3O_6^{2-} + 1/8\ S_8 \quad (1.37)$$

This redox reaction may even be used to prepare trithionate.

1.4.3 Dithionites and dithionous acid

Reduction of aqueous SO_2 with strong reducing agents like sodium amalgam, zinc dust, formic acid or electrochemically yields the corresponding dithionites of which the commercial product $Na_2S_2O_4 \cdot 2\,H_2O$ is best known:

$$2\,HSO_3^- + 2\,e^- \rightleftharpoons S_2O_4^{2-} + 2\,OH^- \qquad (1.38)$$

In reactions where dithionite is used as a reductant the reverse reaction will take place. Aqueous dithionite rapidly reacts with molecular oxygen and therefore has to be handled in an inert atmosphere. This high reactivity is due to the partial dissociation of the dithionite ion:

$$S_2O_4^{2-} \rightleftharpoons 2\,SO_2^{\bullet -} \qquad (1.39)$$

In water this equilibrium is on the left side although the radical anions have been detected by ESR spectroscopy. However, in polar organic solvents the dissociation is almost complete.

Acidification of aqueous dithionite solutions results in decomposition which proceeds at 20 °C and pH values near 6 basically according to the following equation (Drozdova *et al.* 1998):

$$2\,S_2O_4^{2-} + H_2O \rightarrow 2\,HSO_3^- + S_2O_3^{2-} \qquad (1.40)$$

Consequently, the species $HS_2O_4^-$ and $H_2S_2O_4$ are unknown in the form of pure compounds. Since thiosulfate is also unstable in acidic solutions and decomposes to sulfur, SO_2 and polythionate anions, these products will also be observed in such solutions.

1.4.4 Dithionates and dithionic acid

When aqueous sulfite is oxidized by manganese(IV) dioxide, the formation of dithionate and sulfate is observed:

$$3\,HSO + 2\,MnO_2 \rightarrow MnS_2O_6 + MnSO_4 + 3\,OH^- \qquad (1.41)$$

The connectivity of dithionate anions is $[O_3S{-}SO_3]^{2-}$. Iron(III) is also able to oxidize sulfite to dithionate. This reaction occurs in flue gas scrubbers in which gypsum is produced from limestone because traces of Fe^{3+} are always present. When barium dithionate is treated with sulfuric acid aqueous dithionic acid can be prepared (Fehér 1975):

$$BaS_2O_6 + H_2SO_4 \rightarrow H_2S_2O_6 + BaSO_4\downarrow \qquad (1.42)$$

which decomposes above 50 °C as follows:

$$H_2S_2O_6 \rightarrow H_2SO_4 + SO_2 \qquad (1.43)$$

1.5 POLYTHIONATES AND HYDROPHILIC SULFUR SOLS

1.5.1 Polythionates and polythionic acids

Polythionates are actually polysulfane disulfonates: $[O_3S{-}S_x{-}SO_3]^{2-}$. The corresponding sodium and potassium salts with x = 1...4 have been prepared by various syntheses (Fehér 1975):

trithionate:	$SCl_2 + 2HSO_3^- \rightarrow S_3O_6^{2-} + 2\,HCl$	(1.44)
tetrathionate:	$2\,S_2O_3^{2-} + I_2 \rightarrow S_4O^- + 2\,I^-$	(1.45)
pentathionate:	$SCl_2 + 2\,HS_2O_3^- \rightarrow S_5O_6^{2-} + 2\,HCl$	(1.46)
hexathionate:	$S_2Cl_2 + 2\,HS_2O_3^- \rightarrow S_6O_6^{2-} + 2\,HCl$	(1.47)

These salts are water soluble and metastable at acidic pH values. However, the pure polythionic acids (polysulfane disulfonic acids) are unstable and therefore not known as pure materials.

In alkaline solutions and in the presence of strong nucleophiles like sulfide or sulfite anions polythionates are unstable and are degraded, e.g.:

$$S_4O_6^{2-} + OH^- \rightarrow HSSSO_3^- + SO_4^{2-} \qquad (1.48)$$

$$S_4O_6^{2-} + HS^- \rightarrow HSSSO_3^- + S_2O_3^{2-} \qquad (1.49)$$

The sulfane monosulfonates then decompose as outlined in Section 1.4.2. Dithionite reduces polythionates to thiosulfate and sulfane monosulfonates (Münchow and Steudel 1994):

$$S_nO_6^{2-} + S_2O_4^{2-} + 2\,H_2O \rightarrow HSS_{n-3}O_3^- + HS_2O_3^- + 2\,HSO_3^- \qquad (1.50)$$

In other words, polythionates can only exist at acidic pH values and in the absence of strong nucleophiles and reducing agents. Strong oxidizing agents like hydrogen peroxide and elemental chlorine convert polythionates into sulfate.

Polythionates can be detected by their characteristic absorption bands in the infrared spectrum. They can be separated analytically by ion-pair chromatography and all species with chain-lengths up to 22 sulfur atoms have been identified (Steudel and Holdt 1986; Steudel *et al.* 1987).

1.5.2 Hydrophilic sulfur sols (RAFFO and SELMI sols)

Chemically prepared hydrophilic sulfur sols do not consist of elemental sulfur; instead polythionates are their main constituents. Depending on the method of preparation two types of sol are defined: RAFFO (or LAMER) sols are made by decomposition of concentrated sodium thiosulfate solution with concentrated sulfuric acid (Steudel *et al.* 1988). SELMI sols are obtained from the reaction of aqueous sulfite (or SO_2) with sulfide (or H_2S) (Steudel *et al.* 1989b).

When a RAFFO sol is prepared from aqueous thiosulfate and concentrated H_2SO_4 the reactions occurring are similar to those outlined in Section 1.4.2. However, the condensation and/or oxidation of sulfane monosulfonate anions to polythionates occurs in addition:

$$2\ H\text{–}S_x\text{–}SO_3^- \rightarrow H_2S + {}^-O_3S\text{–}S_{2x-1}\text{–}SO_3^- \qquad (1.51)$$

$$2\ H\text{–}S_x\text{–}SO_3^- + [O] \rightarrow H_2O + {}^-O_3S\text{–}S_{2x}\text{–}SO_3^- \qquad (1.52)$$

The concentrated sulfuric acid or other sulfur oxyanions may serve as oxidants. By salting-out with sodium chloride, a yellow precipitate is obtained which looks like sulfur but is mainly composed of long-chain polythionates together with a little elemental sulfur in the form of the homocycles from S_6 to S_{14}. The total sulfur content is ca. 85% but only 17% of this sulfur is present as elemental sulfur. Therefore, this material should not be termed "elemental sulfur" or "S^0", in particular because its reactivity will be quite different from that of α-S_8. The evaporated sol particles dissolve in water forming a colloidal solution (peptization) but the sol can be precipitated again by NaCl. It is believed that the long-chain polythionate anions form micelle-like particles in the core of which some elemental sulfur can be dissolved. The sulfonate groups will cover the surface of the micelles making the particles hydrophilic despite their high sulfur content (Figure 1.5). The particle size is in the range 0.1-0.5 μm; the sulfur content of the aqueous sol can be very high. By evaporation of some of the solvent it has been possible to obtain sulfur concentrations of up to 600 g L^{-1}. Ion-pair chromatography and reversed-phase HPLC are suitable methods to analyse such a sol.

SELMI sols are prepared from H_2S and SO_2 at acidic pH values. Their

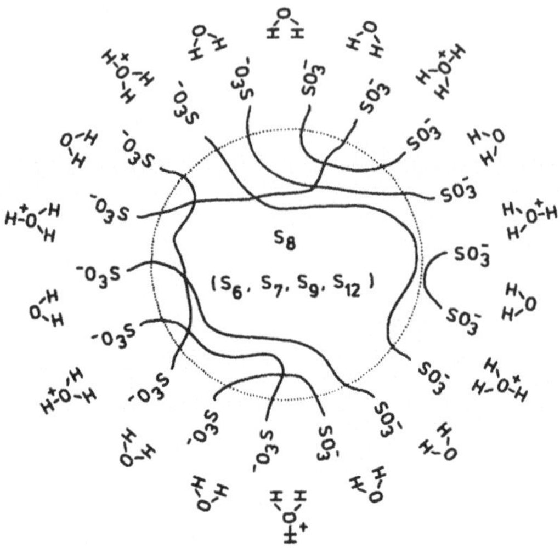

Figure 1.5. Micelle model of the particles in aqueous RAFFO sols. The main constituents are long-chain polythionate anions of average chain-length 29. The hydrophilic surface of the particles will be hydrated and various cations will be present as well. The hydrophobic core of the micelle contains some elemental sulfur in the form of various homocycles (mainly S_8).

composition and properties are quite similar to those of the RAFFO sols but the average chain-length of the polythionate anions is smaller. The reactions resulting in these products are very complex. Both sols are thermodynamically unstable and slowly decompose, the SELMI sol more rapidly than the RAFFO sol. This ageing process involves the conversion of long-chain polythionates into shorter ones with simultaneous formation of S_8. In other words, the micelles are destroyed and elemental sulfur precipitates during several days:

$$^{-}O_3S–S_x–SO_3^- \rightarrow {}^{-}O_3S–S_{x-8}–SO_3^- + cyclo\text{-}S_8 \qquad (1.53)$$

From the model in Figure 1.5 it follows that the particles of hydrophilic sulfur sols are negatively charged. This equal charge results in repulsion which stabilizes the sol. However, when multivalent cations are added, the repulsion is overcome since the electric field is shielded by these cations and since a kind of insoluble "salt" is formed by the attraction of neighbouring sol particles to the same cation. Therefore, even small concentrations of two- or three-valent cations initiate precipitation (Steudel 1999). Compounds or pH values that destroy polythionates (see Section 1.5.1) also destroy RAFFO and SELMI sols.

Hydrophilic sulfur sols are also produced by enzymatic oxidation of sulfide ions or other reduced sulfur compounds by a variety of sulfur bacteria (Brune 1989; Takakuwa 1992). However, the physical properties and chemical composition of this biologically produced sulfur (often abbreviated as S^0) are different from the hydrophilic sulfur sols discussed in this Section (Steudel *et al.* 1987; Janssen 1996; Dahl 1999; Prange *et al.* 1999). Bacterial sulfur globules produced by phototrophic bacteria consist of chain-like sulfur-rich compounds and do not contain substantial amounts of S_8 (Prange *et al.* 1999). Therefore they should not be termed "elemental sulfur". Hydrophilic sulfur particles produced by mixed cultures of *Thiobacilli* have been shown to be composed of a core of elemental sulfur covered by a layer of polymers (presumably proteins) which are equally charged resulting in Coulomb repulsion (Janssen *et al.* 1999).

1.6 SULFURIC ACID AND SULFATES

Sulfuric acid is commercially produced on a huge scale. Concentrated H_2SO_4 contains 4% water. This liquid as well as the 100% acid are oxidants, especially when hot. They are also dehydrating agents since the hydrate $H_2O{\cdot}H_2SO_4$ forms in a strongly exothermic reaction; it has the structure $[H_3O^+][HSO_4^-]$. The pK_a values of H_2SO_4 are –2.8 for the first deprotonation step and +1.92 for the second. Therefore, in dilute solution the acid is completely ionized in the first step but only partly in the second step. At pH = 1 the concentration ratio of SO_4^{2-} to HSO_4^- is ca. 0.1. The sulfate anion is rather inert chemically and needs to be activated for biological reduction at ambient temperatures: APS and PAPS are activated forms containing the sulfate anion attached to a phosphate unit as a mixed anhydride (Figure 1.6).

The energy gained on hydrolysis of the P–O–S bridge promotes any reactions of the liberated sulfate ion.

1.7 DISPROPORTIONATION OF ELEMENTAL SULFUR IN HOT WATER

As shown in Figure 1.1 elemental sulfur is thermodynamically unstable in water at pH values greater than 7. However, the corresponding reactions are slow at 20 °C and in the absence of catalysts. At higher temperatures, however, several equilibrium reactions can be established. The main reactions in *basic solutions* are the following (McCampbell Hamilton 1991):

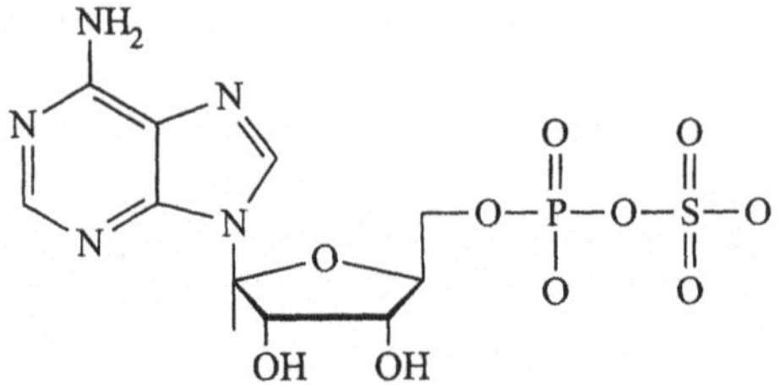

Adenosine Phosphosulfate (APS)

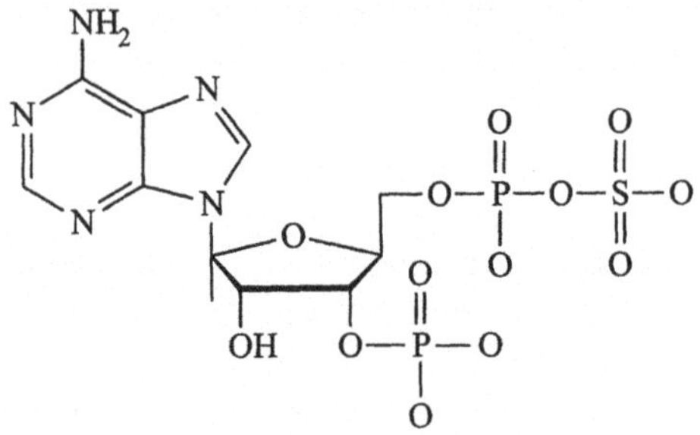

3′-Phosphoadenosine Phosphosulfate (PAPS)

Figure 1.6. Structures of the phosphosulfates APS and PAPS.

$$1/2\ S_8 + 4\ OH^- \rightleftharpoons S_2O_3^{2-} + 2\ HS^- + H_2O \tag{1.54}$$

At pH > 11.5 this reversible reaction occurs at 20 °C. At pH = 7.6, a temperature of 80 °C is necessary. At 100-120 °C the equilibrium is established rapidly. Secondary reactions are the formation of polysulfides:

$$n/8\ S_8 + HS^- + OH^- \rightleftharpoons S_{n-1}^{2-} + H_2O \tag{1.55}$$

the disproportionation of polysulfides, e.g.:

$$S_5^{2-} + 3\ OH^- \rightleftharpoons S_2O_3^{2-} + 3\ HS^- \tag{1.56}$$

and the hydrolysis of thiosulfate:

$$S_2O_3^{2-} + OH^- \rightleftharpoons SO_4^{2-} + HS^- \tag{1.57}$$

The latter two reactions require temperatures above 160 °C. As all of these reactions result in a drop of the pH value (consumption of OH^-), the whole process comes to a standstill unless a buffer is present. The final products are sulfide and sulfate, as shown in Figure 1.1.

In *acidic and neutral solutions* the disproportionation of sulfur proceeds according to the following equation:

$$1/2\ S_8 + 4\ H_2O \rightleftharpoons 3\ H_2S + HSO + H^+ \quad (1.58)$$

This endergonic equilibrium ($\Delta G^\circ > 0$) is established above 150 °C. After some sulfuric acid has been formed, the following secondary equilibrium will be established at this temperature:

$$1/8\ S_8 + 2\ H_2SO_4 \rightleftharpoons 3\ SO_2 + 2\ H_2O \quad (1.59)$$

1.8 ORGANIC DERIVATIVES OF TYPE R–S_n–R (ORGANOPOLYSULFANES)

1.8.1 Synthetic polysulfanes

Bis-organopolysulfanes are often termed *polysulfides* although they are not ionic like, for instance, sodium polysulfides. The recommended nomenclature is *polysulfanes* since they are derivatives of the unsubstituted sulfanes H_2S_n. These in turn are analogues to the alkanes C_nH_{2n+2}. Organic polysulfanes are much more stable thermally than the parent species H–S_n–H. Consequently, many derivatives have been prepared. These compounds are either chain-like or cyclic and *n* can reach values of up to 35 at least. The molecular structures of organic polysulfanes with up to 11 sulfur atoms have been elucidated by X-ray crystallography and many reactions have been studied (for a comprehensive review see Steudel and Kustos 1994).

1.8.2 Naturally occurring polysulfanes

Although it has been known for a long time that organic mono- and disulfanes like methionine and cystine occur in organisms, the widespread natural occurrence of tri- and higher polysulfanes with organic substituents has only relatively recently been demonstrated (Steudel and Kustos 1994). The reason may be that mono- and disulfanes are much more stable chemically and thermally than higher polysulfanes as can be seen from the following dissociation enthalpies of the central sulfur-sulfur bonds (Benson 1978):

Me–S–S–Me	310 kJ mol^{-1}
Me–S–S–S–Me	226 kJ mol^{-1}
Me–S–S–S–S–Me	141 kJ mol^{-1}

Like other compounds with S–S bonds the organic polysulfanes are attacked by nucleophiles. For example, hydrogensulfide cleaves cystine with formation of thiolate (cysteinate) and "persulfide" anions:

$$R–S–S–R + HS^- \rightleftharpoons R–S–S^- + R–S^- + H^+ \tag{1.60}$$

In the presence of O_2 the persulfide is oxidized to the corresponding cysteine-S-sulfonate which is also termed the Bunte salt of cysteine (Steudel and Albertsen 1992):

$$R–S–S^- + 3/2\,O_2 \rightarrow R–S–SO \tag{1.61}$$

Tables 1.4 and 1.5 summarize some compounds containing three or more sulfur atoms linked together which have been isolated from various sources like algae, mushrooms, higher plants or animals. These compounds must have some function beneficial to the particular organism but in most cases this function is unknown. It has been speculated that deterrence of predators is the main function. In fact, varacin has strong bacteriostatic properties. The beneficial properties of the organosulfur compounds contained in onions, garlic and other *Allium* species to human health is also well known (Block 1992).

More recently, unusual peptides have been isolated from genetically engineered bacteria which contain trisulfane units between cysteine residues instead of disulfane units. For example, the human growth hormone with one or both disulfane units replaced by trisulfane groups was biosynthesized by *E. coli* after genetic engineering (Canova-Davis *et al.* 1996).

Table 1.4. Naturally occurring chain-like bis-organyl polysulfanes $R^1–S_n–R^2$ ($n \geq 3$)

R^1	R^2	n	Source
methyl	methyl	4	*Lentinus edodes* (Shiitake mushroom)
methyl	methyl	3	*Lentinus edodes* (Shiitake mushroom); oil made from *Ferula asafoetida* (Afghanistan)
methyl	2-butyl	3	oil made from *Ferula asafoetida* (Afghanistan)
2-butyl	2-butyl	3, 4	oil made from *Ferula asafoetida* (Afghanistan)
allyl	allyl	3–6	garlic oil
alanyl	alanyl	3, 4	wool hydrolysate (acidic)
3-oxoundecyl	3-oxoundecyl	3, 4	*Dictyopteris plagiogramma* (Hawaiian algae)
methyl	complex structure	3	Calichemicin (*Micromonospora echinospora*)

Table 1.5. Naturally occurring cyclic polysulfanes with three or more sulfur atoms connected to each other

Compound	Source
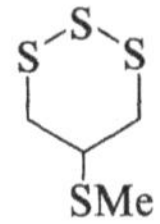 5-Methylthio-1,2,3-trithiane	*Chara globularia* (green algae)
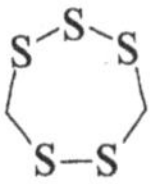 Lenthionine	*Chondria californica* (red algae) *Parkia speciosa* (Mimosaceae species) *Lentinus edodes* (Shiitake mushroom)
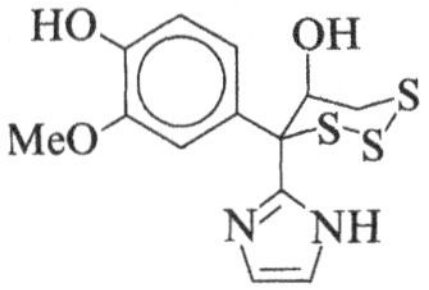	*Aplidium* sp. D (ascidian)
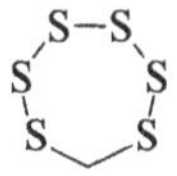 Hexathiepane 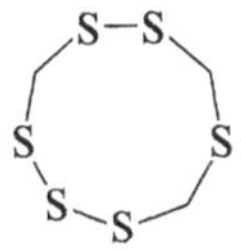 Hexathionane 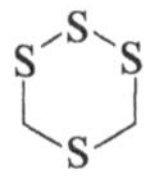 1,2,3,5-Tetrathiane	*Lentinus edodes* (Shiitake mushroom)
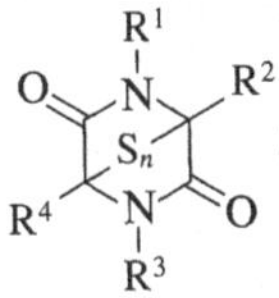 3,6-Epipolythiopiperazine-2,5-dione	
n = 3; $R^1 = R^3 = Me$; $R^2 = CH_2OH$; $R^4 = CH_2Ph$: Sporidesmine E	Fungus *Pithomyces chartarum*
n = 4; $R^1 = R^3 = Me$; $R^2 = CH_2OH$; $R^4 = CH_2Ph$: Sporidesmine G	*Hyalodendron* sp. (fungus); *Penicillium turbatum* *Pithomyces chartarum*
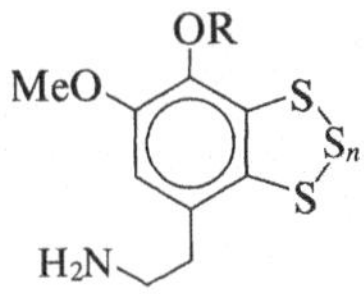 n = 1; R = H Lissoclinotoxin A	*Lissoclinum perforatum* (ascidian)
n = 3; R = Me Varacin	*Lissoclinum vareau* (ascidian)

1.9 REFERENCES

Barton, L. E. (ed.) (1995) *Sulfate Reducing Bacteria*, Plenum, New York.

Benson, S. W. (1978) Thermochemistry and Kinetics of Sulfur-Containing Molecules and Radicals. *Chem. Rev.* **78**, 23-35 (review).

Block, E. (1992) Organosulfur Chemistry of the Species *Allium* and their Importance for the Organic Chemistry of Sulfur. *Angew. Chem. Int. Ed.* **31**, 1135 (review).

Boulègue, J. (1978) Solubility of elemental sulfur in water at 298 K. *Phosphorus and Sulfur* **5**, 127-128.

Brune D. C. (1989) Sulfur oxidation by phototrophic bacteria. *Biochim. Biophys. Acta* **975**, 189-221.

Canova-Davis, E., Baldonado, I. P., Chloupek, R. C., Ling, V. T., Gehant, R., Olson, K. and Gillece-Castro, B. L. (1996) Confirmation by Mass Spectrometry of a Trisulfide Variant in Methionyl Human Growth Hormone Biosynthesized in *Echerichia coli*. *Anal. Chem.* **68**, 4044-4051.

Chen, K. Y. and Morris, J. C. (1972) Kinetics of Oxydation of Aqueous Sulfide by O_2. *Environm. Sci. Technol.* **6**, 529-537.

Chiu, G. and Meerhan, E. J. (1977) *J. Coll. Interface Sci.* **62**, 1.

Chivers, T. and Drummond, I. (1972) Characterization of the trisulfur radical anion S_3^- in blue solutions of alkali polysulfides in hexamethylphosphoramide. *Inorg. Chem.* **11**, 2525-2527.

Connick, R. E., Zhang, Y.-X., Lee, S., Adamic, R. and Chieng, P. (1995) Kinetics and Mechanism of the Oxidation of HSO by O_2. 1. The Uncatalyzed Reaction. *Inorg. Chem.* **34**, 4543-4553.

Connick, R. E. and Zhang, Y.-X. (1996) Kinetics and Mechanism of the Oxidation of HSO by O_2. 2. The Manganese(II)-Catalyzed Reaction. *Inorg. Chem.* **35**, 4613-4621.

Dahl, C. (1999) In *Biochemical Principles and Mechanisms of Biosynthesis and Biodegradation of Polymers* (Steinbüchel, A., ed.), Wiley-VCH, Weinheim, p. 27.

Degrand, C. and Lund, H. (1979) Electrochemical Synthesis of Sulfinic Acids. *Acta Chem. Scand. Part B* **33**, 512-514.

Drozdova, Y., Steudel, R., Hertwig, R. H., Koch, W. and Steiger, T. (1998) Structures and Energies of Various Isomers of Dithionous Acid, $H_2S_2O_4$, and of Its Anion $HS_2O_4^-$. *J. Phys. Chem. Part A* **102**, 990-996 references cited therein.

Fehér, F. (1975) In *Handbuch der Präparativen Anorganischen Chemie*, 3rd edn, vol. 1 (Brauer, G., ed.), Enke, Stuttgart, pp. 356, 362, 395/6 and 398.

Garrels, R. M. and Naeser, C. R. (1958) Equilibrium distribution of dissolved sulphur species in water at 25°C and 1 atm total pressure. *Geochim. Cosmochim. Acta* **15**, 113-130.

Giggenbach, W. F. (1974) Equilibria Involving Polysulfide Ions in Aqueous Sulfide Solutions up to 240°C. *Inorg. Chem.* **13**, 1724-1729.

Gmelin (1960) *Handbuch der anorganischen Chemie*, Schwefel, Teil B, Lieferung 2, VCH, Weinheim, pp. 532-538 and 895-896.

Hahn, J. (1985) Zusammensetzung von Rohsulfan, Nachweis der Sulfane H_2S_9 bis H_2S_{35}. *Z. Naturforsch. Part B.***40**, 263-272.

Holleman-Wiberg (1985) *Lehrbuch der Anorganischen Chemie*, W. de Gruyter, Berlin.

Horner, D. A. and Connick, R. E. (1986) Equilibrium Quotient for the

Isomerization of Bisulfite Ion from HSO to SO_3H^-. *Inorg. Chem.* **25**, 2414-2417.
Janssen, A. J. H. (1996) Formation and colloidal behaviour of elemental sulphur produced from the biological oxidation of hydrogensulphide. Doctoral Dissertation, Landbouwuniversiteit Wageningen (NL).
Janssen, A. J. H., Lettinga, G., De Keizer, A. (1999) Removal of hydrogen sulfide from wastewater and waste gases by biological conversion to elemental sulfur. *Colloids Surfaces A: Physicochemical and Engineering Aspects,* **151**, 389-397.
Karchmer, J. H. (ed.) (1970) *The Analytical Chemistry of Sulfur and its Compounds*, parts I and II, Wiley, New York.
Mäusle, H.-J. and Steudel, R. (1981) Molecular Composition of Liquid Sulfur. Part 3: Quantitative Analysis in the Temperature Region 115-350°C. *Z. Anorg. Allg. Chem.* **478**, 177-190.
McCampbell Hamilton, E. (1991) The Chemistry of Low Valent Sulfur Compounds in the Sulfur-Water System, Ph.D. dissertation, University of Minnesota.
Müller, A. and Krebs, B. (1984) *Sulfur - Its Significance for Chemistry, for the Geo-, Bio- and Cosmosphere and Technology*, Elsevier, Amsterdam.
Müller, A. and Diemann, E. (1987) Polysulfide Complexes of Metals. *Adv. Inorg. Chem.* **31**, 89 (review).
Münchow., V. and Steudel, R. (1994) The Decomposition of Aqueous Dithionite and its Reactions with Polythionates $S_nO_6^{2-}$ (n = 3-5) Studied by Ion-Pair Chromatography. *Z. Anorg. Allg. Chem.* **620**, 121-126.
Nickless, G. (ed.) (1968) *Inorganic Sulphur Chemistry*. Elsevier, Amsterdam.
O'Brien, D. J. and Birkner, F. B. (1977) Kinetics of Oxygenation of Reduced Sulfur Species in Aqueous Solution. *Environm. Sci. Technol.* **11**, 1114-1120.
Otto A. H. and Steudel, R. (1999) *The Gas Phase Acidities of the Sulfanes* H_2S_n (n = 1-4). *Eur. J. Inorg. Chem.*, submitted.
Prange, A.; Arzberger, I., Engemann, C., Modrow, H., Schumann, O., Trüper, H. G., Steudel, R., Dahl, C. and Hormes, J. (1999) In situ analysis of sulfur in the sulfur globules of phototrophic sulfur bacteria by X-ray absorption near edge spectroscopy. *Biochim. Biophys. Acta*, accepted.
Rau, H., Kutty, T. R. N. and Guedes de Carvalho, J. R. F. (1973) Thermodynamics of sulphur vapour. *J. Chem. Thermodyn.* **5**, 833-844.
Schmidt, M. and Siebert, W. (1973) Sulphur. In *Comprehensive Inorganic Chemistry*, Vol. 2, Pergamon, Oxford, pp. 795-933 (review).
Schwarzenbach, G. and Fischer, A. (1960) Die Acidität der Sulfane und die Zusammensetzung wässriger Polysulfidlösungen. *Helv. Chim. Acta.* **43**, 1365-1391.
Steliou, K. (1991) Diatomic Sulfur. *Acc. Chem. Res.* **24**, 341-350 (review).
Steudel, R. and Mäusle, H.-J. (1981) Molecular Compostion of Liquid Sulfur. Part 2: Qualitative Analysis and Preparation of S_7, S_{12}, α-S_{18} and S_{20} from S_8. *Z. Anorg. Allg. Chem.* **478**, 156-176.
Steudel, R. (1982) Homocyclic Sulfur Molecules. *Top. Curr. Chem.* **102**, 149-176 (review).
Steudel, R., Strauss, R. and Koch, L. (1985) Quantitative HPLC Analysis and Thermodynamics of Sulfur Melts. *Angew. Chem. Int. Ed.* **24**, 59-60.
Steudel, R. and Holdt, G. (1986) Ion-pair chromatographic separation of polythionates $S_nO_6^{2-}$ with up to thirteen sulphur atoms. *J. Chromatogr.* **361**, 379-384.
Steudel, R., Holdt, G. and Nagorka, R. (1986) On the Autoxidation of Aqueous

Sodium Polysulfide. *Z. Naturforsch Part B* **41**, 1519-1522 and references cited therein.

Steudel, R. (1987) Sulphur Homocycles. In *The Chemistry of Inorganic Homo- and Heterocycles*, (Haiduc, I. and Sowerby, D. B., eds.), vol 2, Academic Press, London, p. 737-768 (review).

Steudel, R., Holdt, G., Göbel, T. and Hazeu, W. (1987) Chromatographic Separation of Higher Polythionates S_nO^- (n = 3...22) and Their Detection in Cultures of *Thiobacillus ferrooxidans;* Molecular Composition of Bacterial Sulfur Secretions. *Angew. Chem. Int. Ed.* **26**, 151-153.

Steudel, R., Göbel, T. and Holdt, G. (1988) The Molecular Composition of Hydrophilic Sulfur Sols Prepared by Acid Decomposition of Thiosulfate. *Z. Naturforsch. Part B* **44**, 203-218.

Steudel, R. and Holdt, G. (1988) Solubilization of Elemental Sulfur in Water by Cationic and Anionic Surfactants. *Angew. Chem. Int. Ed.* **27**, 1358-1359.

Steudel, R. and Holz, B. (1988) Detection of Reactive Sulfur Molecules (S_6, S_7, S_9, S_μ) in Commercial Sulfur, in Sulfur Minerals, and in Sulfur Melts Slowly cooled to 20°C. *Z. Naturforsch. Part B* **43**, 581-589.

Steudel, R., Holdt, G. and Göbel, T. (1989a) Ion-pair chromatographic separation of inorganic sulphur anions including polysulphide. *J. Chromatogr.* **475**, 442-446.

Steudel, R., Göbel, T. and Holdt, G. (1989b) The Molecular Nature of the Hydrophilic Sulfur Prepared from Aqueous Sulfide and Sulfite (Selmi Sulfur Sol). *Z. Naturforsch Part B* **44**, 526-530.

Steudel, R., Albertsen, J. and Zink, K. (1989c) Photoinduced ESR Signals in Orthorhombic cyclo-Octasulfur (α-S_8) and in Quenched Liquid Sulfur at Low Temperatures. *Ber. Bunsenges. Phys. Chem.* **93**, 502-509.

Steudel, R. and Prenzel, A. (1989) Raman Spectroscopic Discovery of the Hydrogenthiosulfate Anion HSSO in Solid $NH_4HS_2O_3$. *Z. Naturforsch. Part B* **44**, 1499-1502.

Steudel, R. and Albertsen, A. (1992) Determination of cysteine-S-sulfonate by ion-pair chromatography and its formation by autoxidation of cysteine persulfide. *J. Chromatogr.* **606**, 260-263.

Steudel, R. and Kustos, M. (1994) Sulfur: Organic Polysulfanes. In *Encyclopedia of Inorganic Chemistry* (King, R. B., ed.), Vol. 7, Wiley, Sussex, p. 4009-4038 (review).

Steudel, R. (1996) Mechanism for the Formation of Elemental Sulfur from Aqueous Sulfide in Chemical and Microbiological Desulfurization Processes. *Ind. Eng. Chem. Res.* **35**, 1417-1423 and references cited therein.

Steudel, R., Schumann, O., Buschmann, J. and Luger, P. (1998) A New Allotrope of Elemental Sulfur: Convenient Preparation of *cyclo*-S_{14} from S_8. *Angew. Chem. Int. Ed.* **37**, 2377-2378.

Steudel, R. (1998) *Chemie der Nichtmetalle,* 2 edn, W. de Gruyter, Berlin.

Steudel, R. (1999) The Chemistry of Aqueous Sulfur Sols - Models for Bacterial Sulfur Globules? In *Biochemical Principles and Mechanisms of Biosynthesis and Biodegradation of Polymers* (Steinbüchel, A., ed.), Wiley-VCH, Weinheim, p. 17-25.

Szekeres, L. (1974) The Analytical Chemistry of Sulfur Acids. *Talanta* **21**, 1-44 (review).

Szynkarczuk, J., Komorowski, P. G. and Donini, J. C. (1994) Redox Reactions of Hydrosulphide Ions on the Platinum Electrode - I. The Presence of Intermediate Polysulfide Ions and Sulphur Layers. *Electrochim. Acta* **39**, 2285-2289.

Takakuwa S. (1992) Biochemical Aspects of Microbial Oxidation of Inorganic Sulfur Compounds. In *Organic Sulfur Chemistry: Biochemical Aspects* (Oae, S., Okuyama, T. eds.), CRC Press, Tokyo, p. 1.

Tardif, S. L., Rys, A. Z., Abrams, C. B., Abu-Yousef, I. A., Lesté-Lassere, P. B. F., Schultz, E. K. V. and Harpp, D. N. (1997) Recent Chemistry of the Chalcogen Diatomics. *Tetrahedron* **36**, 12225-12236.

Tebbe, F. N., Wasserman, E., Peet, W. G., Vatvars, A. and Hayman, A. C. (1982) *J. Am. Chem. Soc.* **104**, 4971.

Teder, A. (1971) The Equilibrium Between Elementary Sulfur and Aqueous Polysulfide Solutions. *Acta Chem. Scand.* **25**, 1722-1728.

Ueyama, N. and Nakamura, A. (1992) In *Organic Sulfur Chemistry: Biochemical Aspects* (Oae, S. and Okuyama, T., eds.), CRC Press, Boca Raton, p.45.

Williamson, M. A. and Rimstidt, J. D. (1992) Correlation between structure and thermodynamic properties of aqueous sulfur species. *Geochim. Cosmochim. Acta* **56**, 3867-3880.

Zhu, J., Petit, K., Colson, A. O., DeBolt, S. and Sevilla, M. D. (1991) Reactions of $HS^{\bullet}$ and $S^{\bullet-}$ with Molecular Oxygen, H_2S, HS^-, and S^{2-}: Formation of SO^-, $HSSH^{\bullet-}$, $HSS^{\bullet 2-}$ and $HSS^{\bullet}$. *J. Phys. Chem.* **95**, 3676-3681.

2

The geochemical sulfur cycle

Jack J. Middelburg

2.1 INTRODUCTION

Sulfur, the 14th most abundant element in the Earth's crust, occurs in valence states ranging from +6 in sulfate to –2 in sulfides. The reduced states (-2 and –1) are found in metal and metalloid sulfides (Ag, Fe, Cd, Hg, Mn, Ca, Te, Se, As, Sn, Cu, Pb, Pt, Sb, Co, Ni, Mo, Rn, W) of which pyrite (FeS_2) is the most abundant. The intermediate valencies (0, +2, +4) have only a transitory existence in the geochemical cycle and are usually reaction intermediates (e.g. elemental sulfur in sulfide oxidation). Sulfate is the dominant oxidized (+6) form and it is the second most abundant anion in seawater (about 2700 mg/l; after chloride) and in rivers (about 11 mg/l; after bicarbonate). The stability of aqueous sulfur species can be presented in an Eh-pH diagram (often named Pourbaix diagram (Pourbaix and Pourbaix 1992; Sato 1992)) in which the thermodynamic favourable forms are shown as a function of the environmental redox and acidity conditions. The upper and lower limits of the aqueous S species are constrained by the stability of water (Figure 2.1). Sulfate is the dominant aqueous sulfur species in fresh

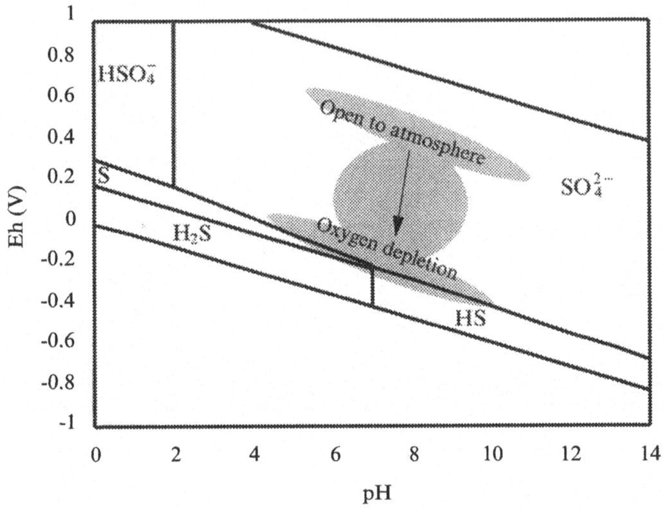

Figure 2.1. Eh-pH diagram for the aqueous sulfur cycle (at a total sulfur concentration of 10^{-2} mole). The distribution of Eh-pH values in natural systems has been shaded. The arrow indicates the evolution of Eh-pH values when oxygen concentrations decrease owing to the limited supply of atmospheric oxygen.

and marine environments. It is only at low redox conditions (less than about –200 mV at pH 7) or at low pH values (pH < 2) that other dissolved sulfur species become important.

The two dominant solid phase appearances of sulfur are in the form of sulfates and sulfides (Table 2.1) although sulfur is also found as elemental sulfur (biogenic as well as volcanic) and in other minerals (e.g. carbonates) and bound in organic matter.

Table 2.1. Major sulfur minerals

Mineral	Formula	Crystal Structure	Specific Gravity (g cm^{-3})	Hardness
Sulfides				
Pyrite	FeS_2	Cubic	4.95-5.03	6-6.5
Marcasite	FeS_2	Orthorhombic	4.89	6-6.5
Pyrrhotite	Fe_7S_8-FeS	Monoclinic	4.6	3.5-4.5
Chalcopyrite	$CuFeS_2$	Tetragonal	4.1-4.3	3.5-4
Sphalerite	ZnS	Cubic	4.1	3.5-4
Galena	PbS	Cubic	7.5-7.6	2.5
Sulfates				
Barites	$BaSO_4$	Orthorhombic	4.5	2.5-3.5
Celestine	$SrSO_4$	Orthorhombic	3.96	3-3.5
Gypsum	$CaSO_4\cdot 2H_2O$	Monoclinic	2.3-2.37	2
Anhydrite	$CaSO_4$	Orthorhombic	2.9-3	3-3.5

2.2 GLOBAL SULFUR RESERVOIRS AND FLUXES

2.2.1 The global sulfur cycle

The original pool of sulfur was contained in igneous rocks, largely as pyrite (FeS_2). Volcanic degassing and weathering under aerobic conditions resulted in transfer of large amounts of S (of the order of 10^{21} moles) from igneous rocks to the hydrosphere (freshwater and ocean). The major global reservoirs of S are currently sediments hosting pyrite and evaporites ($CaSO_4 \cdot 2H_2O$) and the ocean as sulfate (Figure 2.2).

Uplift of rocks results in the weathering of pyrite and evaporites. Pyrite weathering is mainly due to oxidation that results in the production of sulfuric acid. Weathering of gypsum ($CaSO_4 \cdot 2H_2O$), anhydrite ($CaSO_4$) and other sedimentary sulfate minerals occurs through simple dissolution, the rate depending on hydrological conditions. Sulfur is transported back to the ocean by rivers, principally as dissolved sulfate. About half of the riverine sulfate loading is due to atmospheric deposition. Removal of sulfur from the ocean takes place via deposition of sulfate minerals, formation of sedimentary pyrite, sea-air transfer and hydrothermal processes causing deposition of massive sulfide deposits (e.g. black smokers). Hydrothermal sulfide deposition is confined to geologically active zones of the Earth, including mid-ocean ridges and collision zones. Hydrothermal fluids with temperatures up to 350 °C and rich in sulfide (millimolar range) mix with seawater, and upon mixing $CaSO_4$ and iron sulfides precipitate. Hydrothermal vent sulfides are important ore deposits for metals and metalloids. The residence time of sulfate in the ocean is of the order of 11 million years.

2.2.2 The atmospheric sulfur cycle

The atmospheric component of the global sulfur is highly dynamic with residence times of most gases of the order of a few days, because of rapid oxidation to sulfate. The release of sulfur gases is due to volcanic (mainly SO_2) and biological processes with autotrophic organisms (e.g., algae in the ocean, higher plants on land) as well as heterotrophs (bacteria) involved (see chapter 3). The dominant gas emitted from terrestrial systems (freshwater wetlands and anaerobic soils) is hydrogen sulfide, with dimethyl sulfide (DMS) and carbonyl sulfide (COS) being less important. The ocean is a significant source of sulfate containing aerosols and of the biogenic gases DMS and COS. The reduced sulfur gases DMS and H_2S are efficiently oxidized to sulfate in the troposphere. COS is rather inert in the

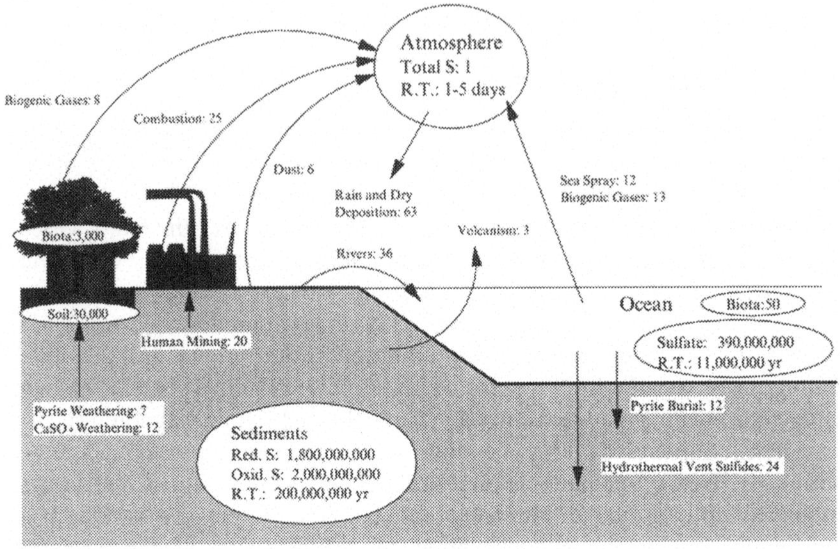

Figure 2.2. Global biogeochemical cycle of sulfur. All reservoirs are given in 10^{11} moles S and all fluxes (arrows) in 10^{11} moles S per year. R.T. is residence time. Based on Schlesinger (1991) and Mackenzie *et al.* (1993).

troposphere, and consequently has a relative long residence time (5 years) and can enter the stratosphere. In the stratosphere, COS is destroyed by a photochemical reaction involving the OH radical, producing sulfate and sulfate aerosols. The oxidation of atmospheric sulfur gases can affect the formation of sulfate aerosol particles that act as cloud-condensation nuclei.

The global geochemical cycle of sulfur has been perturbed by human activities in two ways. Burning of fossil fuels and smelting of sulfide ores has caused elevated release of SO_2 and SO_3 to the atmosphere. Oxidation and hydrolysis of SO_2 causes acid substances that are deposited by rainfall (wet deposition) or in the form of dry deposition. Acid deposition has significantly affected the health of numerous ecosystems and may induce the mobilization of aluminum and other potential toxic metals from soils and sediments (Likens *et al.* 1979).

The second perturbance of the geochemical sulfur cycle is due to elevated emission rates of S gases. Eutrophication of the coastal zone may have enhanced DMS emission from coastal marine areas, and terrrestrial biomass burning has likely affected COS emission. These enhanced emission rates may contribute to climatic change (Mackenzie *et al.* 1993).

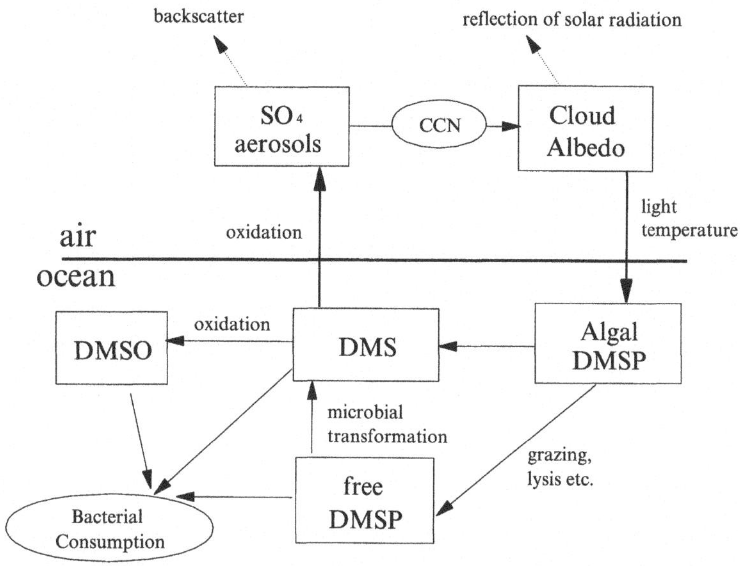

Figure 2.3. Schematic of the processes that link the biogeochemical cycle of dimethylsulfide (DMS) with climate. DMSP: dimethylsulphoniopropionate; DMSO: dimethylsulfoxide; CCN: cloud condensation nuclei.

2.2.3 Sulfur-climate interactions

Various species of phytoplankton produce dimethylsulphoniopropionate (DMSP) as an osmolyte that can alleviate salt stress and prevent freezing. DMSP is converted to dimethylsulfide (DMS) by the enzyme DMSP lyase, in particular when DMSP is liberated from the algae by virus infection or zooplankton grazing (Figure 2.3). Part of the DMS is consumed by bacteria or oxidized to dimethylsulfoxide (DMSO), another part is exchanged to the atmosphere (Malin *et al.* 1992).

In the atmosphere, DMS is oxidized in a sequence of reactions ultimately resulting in sulfate. Atmospheric sulfate contributes to aerosol formation. Sulfate aerosol particles not only cause direct backscattering of solar radiation, but they are also a major source of cloud condensation nuclei, which can form cloud droplets that are important scatterers of solar radiation. Accordingly, DMS-derived aerosols have the potential to affect cloud albedo and consequently the heat and radiation balance of the Earth (Charlson *et al.* 1987; Andreae and Crutzen 1997).

Enhanced cloud albedo due to enhanced DMS emission could lead to cooling because a larger proportion of solar radiation is returned to space. The consequent global cooling and reduced light availability would reduce photosynthesis, hence the production and emission of DMS. This feedback between climate and DMS suggests an active role of organisms in maintaining the climate of our planet (Lovelock 1997).

2.3 THE SEDIMENTARY SULFUR CYCLE

The two major sedimentary sulfur pools are sulfates and sulfides, which are distinct in terms of mineralogy (Table 2.1), weathering, depositional environments and biogeochemistry. It is therefore instructive to discuss them separately.

2.3.1 Sedimentary sulfates

Gypsum is the most important sedimentary sink of sulfate. The oceans are under-saturated with respect to gypsum. However, concentration of seawater by evaporation in restricted basins (and mid-ocean ridges) may result in the precipitation of gypsum. Gypsum starts to precipitate at a concentration factor of about 3.4, i.e. in strong evaporitic conditions as found in salt pans and inland sebkhas (dried up desert lakes). During evaporative concentration, only part of the sulfate can precipitate because initial sulfate concentrations are about 2.8 times those of calcium and part of the calcium has been removed before by calcium carbonate precipitation. Further evaporation of seawater will eventually lead to the formation of potassium- and magnesium-bearing sulfates (Holland 1984). Gypsum ($CaSO_4{\cdot}2H_2O$) can become dehydrated to form anhydrite ($CaSO_4$) if the activity of water decreases.

Formation of barite ($BaSO_4$) and celestine ($SrSO_4$; also known as celestite) does not require extreme evaporitic conditions, because they are just slightly under-saturated in seawater. Barite and celestine have been reported to form during oxidative weathering of sulfide bearing rocks and sediments (see, for example, Van Os *et al.* 1996). Similarly, slightly elevated sulfate concentrations in marine environments (due to slight evaporation or as a result of extensive oxidation of sedimentary sulfides) may result in the formation of barite and celestine.

Barite and celestine also have biological sources. Acantharians, planktonic protozoans that are widespread in the oceans, have skeletons and cysts that consist of celestine. Acantharian-derived celestine contains a variety of trace elements including barium (Bernstein *et al.* 1992). The mechanism for barite formation is uncertain, but includes gravitropism (to regulate the density of organisms relative to seawater) and formation in micro redox environments in decaying organic matter (Dehairs *et al.* 1987).

The sedimentary barium content, i.e. the barite content, is a widely used measure to derive the carbon flux to marine sediments. The premise behind this palaeoceanographic tool is the strong correlation between barite fluxes and organic fluxes to the sediment, and the high preservation potential of barite while most of the organic carbon is removed over time by oxidation to carbon dioxide (Dymond *et al.* 1992).

2.3.2 Iron sulfide formation and organic matter degradation

The sedimentary cycles of organic carbon and reduced sulfur are intimately linked and therefore discussed together.

Part of the organic matter transported from the continent to the ocean or produced in the photic zone of the ocean is being degraded in the water column, the remainder being delivered to the sediments. The flux of organic matter to the sediments depends mainly on the water depth: greater water depths are accompanied with lower carbon inputs. Most of the organic matter arriving at the sediment-water interface is degraded with only a small fraction (a few percent) being buried (Middelburg *et al.* 1993).

Degradation of organic matter is an oxidation reaction because zero-valent carbon in organic matter is oxidized to tetravalent carbon in carbon dioxide, the ultimate reaction product. It is a biologically mediated conversion though photochemical reactions may contribute to organic matter alteration in the surface layer of the ocean. The major electron acceptors for microbial respiration are oxygen, nitrate, manganese oxide, iron oxides, and the oxygen bound in organic matter itself (methanogenesis). The major oxidants are utilized not simultaneously, but sequentially as represented in Table 2.2. The sequence is set by factors such as thermodynamic energy yield (ΔG, Table 2.2), oxidant availability, process kinetics and bacterial physiology.

Table 2.2. Bacterial oxidation reactions of organic matter

Overall reaction	Process	ΔG (kJ/mol)
$CH_2O + O_2 \rightarrow CO_2 + H_2O$	Oxic	-470
$5\ CH_2O + 4NO_3^- \rightarrow 2N_2 + 4HCO_3^- + CO_2 + 3\ H_2O$	Denitrification	-448
$CH_2O + 3\ CO_2 + H_2O + 2MnO_2 \rightarrow 2Mn^{2+} + 4HCO_3^-$	Mn-reduction	-349
$CH_2O + 7\ CO_2 + 4Fe(OH)_3 \rightarrow 4Fe^{2+} + 8HCO_3^- + 3\ H_2O$	Fe-reduction	-114
$CH_2O + SO_4^{2-} \rightarrow H_2S + 2HCO_3^-$	S-reduction	-77
$CH_2O \rightarrow CO_2 + CH_4$	Methanogenesis	-48

Table 2.3. Simplified re-oxidation reactions involving sulfur

$SO_4^{2-} + CH_4 \rightarrow 2HCO_3^- + H_2S$	Anaerobic methane oxidation
$2\,O_2 + H_2S/FeS_2 \rightarrow H_2SO_4$	Aerobic sulfide oxidation
$NO_3^-/MnO_2/FeOOH + H_2S \rightarrow SO_4^{2-}$ + red. prod.	Anaerobic sulfide oxidation

As a consequence of this sequential utilization, there is a distict zonation with depth in sediment: oxic mineralization in the top layer, sulfate reduction at depth and denitrification and metal oxide reduction in between. Similarly, the relative importance of these oxidants in organic matter mineralisation depends strongly on the organic carbon loading (i.e. total degradation rate). At low carbon loadings (< 0.2 mmol m^{-2} yr^{-1}), oxygen supply can keep up with oxygen demand and oxic mineralization dominates. Sulfate reduction dominates at high carbon loadings (Figure 2.4).

The reaction products of mineralization at depth may diffuse upwards and react directly with oxidants in the surface layer (Table 2.3). For instance, methane produced by methanogenesis may diffuse upwards and react with sulfate so that sulfate is reduced to sulfide and methane becomes oxidized to carbon dioxide (anaerobic methane oxidation: Hoehler *et al.* 1994). Similarly, upwardly migrating sulfide may reduce oxygen, nitrate, manganese and iron oxides. In coastal sediments, oxygen consumption due to sulfide oxidation is about as important as oxygen consumption due to oxic mineralization (Jørgensen 1983). Many of these re-oxidation reactions are microbially mediated but the mechanisms and organisms involved have not yet been fully identified.

Sulfate reduction, whether coupled directly to organic matter oxidation (Table 2.2) or indirectly via methane oxidation (Table 2.3), results in the formation of hydrogen sulfide and bisulfide. Free sulfide (S^{2-}) is not a major species in natural ecosystems (Figure 2.1). Hydrogen sulfide and bisulfide may either diffuse upwards or downwards from the place of formation, or react with iron or organic matter.

Hydrogen sulfide, bisulfide or oxidation intermediates (e.g. elemental sulfur and polysulfide) may react with functional groups of organic matter (Luther and Church 1992). Because of the incorporation of sulfur into the organic skeleton, these components may become less labile (Sinninghe Damsté and de Leeuw 1990). In most sediments the availability and reactivity of iron are much higher than that of organic matter. Consequently, organic matter represents an insignificant sink of reduced sulfur.

The reactivity of iron containing minerals towards hydrogen sulfide varies widely, i.e. eight orders of magnitude (Table 2.4). In sediments

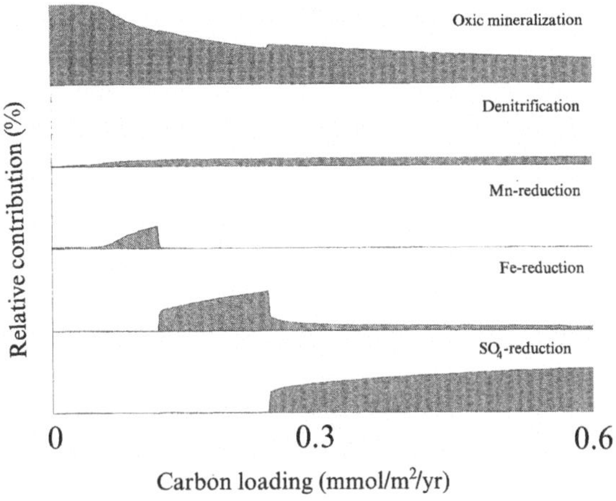

Figure 2.4. Relative contribution of various oxidation pathways to organic matter mineralization in marine sediments as a function of carbon loading (After: Wijsman *et al.* 1999). The contribution of methanogenesis is very small and therefore not shown.

containing the iron minerals ferrihydrite, lepidocrocite, goethite and hematite dissolved sulfide readily precipitates as iron sulfide and dissolved sulfide levels are maintained at a low level. In sediments with the less reactive iron phases or with low iron content, dissolved sulfide may reach high levels and organic sulfur compounds may form.

The first reaction product formed is iron monosulfide rather than the thermodynamic stable form pyrite. Pyrite formation requires either reaction of iron monosulfide with elemental sulfur ($FeS + S_0 \rightarrow FeS_2$; Berner 1984) or with hydrogen sulfide ($FeS + H_2S \rightarrow FeS_2 + H_2$; Rickard 1997). These two pyrite formation pathways have different kinetics. The first reaction requires initially oxidation of hydrogen sulfide to elemental sulfur, whereas hydrogen sulfide acts as the oxidant in the second reaction. Sedimentary bacteria readily consume hydrogen.

Table 2.4. Rate constants (per year) of sedimentary iron minerals with respect to their reaction with dissolved sulfide (From: Canfield *et al.* 1992)

Iron mineral	Rate constant	Iron mineral	Rate constant
Ferrihydrite	2200	Magnetite	6.6×10^{-3}
Lepidocrocite	> 85	Reactive silicates	3×10^{-3}
Goethite	22	Sheet silicates	8×10^{-6}
Hematite	12	Ilmenite, augite	$< 8 \times 10^{-6}$

Iron monosulfide is soluble in hydrochloric acid and is often referred to as acid volatile sulfur (AVS). Pyrite is not soluble in hydrochloric acid and usually determined by reduction with chromium (chromium reducible sulfur, CRS). These sulfur pools represent important sinks for trace elements and their cycling has received much attention from environmental scientists (Huerta-Diaz and Morse 1992).

2.3.3 Carbon-sulfur-iron relationships

Sedimentary pyrite formation can be limited by (1) the availability of sulfate, (2) the amount and lability of organic matter or (3) the amount and reactivity of iron minerals. Relationships between sedimentary concentrations of reduced sulfur, iron and organic carbon can be used to characterize pyrite formation and/or palaeoenvironmental conditions (Figure 2.5). These relationships are usually investigated using sulfur versus organic carbon plots and the degree of pyritization (DOP), where DOP is given by the ratio between pyrite iron and total reactive iron (Berner 1984).

The sulfur versus organic carbon plot can be used to distinguish between marine and freshwater sediments, because pyrite formation in freshwater sediments is primarily limited by the availability of sulfate. The S C plot can also be used to distinguish between marine sediments deposited under oxic water where concentrations of sulfur and organic carbon are closely

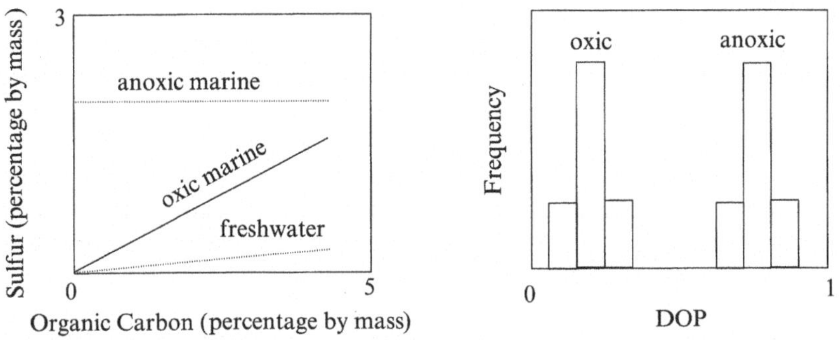

Figure 2.5. Environmental reconstructions using C-S-Fe relationships. Left: The sulfur versus organic carbon plot allows distinction between normal marine sediments being deposited under oxic marine water on the one hand and freshwater sediments or sediments being deposited in anoxic marine water on the other. Right: Frequency distributions of the degree of pyritization values are indicative of sediments deposited under oxic (low DOP) and anoxic (high DOP) water, respectively.

coupled (S=0.36*Organic Carbon), and those deposited in anoxic water where sulfur concentrations are higher and independent of organic carbon concentrations. Moreover, DOP values of sediments being deposited in oxic water (less than 0.42) are significantly lower than those deposited in anoxic water (more than 0.55), partly because of pyrite formation in the anoxic water column (Passier *et al.* 1997).

2.4 SULFUR ISOTOPES

There are nine known isotopes of sulfur of which four are stable (i.e. non-radioactive). The relative crustal abundance of these are ^{32}S (95%), ^{33}S (0.76%), ^{34}S (4.22%) and ^{36}S (0.014). Since the absolute abundance of a stable isotope is not measured readily and extreme small differences in isotope ratios between a standard and a sample can be measured precisely with isotope-ratio mass spectrometers, the delta notation has been introduced. The standard, Canyon Diablo Troilite (CDT, a meteorite), has been chosen so as to be close to the isotopic composition as it was at the time of the formation of the Earth. The delta (δ) notation relates the ^{34}S:^{32}S ratio of a sample to that of CDT and is expressed as parts per thousand using the notation:

$$\delta^{34}S\,(‰) = \left(\frac{(^{34}S/^{32}S)_{sample}}{(^{34}S/^{32}S)_{CDT}} - 1\right) \times 1000$$

2.4.1 Sulfur isotope characteristics and fractionation

The distribution of sulfur isotopes is mainly governed by oxidation and reduction processes, most of which are microbially mediated. Igneous (and lunar) rocks have $\delta^{34}S$ values close to zero, as they are similar to that of sulfides in meteorites. Modern ocean water sulfate has a rather uniform $\delta^{34}S$ value of about + 20‰ and marine evaporites have values close to that of seawater sulfate because little fractionation occurs during mineral formation. There is also little fractionation during formation of sulfide minerals.

Dissimilatory sulfate reduction by bacteria strongly differentiates among the isotopes of sulfur, as a result of a more rapid enzymatic reaction with $^{32}SO_4$. Pure cultures of sulfate-reducing bacteria produce sulfide depleted in ^{34}S by 4 to 46‰, with an average of 18±10‰ (Canfield and Teske 1996). However, sedimentary sulfides show depletions in ^{34}S relative to seawater sulfate varying from 24 to 71‰, with an average of 51±10‰. These large depletions of sedimentary sulfides are thought to result from an initial fractionation by sulfate reducing bacteria, followed by further ^{34}S discrimination during the oxidative part of the sulfur cycle. Intermediates of sulfide oxidation such as elemental sulfur and thiosulfate are subject to disproportionation during which the sulfate becomes enriched in ^{34}S and the sulfide becomes depleted in ^{34}S (Canfield and Thamdrup 1994). Repetitive cycles of sulfide oxidation to elemental sulfur, sulfite or thiosulfate and subsequent disproportionation can result in sulfides strongly depleted in ^{34}S.

2.4.2 Atmospheric oxygen, the global sulfur cycle and sulfur isotopes

The biogeochemical cycles of carbon and sulfur are the main determinants of atmospheric oxygen levels over geological time. Since atmospheric oxygen has been maintained at levels above those critical for aerobic organisms, major changes in the reservoirs of reduced sulfur and organic carbon must have been balanced by those in the oxidized reservoirs (Berner 1987). This means there is a balance between weathering of sedimentary organic carbon and sulfides on land (which consume oxygen):

$$O_2\downarrow + CH_2O \rightarrow CO_2 + H_2O$$
$$15O_2\downarrow + 4FeS_2 + 16CaCO_3 + 8H_2O \rightarrow 2Fe_2O_3 + 16Ca^{2+} + 16HCO_3^- + 8SO_4^{2-}$$

and burial of organic carbon and pyrite in marine sediments (which releases oxygen):

$$CO_2 + H_2O \rightarrow O_2\uparrow + CH_2O$$
$$2Fe_2O_3 + 16Ca^{2+} + 16HCO_3^- + 8SO_4^{2-} \rightarrow 15O_2\uparrow + 4FeS_2 + 16CaCO_3 + 8H_2O$$

These changes have consequences for the sulfur isotopes of marine evaporites. During periods that pyrite deposition in marine sediments is in excess of the oxidative weathering of sulfides on land, there is enhanced transfer of ^{32}S from dissolved sulfate to sedimentary sulfides relative to ^{34}S. The sulfate remaining in seawater becomes enriched in ^{34}S and the $\delta^{34}S$ values of marine evaporites are consequently heavier. This has occurred during the Cambrian (+32‰). Conversely, seawater sulfate was lighter during periods of low pyrite burial in marine sediments (such as the Carboniferous and Permian when net primary production shifted from the ocean to freshwater swamps with little pyrite formation).

The sulfur isotope record of sedimentary sulfides also provides information on the development of atmospheric oxygen levels in the Precambrian (> 540 million years ago). They indicate a two-stage rise in atmospheric oxygen levels (Canfield and Teske 1996; Canfield 1998). The first stage, 2300 to 2000 million years ago, resulted in an increase in seawater sulfate concentrations above the critical limit for sulfate reducers to express isotope fractionation. Sedimentary sulfides formed after this first stage showed a range of isotopes values consistent with sulfate reduction. The second stage, 1000 to 540 million years ago, resulted in the development of fully oxygenated oceanic water and the initiation of the oxidative sulfur cycle. This was accompanied with strongly depleted sedimentary sulfides that have pertained ever since.

2.5 REFERENCES

Andreae, M.O. and Crutzen, P.J. (1997) Atmospheric aerosols: biogeochemical sources and role in atmospheric chemistry. *Science* **276**, 1052-1058.

Berner, R.A. (1984) Sedimentary pyrite formation: an update. *Geochim. Cosmochim. Acta* **48**, 605-615.

Berner, R.A. (1987) Models for carbon and sulfur cycles and atmospheric oxygen: application to paleozoic geological history. *Amer. J. Sci.* **287**, 177-196.

Bernstein, R.E., Byrne, R.H., Betzer, P.R. and Greco, A.M. (1992) Morphologies and transformations of celestite in seawater: The role of acantharians in strontium and barium geochemistry. *Geochim. Cosmochim. Acta* **56**, 3273-3281.

Canfield, D.E. (1998) A new model of Proterozoic ocean chemistry. *Nature* **396**, 450-453.

Canfield, D.E. and Thamdrup, B. (1994) The production of ^{34}S depleted sulfide during bacterial disproportionation of elemental sulfur. *Science* **266**, 1973-1975.

Canfield, D.E. and Teske, A. (1996) Late Proterozoic rise in atmospheric oxygen concentration inferred from phylogenetic and sulphur-isotope studies. *Nature* **382**, 127-132.

Canfield, D.E., Raiswell, R. and Bottrell, S. (1992) The reactivity of sedimentary iron minerals towards sulfide. *Amer. J. Sci.* **292**, 659-683.

Charlson, R.J., Lovelock, J.E., Andreae, M.O. and Warren S.G. (1987) Oceanic phytoplankton, atmospheric sulphur, cloud albedo and climate. *Nature* **326**, 655-661.

Dehairs, F., Lambert, C.E., Chesselet, R. and Risler, N. (1987) The biological production of marine suspended barite and the barium cycle in the western Mediterranean Sea. *Biogeochemistry* **4**, 119-139.

Dymond, J., Suess, E. and Lyle, M. (1992) Barium in deep-sea sediment: A Geochemical proxy for paleoproductivity. *Paleoceanography* **7**, 163-182.

Hoehler, T.M., Alperin, M.J., Albert, D.B., and Martens C.S. (1994) Field and laboratory studies of methane oxidation in an anoxic marine sediment: evidence for a methanogen-sulfate reducer consortium. *Glob. Biogeoch. Cycles* **8**, 451-464.

Holland, H.D. (1984) The chemical evolution of the atmosphere and oceans. Princeton Univ. Press, 582 pp, Princeton, USA.

Huerta-Diaz, M.A. and Morse J.W. (1992) Pyritization of trace metals in anoxic marine sediments. *Geochim. Cosmochim. Acta* **56**, 2681-2702.

Jørgensen, B.B. (1983) Processes at the sediment-water interface In *The major biogeochemical cycles and their interactions*, SCOPE, editors Bolin, B and Cook R.B., J. Wiley & Sons, p.477-515.

Likens, G.E., Wright, R.F., Galloway, J.N., Butler, T.J. (1979) Acid rain. *Scient. Amer.* **241**, 43-51.

Lovelock J. (1997) A geophysiologist's thoughts on the natural sulphur cycle. *Phil. Trans. R. Soc. Lond.* B **352**, 143-147.

Luther, G.W. III and Church, T.M. (1992) An overview of the environmental chemistry of sulphur in wetland systems. In *Sulfur cycling on the continents*, SCOPE, editors Howarth, R.W., Stewart, J.W.B. and Ivanov, M.V., J. Wiley & Sons, p. 125-142.

Malin, G., Turner, S.M. and Liss, P.S. (1992) Sulfur: the plankton/climate connection. *J. Phycol.* **28,** 590-597.

Mackenzie, F.T., Ver, L.M., Sabine, C., Lane, M. and Lerman, A. (1993) C, N, P, S global biogeochemical cycles and modeling of global change. In *Interactions of C, N, P and S biogeochemical cycles and global change*, edited by R. Wollast, F.T. Mackenzie and L. Chou, Springer-Verlag, p. 1-61.

Middelburg, J.J., Vlug, T. and van der Nat, F.J.W.A. (1993) Organic matter mineralization in marine sediments. *Glob. Planet. Change* **8**, 47-58.

Passier, H.F., Middelburg, J.J., de Lange, G.J. and Böttcher, M.E. (1997) Pyrite contents, microtextures, and sulfur isotopes in relation to formation of the youngest eastern Meditterenean sapropel. *Geology* **25**, 519-522.

Pourbaix, M. and Pourbaix, A. (1992) Potential-pH equilibrium diagrams for the system S-H_2O from 25 to 150°C: Influence of access of oxygen in sulphide solutions. *Geochim. Cosmochim. Acta* **56**, 3157-3178.

Rickard, D.T. (1997) Kinetics of pyrite formation by the H_2S oxidation of Fe(II) monosulfide in aqueous solutions between 25 and 125°C: the mechanism. *Geochim. Cosmochim. Acta* **61**, 135-147.

Sato, M. (1992) Persistency-field Eh-pH diagrams for sulfides and their application to supergene oxidation and enrichment of ore bodies. *Geochim. Cosmochim. Acta* **56**, 3133-3156.

Schlesinger, W.H. (1991) Biogeochemistry. Academic Press, 443pp, San Diego, USA.

Sinninghe Damsté, J.S. and de Leeuw, J.W. (1990) Analysis, structure and geochemical significance of organically-bound sulfur in the geosphere: state of the art and future research. *Organic Geochem.* **16**, 1077-1101.

Van Os, B.J.H., Middelburg, J.J. and de Lange, G.J. (1996) Extensive degradation and fractionation of organic matter during subsurface weathering. *Aquat. Geochemistry* **1**, 302-312.

Wijsman, J.W.M., Herman P.M.J., Middelburg, J.J. and Soetaert, K. (1999) A model for early diagenetic processes in sediments of the continental shelf of the Black Sea. *Estuar. Coast. Shelf Sci.* (in press).

3

The biological sulfur cycle†

Thomas Brüser, Piet N.L. Lens and Hans G. Trüper

3.1 INTRODUCTION

Sulfur is abundant in all organisms, appearing in many organic compounds such as amino acids, (poly-)peptides, enzyme cofactors, antibiotics, lipids or carbohydrates. In the various organic compounds sulfur can have either catalytical, structural or regulatory functions. In contrast, the biological roles of inorganic sulfur compounds are rather restricted: either they serve as sources for sulfur assimilation and incorporation into the above mentioned organic compounds, or they are employed as donors or acceptors of electrons for dissimilatory electron transport. The latter pathways establish an electrochemical membrane potential which in turn can be used for ATP synthesis, NAD reduction or other energy consuming purposes.

The biochemical oxidations and reductions of sulfur compounds constitute the biological sulfur cycle which is schematically shown in Figure 3.1. While the assimilatory reduction of sulfate is very common in prokaryotes,

† Dedicated to the memory of Harry D. Peck Jr., who died on 20 November, 1998.

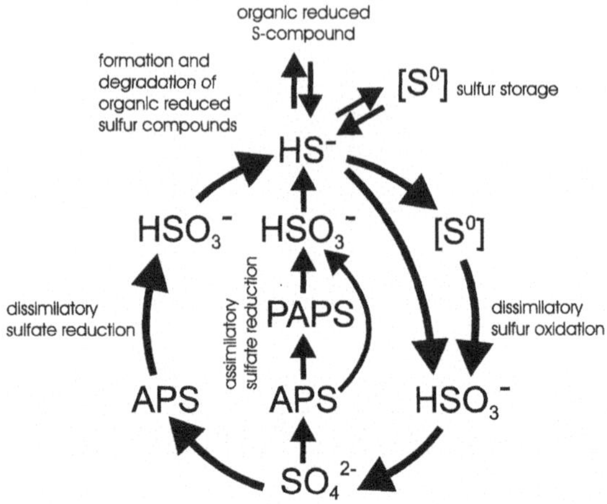

Figure 3.1. The biological sulfur cycle. $[S^0]$ denotes compounds in which sulfur of the oxidation state S^0 occur, like polythionates, polysulfanes, polysulfides or cyclic S^0-sulfur. Note that stored $[S^0]$-sulfur probably has to be reduced back to the sulfide level before further oxidation. APS: adenosine-5'-phosphosulfate, PAPS: 3'-phosphoadenosine 5'-phosphosulfate.

plants and fungi, the dissimilatory pathways are restricted to eubacteria and archaebacteria.

In natural habitats the pathways of the above sulfur cycle are usually interconnected. Sulfur compound oxidizing and reducing bacteria may form a syntrophical community which Baas-Becking (1925) termed "sulfuretum". In most sulfureta a "large sulfur cycle" between sulfide and sulfate can be observed. A high turnover rate of sulfur compounds is typical for such communities. When sulfide is non-biologically oxidized to elemental sulfur near the redoxcline of aquatic systems, a "small sulfur cycle" between elemental sulfur and sulfide can exist. Pfennig and Biebl (1976) described such a synthrophic community of the sulfur reducer *Desulfuromonas* and sulfide oxidizing Chlorobiaceae. Indeed most H_2S produced by bacteria is reoxidized in the same ecosystem. Ingvorsen and Jørgensen (1982) showed this for Danish estuaries where less than 1% of the H_2S formed during one year was released to the atmosphere.

This review summarizes the current knowledge about the various parts of the above cycle, thereby emphasizing the composition and intracellular localization of the various pathways. For further details about particular aspects, the reader is referred to reviews (dissimilatory S-reduction: Hamilton (1998), Hansen (1994), Widdel and Hansen (1992), Fauque *et al.*

(1991), Gibson (1990); dissimilatory S-oxidation: Kelly (1999), Friedrich (1998), Kelly *et al.* (1997), Brune (1995); assimilatory S-reduction: Bick and Leustek (1998), Marzluf (1997), Kredich (1996), LeFaou *et al.* (1990)).

3.2 THE INORGANIC SULFUR COMPOUNDS OF THE BIOLOGICAL SULFUR CYCLE

Sulfide, polysulfides, thiosulfate, polythionates, elemental sulfur, bisulfite and sulfate are the most common inorganic sulfur compounds that occur in natural environments. They can be used by many organisms for dissimilatory or assimilatory purposes. Sulfur compounds of intermediate oxidation states may serve either as electron acceptors or as electron donors in redox processes. In contrast, sulfide and sulfate cannot be further reduced or oxidized, respectively. Sulfide and sulfate are therefore the final products of most sulfur compound reduction or oxidation pathways. The oxidation states of relevant compounds are listed in Table 3.1.

In dissimilatory pathways, the turnover of sulfur compounds is high and the final product is released into the environment. However, truncated redox-pathways also occur, which result in the formation and secretion of sulfur compounds of intermediate oxidation states. The turnover of assimilatory pathways is comparatively low and the sulfur compounds are incorporated into organic compounds mainly - but not exclusively - on the sulfide level.

It is important to recall that free inorganic sulfur compounds are not only produced by biological processes: abiotic processes and the biological sulfur cycle are interconnected. The abiotic formation of inorganic sulfur

Table 3.1. Inorganic sulfur compounds of biological relevance

Compound	Chemical formula	Sulfur oxidation state
sulfide	HS^-	-2
polysulfides	$^-S(S)_nS^-$	-1 (terminal S) / 0 (inner S)
thiosulfate	$S_2O_3^{2-}$	-1 (sulfane S) / +5 (sulfone S)
polythionates	$^-O_3S(S)_nSO_3^-$	0 (inner S) / +5 (sulfone S)
elemental sulfur	S_n rings	0
bisulfite	HSO_3^-	+4
sulfate	SO_4^{2-}	+6

* In analogy to sulfate, the oxidation states -2 for sulfane- and +6 for sulfone sulfur are sometimes used in literature. However, the S-S bond in thiosulfate has much less double bond character than the S-O bonds. Therefore there is no reason to deviate from the formal numbers given in Table 3.1.

substrates can strongly influence the species composition in natural habitats. At this point we will summarize briefly the various known origins of biologically important inorganic sulfur compounds.

Sulfide is formed as the main product of sulfate respiration by bacteria that dominate under anaerobic conditions when organic matter and sulfate are not limiting. Sulfur compounds other than sulfate (e.g. thiosulfate, elemental sulfur) can also serve as electron acceptors for respirations that lead to sulfide. Sulfide is also released by desulfuration of organic compounds that contain reduced sulfur (e.g. proteins). Further sources of sulfide are minerals in soils and rocks or sulfide enriched springs. In this context the bioleaching process has to be mentioned, where metal sulfide minerals (e.g. pyrite) are oxidized by chemolithoautotophic bacteria to sulfuric acid and dissolved heavy metal ion sulfates (Schippers and Sand 1999).

Thiosulfate can be formed by non-biological reaction of sulfite with sulfur or polythionates (Schmidt 1984). It has been proposed that this process may lead to thiosulfate after incomplete oxidation of sulfide (Tuttle and Jannasch 1973). A further origin of thiosulfate is the leaching process where thiosulfate is formed by oxidation of sulfidic minerals like FeS_2, MoS_2 or WS_2 (Schippers *et al.* 1996; Schippers and Sand 1999). In addition, biological thiosulfate formation is known. Release of thiosulfate as a product of taurin fermentation or oxidative sulfur metabolism was observed in rare cases (Denger *et al.* 1997; Brune 1989). However, it is unclear whether the known pathways for biological thiosulfate formation account for a substantial amount of free thiosulfate in natural habitats. In any case, thiosulfate seems to play a major role in the biological sulfur cycle, because most sulfur oxidizing bacteria and also many sulfide producing bacteria accept thiosulfate as a substrate for dissimilatory metabolism (Barrett and Clark 1987).

Elemental sulfur can be formed by sulfide oxidizing bacteria. In some species elemental sulfur is the final oxidation product and therefore it can be further metabolized by syntrophic bacteria. In the Chlorobiaceae, Chloroflexaceae and Ectothiorhodospiraceae extracellular sulfur globules can be formed as intermediates during sulfur compound oxidation to sulfate. The chemistry of this sulfur is not known yet, in contrast to the periplasmic sulfur globules of Chromatiaceae which consist of organic polysulfane sulfur chains (Prange *et al.* 1999). In some thiobacilli (e.g. *Thiobacillus thioparus*, *Thermithiobacillus tepidarius*) elemental sulfur precipitates during growth in batch cultures when oxygen supply is limited (Kelly and Harrison 1989). In those cases, elemental sulfur may originate from non-biological oxidation of polysulfides which accumulate outside the cells when no rapid oxidation is possible. However, in chemostat cultures such phenomena can be minimized (Kelly and Harrison 1989) and it is not known whether these species produce elemental sulfur in natural habitats. Finally it must be mentioned that in some habitats accessable sulfur occurs as a content of soil or rock.

Sulfate is formed under aerobic conditions from most sulfur compounds which are sensitive to oxidation processes. Sulfur compound oxidizing aerobic bacteria (e.g. thiobacilli) accelerate this process. Under anaerobic conditions sulfate is produced by many sulfur compound oxidizing bacteria. These bacteria are either phototrophs or denitrifying chemotrophs. No electron acceptors other than nitrogen oxides of the denitrification pathway are yet known which are employed during anaerobic chemotrophic oxidation of sulfur compounds.

Tetrathionate also can be a natural substrate of sulfur oxidizing bacteria. It originates from incomplete oxidation processes and can be produced from thiosulfate by some bacteria in one oxidation step (Brune 1989; Barrett and Clark 1987).

Polysulfides can be produced either by reductive elemental sulfur ring opening with sulfide or by a partial oxidation of sulfide. The case of *Wollinella succinogenes* is interesting: it can grow on elemental sulfur as an electron acceptor and thereby produces sulfide. In fact, sulfur is not the real acceptor of the electrons but rather polysulfide, which is formed in a reaction of elemental sulfur with sulfide released by this organism (Klimmek *et al.* 1991). Many sulfur reducing species may therefore in fact be polysulfide reducing species.

Sulfite occurs permanently in hot springs where it can be reduced to sulfide by thermophilic archaebacteria such as *Pyrobaculum islandicum*. Sulfite reduction may represent an ancient metabolic pathway. Many bacteria also grow by oxidizing sulfite and some of them seem to be specialized in sulfite oxidation (e.g. *Sulfitobacter* species). Sulfite is an intermediate in most sulfur compound redox pathways which in some cases is released into the medium to some extent. Such a "leakage" could make sulfite available to other bacteria.

3.3 DISSIMILATORY OXIDATION OF REDUCED SULFUR COMPOUNDS

Many eubacteria and some archaebacteria can carry out dissimilatory sulfur oxidations. In the past the term "sulfur bacteria" was used to classify lithotrophic sulfur compound oxidizers taxonomically. "Traditional" sulfur bacteria include green sulfur bacteria (Chlorobiaceae), purple sulfur bacteria (Chromatiaceae and Ectothiorhodospiraceae) and colourless sulfur bacteria which belong either to the proteobacteria (e.g. the genera *Beggiatoa*, *Thiobacillus*, *Thiomicrospira*, *Thioploca*, *Thiosphaera*, *Thiospira*, *Thiothrix* and *Thiovulum*) or to the archaebacteria (*Sulfolobus* and *Acidianus*). However, the ability of lithotrophic growth on reduced sulfur compounds is not restricted to "traditional" sulfur bacteria and in the past many other phototrophic or chemotrophic species with this property were found. On the other hand, numerous traditional sulfur bacteria can grow organoheterotrophically (Brune 1989). Therefore the term "sulfur bacteria" is not very useful anymore.

In Table 3.2 some lithotrophic sulfur-oxidizing bacteria are compared. Taxa not yet characterized on the 16S rDNA level are excluded. Examples of the polyphyletic genus "*Thiobacillus*" are given on the species level. Table 3.2 shows the wide metabolic and phylogenetic diversity of sulfur compound oxidizing bacteria. Many differences can be found in details such as the ability to use various sulfur compounds. As a consequence, the sulfur oxidation pathways can be expected and are indeed found to be variable.

For Table 3.2, data were collected from Friedrich (1998), Moreira and Amils (1997), Shively and Barton (1991), Robertson and Kuenen (1992), Brune (1989), Kelly and Harrison (1989 and Murray *et al.* (1989).

3.3.1 Basic models of sulfur compound oxidation pathways

Most physiological and enzymatic studies on the sulfur compound oxidation pathways concentrate on thiosulfate or sulfide oxidizing species. The best studied chemotrophic species belong to the genus *Thiobacillus*. Among the phototrophic sulfur compound oxidizers the studies centered on *Allochromatium vinosum*. Up to now there are three completely different thiosulfate oxidation pathways proposed to exist in various bacteria. The oxidation of sulfide, sulfur or polythionates can be linked to the thiosulfate oxidation pathways and therefore can be included easily into these models. The first pathway involves an initial cleavage of sulfane and sulfone sulfur and the separate oxidation of these sulfur atoms. In the second pathway thiosulfate is converted to sulfate by a periplasmic multi-enzyme complex. In the third pathway all thiosulfate is oxidized to tetrathionate before further oxidation. The following sections will discuss those models in more detail.

Table 3.2. Comparison of bacteria that are capable of lithotrophic growth on reduced sulfur compounds[1]

	Photo-	Chemo-	Auto-	Oxidation of				Pro-
	trophy	trophy	trophy	HS^-	$S_2O_3^{2-}$	$S_4O_6^{4-}$	S^0	duct
EUBACTERIA								
***Hydrogenobacter* & relatives**	—	+	o, T	—	+		+	SO_4^{2-}
Chlorobiaceae	+	—	o, T	+	+	—	+	SO_4^{2-}
Chloroflexaceae	+	+	f, H	+	—	—	—	S^0
α-Proteobacteria								
Rhodobacter and relatives[2]	—	+	f, C	+	+	—	+	SO_4^{2-}
Paracoccus versutus	—	+	f, C	+	+	—	+	SO_4^{2-}
Acidiphilium acidophilum	—[3]	+	f	+	—	+	+	SO_4^{2-}
"*Thiobacillus" novellus*[4]	—	+	f, C		+	+	—	SO_4^{2-}
β-Proteobacteria								
Thiobacillus thioparus	—	+	o, C	+	+	+	+	S^0 SO_4^{2-}
Thiobacillus denitrificans	—	+	o, C	+[5]	+	+	+	SO_4^{2-}
Thiobacillus aquaesulis	—	+	o, C		+			SO_4^{2-}
Thiomonas thermosulfatus	—	+	f		+	+	+	SO_4^{2-}
Thiomonas perometabolis	—	+	f		+	+	+	SO_4^{2-}
Thiomonas cuprinus	—	+	f		—	—	+	SO_4^{2-}
Thiomonas intermedius	—	+	f, C	+	+	+	+	SO_4^{2-}
γ-Proteobacteria[6]								
Chromatiaceae	+	+	f, C	+	+	—	+	SO_4^{2-}
Ectothiorhodospiraceae	+	+	f	+	+	—	+	S^0 SO_4^{2-}
Beggiatoa	—	+	f	+	+			SO_4^{2-}
Thioploca	—	+	f	+				SO_4^{2-}

[1] + : property present, — : property absent, + / — : property present in some strains, f = facultative autotroph, o = obligate autotroph, H = hydroxypropionate pathway, T = reductive tricarboxylic acid cycle, C = Calvin cycle.

[2] The expression "Rhodospirillaceae" is avoided, since the genera *Rhodoferax* and *Rhodocyclus* - which do not oxidize reduced sulfur compounds - belong to the β-subgroup of the Proteobacteria

[3] *Ac. acidophilus* (formerly *Thiobacillus acidophilus*) belongs to the group of aerobic anoxygenic photosynthetic bacteria, which - to our knowledge - cannot grow solely phototrophically (Yurkov and Beatty 1998)

[4] "*Thiobacillus*" is put into quotation-marks in the case of *Tb. novellus*, because the type species *Thiobacillus thioparus* belongs to the β-proteobacteria (see Table 3.2).

[5] Some strains - as *Thiobacillus denitrificans* RT (DSMZ 807) - may not be able to grow on sulfide (Schedel and Trüper 1980)

[6] The genera *Acidithiobacillus, Thermithiobacillus* and *Halothiobacillus* were formerly named *Thiobacillus* (Kelly and Wood 1999).

Table 3.2 (continued)	Photo-trophy	Chemo-trophy	Auto-trophy	Oxidation of HS$^-$	$S_2O_3^{2-}$	$S_4O_6^{4-}$	S^0	Pro-duct
Acidithiobacillus ferrooxidans	—	+	o	+	+	+	+	SO_4^{2-}
Acidithiobacillus caldus	—	+	o		+			SO_4^{2-}
Acidithiobacillus thiooxidans	—	+	o, C	+	+	+	+	SO_4^{2-}
Thermithiobacillus tepidarius	—	+	o	+	+	+	+	SO_4^{2-}
Halothiobacillus hydrothermalis	—	+	o		+			SO_4^{2-}
Halothiobacillus halophilus	—	+	o		+			SO_4^{2-}
Halothiobacillus neapolitanus	—	+	o, C	+	+	+	+	SO_4^{2-}
Thiomicospira	—	+	o, C	+	+	+		SO_4^{2-}
Riftia symbiont	—	+	o, C	+				SO_4^{2-}
Achromatium	—	+		+				SO_4^{2-}
ε-Proteobacteria								
Thiovulum	—	+	o	+				SO_4^{2-}
"*Thiomicrospira*" *denitrificans*[7]	—	+	o, C	+	+	+	+	SO_4^{2-}
Cyanobacteria								
Oscillatoria	+	—	o, C	+	—	—	—	S^0
Gram-positive bacteria								
Sulfobacillus	—	+					+	SO_4^{2-}
ARCHAEBACTERIA								
Crenarchaeota								
Acidianus and *Sulfolobus*	—	+	o/f, C	+		+	+	SO_4^{2-}

3.3.1.1 The branched thiosulfate oxidation pathway

Up to now the branched thiosulfate oxidation pathway is only established for phototrophs of the γ-subclass and for obligate chemotrophs of the β-subclass of the proteobacteria (Figure 3.2). *Allochromatium vinosum*, *Thiocapsa roseopersicina* and *Thiobacillus denitrificans* are known to cleave thiosulfate in the initial step of thiosulfate metabolism (Brune 1989; Schedel and Trüper 1980; Trüper and Pfennig 1966; Smith and Lascelles 1966). However, it is very likely that other Chromatiaceae as well as *Thiobacillus thioparus* and possibly members of the Chlorobiaceae and some *Beggiatoa* species also use this metabolic pathway.

Most studies were done with anaerobically grown cells, although the species using this pathway usually are not obligate anaerobes (with a few

[7] "*Thiomicrospira*" *denitrificans* is written with quotation marks, because it does not belong to the γ-proteobacteria to which the type species *Thiomicrospira pelophila* belongs.

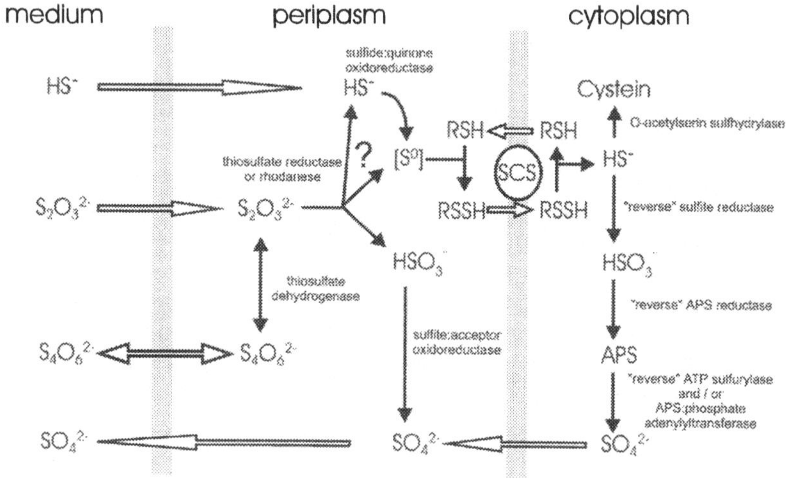

Figure 3.2. Model of the branched thiosulfate oxidation pathway. Black arrows: chemical reactions; white arrows: transport processes; SCS: "sulfide carrier system" involving an unknown transsulfhydrylase on the periplasmic side, a thiol/perthiol transporting system in the membrane and a heterodisulfide reductase on the cytoplasmic side (see text).

exceptions). There are a few studies on aerobically grown *Thiobacillus denitrificans* and *Thiocapsa roseopersicina* and on microaerophilically grown *Chromatiaceae* (see Dahl and Trüper 1989; Kämpf and Pfennig 1980, 1986a,b; Timmer-Ten Hoor 1981; Gibson 1967; Breuker 1964). The following discussion of the branched thiosulfate oxidation pathway will be based on results obtained mainly from studies on Chromatiaceae and *Thiobacillus denitrificans* grown under anaerobic conditions.

In experiments with *Thiobacillus denitrificans*, *Allochromatium vinosum* and *Thiocapsa roseopersicina* cell suspensions, the thiosulfate cleavage was studied with radioactively labelled sulfane or sulfone sulfur (Schedel and Trüper 1980; Trüper and Pfennig 1966; Smith and Lascelles 1966). It could be shown in all cases that the sulfane sulfur and sulfone sulfur were further oxidized by different pathways. The sulfane sulfur accumulates as [S^0] before further oxidation, whereas the sulfone sulfur is rapidly converted into sulfate (Schedel and Trüper 1980; Smith and Lascelles 1966). However, sulfur and sulfate are not the initial products of thiosulfate cleavage (Figure 3.2). Rather the cleavage of thiosulfate occurs by the action of a thiosulfate reductase or a rhodanese (Brune 1989).

Rhodanese activity (thiosulfate sulfur transferase) is measured as the thiocyanate formation in an assay with thiosulfate and cyanide. The enzyme structure seems to be complex and varies in various species. In *Thiobacillus denitrificans* rhodanese is reported to be an 38 kDa enzyme consisting of four subunits. However, rhodanese activities were also found for 150,000 Da proteins and even for proteins with a molecular mass higher than 500,000 Da. It is proposed that tetrameric aggregates of rhodanese can be formed (Bowen *et al.* 1965). This aggregation may be due to cystine bridge formation which may influence the activity of the enzyme (Schedel and Trüper 1980). Rhodanese is also present in other bacterial species with a branched thiosulfate oxidation pathway (Smith and Lascelles 1966, Yoch and Lindstrom 1971). Because organic thiols can substitute for cyanide as sulfane sulfur acceptors, rhodanese may transfer sulfane sulfur to organic carriers like glutathione or to polysulfane chains. However, no function of any rhodanese is unequivocally established yet and the occurence of this activity in bacteria that do not oxidize sulfur compounds (e.g. *Escherichia coli*, Alexander and Volini 1987) suggests that rhodanese may have other functions.

Thiosulfate reductase is currently the most probable candidate for the thiosulfate cleaving step. It has been detected in many organisms, among them *Thiobacillus denitrificans* (Schedel and Trüper 1980). The enzyme is neither purified nor is its gene cloned yet from any sulfur compound oxidizing organism. Thiosulfate reductase activity is defined as the production of sulfide and sulfite from thiosulfate. In some cases the reaction appears to depend on the presence of organic thiols (e.g. glutathione or dihydrolipoate) and then the distinction from rhodanese becomes unclear (*Thiobacillus thioparus*, Peck 1960; *Tc. roseopersicina*, Petushkova and Ivanowskii 1976). Because thiosulfate reductases from sulfide producing organisms appear to be periplasmic or membrane-bound with a catalytic subunit directed towards the periplasm, we suggest this localization also for thiosulfate oxidizing species (Heinzinger *et al.* 1995; see Figure 3.2). Another line of evidence for this localization is the fact that during growth on thiosulfate, sulfur globules are formed inside the periplasm of *Allochomatium vinosum* (Pattaragulwanit *et al.* 1998). If thiosulfate would be cleaved in the cytoplasm, then the sulfane sulfur had to be transported for storage back into the periplasm and again transported into the cytoplasm for further oxidation (discussed below). Therefore the periplasmic sulfur globules are strong evidence for a periplasmic thiosulfate cleavage, at least in *Allochomatium vinosum*.

Thiosulfate reductase (or rhodanese) converts the sulfone sulfur to sulfite which can be further oxidized to sulfate by a **sulfite:acceptor oxidoreductase**.

Sulfite:acceptor oxidoreductase from *Thiobacillus denitrificans* is membrane associated and coupled to *c*-type cytochrome reduction (Aminuddin and Nicholas 1974). Similar coupling to *c*-type cytochromes is reported from other sulfite:acceptor oxidoreductases (Hooper and DiSpirito 1985). The 48-fold enriched sulfite reductase from *Tb. denitrificans* exhibited cytochrome *c* spectral properties and slowly reduced yeast and horse heart cytochrome *c* (Aminuddin and Nicholas 1974). Endogenous cytochromes of *Tb. denitrificans* were not tested but could possibly function better as electron acceptors. The purified soluble enzyme of *Thiobacillus novellus* contains heme *c* and molybdopterin (Southerland and Toghrol 1983; Toghrol and Southerland 1983). Since *c*-type cytochromes are periplasmic, sulfite:acceptor oxidoreductases are proposed to be located inside the periplasm (Hooper and DiSpirito 1985). The sulfite produced by thiosulfate reductase (or rhodanese) therefore probably does not have to enter the cytoplasm and can be oxidized directly in the periplasm.

What happens with the sulfide that is formed inside the periplasmic space from the thiosulfate sulfane sulfur by thiosulfate reductase? It can be oxidized by the same pathway that is also employed for the oxidation of exogenous sulfide. In a first step, all sulfide is oxidized to sulfur chains which are usually termed [S^0]-sulfur although this sulfur is more reactive than circular S^0-sulfur (Schedel and Trüper 1980). This sulfur is usually stored until no free sulfide is available inside the periplasm or in the surrounding medium. The stored sulfur was recently analysed in Chromatiaceae and found to be polysulfane sulfur of variable chain length (Prange *et al.* 1999). The bacteria benefit from the storage of sulfur, because they can further oxidize it in times when sulfide or thiosulfate is limiting. Chromatiaceae - which grow best phototrophically - can use this [S^0]-sulfur also as an electron acceptor under anoxic conditions in the dark when they oxidize organic compounds. Under oxic conditions in the dark they behave as chemotrophic sulfur oxidizers, which is very interesting considering the relatedness of the sulfur compound oxidation pathway to the chemotroph *Thiobacillus denitrificans.*

The enzyme that is likely to be responsible for the oxidation of sulfide to periplasmic storage sulfur is **sulfide:quinone oxidoreductase (SQR)**. The exact product of this enzyme is not known yet, but it is proposed to be disulfide (Reinartz *et al.* 1998). SQR is an integral membrane protein which transfers electrons from sulfide to the quinone pool of the membrane. This enzyme activity has been described for *Oscillatoria limnetica*, *Chlorobium limicola*, *Rhodobacter capsulatus* and *Paracoccus denitrificans* (Arieli *et al.* 1994; Shahak *et al.* 1992a/b; Schütz *et al.* 1997, 1998) and very recently for

Allochromatium vinosum (Reinartz *et al.* 1998). In *Allochromatium vinosum* and *Chlorobium limicola* f. *thiosulfatophilum* there exists also a second periplasmic sulfide-oxidizing enzyme, a **flavocytochrome *c*** (Brune 1989). However, at least in *Allochromatium vinosum* it is shown that this enzyme is not responsible for periplasmic sulfide oxidation *in vivo* and therefore SQR seems to play that role (Reinartz *et al.* 1998).

The periplasmic stored sulfur has to be transported into the cytoplasm for further oxidation (Pott and Dahl 1998). It is likely that the transport across the membrane is done via organic thiols which can accept a terminal thiol group from polysulfane chains thereby forming a perthiol. Glutathione derivatives are the first candidates for such a function. Indeed, glutathioneamide persulfide was detected in large amounts in sulfur oxidizing *Allochromatium vinosum* (Bartsch *et al.* 1996). An enzyme that transfers free sulfhydryl groups from polysulfane chains onto the organic thiol is not known but has to be postulated. After transport into the cytoplasm sulfide has to be reductively released from the organic perthiol by a kind of **heterodisulfide reductase**. The complete system of sulfide transport from periplasmic polysulfanes or polysulfides to the cytoplasm can be termed sulfide carrier system (SCS). This system includes a thiol export / perthiol import system and at least two enzymatic activities, one sulfhydryltransferase on the periplasmic side and a heterodisulfide reductase on the cytoplasmic side of the membrane. The SCS model can explain how sulfide can be transported to the cytoplasm without any sulfide generation from stored sulfur inside the periplasm. Therefore, in the SCS model the presence of periplasmic sulfide could be the signal which leads to sulfur storage. If sulfide is used up inside the periplasm, this is likely to be the signal for the oxidation of the stored sulfur. Older models of sulfur rereduction inside the periplasm could not explain the observation that [S^0]-sulfur oxidation starts when sulfide is depleted.

Cytoplasmically released sulfide serves as the substrate for a reverse acting **sulfite reductase**. This sulfite reductase has been purified from *Allochomatium vinosum* and *Thiobacillus denitrificans* (Schedel *et al.* 1979; Schedel and Trüper 1979). The enzyme has been reported to be a 160,000 Da enzyme of a proposed $\alpha_2\beta_2$ structure in *Thiobacillus denitrificans* and a 280,000 Da enzyme of proposed $\alpha_4\beta_4$ structure in *Allochromatium vinosum*. This enzyme seems to be very similar to sulfite reductases which function in the reductive direction, e.g. from *Desulfovibrio vulgaris* (Steuber *et al.* 1995). The enzyme contains [4Fe-4S]-clusters and siroheme. In the sulfate reducing *Desulfovibrio vulgaris* the enzyme has a third only loosely bound small γ-"subunit" whose gene is also present in the *Allochromatium vinosum* "*dsr*"-

operon which contains the structural and accessory genes of dissimilatory sulfite reductase (Pott and Dahl 1998, see also Steuber *et al.* 1995 and section 3.4). Interestingly a gene coding for a heterodisulfide reductase is found in the same operon (Pott and Dahl 1998), supporting the SCS model described above. Reverse sulfite reductase is essential for oxidation of stored sulfur in *Allochromatium vinosum* (Pott and Dahl 1998). Since sulfite is produced from reverse sulfite reductase, it can be concluded that all sulfane sulfur from thiosulfate is oxidized to sulfite inside the cytoplasm and not inside the periplasm. This could possibly be the main reason for the existence of a second sulfite oxidation pathway inside the cytoplasm in addition to the mentioned periplasmic pathway.

The cytoplasmic sulfite oxidation pathway is AMP-dependent and consists of two steps: the formation of adenosine-5'-phosphosulfate (= adenylylsulfate or APS) and the release of sulfate from adenylylsulfate. APS is formed by a reverse acting **APS reductase**, a FAD and [4Fe-4S] cluster containing enzyme which occurs membrane-bound or soluble in phototrophic species of the Chromatiaceae and only soluble in the cytoplasm of *Thiobacillus denitrificans*, *Thiobacillus thioparus* and *Beggiatoa* MS-81-1c (Trüper and Rogers 1971; Trüper and Peck 1970; Lyric and Suzuki 1970; Bowen *et al.* 1966; Hagen and Nelson 1997). The enzyme consists of two different subunits of a proposed $\alpha_2\beta_2$ structure. Molecular masses are reported ranging from 170,000 to 180,000 Da. The structural genes of this enzyme were cloned from *Allochromatium vinosum* (Dahl 1996, Hipp *et al.* 1997). The "*in vivo*" electron acceptors for "reverse" APS reductase are not yet known, but probably the electrons are fed from APS reductase somehow into the quinone pool.

Sulfate is released from APS either by the reverse **ATP sulfurylase** or by an **adenylylsulfate:phosphate adenylyltransferase (APAT)**. ATP sulfurylase transfers the AMP moiety of APS onto pyrophosphate and therefore produces ATP and sulfate. APAT transfers AMP from APS onto phosphate and therefore releases ADP and sulfate as reaction products. In contrast to the opinion that an "ADP sulfurylase" exists, we found that the enzyme catalyses the reaction strictly unidirectionally and therefore this enzyme has to be named correctly adenylylsulfate:phosphate adenylyltransferase or APAT (Brüser *et al.*, submitted). APAT exists so far it is known only together with ATP sulfurylase in APS reductase containing organisms. This includes *Thiobacillus denitrificans* and *Thiobacillus thioparus* (Peck 1960; T. Brüser, unpublished results). Therefore new models for the function of APAT had to be developed which could explain the existence of two sulfate releasing enzymes, APAT and ATP sulfurylase, in the same compartment. We propose that under pyrophosphate limiting conditions APAT serves to ensure a high turnover of APS (Brüser *et al.*, submitted). Without APAT, APS and toxic sulfite would accumulate cytoplasmically. In the absence of APAT a sulfite extrusion mechanism probably exists. Sulfite may be further oxidized inside the periplasm by sulfite:acceptor oxidoreductase (see above). This sulfite extrusion mechanism might work in *Allochromatium vinosum*, where the knock-out of APS reductase via mutation had no significant effect on the viability of this organism, despite the probably continued production of sulfite by sulfite reductase inside the cytoplasm of the mutant (Dahl 1996).

Sulfate produced cytoplasmically is exported and released into the surrounding medium together with the sulfate produced inside the periplasmic space. The outer membrane of Gram-negative bacteria is not a barrier for ions of this size (as it is not a barrier for the sulfur compound substrates).

The role of thiosulfate dehydrogenase in bacteria of the branched thiosulfate oxidation pathway is not clear (see Figure 3.2). In *Allochromatium vinosum* tetrathionate is produced from thiosulfate under mildly acidic conditions (Smith 1966). The responsible enzyme is probably a *c*-type cytochrome which is proposed to use a flavocytochrome c_{552} as an electron acceptor (Schmitt *et al.* 1981). Therefore tetrathionate can be assumed to be formed inside the periplasmic space. However, tetrathionate is not further metabolized in *Allochromatium vinosum*. In other cases tetrathionate may be reduced to thiosulfate by the same enzyme acting in the reverse direction (Barrett and Clark 1987). Thiosulfate may then be further metabolized as described above. However, this is not yet shown for organisms of the branched thiosulfate oxidation pathway.

3.3.1.2 The multi-enzyme-complex pathway

This pathway was proposed by Kelly and coworkers for *Paracoccus versutus* (formerly "*Tb. versutus*" or "*Tb. A2*"; Lu and Kelly 1983a,b). It is also shown to exist in *Paracoccus denitrificans* (Wodara *et al.* 1994) and proposed to be possibly present in "*Thiobacillus*" *novellus* and *Xanthobacter* ssp. (Kelly 1997). The pathway has been termed "*Paracoccus* sulfur oxidation pathway" or PSO pathway, whereas the enzyme system has been termed "thiosulfate-oxidizing multi-enzyme system" or TOMES (Kelly *et al.* 1997). Up to now all examples for bacteria with this pathway come from facultative autotrophic α-proteobacteria.

The TOMES of *Paracoccus versutus* consists of five different polypeptides, one 16 kDa thiosulfate binding enzyme, one 59 kDa protein

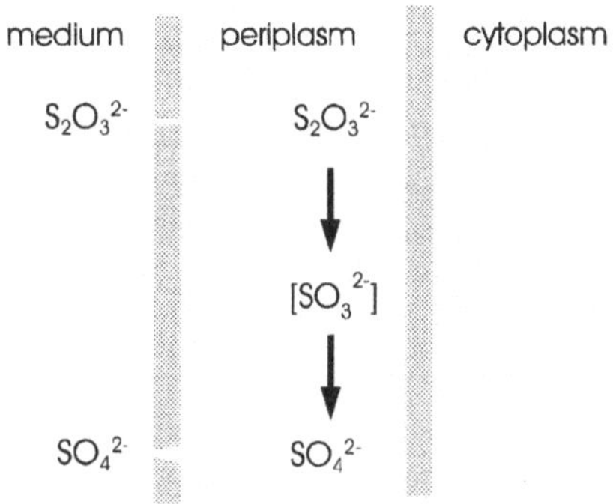

Figure 3.3. Model of the multi-enzyme-complex pathway of thiosulfate oxidation in *Paracoccus* species.

of unknown function which contains binuclear manganese sites, a 56 kDa homodimeric cytochrome $c_{552.5}$, a hexameric 260 kDa cytochrome c_{551} and a 44 kDa molybdopterin cofactor containing sulfite dehydrogenase (Friedrich 1998). The cytochromes indicate a periplasmic localization which is confirmed by gene sequence data available for the similar *Paracoccus denitrificans* TOMES. These data reveal signal peptides for a targeting of the TOMES subunits into the periplasm (Wodara *et al.* 1994). The complex oxidizes thiosulfate to sulfate without the transient release of reaction intermediates. Therefore *Paracoccus* species seem to have the ability to oxidize both sulfane and sulfone sulfur of thiosulfate in the periplasmic space (Friedrich 1998). It was suspected that the manganese containing enzyme (SoxB) could be the key enzyme for the oxidation of sulfane sulfur to the sulfite-level, and some experimental evidence supported this hypothesis (Schneider and Friedrich 1994). Confusingly, two other sulfide oxidizing proteins may exist in the periplasm of *Paracoccus* species. A sulfide quinone oxidoreductase was detected in *Paracoccus denitrificans* membranes (Schütz *et al.* 1998). In addition, in the same organism a gene coding for a flavoprotein with homology to the *Allochromatium vinosum* flavocytochrome *c* was detected downstream of the genes coding for the multi-enzyme complex. Mainly for those reasons, Friedrich recently questioned the function of SoxB for sulfide oxidation within the multi-enzyme-complex (Friedrich 1998). However, it seems unlikely that SQR is involved in sulfane sulfur oxidation when it is true that no reaction intermediate is released by the enzyme complex. Rather the flavoprotein could serve this function if it is associated with the complex. *Paracoccus denitrificans* can grow anaerobically on sulfide and therefore SQR may play a role in a sulfide oxidation pathway which may be distinct from the thiosulfate oxidation pathway in this organism. The only conclusion which can be made about the mechanism of thiosulfate oxidation in *Paracoccus* thus far seems to be that sulfite must be an intermediate of the reaction since a sulfite dehydrogenase is associated with the complex (see Figure 3.3). What happens with the thiosulfate sulfane sulfur remains to be revealed.

3.3.1.3 The tetrathionate pathway

For many years, there has been controversy about whether thiosulfate can be oxidized to sulfate via polythionates as intermediates (Figure 3.4). The first proposal of a thiosulfate oxidation via tetrathionate was made in 1952 for *Thiobacillus thioparus* (Vishniac 1952). However, this organism has both the AMP-dependent and the AMP-independent sulfite oxidation pathways and is very closely related to *Thiobacillus denitrificans*. For *Thiobacillus thioparus* this strongly suggests a branched thiosulfate oxidation pathway as described above (section 3.3.1.1). There is also a glutathione-dependent thiosulfate reductase reported from *Tb. thioparus* (Peck 1960). This and other data have been taken as a strong evidence against a tetrathionate pathway in this organism (Peck 1960). Nevertheless, there are aerobic obligate autotrophic thiobacilli that belong to the γ-proteobacteria, including *Thermithiobacillus tepidarius*, *Halothiobacillus neapolitanus*, *Acidithiobacillus ferrooxidans*, *Acidithiobacillus caldus* and *Acidithiobacillus thiooxidans* and a facultative autotrophic α-proteobacterium, *Acidiphilium acidophilum*, which possibly oxidize thiosulfate via tetrathionate (Lu and Kelly 1988a, Kelly *et al.* 1997).

The most striking evidence for this pathway is, that thiosulfate seems to be quantitatively converted to tetrathionate before further oxidation. The responsible enzyme is a thiosulfate dehydrogenase which has been purified from various "*Thiobacillus*"-species with a postulated tetrathionate pathway. In "*Thiobacillus*" strain W5 thiosulfate dehydrogenase is probably a $\alpha_2\beta_2$-tetramer comprising 33 kDa and 28 kDa subunits (Visser *et al.* 1997a). However, the oxidation-pathway of tetrathionate is obscure. From inhibition studies Lu and Kelly (1988b) concluded that tetrathionate oxidation by *Thermithiobacillus tepidarius* may take place inside the cytoplasm. Similar results were obtained with *Acidithiobacillus caldus* (Hallberg *et al.* 1996). In contrast, in the acidophiles *Acidiphilium acidophilum* and *Acidithiobacillus ferrooxidans*, it is proposed that tetrathionate is metabolized inside the periplasm (Meulenberg *et al.* 1993; Hazeu *et al.* 1988). However, in none of these cases has the localization been unequivocally established.

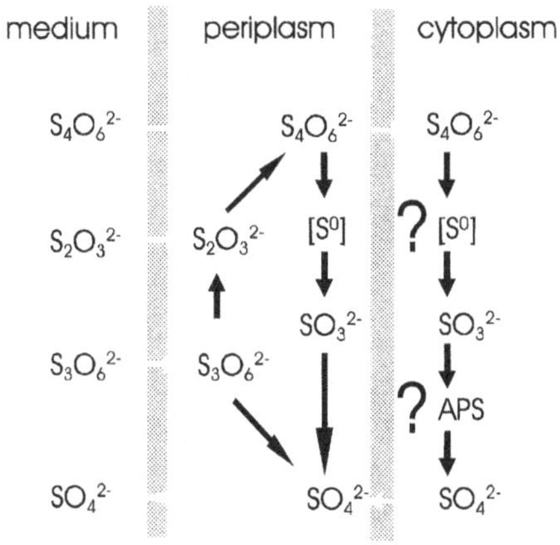

Figure 3.4. Model of the tetrathionate pathway of thiosulfate oxidation in obligate aerobic *Thiobacillus* species. Note that the tetrathionate oxidation pathway is highly hypothetical.

It has been proposed that tetrathionate is hydrolyzed to disulfane-monosulfonic acid and sulfuric acid (Equation 3.1, Pronk *et al.* 1990):

$$S_4O_6^{2-} + H_2O \rightarrow S_3O_3^{2-} + H_2SO_4 \qquad (3.1)$$

More recent experimental data support that thiosulfate, elemental sulfur, sulfate and protons are formed from tetrathionate in *Acidiphilus acidophilus* (Equation 3.2, Meulenberg *et al.* 1993, DeJong *et al.* 1997):

$$S_4O_6^{2-} + H_2O \rightarrow S^0 + S_2O_3^{2-} + H_2SO_4 \qquad (3.2)$$

Oxidation of tetrathionate and elemental sulfur may be coupled to respiratory energy conservation. Also the possibly periplasmic release of protons as reaction products may contribute to the electrochemical membrane potential. The proposed cytoplasmic oxidation of tetrathionate to sulfate in *Thermithiobacillus tepidarius* would strongly acidify the cytoplasm (Kelly *et al.* 1997), which would make this pathway energetically less effective. If tetrathionate is metabolized according to Equation (3.2), then thiosulfate is produced which again can be oxidized to tetrathionate. Sulfate, the second product, can be released into the medium. Only sulfur remains to be oxidized by a pathway that is not known yet (Figure 3.5).

The tetrathionate metabolizing enzyme (**tetrathionate hydrolase**) has recently been purified and characterized (DeJong *et al.* 1997). Indeed

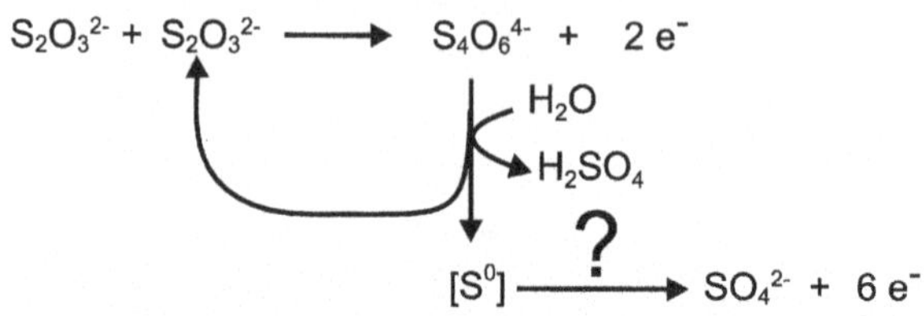

Figure 3.5. Proposed tetrathionate oxidation pathway of *Acidiphilium acidophilum* (After: Meulenberg *et al.* 1993).

thiosulfate, sulfur and sulfate have been detected as products of its enzymatic reaction.

Because trithionate is not an intermediate in the pathway, it remains unclear why trithionate hydrolase is formed. This enzyme has been purified from *Acidiphilium acidophilum* (Meulenberg *et al.* 1992) and is reported to exist also in other species with the tetrathionate pathway (Pronk *et al.* 1990, Kelly *et al.* 1997). In an older "cyclic" model it was proposed that from tetrathionate a disulfane-monosulfonic acid is formed which could be oxidized to trithionate. Trithionate in turn could have been hydrolysed to thiosulfate and sulfate (Pronk *et al.* 1990). This idea has been adopted by Schippers *et al.* (1996) for the pyrite leaching process. Confusingly, trithionate has been reported to be formed as an intermediate during thiosulfate oxidation in *Halothiobacillus neapolitanus* (Kelly *et al.* 1997). Possibly, the tetrathionate hydrolase pathway may be specific for acidophilic thiobacilli.

The pathways by which elemental sulfur may be oxidized by thiobacilli have been recently discussed by Kelly (1999). It is very unlikely that sulfur is oxidized by the use of an oxygenase, as the bacteria would not profit energetically from such an oxidation. Further, there is accumulating evidence for membrane associated electron transport being involved in elemental sulfur oxidation (Kelly 1999). Elemental sulfur oxidizing activities have been purified in rare cases from species with a tetrathionate pathway. From *Acidithiobacillus ferrooxidans* a sulfur:ferric iron oxidoreductase has been purified to apparent homogeneity (Sugio *et al.* 1987). It is a homodimer of 23,000 Da subunits. The enzyme required glutathione for activity. The initial product of sulfur oxidation was sulfite. Thiosulfate was formed slowly, probably in an uncatalysed reaction of sulfite with sulfur. Since ferric iron reduction is not a common property of all thiobacilli with a tetrathionate-pathway, this enzyme may represent some specialization of *Acidithiobacillus ferrooxidans*. A glutathione dependent sulfur metabolizing enzyme has also been described for *Acidithiobacillus thiooxidans* (Suzuki and Silver 1966). In this case oxygen has been postulated as the electron acceptor. For about 40 years it has been proposed that glutathione or other organic thiols are involved in sulfur oxidation of various thiobacilli (references in Trudinger and Loughlin 1981). Trudinger and Loughlin (1981) listed evidence for a glutathione dependent soluble or alternatively a glutathione-independent but thiol reactive membrane-bound sulfur oxidizing system. They supported the opinion that a reaction of sulfur with a thiol initiates the sulfur oxidation. Regarding the older ideas about elemental sulfur oxidation (Trudinger and Loughlin 1981), there seems not to be much difference detectable to the branched thiosulfate oxidation pathway discussed above. However, the sulfur formed from thiosulfate by acidophilic thiobacilli is reported to consist of polythionates and some elemental sulfur (Steudel *et al.* 1987; Pronk *et al.* 1990). Recall that in bacteria with a branched thiosulfate oxidation pathway storage sulfur (from sulfide grown cells) consists of organic polysulfanes in all cases studied so far (Prange *et al.* 1999). This suggests that the sulfur oxidation in acidophilic thiobacilli may differ significantly from that of species with a branched thiosulfate oxidation pathway. The assessment of the occurence of enzymes typical for a branched thiosulfate oxidation pathway in "tetrathionate-pathway" thiobacilli is therefore interesting.

Two sulfide dehydrogenases are purified from membranes of autotrophic chemolithotrophic sulfur oxidizers (Visser *et al.* 1997b, Sorokin *et al.* 1998). Both enzymes are reported to be *c*-type cytochromes. Therefore sulfide oxidation - like in the case of the branched thiosulfate oxidation pathway - appears to occur inside the periplasm. Also in "*Thiobacillus*" *concretivorus* sulfide oxidation is associated with the membranes (Moriarty and Nicholas 1969). Information about the role of thiosulfate reductases or rhodaneses in these bacteria is insufficient, although a rhodanese is reported from *Acidithiobacillus ferrooxidans* (Tabita *et al.* 1969, Tuovinen *et al.* 1975). No clear evidence exists so far for a reverse sulfite reductase in obligate chemotrophic γ-proteobacteria. In *Acidithiobacillus thiooxidans* the first product of glutathione-dependent sulfur oxidation is reported to be sulfite, which had been trapped with formaldehyde (Suzuki and Silver 1966; Silver and Lundgren 1968). The glutathione dependence may indicate some sulfur transferase mechanism. Many reports about sulfite-oxidizing activities in bacteria with the postulated tetrathionate pathway strongly suggest that sulfite is indeed an intermediate of sulfur oxidation in these species. AMP-dependent sulfite oxidation was reported for one strain of *Acidithiobacillus ferrooxidans* (Tuovinen *et al.* 1976). The *Acidithiobacillus ferrooxidans* culture in those experiments was not pure and therefore the results require confirmation. APS reductase activity has also been detected in *Acidithiobacillus thiooxidans* crude extracts (Peck 1961). The activity in *Acidithiobacillus thiooxidans* was considerably lower than in thiobacilli of the β-proteobacteria (13-25%, Peck 1961), but APS reductase was proposed to generally play a role for sulfite reduction in thiobacilli (Peck 1961, Peck and Fisher 1962). APS reductase has never been purified from "tetrathionate-pathway" thiobacilli. However, regarding the data from *Acidithiobacillus ferrooxidans* and *Acidithiobacillus thiooxidans* and the existence of APS reductase in other γ-proteobacteria (e.g. *Chromatiaceae*, see section 3.3.1.1) it appears possible that this enzyme also plays a role in "tetrathionate-pathway" thiobacilli. Surely most - if not all - "*Thiobacillus*" species contain an AMP-independent sulfite dehydrogenase. It is proposed that sulfite oxidation takes place inside the cytoplasm in the case of acidophilic thiobacilli like *Acidithiobacillus thiooxidans*, because neutral pH values are required for sulfite oxidation by crude extracts (Takeuchi and Suzuki 1994). Since other AMP-independent sulfite dehydrogenases are proposed to be periplasmic, the characterization of sulfite dehydrogenase from this species would be of interest. An unusual sulfite dehydrogenase has also been reported from one strain of *Acidithiobacillus ferrooxidans*, which is a 650,000 Da membrane-bound sulfite:ferric iron oxidoreductase (Sugio *et al.* 1992).

The pathway described above may cover thiosulfate, polythionate, sulfur and sulfite oxidation in these species (Figure 3.4). Sulfite has to be included in the scheme of Meulenberg *et al.* (1993) as an intermediate in sulfur oxidation. How sulfur itself is oxidized in this pathway remains an open question. If polysulfides are formed from sulfur, the pathway could resemble that one of the branched thiosulfate oxidizing pathway (section 3.3.1.1). This could explain the existence of cytoplasmic sulfite oxidation systems like the AMP-dependent one. Additional periplasmic sulfite reductases could indicate that sulfite is formed in the periplasm or transported into the periplasm.

Cells of various tetrathionate-pathway-thiobacilli are also able to oxidize sulfide rapidly (Pronk *et al.* 1990). A membrane-bound flavocytochrome *c* sulfide dehydrogenase which is possibly responsible for sulfide oxidation has been purified from *Thiobacillus* W5, a strain with similarities to "*Tb. neapoitanus*" (Visser *et al.* 1997b). It remains an open question whether sulfide and thiosulfate oxidation share parts of the same enzymatic pathway in these organisms. If a thiosulfate reductase is involved in thiosulfate oxidation, a sulfide dehydrogenase would be required for growth on thiosulfate.

In the tetrathionate pathway described above, more questions than answers are given. Kelly, who contributed a lot to the model of the tetrathionate pathway, recently questioned the main idea of this model when he noted that there exists uncertainty about "the centrality of the role of the conversion of thiosulfate to tetrathionate" (Kelly 1999). Indeed, if one would not assume tetrathionate-formation as required for thiosulfate oxidation, then the formation of polysulfanes or polysulfides with the aid of glutathione or thiol-dependent enzymes would indicate that the pathway in the chemotrophic γ-proteobacteria could be at least very similar to that in the phototrophic γ-proteobacteria. Clearly more studies are required to answer the open questions.

3.3.1.4 Sulfur oxidation in archaebacteria

Members of the Sulfolobales characteristically grow on sulfur either anaerobically or aerobically, thereby producing sulfide or sulfuric acid, respectively (Segerer *et al.* 1985). The sulfur metabolizing enzyme has been identified in *Acidianus ambivalens* (formerly *Desulfurolobus ambivalens*) as a sulfur oxygenase/reductase, since sulfite, thiosulfate and sulfide are formed from sulfur (Kletzin 1989). Thiosulfate formation has been attributed to a non-enzymatical reaction of sulfite with the sulfur substrate (Kletzin 1992). Also, the sulfide formation was tentatively explained by a nonenzymatical mechanism (Kletzin 1989). Therefore is not sure whether sulfur oxygenase/reductase of *Acidianus ambivalens* is a sulfite-producing sulfur oxygenase only. Such a sulfur oxygenase has been reported from *Acidianus brierleyi* to be a 560 kDa enzyme (Emmel *et al.* 1986). The sulfur oxygenase of *Acidianus brierleyi* is probably very similar to the sulfur oxygenase/reductase of *Acidianus ambivalens*. The latter soluble 550 kDa cytoplasmic enzyme has been purified and the corresponding gene has been analysed (Kletzin 1992). Sulfur oxygenase/reductase is an oligomer consisting of 35 kDa subunits. The enzyme is proposed to contain no cofactors and to function similar to rhodanese with active thiols only (Kletzin 1989). Putative active thiols have been identified in the amino acid sequence (Kletzin 1992). Further studies are required to elucidate the mechanism of this enzyme.

If sulfite would be the only product of an oxygenase reaction with sulfur, then the oxidation of sulfite would be the only possible electron source for aerobic lithotrophic growth. If sulfide is formed as a further product, sulfide oxidation may contribute a lot to energy conservation. The situation then is similar to the branched thiosulfate oxidation pathway, where sulfide and sulfite can be formed from thiosulfate by the reaction of thiosulfate reductase.

The enzymes that oxidize sulfide or sulfite in *Acidianus* species have not yet been identified. There remain many questions about the archaebacterial sulfur oxidation pathway. It has to be noted that these species are thermophilic acidophilic organisms, which obviously are difficult to study.

3.4 DISSIMILATORY REDUCTION OF OXIDIZED SULFUR COMPOUNDS

In the microbial sulfur cycle, sulfate is converted into sulfide by sulfate-reducing bacteria (SRB) via dissimilatory sulfate reduction. This process of bacterial respiration occurs under strictly anaerobic conditions and uses sulfate as the terminal electron acceptor. Electron donors are usually organic compounds or hydrogen (see chapters 7 and 14).

SRB comprise the traditional sulfate-reducing genera *Desulfovibrio* and *Desulfotomaculum* in addition to the morphologically and physiologically different genera *Thermodesulfobacterium, Desulfobacter, Desulfobulbus, Desulfococcus, Desulfonema* and *Desulfosarcina* and others, whose name usually begins with "*Desulfo-*" (Widdel and Hansen 1992). In the presence of sulfate, SRB are able to use several intermediates of the anaerobic mineralisation process. Besides the direct methanogenic substrates molecular hydrogen (H_2), formate, acetate and methanol, they can also use propionate, butyrate, higher and branched fatty acids, lactate, ethanol and higher alcohols, fumarate, succinate, malate, alkanes and aromatic compounds. In sulfidogenic breakdown of volatile fatty acids, two oxidation patterns can be distinguished. Some SRB are able to completely oxidize volatile fatty acids to CO_2 and sulfide as end products. Other SRB lack both the tricarboxylic acid cycle and the CO dehydrogenase dependent acetyl-CoA pathway and carry out an incomplete oxidation of volatile fatty acids with acetate and sulfide as the endproducts.

In addition to the reduction of sulfate, reduction of sulfite and thiosulfate is also very common among SRB. *Desulfovibrio* strains have been reported to be able to reduce di-, tri- and tetra-thionate. A unique ability of some SRB, e.g. *Desulfovibrio dismutans* and *Desulfobacter curvatus*, is the dismutation of sulfite (Equation 3.3) or thiosulfate (Equation 3.4):

$$4\ SO_3^{2-} + H^+ \rightarrow 3\ SO_4^{2-} + HS^- \quad (3.3)$$

$$S_2O_3^{2-} + H_2O \rightarrow SO_4^{2-} + HS^- + H^+ \quad (3.4)$$

Some SRB were found to be able to respire oxygen, despite being classified as strict anaerobic bacteria. Thus far, however, aerobic growth of pure cultures of SRB has not been demonstrated. The ability of SRB to carry out sulfate reduction under aerobic conditions remains nevertheless intriguing and is of significance for micro-scale sulfur cycles.

In the absence of an electron-acceptor, most SRB are able to grow through a fermentative or acetogenic reaction (Widdel and Hansen 1992). Pyruvate, lactate and ethanol are easily fermented by many SRB. An interesting feature of SRB is their ability to perform acetogenic oxidation in syntrophy with hydrogenotrophic methanogenic bacteria, as described for co-cultures of hydrogenotrophic methanogenic bacteria with *Desulfovibrio* sp. using lactate and ethanol or with *Desulfobulbus*-like bacteria using propionate. In the presence of sulfate, however, these bacteria behave as true SRB and metabolize propionate as electron-donor for the reduction of sulfate.

3.4.1 Pathways of dissimilatory sulfur compound reductions

3.4.1.1 Sulfate reduction pathways

In dissimilatory sulfate reduction three enzymes are generally employed, all of which are located inside the cytoplasm to which sulfate is transported in an active process (Hansen 1994; Kreke and Cypionka 1995). The direct reduction of sulfate to bisulfite would require unphysiologically negative redox potentials (E_o' = -516 mV, Thauer *et al.* 1977). Therefore sulfate is converted to a mixed anhydride which can be more easily reduced. **ATP sulfurylase** catalyses this activation of sulfate in a reaction with ATP, thereby forming adenosine-5'-phosphosulfate (= adenylylsulfate or APS) and inorganic pyrophosphate. APS formation from ATP and sulfate is an endergonic process ($\Delta G^{o\prime}$ = + 46 kJ mol^{-1}, Moodie and Ingledew 1990) and requires the hydrolysis of the second reaction product pyrophosphate ($\Delta G^{o\prime}$ = - 33 kJ mol^{-1}) and the subsequent reduction of APS ($\Delta G^{o\prime}$ = - 60 kJ mol^{-1}) for high efficiency. ATP sulfurylase is also found in sulfite oxidation pathways (see section 3.3.1) and in assimilatory sulfate reduction (section 3.5). ATP sulfurylase from the dissimilatory sulfate reduction pathway has been purified and characterized from *Archaeoglobus fulgidus* (Dahl *et al.* 1990, 1994). Recently the corresponding gene has been analysed and expressed (Sperling *et al.* 1998). Interestingly, ATP sulfurylases from dissimilatory and assimilatory pathways are related on the amino acid sequence level (Sperling *et al.* 1998). The recombinant active enzyme from *Archaeoglobus fulgidus* consists of two identical 53 kDa subunits (Sperling *et al.* 1998).

The reduction of APS is carried out by a FAD and [4Fe-4S]-cluster containing enzyme called **APS reductase**. This is essentially the same enzyme as the "reverse" acting APS reductase of sulfur oxidizing bacteria (section 3.3.1.1). However, in contrast to some membrane-bound APS reductases of Chromatiaceae, APS reductase seems generally to be a soluble enzyme in sulfate reducing organisms. Sulfite and AMP are the products of the two electron reduction of APS by APS reductase ($E_o' = -60$ mV, Thauer *et al.* 1977). APS reductase has been purified and characterized from several sulfate reducers, e.g. *Desulfovibrio vulgaris* (Bramlett and Peck 1975), *Desulfovibrio gigas* (Lampreia *et al.* 1990) and *Archaeoglobus fulgidus* (Speich and Trüper 1988; Lampreia *et al.* 1991; Speich *et al.* 1994; Dahl *et al.* 1994). It is composed of a large flavin subunit (approx. 70 - 80 kDa) and a small ferredoxin subunit (approx. 18 - 25 kDa) which together may form $\alpha\beta$-dimers or $\alpha_2\beta_2$- or possibly $\alpha_4\beta_4$-oligomers. Although, among others, NADH has been proposed as an electron donor for APS reductase (Chen *et al.* 1994), this is not unequivocally clear.

Sulfite is further reduced by dissimilatory **sulfite reductase**, a siroheme and iron-sulfur cluster containing enzyme. This enzyme is related to the "reverse" sulfite reductase of the branched thiosulfate oxidation pathway (section 3.3.1.1). Siroheme sulfite reductase of a different type exists in assimilatory pathways (section 3.3.3). Dissimilatory sulfite reductases have been isolated from many sulfate reducers and their classification has been based on spectral differences, leading to trivial names like desulforubidin, desulfofuscidin or desulfoviridin. However, all dissimilatory sulfite reductases seem to be structurally related. A comparison of sulfite reductases from archaebacteria and eubacteria is given by Dahl *et al.* (1993). Sulfite reductases are ca. 200 kDa enzymes, consisting of two or three subunits of a proposed $\alpha_2\beta_2$ or $\alpha_2\beta_2\gamma_2$ structure. The γ-subunit was dicovered with *Desulfovibrio vulgaris* desulfoviridin (Pierik *et al.* 1992). This small subunit may be only loosely associated with the enzyme in some cases and the number of γ-subunits per enzyme has been proposed to be variable (Steuber *et al.* 1995). All known sulfite reductases are soluble with the exception of a desulfoviridin-type sulfite reductase from *Desulfovibrio desulfuricans* which is reported to be membrane-bound (Steuber *et al.* 1994). Since the same organism contains a spectroscopically identical soluble sulfite reductase which has identical N-termini in all subunits (Steuber *et al.* 1995), the membrane-bound and soluble enzymes may differ structurally only in the existence of a membrane anchor. However, both enzymes are reported to have considerably different catalytic properties and a genetic approach may help to define relevant structural and possibly physiological differences.

There exists some confusion about the product of sulfite reductase (Jones and Skyring 1975). Depending on the assay conditions, thiosulfate, trithionate, tetrathionate or sulfide can be detected as products of sulfite reduction. Sulfide is produced in assays with reduced methyl viologen as an electron donor. However, the experimental determination of the *in vivo* product may require the use of physiological electron donors (Fritz-Steuber 1996). Such experiments were carried out with membrane-bound desulfoviridin-type sulfite reductases of *Desulfovibrio desulfuricans* (Steuber *et al.* 1994) and *Desulfovibrio gigas* (Chen *et al.* 1993). In both studies sulfite reductase reduced bisulfite to mainly sulfide when hydrogenase was used as an electron delivering system. Cytochrome c_3 and a flavoprotein were successfully used to mediate electrons between hydrogenase and sulfite reductase *in vitro* (Steuber *et al.* 1994, Chen *et al.* 1993). Some trithionate or thiosulfate formation in such assays is currently believed to be artificial, at least in the case of *Desulfovibrio* species (Steuber *et al.* 1994). The relevance of a trithionate reducing system in *Desulfovibrio vulgaris* for a possible sulfide formation from bisulfite pathway via trithionate and thiosulfate as intermediates (Kim and Akagi 1985) remains to be clarified (Hansen 1994).

In *Desulfotomaculum* species trithionate is believed to be the product of sulfite reductase (Akagi *et al.* 1973,1974; Fauque *et al.* 1991). In this case a more complex pathway is proposed to exist. Trithionate may be reduced to thiosulfate and bisulfite by **trithionate reductase**. Thiosulfate is further reduced by **thiosulfate reductase** to bisulfite and sulfide. The bisulfite produced by trithionate reductase or thiosulfate reductase may serve again as a substrate for bisulfite reductase. However, the relevance of this pathway has been doubted (Widdel and Hansen 1992). Thiosulfate and trithionate may be formed abiotically under assay conditions in a reaction of sulfite with sulfide (Widdel and Hansen 1992). The two enzymes trithionate reductase and thiosulfate reductase may serve to scavenge side products of sulfite reductase or alternatively they may be used to metabolize external trithionate or thiosulfate (Widdel and Hansen 1992).

3.4.1.2 Sulfur reduction pathways

Some bacteria are known which can grow on sulfur as an electron acceptor during heterotrophic or lithotrophic respirations. The best studied examples for sulfur reducing organisms are *Desulfuromonas acetoxidans* and *Wollinella succinogenes*. The latter, a spirilloid ε-proteobacterium, produces sulfide during growth which can react uncatalysed with sulfur, thereby producing polysulfide, the actual substrate for "sulfur" respiration (Klimmek *et al.* 1991). A membrane-bound **polysulfide reductase** is responsible for polysulfide reduction to sulfide. Polysulfide is proposed to be delivered to polysulfide reductase by the periplasmic Sud protein (Klimmek *et al.* 1998). Polysulfide reductase has been characterized and the corresponding genes have been analysed (Schröder *et al.* 1988; Krafft *et al.* 1992). The enzyme is a heterotrimer, consisting of a catalytical 81 kDa molybdenum cofactor containing subunit, a 21 kDa ferredoxin subunit and a 34 kDa membrane anchor subunit (Krafft *et al.* 1992). The active site of polysulfide reductase probably faces the periplasm, since the catalytic subunit is synthesized with a signal peptide which may serve for membrane translocation. In other sulfur reducing species the reduction mechanism may be similar.

A different polysulfide reductase has been purified from *Pyrococcus furiosus* (Ma and Adams 1994). This enzyme is cytoplasmic and catalyses sulfide formation from polysulfides with NADPH as the electron donor. It is a heterodimer of 52 and 29 kDa subunits.

Some archaebacteria are also able to grow on sulfur. Sulfolobales can reduce sulfur to sulfide under anaerobic conditions (Segerer *et al.* 1985). The chemolithoautotrophic *Pyrodictium abyssi* contains a membrane-bound 520 kDa **hydrogen:sulfur oxidoreductase** complex (Dirmeier *et al.* 1998). Many different subunits build up this complex and - as expected - nickel and iron-sulfur clusters are present in this enzyme. Also *b*-type and *c*-type cytochromes could be detected, suggesting that the complex may possess subunits at the outer side of the cytoplasmic membrane. Surprisingly no molybdenum was found. Instead copper was shown to be in the preparation. Possibly this recently enriched enzyme complex represents a new type of "sulfur reductase". A similar membrane-bound hydrogenase-coupled sulfur reductase may also be present in *Pyrodictium brockii* (Pihl *et al.* 1992).

A different sulfur reductase is reported from a moderately thermophilic prokaryote strain "TI1". This enzyme is a cytoplasmic 86 kDa homodimer which uses NADH to reduce sulfur to sulfide (Sugio *et al.* 1998). Tetrathionate, thiosulfate or sulfite were not substrates of this enzyme.

3.4.1.3 Thiosulfate and sulfite reduction

Thiosulfate reductase - possibly the key enzyme of thiosulfate respiration - has been mentioned already in the discussion of oxidative pathways (see 3.3.1.1) and of sulfate respiration (3.4.1.1). As mentioned above, sulfide and bisulfite are the products of this enzyme. Regarding the abundance of thiosulfate reductase, it is understandable that there are many thiosulfate-reducing bacteria among sulfate-reducing species. However, also several organisms that cannot carry out dissimilatory sulfate reduction can grow as thiosulfate reducers. Examples of such organisms are some eubacteria like *Salmonella typhimurium, Citrobacter freundii* and *Proteus mirabilis* or archaebacteria like *Pyrobaculum islandicum.* A detailed discussion of this pathway is given by Barrett and Clark (1987). Thiosulfate reductase has been shown to be membrane-bound in *Citrobacter* and *Proteus* (Barrett and Clark 1987). In *Salmonella typhimurium* thiosulfate reductase has been cloned and sequenced (Heinzinger *et al.* 1995). It is a membrane-bound, probably the periplasm-facing molybdenum enzyme, very similar to the polysufide reductase of *Wollinella succinogenes* (see section 3.4.1.2).

How is bisulfite reduced to sulfide in thiosulfate reducers? One would expect that there are siroheme sulfite reductases present in thiosulfate reducers. Indeed this seems to be the case. Recently, Molitor *et al.* (1998) reported the purification of a siroheme enzyme from thiosulfate grown *Pyrobaculum islandicum* which most probably represents a bisulfite reductase. Also in enterobacteria there is evidence that sulfide production from sulfite depends on siroheme enzymes (Cole *et al.* 1980). It is very likely that sulfite reduction occurs in thiosulfate reducers by a pathway identical to the one described above for sulfate reducing species (see 3.4.1.1). In the cases of *Clostridium pasteurianum* and *Pyrobaculum islandicum,* these species may be considered as facultative dissimilatory sulfite reducers. From *Clostridium pasteurianum* an unusual 83.6 kDa dissimilatory sulfite reductase has been purified from which no siroheme could be isolated (Harrison *et al.* 1984).

3.4.1.4 Tetrathionate reduction

Tetrathionate reductase activity is suggested as a second activity of some thiosulfate reductases and of a reversible thiosulfate dehydrogenase (Oltmann *et al.* 1974, Barrett and Clark 1987). A tetrathionate reductase coding operon has been cloned and sequenced from *Salmonella typhimurium* (Hensel *et al.* 1999). Like thiosulfate reductase of the same organism or polysulfide reductase from *Wollinella succinogenes,* this enzyme seems to be a membrane-bound heterotrimeric protein. The large catalytic subunit is hydrophilic and probably contains a molybdopterin guanine dinucleotide cofactor. A ferredoxin subunit may play a role for electron transport and a membrane subunit contains a quinol oxidation site. Apparently the same cofactors and similar proteins are employed for the reduction of different sulfur compounds. This could explain why indeed multifunctional enzymes exist which can reduce tetrathionate, thiosulfate and possibly also polysulfides (Oltmann *et al.* 1974).

3.5 SULFUR ASSIMILATION

L-cysteine is a common precursor for many organic sulfur compounds like methionine, lipoic acid, thiamine, coenzyme A, molybdopterin, iron-sulfur clusters and many other sulfur compounds within cells. Sulfatations of sugars, proteins or lipids usually require activated sulfate-phosphate mixed anhydrides like 3'-phosphoadenosine 5'-phosphosulfate (PAPS). For *de novo* synthesis of cysteine or methionine usually sulfide (in some cases also thiosulfate) can function as direct sulfhydryl donor. Up to now three different systems for sulfide incorporation are known. The enzymes that catalyse the cysteine formation from sulfide or thiosulfate and O-acetylserine in *E. coli* are **O-acetylserin sulfhydrylases**. Cystathionine, a methionine precursor, is generated from cysteine and O-succinylhomoserine (Figure 3.6 A). In fungi and some bacteria (e.g. *Clostridium acetobutylicum*, *Leptospira meyeri* and *Brevibacterium flavum*) there exists an alternative pathway where a **O-acetylhomoserine sulfhydrylase** converts O-acetylhomoserine to homocysteine, a precursor for cysteine and methionine (Figure 3.6 B; Grundy and Henkin 1998; Belfaiza *et al.* 1998; Ozaki and Shiio 1982). In *Pseudomonas aeruginosa* O-succinylhomoserine is directly sulfhydrylated by a **O-succinylhomoserine sulfhydrylase** to homocysteine which then may serve as a precursor of cysteine and methionine (Figure 3.6C; Foglino *et al.* 1995).

In case there is no external reduced sulfur compound available that can be incorporated into the organic matter of the organism, sulfate reduction to sulfide is required. Some bacteria are very much specialized for living in habitats with reduced sulfur compounds and such bacteria may lack a sulfate reduction pathway (e.g. Chlorobiaceae). Usually assimilatory sulfate reduction is a highly regulated process. Most studies have been done on eucaryotes, but it is also known from bacteria that assimilatory sulfate reduction pathways are induced under sulfur-limted conditions (Kredich 1996; Marzluf 1997).

At least two assimilatory pathways are known by which sulfate can be reduced to sulfide. In one pathway APS, PAPS and sulfite are believed to be intermediates, the other pathway does not include the formation of PAPS and APS is directly reduced to sulfite. In both cases sulfate is actively transported into the cytoplasm (Sirko *et al.* 1995; Warthmann and Cypionka 1996) and activated by an **ATP sulfurylase**, as already discussed for dissimilatory ATP sulfurylases (section 3.4.1.1). However, ATP sulfurylase from *E. coli* has been found to be also a guanosine triphosphatase (GTPase) and GTP hydrolysis triggers APS formation from ATP and sulfate (Liu *et al.* 1998). The high intracellular pyrophosphate

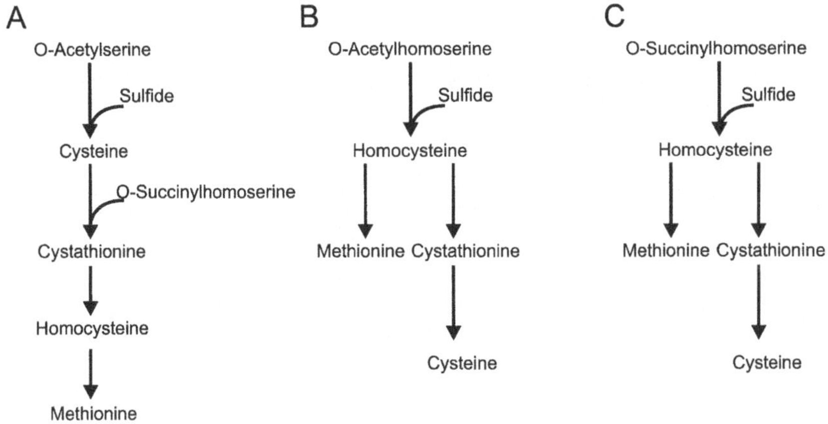

Figure 3.6. Pathways of introduction of sulfide into organic compounds.

concentration in *E. coli* is suggested as the reason for the need of a GTPase-activated ATP sulfurylase. In other organisms studied, no GTPase activation of ATP sulfurylase was required (Cooper and Trüper 1985).

APS can be phosphorylated to PAPS by an ATP-dependent **APS kinase**. ATP sulfurylase and APS kinase are in some cases two functions of one protein (Schwartz *et al.* 1998). In these cases a substrate channeling occurs and PAPS is released as the reaction product. Whether such a substrate channeling can also occur in cases where ATP sulfurylase and APS kinase are functions of two different polypeptides (e.g. in *E. coli*) is not known yet.

PAPS can serve as substrate for sulfatation reactions or its sulfate moiety is reduced to free sulfite by an enzyme called **PAPS reductase**. This enzyme is a 56 kDa homodimer in *E. coli*. A similar PAPS reductase is present in *Thiocapsa roseopersicina* and some cyanobacteria (Haverkamp and Schwenn 1999).

In some cases it has been reported that PAPS formation is circumvented and that sulfite is directly formed from APS in the assimilatory pathway by an **APS reductase** (Imhoff 1982). This PAPS circumventing pathway has recently been questioned (Haverkamp and Schwenn 1999), but still there is experimental evidence lacking against the direct assimilatory formation of sulfite from APS. Recent studies on plant sulfate assimilation mechanisms suggest that some plants may also circumvent PAPS formation (Wray *et al.* 1998). It is suggested that in plants PAPS may serve as a substrate for sulfotransferases or as a storage compound for activated sulfate.

Sulfite is reduced to sulfide by an assimilatory **sulfite reductase**. The *E. coli* enzyme is NADPH-dependent and consists of two different polypeptides of 66 and 64 kDa which form an $\alpha_8\beta_4$ oligomer. The α-subunit is a flavoprotein containing one FMN and one FAD. The β-subunit contains siroheme and an iron-sulfur cofactor and is regarded as the catalytic subunit. The structure of the catalytic subunit has been revealed by X-ray analysis (Crane *et al.* 1995). In the active site, siroheme is coupled to an [4Fe-4S]-cluster. Sulfite binds to siroheme via its sulfur atom. A different low molecular weight type of assimilatory sulfite reductases has been isolated from several anaerobic bacteria (Moura *et al.* 1986). The only cofactors of this enzyme are siroheme and an iron-sulfur cluster. The gene sequence of this enzyme was obtained from *Desulfovibrio vulgaris* and revealed some similarity to the heme subunit of the *E. coli* assimilatory sulfite reductase (Tan *et al.* 1991, 1994). Siroheme appears to be an ancient cofactor optimized for sulfite reduction.

3.6 ACKNOWLEDGEMENTS

We thank the Deutsche Forschungsgemeinschaft for grants to H.G. Trüper and C. Dahl. This work was supported by the Fonds der Chemischen Industrie. We are grateful to Dr. Christiane Dahl and co-workers for many helpful discussions.

3.7 REFERENCES

Akagi, J.M. and Adams, V. (1973) Isolation of a bisulfite reductase activity from *Desulfotomaculum nigrificans* and its identification as the carbon monoxide-binding pigment P582. *J. Bacteriol.* **116**, 392-396

Akagi, J.M., Chan, M. and Adams, V. (1974) Observations on the bisulfite reductase (P582) isolated from *Desulfotomaculum nigrificans*. *J. Bacteriol.* **120**, 240-244

Alexander, K. and Volini, M. (1987) Properties of *Escherichia coli* rhodanese. *J. Biol. Chem.* **262**, 6595-6604

Aminuddin, M. and Nicholas, D.J.D. (1974) An AMP-independent sulphite oxidase from *Thiobacillus denitrificans*: purification and properties. *J. Gen. Microbiol.* **82**, 103-113

Arieli, B., Shahak, Y., Taglicht, D., Hauska, G. and Padan, E. (1994) Purification and characterization of sulfide-quinone reductase, a novel enzyme driving anoxygenic photosynthesis in *Oscillatoria limnetica*. *J. Biol. Chem.* **269**, 5705-5711

Baas-Becking, L.G.M. (1925) Studies on suphur bacteria. *Annals of Botany* **39**, 613-650

Barrett, E. and Clark, M.A. (1987) Tetrathionate reduction and production of hydrogen sulfide from thiosulfate. *Microbiological Reviews* **51**, 192-205

Bartsch, R.G., Newton, G.L., Sherrill, C. and Fahey, R.C. (1996) Glutathione amide and its perthiol in anaerobic sulfur bacteria. *J. Bacteriol.* **178**, 4742-4746

Belfaiza, J., Martel, A., Margarita, D. and Saint Girons, I. (1998) Direct sulfhydrylation for methionine biosynthesis in *Leptospira meyeri. J. Bacteriol.* **180**, 250-255

Bick, J.A. and Leustek, T. (1998) Plant sulfur metabolism--the reduction of sulfate to sulfite. *Curr. Opin. Plant Biol.* **1**, 240-244

Bowen, T.J. Happold, F.C. and Taylor, B.F. (1965) Some properties of the rhodanese system of *Thiobacillus denitrificans. Biochem. J.* **97**, 651-657

Bowen, T.J. Happold, F.C. and Taylor, B.F. (1966) Studies on adenosine-5'-phosphosulfate reductase from *Thiobacillus denitrificans. Biochim. Biophys. Acta* **118**, 67-73

Bramlett, R.N. and Peck, H.D. Jr. (1975) Some physical and kinetic properties of adenylyl sulfate reductase from *Desulfovibrio vulgaris. J. Biol. Chem.* **250**, 2979-2986

Breuker, E. (1964) Die Verwertung von intrazellulärem Schwefel durch Chromatium vinosum im aeroben und anaeroben Licht- und Dunkelstoffwechsel. *Zentralblatt f. Bakteriologie, Parasitenkunde, Infektionskrankheiten u. Hygiene, II.Abt.* **118**, 561-579

Brune, D. C. (1989) Sulfur oxidation by phototrophic bacteria. *Biochim. Biophys. Acta* **975**, 189-221

Brune, D. C. (1995) Sulfur compounds as photosynthetic electron donors, in *Anoxygenic Photosynthetic Bacteria* (Eds.: Blanenship, R.E., Madigan, M.T., Bauer, C.E.) pp.847-870, Kluwer Academic Press, Niederlande

Brüser, T., Selmer, T. and Dahl, C. (1999) Biochemical and genetic analyses identify "ADP sulfurylase" from *Thiobacillus denitrificans* as an unidirectionally acting adenylylsulfate:phosphate adenylyltransferase. submitted

Chen, L., Le Gall, J. and Xavier, A.V. (1994) Purification, characterization and properties of an NADH oxidase from Desulfovibrio vulgaris (Hildenborough) and its coupling to adenylyl phosphosulfate reductase. *Biochem. Biophys. Res. Commun.* **203**, 839-844

Chen, L., Liu, M.-Y. and LeGall, J. (1993) Isolation and characterization of flavoredoxin, a new flavoprotein that permits in vitro reconstitution of an electron transfer chain from molecular hydrogen to sulfite reductionin the bacterium *Desulfovibrio gigas. Arch. Biochem. Biophys.* **303**, 44-50

Cole, J.A., Newman, B.M. and White, P. (1980) Biochemical and genetic characterization of nirB mutants of *Escherichia coli* K 12 pleiotropically defective in nitrite and sulphite reduction. *J. Gen. Microbiol.* **120**, 475-483

Cooper, B.P. and Trüper, H.G. (1985) Sulfate activation in *Rhodobacter sulfidophilus* and other species of Rhodospirillaceae. *Arch. Microbiol.* **141**, 384-391

Crane, B.R., Siegel, L.M. and Getzoff, E.D. (1995) Sulfite reductase structure at 1.6 A: evolution and catalysis for reduction of inorganic anions. *Science* **270**, 59-67

Dahl, C. (1996) Insertional gene inactivation in a phototrophic sulphur bacterium: APS-reductase-deficient mutants of *Chromatium vinosum. Microbiology* **142**, 3363-3372

Dahl, C., Koch, H.-G., Keuken, O. and Trüper, H.G. (1990) Purification and characterization of ATP sulfurylase from the extremely thermophilic archaebacterial sulfate-reducer, *Archaeoglobus fulgidus*. *FEMS Microbiol. Lett.* **67**, 27-32

Dahl, C., Kredich, N.M., Deutzmann, R. and Trüper, H.G. (1993) Dissimilatory sulphite reductase from *Archaeoglobus fulgidus*: physico-chemical properties of the enzyme and cloning, sequencing and analysis of the reductase genes. *J. Gen. Microbiol.* **139**, 1817-1828

Dahl, C., Speich, N. and Truper, H.G. (1994) Enzymology and molecular biology of sulfate reduction in extremely thermophilic archaeon *Archaeoglobus fulgidus*. *Methods Enzymol.* **243**, 331-349

Dahl, C. and Trüper, H. G. (1989) Comparative enzymology of sulfite oxidation in *Thiocapsa roseopersicina* strains 6311, M1 and BBS under chemotrophic and phototrophic conditions. *Z. Naturforsch.* **44c**, 137-142

De Jong, G.A., Hazeu, W., Bos, P. and Kuenen, J.G. (1997) Isolation of the tetrathionate hydrolase from *Thiobacillus acidophilus*. *Eur. J. Biochem.* **243**, 678-683

Denger, K., Laue, H. and Cook, A.M. (1997) Thiosulfate as a metabolic product: the bacterial fermentation of taurine. *Arch. Microbiol.* **168**, 297-301

Dirmeier, R., Keller, M., Frey, G., Huber, H. and Stetter, K.O. (1998) Purification and properties of an extremely thermostable membrane-bound sulfur-reducing complex from the hyperthermophilic *Pyrodictium abyssi*. *Eur. J. Biochem.* **252**, 486-491

Emmel, T., Sand, W., König, W.A. and Bock, E. (1986) Evidence for the existence of a sulfur oxygenase in *Sulfolobus brierley*. *J. Gen. Microbiol.* **132**, 3415-3420

Fauque, G., LeGall, J. and Barton, L.L. (1991) Sulfate-reducing and sulfur-reducing bacteria, In: *Variations in Autotrophic Life*. (Eds.: Shively, J.M., Barton, L.L.) pp.271-338, Academic Press, London

Foglino, M., Borne, F., Bally, M., Ball, G. and Patte, J.C. (1995) A direct sulfhydylation pathway is used for methionine biosynthesis in *Pseudomonas aeruginosa*. *Microbiology* **141**, 431-439

Friedrich, C.G. (1998) Physiology and genetics of sulfur-oxidizing bacteria. *Adv. Microbial Physiol.* **39**, 235-289

Fritz-Steuber, J. (1996) Molecular and kinetic properties of desulfoviridin, the dissimilatory ferrocytochrome c_3:sulfite oxidoreductase from *Desulfovibrio desulfuricans* (Essex). Ph.D. thesis, Konstanz (Germany)

Gibson, J. (1967) Aerobic metabolism of *Chromatium* sp. strain *D*. *Arch. Mikrobiol.* **59**, 104-112

Gibson, G.R. (1990) Physiology and ecology of the sulphate-reducing bacteria. *J. Appl. Bacteriol.* **69**, 769-797

Grundy, F.J. and Henkin, T.M. (1998) The S box: a new global trascription termination control system for methionine and cysteine biosynthesis genes in Gram-positive bacteria. *Mol. Microbiol.* **30**, 737-749

Hagen, K.D. and Nelson, D.C. (1997) Use of reduced sulfur compounds by *Beggiatoa* spp.: enzymology and physiology of marine and freshwater strains in homogenous and gradient cultures. *Appl. Environm. Microbiol*. **63**, 3957-3964

Hallberg, K.B., Dopson, M. and Lindström, E.B. (1996) Reduced sulfur compound oxidation by *Thiobacillus caldus*. *J. Bacteriol.* **178**, 6-11

Hamilton, W.A. (1998) Bioenergetics of sulfate-reducing bacteria in relation to their environmental impact. *Biodegradation* **9**, 201-212

Hansen, T.A. (1994) Metabolism of sulfate-reducing prokaryotes. *Antonie van Leeuwenhoek* **66**, 165-185

Harrison, G., Curle, C. and Laishley, E.J. (1984) Purification and characterization of an inducible dissimilatory type sulfite reductase from *Clostridium pasteurianum*. *Arch. Microbiol.* **138**, 72-78

Haverkamp, T. and Schwenn, J.D. (1999) Structure and function of a *cysBJIH* gene cluster in the purple sulphur bacterium *Thiocapsa roseopersicina*. *Microbiology* **145**, 115-125

Hazeu, W., Batenburg-van der Vegte, W.H., Bos, P., van der Pas, R.K. and Kuenen, J.G. (1988) The production and utilization of intermediary elemental sulfur during the oxidation of reduced sulfur compounds by *Thiobacillus ferrooxidans*. *Arch. Microbiol.* **150**, 574-579

Heinzinger, N.K., Fujimoto, S.Y., Clark, M.A., Moreno, M.S. and Barrett,E.L. (1995) Sequence analysis of the *phs* operon in *Salmonella typhimurium* and the contribution of thiosulfate reduction to anaerobic energy metabolism. *J. Bacteriol.* **177**, 2813-2820

Hensel, M., Hinsley, A.P., Nikolaus, T., Sawers, G. and Berks, B.C. (1999) The genetic basis of tetrathionate respiration in *Salmonella typhimurium*. *Mol. Microbiol.* **32**, 275-287

Hipp, W.M., Pott, A.S., Thum-Schmitz, N., Faath, I., Dahl, C. and Trüper, H.G. (1997) Towards the phylogeny of APS reductase and sirohaem sulfite reductases in sulfate-reducing and sulfur-oxidizing prokaryotes. *Microbiology* **143**, 2891-2902

Hooper, A.B. and DiSpirito, A.A. (1985) In bacteria which grow on simple reductants, generation of a proton gradient involves extracytoplasmic oxidation of substrate. *Microbiol. Rev.* **49**, 140-157

Imhoff, J.F. (1982) Occurence and evolutionary significance of two sulfate assimilation pathways in the Rhodospirillaceae. *Arch. Microbiol.* **132**, 197-203

Ingvorsen, K. and Jørgensen, B.B. (1982) Seasonal variation in H_2S emission from internal sediments in Denmark. *Atmospheric Environment* **16**, 855-865

Jones, H.E. and Skyring, G.W. (1975) Effect of enzymic assay conditions on sulfite reduction catalysed by desulfoviridin from *Desulfovibrio gigas*. *Biochim. Biophys. Acta* **377**, 52-60

Kämpf, C. and Pfennig, N. (1980) Capacity of Chromatiaceae for chemotrophic growth. Specific respiration rates of *Thiocystis violacea* and *Chromatium vinosum*. *Arch. Microbiol.* **127**, 125-135

Kämpf, C. and Pfennig, N. (1986a) Isolation and characterization of some chemoautotrophic Chromatiaceae. *J. Basic Microbiol.* **9**, 507-515

Kämpf, C. and Pfennig, N. (1986b) Chemoautotrophic growth of *Thiocystis violacea, Chromatium gracile* and *A. vinosum* in the dark at various O_2-concentrations. *J. Basic Microbiol.* **9**, 517-531

Kelly, D.P. (1999) Thermodynamic aspects of energy conservation by chemolithotrophic sulfur bacteria in relation to the sulfur oxidation pathways. *Arch. Microbiol.* **171**, 219-229

Kelly, D.P. and Harrison, A.P. (1989) Genus *Thiobacillus*. In: *Bergeys manual of systematic bacteriology*. (Eds.: Murray, R.G.E., Brenner, D.J., Holt, J.G., Krieg, N.R., Moulder, J.W., Pfennig, N.,Sneath, P.H.A., Staley, J.T., Williams, S.T.) Williams and Wilkins. Baltimore

Kelly, D.P., Shergill, J.K., Lu, W.-P. and Wood, A.P. (1997) Oxidative metabolism of inorganic sulfur compounds by bacteria. *Antonie van Leeuwenhoek* **71**, 95-107

Kelly, D.P. and Wood, A.P. (1999) Reclassification of some species of *Thiobacillus* to the newly designated genera *Acidithiobacillus* gen. nov., *Halothiobacillus* gen. nov. and *Thermothiobacillus* gen. nov.. *Int. J. System. Bacteriol.*, in press

Kim, J.H. and Akagi, J.M. (1985) Characterization of a trithionate reductase system from *Desulfovibrio vulgaris*. *J.Bacteriol.* **163**, 472-475

Kletzin, A. (1989) Coupled enzymatic production of sulfite, thiosulfate, and hydrogen sulfide from sulfur: purification and properties of a sulfur oxygenase reductase from the facultatively anaerobic archaebacterium *Desulfurolobus ambivalens*. *J. Bacteriol.* **171**, 1638-1643

Kletzin, A. (1992) Molecular characterization of the sor gene, which encodes the sulfur oxygenase/reductase of the thermoacidophilic archaeum *Desulfolobus ambivalens*. *J. Bacteriol.* **174**, 5854-5859

Klimmek, O., Kreis, V., Klein, C., Simon, J., Wittershagen, A. and Kröger, A. (1998) The function of the periplasmic Sud protein in polysulfide respiration of *Wolinella succinogenes*. *Eur. J. Biochem.* **253**, 263-269

Klimmek, O., Kröger, A., Steudel, R. and Holdt, G. (1991) Growth of *Wolinella succinogenes* with polysulphide as terminal acceptor of phosphorylative electron export. *Arch. Microbiol.* **155**, 177-182

Krafft, T., Bokranz, M., Klimmek, O., Schröder, I., Fahrenholz, F., Kojro, E. and Kröger, A. (1992) Cloning and nucleotide sequence of the psrA gene of *Wolinella succinogenes* polysulphide reductase. *Eur. J. Biochem.* **206**, 503-510

Kredich, N.M. (1996) Biosynthesis of cysteine. In: *Escherichia coli and Salmonella typhimurium: cellular and molecular biology.* 2nd ed. (Eds.: Neidhardt, F.C., Curtiss, R., Ingraham, J.L., Lin, E.C.C., Low, K.B., Magasanik, B., Reznikoff, W.S., Riley, M., Schaechter, M., Umbarger, H.E.) pp.514-527, ASM Press, Washington, D.C.

Kreke, B. and Cypionka, H. (1995) Energetics of sulfate transport in Desulfomicrobium baculatum. *Arch. Microbiol.* **163**, 307-309

Lampreia, J., Fauque, G., Speich, N., Dahl, C., Moura, I., Trüper, H.G. and Moura, J.J.G. (1991) Spectroscopic studies on APS reductase isolated from the hyperthermophilic sulfate-reducing archaebacterium *Archaeoglobus fulgidus*. *Biochem. Biophys. Res. Commun.* **181**, 342-347

Lampreia, J., Moura, I., Teixeira, M., Peck, H.D. Jr., Legall, J., Huynh, B.H. and Moura, J.J. (1990) The active centers of adenylylsulfate reductase from *Desulfovibrio gigas*. Characterization and spectroscopic studies. *Eur. J. Biochem* **188**, 653-664

Liu, C., Wang, R., Varlamova, O. and Leyh, T.S. (1998) Regulating energy transfer in the ATP sulfurylase-GTPase system. *Biochemistry* **37**, 3886-3892

LeFaou, A., Rajagopal, B.S., Daniels, L. and Fauque, G. (1990) Thiosulfate, polythionates and elemental sulfur assimilation and reduction in the bacterial world. *FEMS Microbiol. Rev.* **75**, 351-382

Lu, W.P. and Kelly, D.P. (1983a) Partial purification and resolution of a thiosulfate-oxidizing system from *Thiobacillus* A2. *J. Gen. Microbiol.* **129**, 1673-1681

Lu, W.P. and Kelly, D.P. (1983b) Purification and some properties of two principal enzymes of the thiosulfate-oxidizing multi-enzyme system from *Thiobacillus* A2.

J. Gen. Microbiol. **129**, 3549-3564

Lu, W.P. and Kelly, D.P. (1988a) Kinetic and energetic aspects of inorganic sulphur compound oxidation by *Thiobacillus tepidarius*. *J. Gen. Microbiol.* **134**, 865-876

Lu, W.P. and Kelly, D.P. (1988b) Cellular location and partial purification of the 'thiosulphate-oxidizing enzyme' and 'trithionate hydrolase' from *Thiobacillus tepidarius*. *J. Gen. Microbiol.* **134**, 877-885

Lyric. M.R. and Suzuki, I. (1970) Enzymes involved in the metabolism of thiosulfate by *Thiobacillus thioparus* . II. Properties of APS reductase. *Can. J. Biochem.* **48**, 344-353

Ma, K. and Adams, M.W. (1994) Sulfide dehydrogenase from the hyperthermophilic archaeon *Pyrococcus furiosus*: a new multifunctional enzyme involved in the reduction of elemental sulfur. *J. Bacteriol.* **176**, 6509-6517

Marzluf, G.A. (1997) Molecular genetics of sulfur assimilation in filamentous fungi and yeast. *Annu. Rev. Microbiol.* **51**, 73-96

Meulenberg, R., Pronk, J.T., Frank, J., Hazeu, W., Bos, P. and Kuenen, J.G. (1992) Purification and partial characterization of a thermostable trithionate hydrolase from the acidophilic sulphur oxidizer *Thiobacillus acidophilus*. *Eur. J. Biochem.* **209**, 367-374

Meulenberg, R., Scheer, E.J., Pronk, J.T., Hazeu, W., Bos, P. and Kuenen, J.G. (1993) Metabolism of tetrathionate in *Thiobacillus acidophilus*. *FEMS Microbiol. Lett.* **112**, 167-172

Molitor, M., Dahl, C., Molitor, I., Schäfer, U., Speich, N., Huber, R., Deutzmann, R. and Trüper, H.G. (1998) A dissimilatory sirohaem-sulfite-reductase-type protein from the hyperthermophilic archaeon *Pyrobaculum islandicum*. *Microbiology* **144**, 529-541

Moodie, A.D. and Ingledew, W.J. (1990) Microbiol anaerobic respiration. *Adv. Microbial. Physiol.* **31**, 225-269

Moreira, D. and Amils, R. (1997) Phylogeny of *Thiobacillus cuprinus* and other mixotrophic thiobacilli: proposal for *Thiomonas* gen. nov.. *Int. J. Syst. Bacteriol.* **47**, 522-528

Moriarty, D.J.W. and Nicholas, D.J.D. (1969) Enzymic sulphide oxidation by *Thiobacillus concretivorus*. *Biochim. Biophys. Acta* **184**, 114-123

Moura, I., Lino, A.R., Moura, J.J., Xavier, A.V., Fauque, G., Peck, H.D. Jr. and LeGall, J. (1986) Low-spin sulfite reductases: a new homologous group of non-heme iron-siroheme proteins in anaerobic bacteria. *Biochem. Biophys. Res. Commun.* **141**, 1032-1041

Murray, R.G.E., Brenner, D.J., Holt, J.G., Krieg, N.R., Moulder, J.W., Pfennig, N.,Sneath, P.H.A., Staley, J.T., Williams, S.T. (eds.) (1989) *Bergeys manual of systematic bacteriology*. Williams and Wilkins. Baltimore

Oltmann, L.F., Schoenmaker, G.S. and Stouthamer, A.H. (1974) Solubilization and purification of a cytoplasmic membrane-bound enzyme catalyzing tetrathionate and thiosulphate reduction in *Proteus mirabilis*. *Arch. Microbiol.* **98**, 19-30

Ozaki, H. and Shiio, I. (1982) Methionine biosynthesis in *Brevibacterium flavum*: properties and essential role of O-acetylhomoserine sulfhydrylase. *Acta. Biochim. Pol.* **23**, 321-324

Pattaragulwanit, K., Brune, D.C., Trüper, H.G. and Dahl, C. (1998) Molecular genetic evidence for extracytoplasmic localization of sulfur globules in *Chromatium vinosum*. *Arch. Microbiol.* **169**, 434-444.

Peck, H.D. Jr. (1960) Adenosine 5'-phosphosulfate as an intermediate in the oxidation of thiosulfate by *Thiobacillus thioparus. Biochemistry* **46**, 1053-1057

Peck, H.D. Jr. (1961) Enzymatic basis for assimilatory and dissimilatory sulfate reduction. *J. Bacteriol.* **82**, 933-939

Peck, H.D. Jr. and Fisher, E. Jr. (1962) The oxidation of thiosulfate and phosphorylation in extracts of *Thiobacillus thioparus. J. Biol. Chem.* **237**, 190-197

Petushkova, I.P., Ivanovskii, R.N. (1976) Enzymes involved in thiosulfate metabolism in *Thiocapsa roseopersicina* under various conditions of growth. (russian) *Mikrobiologiia* **45**, 960-965

Pfennig, N. and Biebl, H. (1976) *Desulfuromonas acetoxidans* gen. nov. and sp. nov. a new anaerobic sulfur-reducing acetate-oxidizing bacterium. *Arch. Microbiol.* **110**, 3-12

Pierik, A.J., Duyvis, M.G., van Heelvort, J.M.L.M., Wolbert, R.B.G. and Hagen, W.R. (1992) The third subunit of desulfoviridin-type dissimilatory sulfite reductases. *Eur. J. Biochem.* **205**, 111-115

Pihl, T.D., Black, L.K., Schulman, B.A. and Maier, R.J. (1992) Hydrogen-oxidizing electron transport components in the hyperthermophilic archaebacterium *Pyrodictium brockii. J.Bacteriol.* **174**, 137-143

Pott, A. and Dahl, C. (1998) Sirohaem sulfite reductase and other proteins encoded by genes at the *dsr* locus of *Chromatium vinosum* are involved in the oxidation of intracellular sulfur. *Microbiology* **144**, 1881-1894

Prange, A., Arzberger, I., Engemann, C., Modrow, H., Schumann, O., Trüper, H.G., Steudel, R., Dahl, C. and Hormes, J. (1999) In situ analysis of sulfur in the sulfur globules of phototrophic sulfur bacteria by X-ray absorption near edge spectroscopy. *Biochim. Biophys. Acta.* (in press)

Pronk, J.T., Meulenberg, R., Hazeu, W., Bos, P. and Kuenen, J.G. (1990) Oxidation of reduced inorganic sulphur compounds by acidophilic thiobacilli. *FEMS Microbiol. Rev.* **75**, 293-306

Reinartz, M., Tschäpe, J., Brüser, T., Trüper, H.G. and Dahl, C. (1998) Sulfide oxidation in the phototrophic sulfur bacterium *Chromatium vinosum. Arch. Microbiol.* **170**, 59-68

Robertson L.A. and Kuenen, J.G. (1992) In: *The Prokaryotes* (Ed.: Barlows, A., Trüper, H.G., Dworkin, M., Harder, W., Schliefer, H.K.) 2nd Ed., pp.385-413, Springer, New York

Santer, M., Margulies, M., Klinman, N. and Kaback, R. (1960) Role of inorganic phosphate in thiosulfate metabolism by *Thiobacillus thioparus. J. Bacteriol.* **79**, 313-320

Schedel, M. and Trüper, H.G. (1979) Purification of *Thiobacillus denitrificans* siroheme sulfite reductase and investigation of some molecular and catalytic properties. *Biochim. Biophys. Acta* **568**, 454-467

Schedel, M. and Trüper, H.G. (1980) Anaerobic oxidation of thiosulfate and elemental sulfur in *Thiobacillus denitrificans. Arch. Microbiol.* **124**, 205-210

Schedel, M., Vanselow, M. and Trüper, H.G. (1979) Siroheme sulfite reductase isolated from *Chromatium vinosum.* Purification and investigation of its molecular and catalytic properties. *Arch. Microbiol.* **121**, 29-36

Schippers, A., Jozsa, P.-G. and Sand, W. (1996) Sulfur chemistry in bacterial leaching of pyrite. *Appl. Environm. Microbiol.* **62**, 3424-3431

Schippers, A. and Sand, W. (1999) Bacterial leaching of metal sulfides proceeds by two indirect mechanisms via thiosulfate or via polysulfides and sulfur. *Appl. Environm. Microbiol.* **65**, 319-321

Schmidt, M. (1984) Chemical problems of flue-gas desulfurization, in: *Sulfur, its significance for chemistry, for the geo-, bio- and cosmosphere and technology* (Ed.: Müller, A. and Krebs, B.) pp.259-276, Elsevier, Amsterdam

Schmitt, W., Schleifer, G. and Knobloch, K. (1981) The enzymatic system thiosulfate:cytochrome c oxidoreductase from photolithoautotrophically grown *Chromatium vinosum. Arch. Microbiol.* **130**, 334-338

Schneider, A. and Friedrich, C. (1994) Sulfide dehydrogenase is identical with the SocB protein of the thiosulfate-oxidizing enzyme system of *Paracoccus denitrificans* GB17. *FEBS Lett.* **350**, 61-65

Schröder, I., Kröger, A. and Macy, J.M. (1988) Isolation of the sulphur reductase and reconstitution of the sulphur respiration of *Wolinella succinogenes. Arch. Microbiol.* **149**, 572-579

Schütz, M., Klughammer, C., Griesbeck, C., Quentmeier, A., Friedrich, C.G. and Hauska, G. (1998) Sulfide-quinone reductase activity in membranes of the chemotrophic bacterium *Paracoccus denitrificans* GB17, *Arch. Microbiol.* **170**, 353-360

Schütz, M., Shahak, Y, Padan, E. and Hauska, G. (1997) Sulfide-quinone reductase from *Rhodobacter capsulatus. J. Biol. Chem.* **272**, 9890-9894

Schwartz, N.B., Lyle, S., Ozeran, J.D., Li, H., Deyrup, A., Ng, K. and Westley, J. (1998) Sulfate activation and transport in mammals: system components and mechanisms. *Chem. Biol. Interact.* **109**,143-151

Segerer, A., Stetter, K.O. and Klink, F. (1985) Two contrary modes of chemolithotrophy in the same archaebacterium. *Nature* **313**, 787-789

Shahak, Y., Arieli, B., Hauska, G., Herrmann, I. and Padan, E. (1992a) Isolation of sufide-quinone reductase (SQR) from Prokaryotes. *Phyton* **32**, 133-137

Shahak, Y., Arieli, B., Padan, E. and Hauska, G., (1992b) Sulfide quinone reductase (SQR) activity in *Chlorobium. FEBS Lett.* **299**, 127-130

Shively, J.M. and Barton, L.L.(eds.) (1991) *Variations in Autotrophic Life*. Academic Press, London

Silver, M. and Lundgren, D.G. (1968) Sulfur-oxidizing enzyme of *Ferrobacillus ferrooxidans* (*Thiobacillus ferrooxidans*). *Can. J. Biochem.* **46**, 457-461

Sirko, A., Zatyka, M., Sadowy, E. and Hulanicka, D. (1995) Sulfate and thiosulfate transport in *Escherichia coli* K-12: evidence for a functional overlapping of sulfate- and thiosulfate-binding proteins. *J. Bacteriol.* **177**, 4134-4136

Smith, A.J. (1966) The role of tetrathionate in the oxidation of thiosulphate by *Chromatium* sp. strain D. *J. Gen. Microbiol.* **42**, 371-380

Smith, A.J. and Lascelles, J. (1966) Thiosulphate metabolism and rhodanese in *Chromatium* sp. strain D. *J. Gen. Microbiol.* **42**, 357-370

Southerland, W.M. and Toghrol, F. (1983) Sulfite oxidase activity in *Thiobacillus novellus. J. Bacteriol.* **156**, 941-944

Speich, N., Dahl, C., Heisig, P., Klein, A., Lottspeich, F., Stetter, K.O. and Trüper, H.G. (1994) Adenylylsulfate reductase from the sulphate-reducing archaeon *Archaeoglobus fulgidus*: cloning and characterization of the genes and comparison of the enzyme with other iron-sulphur flavoproteins. *Microbiology* **140**, 1273-1284

Speich, N. and Trüper, H.G. (1988) Adenylylsulfate reductase in a dissimilatory sulphate-reducing archaebacterium. *J. Gen. Microbiol.* **134**, 1419-1425

Sperling, D., Kappler, U., Wynen, A., Dahl, C. and Trüper, H.G. (1998) Dissimilatory ATP sulfurylase from the hyperthermophilic sulfate reducer *Archaeoglobus fulgidus* belongs to the group of homo-oligomeric ATP sulfurylases. *FEMS Microbiol. Lett.* **162**, 257-264

Steuber, J., Arendsen, A.F., Hagen, F. and Kroneck, P.M.H. (1995) Molecular properties of the dissimilatory sulfite reductase from *Desulfovibrio desulfuricans* (Essex) and comparison of the enzyme with the enzyme from *Desufovibrio vulgaris* (Hildenborough). *Eur. J. Biochem.* **233**, 873-879

Steuber, J., Cypionka, H. and Kroneck, P.M.H. (1994) Mechanism of dissimilatory sulfite reduction by *Desulfovibrio desulfuricans*: purification of a membrane-bound sulfite reductase and coupling with cytochrome c_3 and hydrogenase. *Arch. Microbiol.* **162**, 255-260

Steudel, R., Hold, G., Gobel, T. and Hazeu, W. (1987) Chromatographie separation of higher polythionates $S_nO_6^{2-}$ (n=3...22) and their detection in cultures of *Thiobacillus ferrooxidans*; molecular composition of bacterial sulfur secretions. Angew. Chem. Int. Ed. Engl. 26, 151-153

Sugio, T., Hirose, T., Li-Zhen, Y. and Tano, T. (1992) Purification and some properties of sulfite:ferric ion oxidoreductase from *Thiobacillus ferrooxidans*. *J. Bacteriol.* **174**, 4189-4192

Sugio, T., Mizunashi, W., Inagaki, K. and Tano, T. (1987) Purification and some properties of sulfur:ferric ion oxidoreductase from *Thiobacillus ferrooxidans*. *J. Bacteriol.* **169**, 4916-4922

Sugio, T., Oda, K., Matsumoto, K., Takai, M., Wakasa, S. and Kamimura, K. (1998) Purification and characterization of sulfur reductase from a moderately thermophilic bacterial strain, TI-1, that oxidizes iron. *Biosci. Biotechnol. Biochem.* **62**, 705-709

Suzuki, I. and Silver, M. (1966) The initial product and properties of the sulfur-oxidizing enzyme of thiobacilli. *Biochim. Biophys. Acta.* **122**, 22-33

Tabita, R., Silver, M. and Lundgren, D.G. (1969) The rhodanese enzyme of Ferro*bacillus ferrooxidans* (*Thiobacillus ferrooxidans*). *Can. J. Biochem.* **47**, 1141-1145

Takeuchi, T.L. and Suzuki, I. (1994) Effect of pH on sulfite oxidation by *Thiobacillus thiooxidans* cells with sulfurous acid or sulfur dioxide as a possible substrate. *J. Bacteriol.* **176**, 913-916

Tan, J., Helms, L.R., Swenson, R.P. and Cowan, J.A. (1991) Primary structure of the assimilatory-type sulfite reductase from *Desulfovibrio vulgaris* (Hilenborough): cloning and nucleotide sequence of the reductase gene. *Biochemistry* **30**, 9900-9907

Tan, J., Soriano, A., Lui, S.M. and Cowan, J.A. (1994) Functional expression and characterization of the assimilatory-type sulfite reductase from *Desulfovibrio vulgaris* (Hilenborough). *Arch. Biochem. Biophys.* **312**, 516-523

Thauer, R.K., Jungermann, K. and Decker, K. (1977) Energy conservation in chemotrophic anaerobic bacteria. *Bacteriol. Rev.* **41**, 100-180

Timmer-Ten Hoor, A. (1981) Cell yield and bioenergetics of *Thiomicrospira denitrificans* compared with *Thiobacillus denitrificans*. *Antonie van Leeuwenhoek* **47**, 231-243

Toghrol, F. and Southerland, W.M. (1983) Purification of *Thiobacillus novellus* sulfite oxidase. Evidence for the presence of heme and molybdenum. *J. Biol. Chem.* **258**, 6762-6766

Trudinger, P.A. and Loughlin, R.E. (1981) Metabolism of simple sulphur compounds. In: *Comprehensive Biochemistry* (Eds.: Neuberger, A., van Deenen, L.L.M.) Vol. 19A Amino acid metabolism and sulphur metabolism. pp.165-256 Elsevier, Amsterdam

Trüper, H.G. and Peck, H.D., Jr. (1970) Formation of adenylyl sulfate in phototrophic bacteria. *Arch. Microbiol.* **73**, 125-142

Trüper, H.G. and Pfennig, N. (1966) Sulphur metabolism in Thiorhodaceae. III. Storage and turnover of thiosulphate sulphur in *Thiocapsa floridana* and *Chromatium* species. *Antonie van Leeuwenhoek* **32**, 261-276

Trüper, H.G. and Rogers, L.A. (1971) Purification and properties of adenylyl sulfate reductase from the phototrophic sulfur bacterium, *Thiocapsa roseopersicina*. *J. Bacteriol.* **108**, 1112-1121

Tuovinen, O.H., Kelley, B.C. and Nicholas, D.J.D. (1976) Enzymic comparisons of the inorganic sulfur metabolism in autotrophic and heterotrophic *Thiobacillus ferrooxidans*. *Can. J. Microbiol.* **22**, 109-113

Tuttle, J.H. and Jannasch, H.W. (1973) Dissimilatory reduction of inorganic sulfur by facultatively anaerobic marine bacteria. *J. Bacteriol.* **115**, 732-737

Vishniac, W. (1952) The metabolism of *Thiobacillus thioparus*. I. The oxidation of thiosulfate. *J.Bacteriol.* **64**, 363-373

Visser, J.M., de Jong, G.A.H., Robertson, L.A. and Kuenen, J.G. (1997a) Purification and characterization of a periplasmic Thiosulfate dehydrogenase from the obligately autotrophic *Thiobacillus* sp. W5. *Arch. Microbiol.* **166**, 372-378

Visser, J.M., de Jong, G.A.H., Robertson, L.A. and Kuenen, J.G. (1997b) A novel membrane-bound flavocytochrome c sulfide dehydrogenase from the colourless sulfur bacterium *Thiobacillus* sp. W5. *Arch. Microbiol.* **167**, 295-301

Warthmann, R. and Cypionka, H. (1996) Characteristics of assimilatory sulfate transport in *Rhodobacter sulfidophilus*. *FEMS Microbiol. Lett.* **142**, 243-246

Widdel, F., and Hansen, T.A. (1992) The dissimilatory sulfate- and sulfur-reducing bacteria. In: *The Prokaryotes* (Eds.: Barlows, A., Trüper, H.G., Dworkin, M., Harder, W., Schliefer, H.K.) 2nd Ed., Vol. 1, pp.583-624, Springer, New York

Wodara, C., Kostka, S., Egert, M., Kelly, D.P. and Friedrich, C.G. (1994) Identification and sequence analysis of the soxB gene essential for sulfur oxidation of Paracoccus denitrificans GB17. J. Bacteriol. **176**, 6188-6191

Wray, J.L., Campbell, E.I., Roberts, M.A., Gutierrez-Marcos, J.F. (1998) Redefining reductive sulfate assimilation in higher plants: a role for APS reductase, a new member of the thioredoxin superfamily? *Chem. Biol. Interact.* **109**, 153-167

Yoch, D.C. and Lindstrom ES (1971) Survey of the photosynthetic bacteria for rhodanese (thiosulfate: cyanide sulfur transferase) activity. *J. Bacteriol.* **106**, 700-701

Yurkov, V.V. and Beatty, J.T. (1998) Aerobic anoxygenic phototrophic bacteria. *Microbiol. Mol. Biol. Rev.* **62**, 695-724

4

Catalytic removal of sulfur from diesel oil by hydrotreating

Hank R. Reinhoudt and Jacob A. Moulijn

4.1 INTRODUCTION

In general, hydrotreating is the collection of catalytic refinery processes that aim to remove hetero-atoms (like S, N, O) and metals (like Ni and V) from various mineral oil fractions. In addition, saturation of aromatic compounds and olefins may also be considered as a hydrotreating process. Among these hydrotreating processes the removal of sulfur, or hydrodesulfurisation (HDS), is most widely applied and can therefore be considered as the most important hydrotreating process.

The sulfur content of crude oil may vary significantly with its origin. For example, Nigerian crude oil contains about 0.2 w.% sulfur whereas that in Kuwaiti crude oil may even exceed 2.5 w.%. Independent of the total sulfur content, the distribution of sulfur containing molecules over the boiling point range of the crude oil is similar. Low boiling fractions like naphtha

mercaptans	sulphides	disulphides
R-SH	R-S-R'	R-S-S-R'

thiophenes benzothiophenes dibenzothiophenes

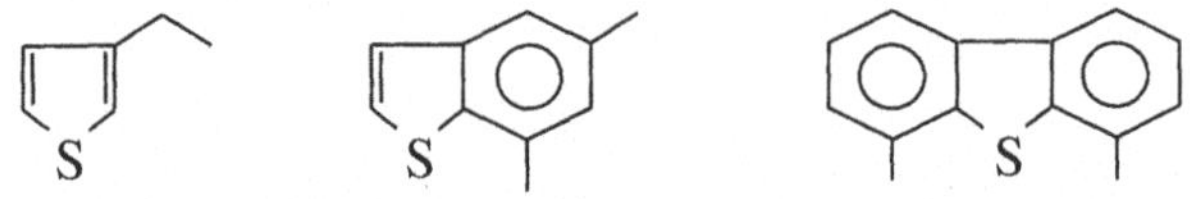

Figure 4.1. Representative sulfur compounds in crude oil.

only contain relatively low amounts of sulfur whereas in high boiling fractions like atmospheric residue the sulfur content is considerably high, in some cases even up to 4.5 w.%. Sulfur in crude oil is present in a wide range of organic sulfur compounds of which the most important representatives are shown in Figure 4.1.

The incentive for HDS of various oil fractions is not always the same. The removal of sulfur may be required for the protection of downstream process equipment or catalysts. For example, the platinum-based catalyst applied in catalytic reforming, which is one of the most important processes in present day refineries, is very sensitive for poisoning with sulfur which necessitates the removal of sulfur to very low concentrations (less than 1 ppm S). Next to technical reasons or improvement of the product quality, the application of HDS processes can also be imposed by environmental legislation. An important and actual example of environmental legislation is the removal of sulfur from diesel oil. In the past decade, the maximum sulfur content in automotive diesel fuel in the European Union has been decreased from 3000 to 500 ppm S (Anabtawi *et al.* 1996). Recently, a further decrease has been announced to 350 ppm S in 2000 to only 50 ppm by the year 2005 (European Union 1998). The main reason for this reduction is the fact that an effective application of a diesel exhaust catalyst for cars is strongly hampered by the presence of sulfur in diesel fuel.

The severe reduction of the maximum sulfur content in diesel fuel puts pressure on the performance of currently applied HDS processes. Instead of a conversion of about 95%, the future diesel fuel specifications require a conversion of sulfur compounds of more than 99.5%. In addition, the remaining sulfur compounds at conversions above 95% show a low reactivity over the conventionally applied HDS catalysts. It will be clear that the transition of conventional HDS to the so-called deep HDS of diesel fuel necessitates the development of improved HDS processes and catalysts.

In this chapter we first go into the reactions, the catalysts and the processes of conventional diesel HDS processes. Gradually, the background of the problem of reaching very low sulfur concentrations in diesel fuel with the existing HDS technology will become clear. Finally, we focus on recent developments in deep HDS processes and catalysts and we discuss the possibilities for refineries to meet the future sulfur limits.

4.2 CONVENTIONAL HDS OF DIESEL FUEL

4.2.1 HDS processes

In general, HDS processes are carried out at reactions temperatures between 573 and 673 K and an overall pressure between 3.0 and 6.0 MPa in the presence of hydrogen and a catalyst. Organic sulfur compounds react with hydrogen to form H_2S, which can be easily separated from the oil after the HDS reactor. HDS of naphtha, a component stream of gasoline, is carried out in the gas phase since naphtha is gaseous at HDS reaction temperatures.

In contrast, in diesel fuel HDS, the oil feed is for a major part in the liquid phase. The latter reaction is carried out in a fixed bed reactor which is operated in the so-called trickle flow regime. In a trickle flow reactor, the liquid diesel feed flows downward through the reactor forming a thin liquid film over the solid catalyst particles (5 mm diameter). The gas phase flows in the same direction in the void space in between the catalyst particles. The trickle flow operation is schematically represented in Figure 4.2.

4.2.2 HDS reactions

In general, desulfurisation of organic sulfur compounds with hydrogen involve the cleavage of a C-S bond by hydrogenolysis. A distinction can be made between compounds where the sulfur atom is a part of a hetero-cyclic ring (like in thiophene or dibenzothiophene, Equation 4.1), and non-cyclic sulfur compounds (like sulfides and mercaptans, Equation 4.2).

$$\text{-C-S-C-} + 2\ \text{H-H} \rightarrow \text{H-S-H} + 2\ \text{C-H} \tag{4.1}$$

$$\text{-C-S} + \text{H-H} \rightarrow \text{H-S-H} + \text{R-C-H} \tag{4.2}$$

When sulfur is located in a hetero-cyclic ring, the direct cleavage of the C-S bond is not likely. In those cases, partial hydrogenation of the thiophenic ring probably precedes the actual cleavage of the C-S bond which results in an increased hydrogen consumption.

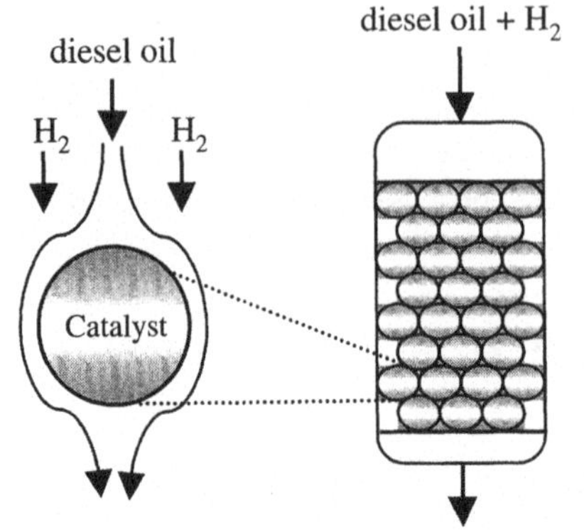

Figure 4.2. Schematic representation of trickle flow operation as applied in conventional diesel oil HDS processes.

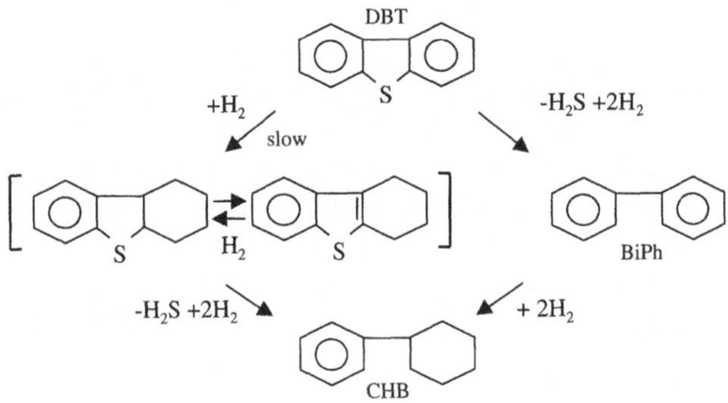

Figure 4.3. Detailed reaction scheme for DBT HDS.

A representative sulfur compound for diesel fuel HDS is dibenzothiophene (DBT). In Figure 4.3, a more detailed reaction scheme for DBT HDS is shown.

At the reaction conditions generally applied in industrial HDS, the two main products are cyclohexylbenzene (CHB) and biphenyl (BiPh). It has been reported that in general the hydrogenation of DBT to form sulfur containing hydrogenated intermediates is slow compared with the direct HDS to BiPh (Houalla *et al.* 1978).

The HDS of organic sulfur compounds is an exothermic reaction and is essentially irreversible under the reaction conditions employed in industrial processes (Girgis and Gates 1991). This is illustrated for methyl-sulfide, thiophene and DBT. Thermodynamic data for DBT (Vrinat 1983) indicate that the HDS of DBT to give BiPh (Figure 4.4) is favoured at temperatures that are representative for industrial deep HDS. However, as was shown in Figure 4.3, the removal of sulfur from for example DBT may either take place with or without preceding hydrogenation of the heterocyclic ring. The pathway involving a hydrogenation step preceding the desulfurisation step may indeed be affected by an unfavourable thermodynamic equilibrium.

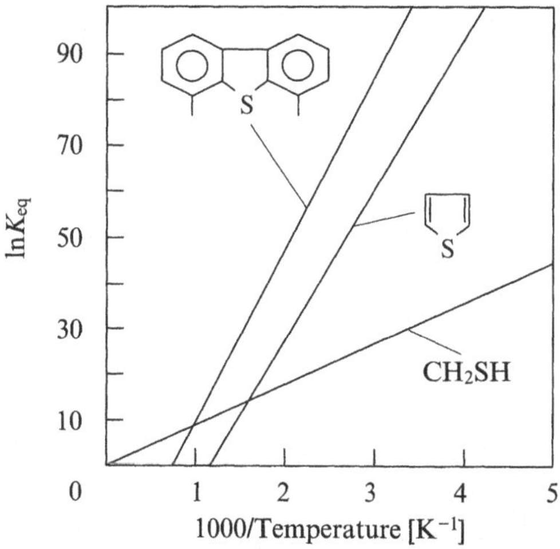

Figure 4.4. Equilibrium data for HDS reactions with different compounds as a function of temperature.

The reactivity of the various sulfur compounds in diesel oil varies considerably. In Figure 4.5, the order and relative reactivity of representative sulfur compounds in diesel fuel is shown (Ma *et al.* 1994).

Based on the relative reactivity (Figure 4.5), it can be expected that at high HDS conversions mainly alkylated DBTs remain. Kilanowski *et al.* (1978) and Houalla *et al.* (1980) were the first to point to the retarding effect of alkyl groups on the 4 and 6 position of DBT on the reactivity in model feed HDS reactions. Also in diesel oil it has been found that especially 4, and 6-alkylated DBTs are most refractory towards desulfurisation (Amorelli *et al.* 1992). Based on these findings it will be clear that the ease of desulfurisation of a particular diesel oil will strongly depend on the type of sulfur compounds that are present. In line with this, the reaction temperature required for deep HDS (< 500 ppm S) has been correlated to

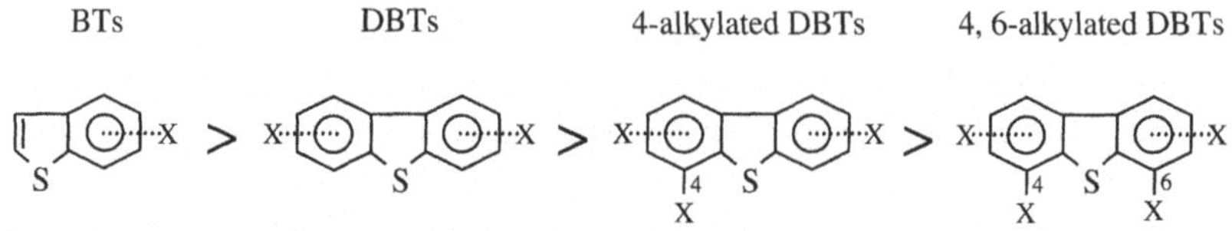

40 : 8 : 4 : 1

Figure 4.5. The order and relative reactivity of representative sulfur compounds in diesel fuel (X denotes the presence of an alkyl-group and BT = benzothiophene).The presence of 4, and 6-alkylated DBTs is a key factor in the difficulty to produce diesel fuels that comply with the future diesel fuel legislation within the European Union (maximum 50 ppm S).

the concentration of sulfur in the boiling range above 589 K (Shih *et al.* 1992). Since the boiling point of DBT is 604 K, this fraction will indeed contain sulfur compounds with a low HDS reactivity.

4.2.3 HDS kinetics

The kinetics of the hydrodesulfurisation reactions of DBT have been discussed rather extensively in the literature. This has however, not let to a clear and unambiguous picture. This is most likely due to the different experimental conditions employed which cover a wide range of temperatures, pressures and catalysts.

In general, the kinetics of DBT HDS are consistent with a pseudo first order so-called Langmuir-Hinshelwood type rate equation (see for example Equation 4.3). In most equations, the competitive adsorption of DBT and H_2S onto the active sites on the catalyst is taken into account. It is important to note that H_2S, which is formed in the HDS reaction lowers the reaction rate. Moreover, the adsorption of hydrogen in not considered to take place on the same site as DBT. In case of conventional hydrotreating catalysts at low hydrogen pressures (P_{H2} < 2.5 MPa), the contribution of the hydrogenation pathway can be neglected and the HDS of DBT may be described by Equation 4.3 (Vrinat 1983):

$$r_{\mathrm{HDS}} = k\frac{K_{\mathrm{DBT}}P_{\mathrm{DBT}}}{1 + K_{\mathrm{DBT}}P_{\mathrm{DBT}} + K_{\mathrm{H_2S}}P_{\mathrm{H_2S}}}\frac{K_{\mathrm{H_2}}P_{\mathrm{H_2}}}{1 + K_{\mathrm{H_2}}P_{\mathrm{H_2}}} \tag{4.3}$$

In Equation 4.3, r_{HDS} is the reaction rate for direct HDS of DBT; k is the reaction rate constant for DBT HDS; K_x is the adsorption constant for either DBT, H_2S or H_2 and P_x is the partial pressure for either DBT, H_2S or H_2. Instead of partial pressures, also concentrations can be used. When high hydrogen pressures are applied (P_{H_2} > 5.0 MPa), the contribution of the hydrogenation pathway can not be neglected anymore and an additional equation is needed to account for this reaction (Vrinat 1983):

$$r_{\mathrm{HYD}} = k\frac{P_{\mathrm{DBT}}P_{\mathrm{H_2}}}{1 + K_{\mathrm{DBT}}P_{\mathrm{DBT}}} \qquad (4.4)$$

In Equation 4.4, r_{HYD} is the reaction rate for hydrogenation of DBT. In Equation 4.4 it is assumed that no competitive adsorption of H_2S or H_2 takes place on the active sites for DBT HDS. At high hydrogen pressures, the kinetics of DBTs are best described by the combination of 4.3 and 4.4. The resulting overall desulfurisation rate r is then given by Equation 4.5:

$$r = r_{\mathrm{HDS}} + r_{\mathrm{HYD}} \qquad (4.5)$$

Whereas for individual sulfur compounds a first-order dependency on the total sulfur concentration is generally observed, the kinetics of HDS in an industrial feed appear to be more complex. This is due to the presence of a wide range of sulfur compounds with a different reactivity. In Figure 4.6 it is illustrated that the total sulfur concentration in a gas oil as a function of the residence time in the reactor initially decreases rapidly, followed by a much slower decrease when which is typical for a higher order dependency on the total sulfur concentration. Indeed, as shown in Figure 4.7, the

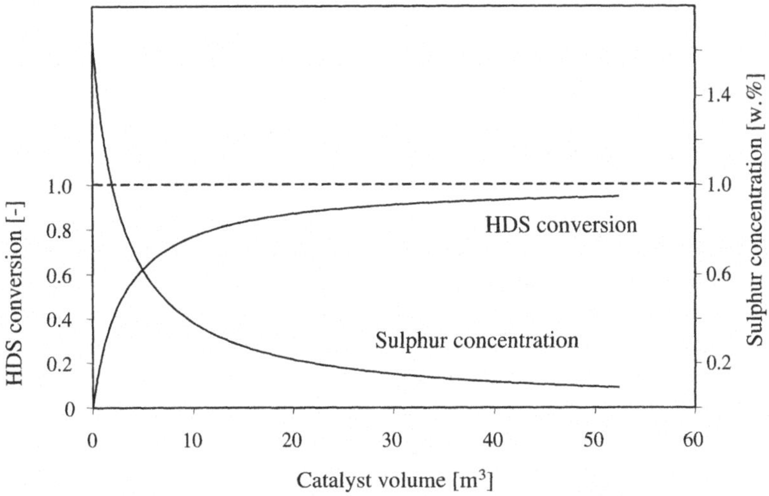

Figure 4.6. The total sulfur concentration and the total HDS conversion in diesel oil (1.6 w.% S initial) as function of the reactor height (total volume 52 m^3), expressed as its cumulative volume.

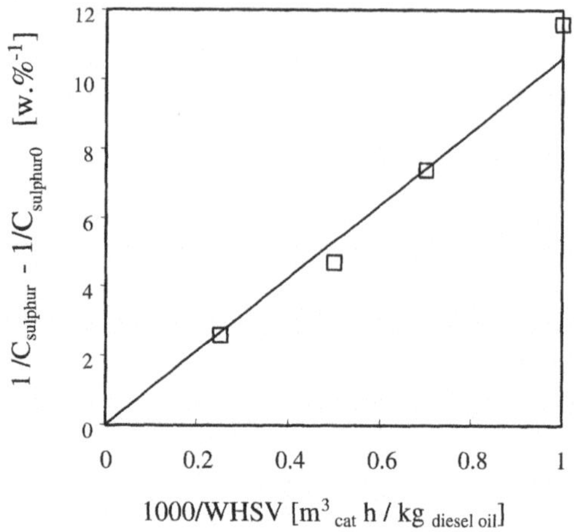

Figure 4.7. Illustration of the second order dependency of the total sulfur concentration in diesel oil. C_{sulfur} denotes the total sulfur concentration, $C_{sulfur0}$ the initial total sulfur concentration and WHSV the diesel Weight Hourly Space Velocity.

conversion of the sum of all sulfur compounds present in the diesel oil can be well described by applying higher order kinetics, in this case second-order kinetics were used.

The second-order behaviour that is observed for real feed HDS has no physical meaning and in fact we deal with apparent second-order kinetics. Because the individual sulfur compounds also in a real feed show first-order kinetics, it appears logical to model real feed HDS kinetics with several first-order reactions with different rate constants. A simulated example of such a model is shown in Figure 4.8.

In this model, it was assumed that three parallel first-order reactions occur simultaneously. The symbols in the figure are simulated second-order data (C_{S0} = 1.64 w.%, k_{2nd} = 10.6 × 10^3 kg$_{oil}$·m$_{cat}^{-3}$ w.% S^{-1} h^{-1}), obtained for the second-order reaction of Figure 4.6 and 4.7. The numbered curves represent first-order reactions of 'lumps' of sulfur-containing compounds which have a similar reactivity (*e.g.*, 1: easily to convert like benzothiophenes, 2: more difficult to convert like dibenzothiophenes and 3: 4, and 6-alkylated DBTs). The sum of these three curves is the curve through the symbols. So, a three-lump model consisting of first-order reactions fits an apparent second-order reaction. According to the applied model, diesel oil HDS can be described with first-order kinetics at very high HDS conversions. However, in practice a reaction order close to two has been observed up to diesel oil sulfur concentrations of less than 500 ppm.

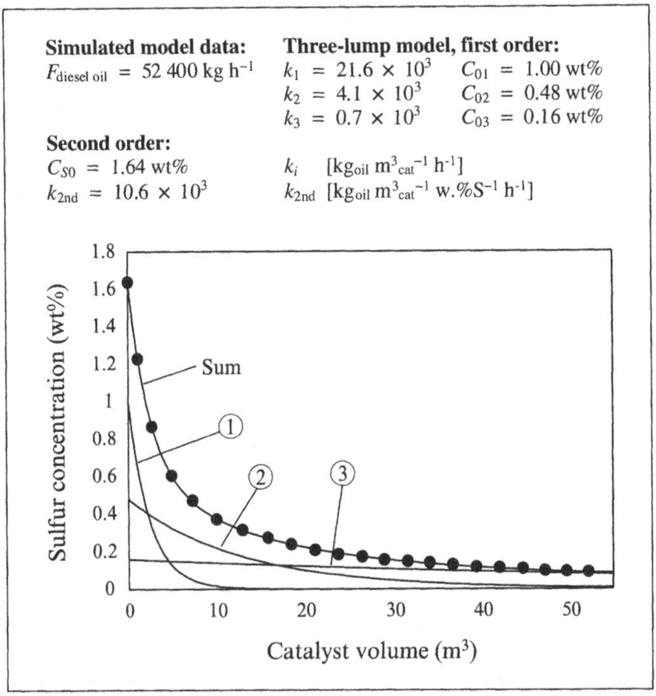

Figure 4.8. Simulated concentration profiles of a model of three lumped, first order HDS reactions. $F_{\text{diesel oil}}$ is the through-put of diesel per hour.

Apparently, the reactivity of the sulfur compounds that remain present at high HDS conversions still differs in such a way that they can not be lumped together.

4.2.4 HDS catalysts

Typical HDS catalysts consist of a mixed metal sulfide of a combination of either cobalt (Co) or nickel (Ni) with either molybdenum (Mo) or tungsten (W), supported on a porous aluminium-oxide carrier (γ-Al_2O_3) with a high surface area (200-400 m^2 g^{-1}). The combination of these four metals emerged from an extensive search for sulfur-tolerant hydrogenation catalysts at IG-Farben Industries in Germany in the 1920s. This company subsequently developed a hydrocracking process for coal and tar on the basis of a Ni/W catalyst. The first industrial application of a Co/Mo catalyst for hydrotreating was already reported in 1943 (Byrns 1943).

Up to now it still has been the preferred catalyst for most hydrotreating reactions. Because of its industrial significance, molybdenum and molybdenum catalysts promoted with either Co or Ni have received massive attention in the literature.

MoS_2 or WS_2 on itself already possess a significant activity for HDS reactions. These compounds are present at the surface of the aluminiumoxide support in the form of a so-called slab (Figure 4.9A). The formation of a mixed sulfide catalyst with either Co or Ni (Figure 4.9B) strongly enhances the activity for HDS. The origin of this so-called promotion effect of Co(Ni), has been extensively discussed in the literature. Currently, the most widely accepted model for the promoting effect of Co and Ni is the so-called CoMoS model of Topsøe and co-workers (Wivel *et al.* 1981; Topsøe *et al.* 1984; Sorensen *et al.* 1985; Topsøe and Topsøe 1984). In this model, the high activity in HDS reactions is directly allocated to a specific Co(Ni)-Mo(W)-S interaction phase. This phase has been identified with Mössbauer spectroscopy and infrared studies. Based on the fact that Co(Ni) promoter atoms are surface exposed on MoS_2 slabs (Figure 4.9), it has been proposed that highly disperse Co(Ni) particles are located at the edges of the MoS_2 (WS_2) slabs (Ratnasamy and Sivasanker 1980) (Figure 4.9B). The removal of sulfur from the MoS_2 (WS_2) edges and the concomitant creation of coordinative unsaturated sites, also referred to as sulfur anion vacancies, is thought to play an important role in HDS catalysis. The importance of sulfur anion vacancies in HDS reactions is supported by the correlation between the HDS activity and the ease of surface sulfur hydrogenation (De Beer *et al.* 1976; Scheffer *et al.* 1989;

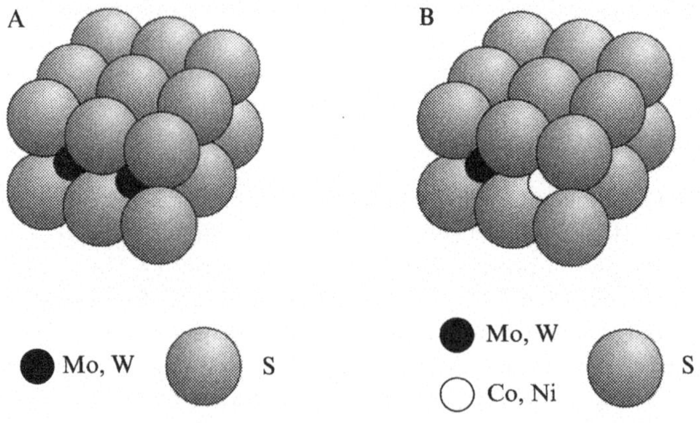

Figure 4.9. Schematic representation of (A) MoS_2 or WS_2 slab, (B) Position of Co, Ni on the edge of a MoS_2, WS_2 slab forming a mixed sulfide.

Scheffer *et al.* 1990). Based on infra-red measurements on adsorbed NO, it could be inferred that the metal-sulfur bond strength in Co-Mo-S is lower compared with that in MoS_2. This could be the background of the promotion effect of Co (Topsøe *et al.* 1983).

4.3 DEEP HDS OF DIESEL FUEL

Now that we have a basic understanding of the most important issues in HDS processes, reactions and catalysts, we can focus on the specific problems that play a part in the production of ultra low sulfur diesel fuel. In general we can speak of deep HDS as opposed to conventional HDS when the sulfur concentration is below 1000 ppm. As was mentioned previously, the key problem in deep HDS is the low reactivity of DBTs with an alkyl group at the 4- and 6-position which make up for a significant part of the sulfur compounds when the total sulfur content has been brought down to concentrations below 1000 ppm. The low reactivity of these compounds is nicely illustrated in Figure 4.10. In this case, a diesel oil was taken which contained only 760 ppm sulfur. After deep HDS at 613 K and 6.0 MPa over a conventional CoMo/γ-Al_2O_3 catalyst it is clearly observed that especially 4, and 6-alkylated DBTs remain in the product. The background of the low reactivity of 4, and 6-alkylated DBTs as compared to other DBTs is probably related to the position of the alkyl groups in relation to the sulfur atom.As shown in paragraph 4.2.2, DBT is preferably converted through direct extrusion of the sulfur atom without preceding hydrogenation to form BiPh. It is generally believed that the preferred mode of adsorption in this reaction is the adsorption via the sulfur atom. Conventional HDS catalysts like CoMo/γ-Al_2O_3 show a high activity for this reaction. However, in case of 4, and 6-alkylated DBTs, adsorption via the sulfur atom is sterically hindered by the presence of the 'bulky' alkyl groups. The direct desulfurisation of 4, and 6-alkylated DBTs is, therefore, more difficult and consequently the HDS reaction rate is much lower. Thus, although conventional mixed sulfide catalysts like CoMo/γ-Al_2O_3 posses a high activity for the conversion of most of the sulfur compounds in diesel oil, they are probably not the most suitable catalysts for deep HDS reactions.

It was shown by Landau *et al.* (1996) that after hydrogenation of one the aromatic rings, the sterical hindrance of the alkyl groups in 4, and 6-alkylated DBTs becomes much lower. Indeed, it was shown by Kabe *et al.* (1993) that the rate of HDS of DBTs with one hydrogenated aromatic ring is virtually independent on the presence and position of alkyl groups. An

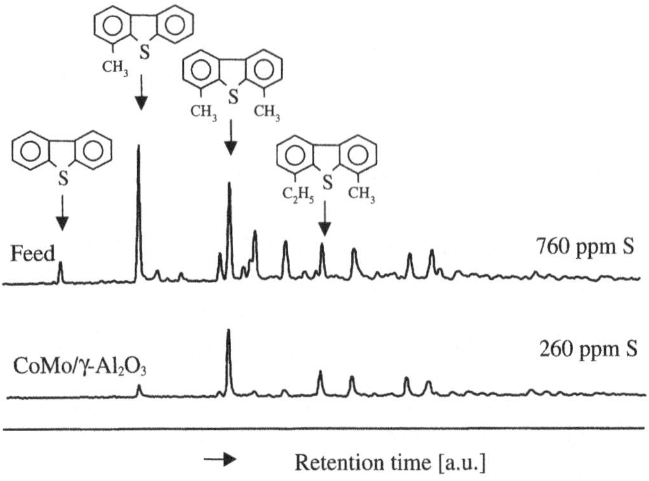

Figure 4.10. Sulfur specific spectra (analysed by GC with sulfur chemiluminescence detector) of a diesel fuel and its desulfurised product after deep HDS (CoMo/γ-Al_2O_3, 613 K and 6.0 MPa).

important conclusion from these findings is that an effective catalyst for deep HDS preferably must have a high activity for aromatic ring hydrogenation. Alternative principle routes for conversion of 4, and 6-alkylated DBTs are isomersation, de-alkylation or cracking. The two latter reactions are however less desired in practice since they will lower the amount of valuable products in the diesel fuel range.

4.3.1 Alternative catalysts for deep HDS of diesel fuel

As was discussed in the previous paragraph, alternative deep HDS catalysts preferably have a high hydrogenation activity. With respect to the hydrogenation activity, noble metal based (Pt, Pd) catalysts may be interesting options. However, it is generally known that noble metals are very sensitive towards poisoning with sulfur compounds which are intrinsically present in HDS reactions. Still, when noble metals are applied on acidic supports like zeolites, it was found that the tolerance for sulfur compounds can be significantly increased (for example: Marécot *et al.* 1993; Homayer *et al.* 1990). Other options for deep HDS catalysts may be found in improved mixed metal sulfide catalysts by modification of the support or by increasing the metal loading. In Figure 4.11, the performance of several conventional and alternative catalysts in deep HDS is shown.

From these results several important conclusions can be drawn. At first, it appears that the application of mixed metal sulfide catalysts with a higher hydrogenation activity than CoMo/γ-Al_2O_3, like NiMo/- and NiW/γ-Al_2O_3 indeed show a better performance in deep HDS of diesel fuel. However, at the same time it can be concluded that the improvements realised with these catalysts are by far insufficient to comply with the future diesel fuel specifications of maximum 50 ppm S. Another important conclusion is that noble metal catalysts on acidic supports, in this case amorphous silica alumina (ASA), show a much better performance than the mixed metal sulfide catalysts. Hence they may have potential for application in deep HDS reactions. Especially the combination of Pt and Pd on ASA shows a superior activity. However, it was reported by Reinhoudt *et al.* (1999), that ASA supported noble metal catalysts can only be successfully applied when the total sulfur concentration in the oil feed is lower than roughly 1000 ppm S. At higher concentrations, the ASA supported noble metal catalysts are still active but they loose their advantage over the more sulfur tolerant mixed metal sulfide catalysts since they are more strongly inhibited by sulfur compounds.

The balance between catalyst properties like activity, costs and the sulfur tolerance on one hand and the required sulfur reduction in the diesel oil on the other hand eventually determines the suitability of a particular catalyst. The relation between the catalyst properties and the deep HDS process lay-out will be discussed in the next paragraph.

4.3.2 The relation between deep HDS process and catalyst

In the literature, various proposals for adapted or new HDS processes have been reported. These proposals can generally be divided into single- and two-stage conventional fixed-bed reactors or alternative reactor concepts.

The application of a single-stage process for the production of low sulfur diesel fuel is the most attractive option since current hydrotreating processes usually are based on single-stage reactors. Solutions to realise high HDS conversions in a single-stage reactor can be sought in a decreased throughput, increased reaction temperature and/or pressure and in more active catalysts. However, most of these possible solutions have obvious disadvantages. By increasing the reaction temperatures, the cycle length of catalyst life will be significantly abbreviated owing to fast deactivation which lowers the economic feasibility. In a single-stage deep HDS process, longer operation cycles can be achieved at the expense of high hydrogen pressures which will require radical revamping of the existing reactors and

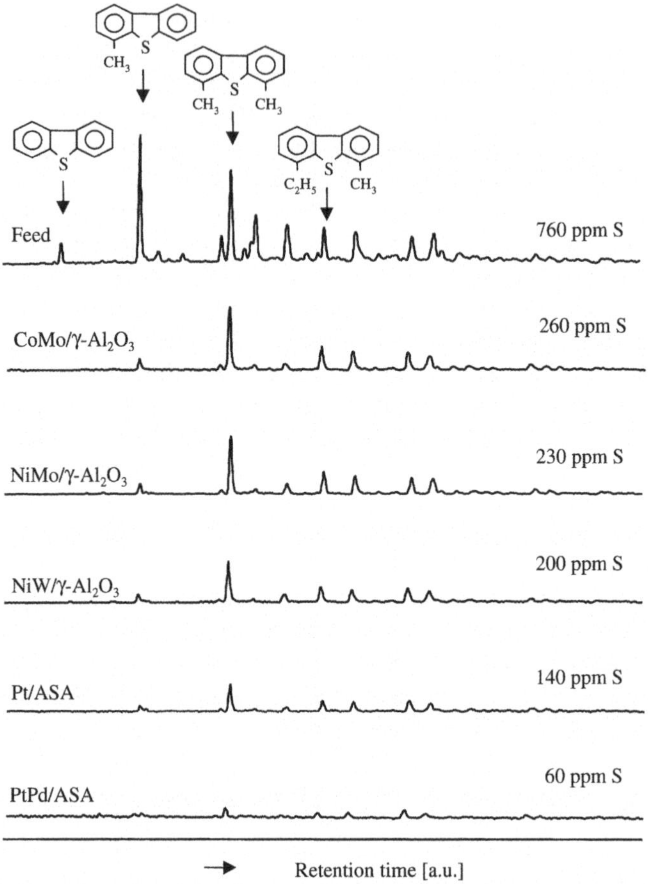

Figure 4.11. The sulfur specific GC spectra of a diesel fuel and of its desulfurised products after deep HDS with various catalysts at 613 K and a pressure of 6.0 MPa.

offer only limited flexibility. The application of more active catalysts is a valid option although large improvements in their performance for deep HDS are not expected. It was shown in the previous paragraph that alternative mixed metal catalysts only promise a moderate improvement in the desulfurisation activity. In addition, noble metal based catalysts cannot be applied in a single-stage reactor since the H_2S level is too high. In general, it can be stated that single-stage processes are still attractive for the production of low sulfur diesel fuel only when the required sulfur content in the product is in the order of 500 ppm S. However, when the sulfur content has to be reduced to 150 ppm or lower, single-stage processes are not a feasible option.

In two-stage processes, the first reactor is generally used to reduce the sulfur- and nitrogen compounds concentration to levels that allow the application of noble metal catalysts in the second stage. Obviously, a thorough interstage removal of H_2S is essential. Most of the two-stage processes that have been reported in literature apply a sulfur tolerant zeolite based noble metal catalysts in the second stage. The main advantage of this approach is the high flexibility of the first stage. Moreover, in addition to a high reduction of the sulfur content also hydrogenation of aromatics can be achieved in the second stage which improve the diesel fuel quality. An example of an integrated two-stage process is the SMDH (Shell Middle Distillates Hydrogenation) of Shell (Minderhoud and Lucien 1988) which currently operates in Sweden.

Instead of changing the process lay-out, an improved diesel fuel production process may also be realised by changing the reactor concept. A concept with potential advantages for deep HDS processing, is a counter-current operation in which the oil flows downward and the hydrogen gas upward through the reactor. The main advantage of counter-current operation for deep HDS processing is the low H_2S and NH_3 partial pressure in that part of the reactor where the most refractory sulfur compounds have to be converted. This approach opens the possibility of applying catalysts with a relatively low sulfur tolerance, like ASA supported noble metal catalysts. Two examples of such reactor concepts have recently been published: the Three Levels of Porosity Reactor (Van Hasselt *et al.* 1997) and the Internally Finned Monolithic Reactor (Sie and Lebens 1998).

4.3.3 Evaluation

The production of diesel fuel of the currently required sulfur concentration (500 ppm S) is in most cases realised in a conventional single-stage reactor at high reaction temperatures (633-653 K) and lower throughputs in combination with the application of low sulfur crude oils. However, the recently proposed legislation for the European Union, which limits the sulfur level in diesel fuel to only 50 ppm in 2005, will necessitate the implementation of improved processes for ultra-clean diesel fuel production. Considering the short time for refiners to realise the production of ultra clean diesel fuel, it is expected that they will rely on conventional trickle bed reactors rather than on new, sophisticated reactor technology.

As was discussed previously, single-stage processes will in most cases not

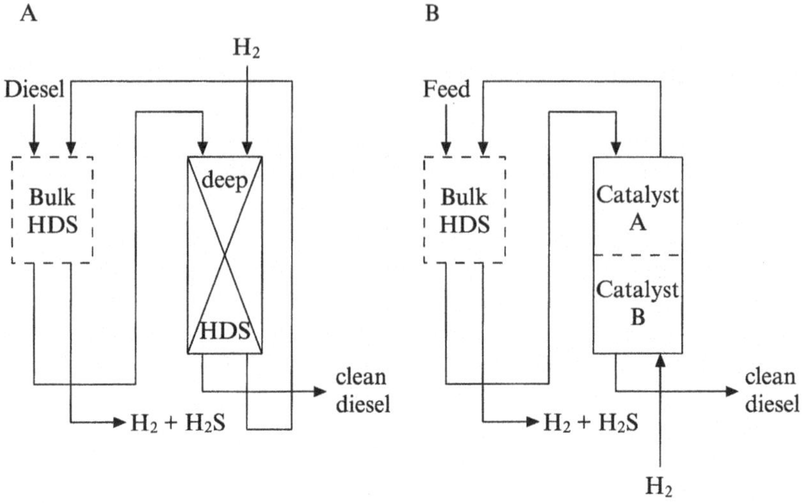

Figure 4.12. Optional process lay-out for the production of ultra low sulfur diesel fuel, (A) a second stage deep HDS reactor in series with the existing HDS reactor, (B) a counter current reactor with catalyst profiling (catalyst A, sulfur tolerant mixed metal catalyst; catalyst B, highly active noble metal catalyst on acidic support).

be capable of reaching these low sulfur concentrations and hence two-stage processes appear to be a promising option for future deep HDS (Figure 4.12A). In the first stage, the existing HDS reactor, the sulfur content has to be brought down to levels in the order of 1000 ppm S, which should be possible with conventional HDS catalysts. In the second, deep HDS stage, the sulfur level can be brought down to the desired level by applying a specific deep HDS catalyst which will need a high specific activity for the conversion of mainly 4, 6 di-alkylated DBTs.

A more sophisticated possibility for the production of ultra low sulfur diesel fuels is the application of a counter current process (Figure 4.12B). As a second stage tail treatment after the existing HDS reactor, counter current operation clearly has advantages. Catalyst profiling could be especially attractive since the low H_2S concentration in the lower part of the reactor (catalyst bed B) is well suited for PtPd/ASA-like catalysts. Considering its lower H_2S sensitivity, NiW/γ-Al_2O_3 would be a suitable catalyst for the upper part of the reactor (catalyst bed A) where a relatively high H_2S concentration prevails. A remark has to be made about the applicability of this type of reactor in the short term. As the technology involved in counter current operation is still relatively new, the application of a second stage trickle bed reactor seems to be the most feasible option for the short term.

4.4 REFERENCES

Amorelli, A., Amos, Y. D., Halsig, C. P., Kosman, J. J., Jonker, R. J., De Wind, M., and Vrieling, J. (1992) Estimate feedstock processability. *Hydrocarbon Processing* **June**, 93-101.

Anabtawi, J. A., Ali, S. A., and Ali, M. A. (1996) Impact of gasoline and diesel specifications on the refining industry. *Energy Sources* **18**, 203-214.

Byrns, A. C., Bradley, W. E., and Less, M. W. (1943) *Ind. Eng. Chem.* **35**, 1160.

De Beer, V. H. J., Van der Aalst, M. J. M., Machiels, C. J., and Schuit, G. C. A. (1976) The CoO-MoO_3 / Al_2O_3 catalyst VII. Influence of the support. *J. Catal.* **43**, 78-89.

European Union (1998) E.U.Directive 98/70/EC.

Girgis, M. J. and Gates, B. C. (1991) Reactivities, reaction networks, and kinetics in high pressure catalytic processing. *Ind. Eng. Chem. Res.* **30**, 2021-2058.

Homayer, S. T., Karpinski, Z., and Sachtler, W. M. H. (1990) Effect of zeolite protons on palladium catalyzed hydrocarbon reactions. *J. Catal.* **123**, 60-73.

Houalla, M., Broderick, D. H., Nag, N. K., Sapre, A. V., De Beer, V. H. J., Gates, B. C., and Kwart, H. (1980) Hydrodesulphurzation of methyl-substituted dibenzothiophenes catalyzed by sulphided CoMo/Al_2O_3. *J. Catal.* **61**, 523-527.

Houalla, M., Nag, N. K., Sapre, A. V., Broderick, D. H., and Gates, B. C. (1978) Hydrodesulphurization of dibenzothiophene catalyzed by sulphided CoO-MoO_3/ Al_2O_3: The reaction network. *AIChE Journal* **24(6)**, 1015-1021.

Kabe, T., Ishihara, A., and Zhang, Q. (1993) Deep Desulphurization of Light Oil. Part 2: hydrodesulphurization of dibenzothiophene, 4-methyldibenzothiophene and 4,6-dimethyldibenzothiophene. *Appl. Catal. A: General* **97**, L1-L9.

Kilanowski, D. R., Teeuwen, H., De Beer, V. H. J., Gates, B. C., Schuit, G. C. A., and Kwart, H. (1978) Hydrodesulphurization of thiophene, benzothiophene, dibenzothiophene, and related compounds catalyzed by sulphided CoO-MoO_3/ Al_2O_3: Low pressure Reactivity Studies. *J. Catal.* **55**, 129-137.

Landau, M. V., Berger, D., and Herskowitz, M. (1996), Hydrodesulfurization of methyl-substituted dibenzothiophenes: fundamental study of routes to deep desulfurization. *J. Catal.* **159**, 236.

Ma, X., Sakanishi, K., and Mochida, I. (1994) Hydrodesulphurization reactivities of various sulphur compounds in diesel fuel. *Ind. Eng. Chem. Res.* **33**, 218-222.

Marécot, P., Mahoungou, J. R., and Barbier, J. (1993) Benzene hydrogenation on platinum and iridium catalysts. Variation of the toxicity of sulphur with the nature of the support. *Appl. Catal. A: General* **101**, 143-149.

Minderhoud, J. K. and Lucien, J. (1988) Process for the hydrogenation of hydrocarbon oils. *Eur. Patent* 303.332.

Ratnasamy, P. and Sivasanker, S. (1980) Structural chemistry of Cobalt-molybdenum-alumina catalysts. *Catal. Rev. -Sci. Eng.* **22**, 401-429.

Reinhoudt, H. R., Troost, R., Van Langeveld, A. D., Van Veen, J. A. R., Sie, S. T., and Moulijn, J. A. (1999) Testing and characterisation of Pt/ASA applied in model reactions for deep HDS. *Fuel Proc. Technol.*, In Press.

Scheffer, B., Dekker, N. J. J., Mangnus, P. J., and Moulijn, J. A.(1990) A temperature programmed reduction study of sulphided $CoMo/Al_2O_3$ hydrodesulphurization catalysts. *J. Catal.* **121**, 31-46.

Scheffer, B., Molhoek, P., and Moulijn, J. A. (1989) Temperature Programmed Reduction of $NiO-WO_3/Al_2O_3$ hydrodesulphurization catalysts. *Appl. Catal. A: General* **46**, 11-30.

Shih, S. S., Mizrahi, S., Green, L. A., and Sarli, M. S. (1992) Deep desulphurization of Distillates. *Ind. Eng. Chem. Res.* **31**, 1232-1235.

Sie, S. T. and Lebens, P. J. M. (1998) Monolithic reactors for counter-current gas-liquid operation. *In: Structured catalysts and* reactors (Cybulski, A. and Moulijn, J. A. Eds.) **11**, 305-321. Marcel Dekker New York.

Sørensen, O., Clausen, B. S., Candia, R., and Topsøe, H. (1985) HREM and AEM studies of HDS catalysts: direct evidence for the edge location of cobalt in CoMoS. *Appl. Catal. A: General* **13**, 363-372.

Topsøe, H., Candia, R., Topsøe, N-Y., and Clausen, B. S. (1984) On the state of the Co-Mo-S model. *Bull. Soc. Chim. Belg.* **93(8-9)**, 783-806.

Topsøe, N-Y. and Topsøe, H. (1983) Characterization of the structures and active sites in sulphided $CoMo/Al_2O_3$ and $NiMo/Al_2O_3$ catalysts by NO chemisorption. *J. Catal.* **84**, 386-401.

Topsøe, N-Y., Topsøe, H., Sørensen, O., Clausen, B. S., and Candia, R. (1984) IR and AEM evidence for the edge location of the promoter atoms in CoMoS type structures. *Bull. Soc. Chim. Belg.* **93(8-9)**, 727-733.

Van Hasselt, B. W., Lindenbergh, D. J., Calis, H. P., Sie, S. T., and Van den Bleek, C. M. (1997) The three-levels-of-porosity reactor. A novel reactor for counter-current trickle-flow processes. *Chem. Eng. Sci.* **52(21/22)**, 3901-3907.

Vrinat, M. (1983) The kinetics of the hydrodesulphurization process - a review. *Appl. Catal. A: General* **6**, 137-158.

Wivel, C., Candia, R., Clausen, B. S., Mørup, S., and Topsøe, H. (1981) On the catalytic significance of a cobalt-molybdenum-sulfur phase in cobalt-molybdenum/alumina hydrodesulfurization catalysts: combined in-situ Mössbauer emission spectroscopy and activity studies. *J. Catal.* **68**, 453.

5

Bioleaching of sulfide minerals

Mieke Boon

5.1 INTRODUCTION

Microbial leaching of sulfide minerals has several applications: the dissolution of metals from low-grade sulfide ores; the generation of acidic ferric sulfate lixiviant which is used as a leaching medium in extractive hydrometallurgical processes; and the oxidation of gold-bearing pyritic ores. Compared with other extractive techniques, biohydrometallurgical processes have the advantage of improved efficiency and less environmental problems (Dutrizac and MacDonald 1974; Brierley 1978; Torma 1977; Torma and Bosecker 1982). On the other hand, bacterial oxidation of pyrite is the major cause of acid mine drainage production which may pollute the environment. Bacterial oxidation in pyrite-containing soils, which occur in coastal areas and originate from anaerobic marine sediments, is also undesirable because it causes acidification which makes soils unsuitable for agricultural use (Arkesteyn 1980). In the case of heavy-metal polluted areas acidification and the production of ferric ions due to the bacterial oxidation of pyrite causes the dissolution of immobilized toxic metals which is also highly undesirable. To either improve bacterial oxidation of sulfide minerals in industrial mining processes or prevent undesired pyrite oxidation in natural environments, the mechanism of bacterial metal sulfide oxidation needs to be known.

Box 5.1. Design and optimization of biohydrometallurgical processes

Two basically different large-scale processes can be applied in bacterial leaching of metals from ore (Figure 5.1). Nowadays heap leaching is the most widely spread bacterial leaching process while aerated slurry reactors find less application (Torma 1988; Kelly *et al.* 1979; Duncan *et al.* 1967). From feasibility studies it was found that the costs of the slurry process for gold production are mainly determined by the aeration, heat transfer (heat is produced from bacterial sulfide oxidation), and the pyrite oxidation rate (Duncan *et al.* 1966; Corrans *et al.* 1972; Acevedo and Gentina 1989). Also, for low-grade ores the oxidation rate per unit of reactor volume is essential. Optimization of bioleaching processes aims at finding optimum pH, temperature, bacterial strain, particle size, slurry density, concentrations of ferric and ferrous iron and biomass, and improvement of process operation. It is crucial to know which sub-processes (e.g. bacterial and chemical oxidation rates, mass transfer between the gas- and liquid phase and between liquid and mineral surface, Figure 5.2) are involved in the bioleaching of minerals.

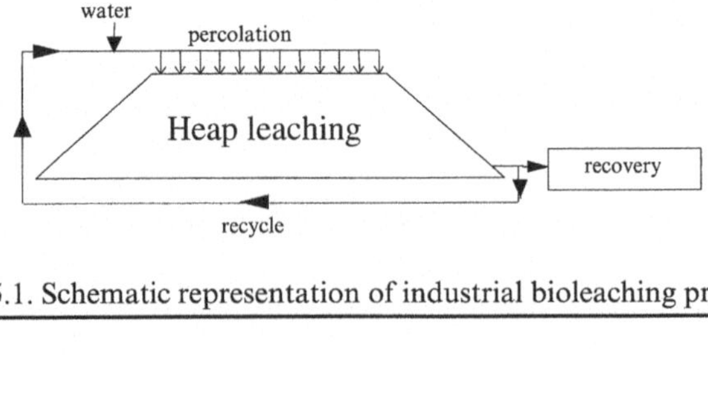

Figure 5.1. Schematic representation of industrial bioleaching processes.

5.2 RATE-DETERMINING STEPS IN BACTERIAL OXIDATION OF SULFIDE MINERALS

5.2.1 Sub-processes in biohydrometallurgical processes

A debate has been going on for several decades between defenders of a so-called *direct* mechanism and those who assume that an *indirect* mechanism explains the bacterial oxidation of sulfide minerals (Dutrizac and MacDonald 1974; Brierley 1978; Torma 1977; Torma and Bosecker 1982). The *indirect* mechanism will occur as soon as ferric iron is available. Sulfide minerals are chemically oxidized with ferric to ferrous iron, sulfate, and metal ions (Figure 5.2). Table 5.1 shows possible intermediate reactions if incomplete chemical oxidation occurs. The role of bacteria in the *indirect* mechanism is to oxidize intermediate reduced sulfur compounds and to regenerate ferric iron. Bacteria that are involved in bioleaching of sulfide minerals are mostly autotrophic (i.e. carbon dioxide is their carbon source) and utilize the chemical reaction energy of oxidation reactions. The bacteria operate at low pH (pH between 1.6 and 2.5). Both mesophyllic bacteria like *Thiobacillus ferrooxidans* and *Leptospirillum ferrooxidans* which operate at moderate temperatures (25–30 °C), as well as thermophilic bacteria like *Sulfolobus* species which have their optimum near 70 °C (LeRoux and Wakerley 1987; Lawrence and Marchant 1987; Norris and Barr 1987; Boogerd *et al.* 1989), are applied. In the bacterial oxidation reactions oxygen is the electron acceptor (Table 5.1). Simultaneously a *direct* mechanism might occur. It is assumed that bacteria need close contact with the mineral surface and 'directly' oxidize the sulfur moiety by means of an enzymatic action (Duncan *et al.* 1967; Arkesteyn 1980). In this reaction oxygen is consumed at the mineral surface (Table 5.1). The simplified overall stoichiometry (i.e. bacterial consumption of CO_2 is neglected, Boon *et al.* 1995; 1998a) is the same for the *direct* and *indirect* mechanism. Protons are produced in bacterial pyrite oxidation, consumed in the case of chalcopyrite, and neutral for sphalerite. Other sub-processes (Figure 5.2) that might play a role in the bacterial oxidation of sulfide minerals are: (1) Galvanic oxidation reactions occur in mixtures of sulfide minerals in which the less noble mineral is electro-chemically oxidized by the more noble mineral (not shown in Figure 5.2). In the presence of ferric iron the stoichiometry of the oxidation of the less noble mineral ($FeS_2 > CuFeS_2 > ZnS$) is the same as for chemical oxidation with ferric iron (Mehta and Murr 1982; Jyothi *et al.* 1988). (2) Besides these oxidation processes the formation of precipitates may occur at the mineral surface due to the

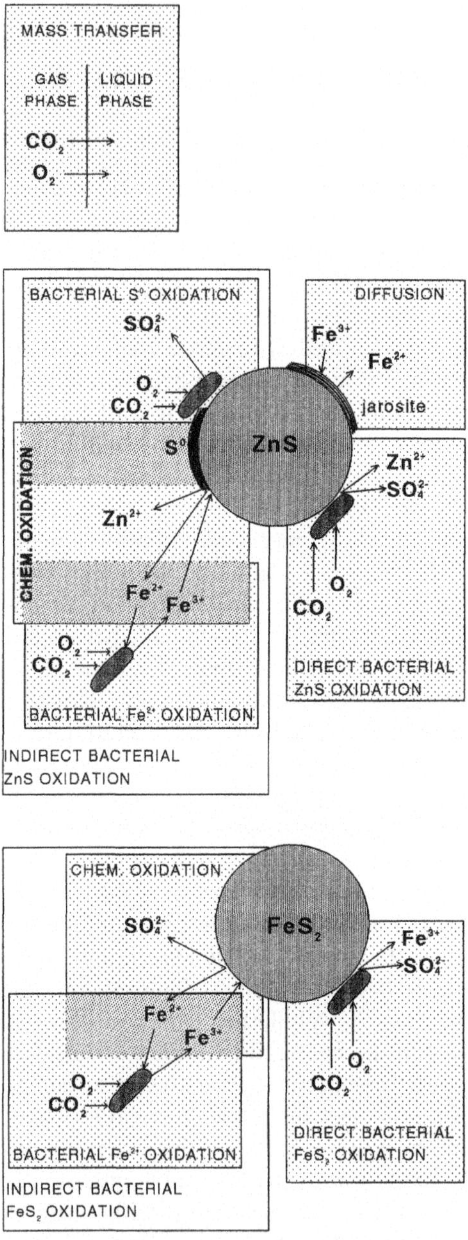

Figure 5.2. Sub-processes in the bacterial oxidation sulfide minerals, e.g. sphalerite (ZnS) and pyrite (FeS_2).

Table 5.1. Stoichiometric equations for bacterial and chemical oxidation of sphalerite (ZnS), chalcopyrite ($CuFeS_2$), pyrite (FeS_2), and intermediates

Bacterial oxidation with O_2		
$ZnS + 2O_2$	→	$ZnSO_4$
$CuFeS_2 + 4.25O_2 + H^+$	→	$Cu^{2+} + 2SO_4^{2-} + Fe^{3+} + 0.5H_2O$
$FeS_2 + 15/4O_2 + 0.5H_2O$	→	$Fe^{3+} + 2SO_4^{2-} + H^+$
$S^0 + 1.5O_2 + H_2O$	→	$SO_4^{2-} + 2H^+$
$S_2O_3^{2-} + 2O_2 + H_2O$	→	$2SO_4^{2-} + 2H^+$
$Fe^{2+} + 1/4O_2 + H^+$	→	$Fe^{3+} + 1/2H_2O$
Incomplete chemical oxidation with Fe^{3+}		
$ZnS + 2Fe^{3+}$	→	$Zn^{2+} + S^0 + 2Fe^{2+}$
$CuFeS_2 + 4Fe^{3+}$	→	$Cu^{2+} + 2S^0 + 5Fe^{2+}$
$FeS_2 + 2Fe^{3+}$	→	$3Fe^{2+} + 2S^0$
$S^0 + 2Fe^{3+} + 1.5H_2O$	→	$0.5S_2O_3^{2-} + 3H^+ + 2Fe^{2+}$
Complete chemical oxidation with Fe^{3+}		
$ZnS + 8Fe^{3+} + 4H_2O$	→	$ZnSO_4 + 8Fe^{2+} + 8H^+$
$CuFeS_2 + 16Fe^{3+} + 8H_2O$	→	$Cu^{2+} + 2SO_4^{2-} + 17Fe^{2+} + 16H^+$
$FeS_2 + 14Fe^{3+} + 8H_2O$	→	$15Fe^{2+} + 2SO_4^{2-} + 16H^+$
$S^0 + 6Fe^{3+} + 4H_2O$	→	$SO_4^{2-} + 6Fe^{2+} + 8H^+$
$S_2O_3^{2-} + 8Fe^{3+} + 5H_2O$	→	$2SO_4^{2-} + 8Fe^{2+} + 10H^+$

elemental sulfur production in chemical oxidation reactions or the precipitation of iron hydroxide (e.g. jarosite). Formation of layers will prevent *direct* bacterial oxidation of the mineral surface. Diffusion of soluble compounds through these layers might become a rate determining process. (3) Finally, the mass transfer of oxygen and carbon dioxide from the aeration air to the liquid is an important sub-process. This review focuses on the mechanism and rate determining sub-processes in bacterial oxidation of sulfide minerals.

5.2.2 *Direct* mechanism

Defenders have used several methods to prove the *direct* mechanism in the oxidation of sulfide minerals with *T. ferrooxidans*. For example, by showing that pyrite oxidation is negligible when pyrite and bacteria are separated by a membrane (Arkesteyn 1980), or by showing that the bacterial growth yield on pyrite cannot be explained when assuming a purely *indirect* mechanism (Arkesteyn 1980; Hazeu *et al.* 1986; 1987), or by showing that bacterial pyrite or chalcopyrite oxidation rates decrease when using selective inhibitors (Duncan *et al.* 1967; Arkesteyn 1980), or by showing the specific surface pattern of pyrite after bacterial oxidation by scanning electron micrographs (Hansford and Drossou. 1987; Natarajan 1988; Rodriquez-Leiva and Tributsch 1988; Bärtels *et al.* 1988; Mustin *et al.* 1992; Vargas *et al.* 1993). Another approach to prove the *direct* mechanism was to show significant oxidation of metal sulfides in bio-oxidation experiments with synthetic or natural metal sulfides under iron-free conditions (Torma *et al.* 1970, 1972; Sanmugasunderam *et al.* 1985; Natarajan 1988), or by showing that bio-oxidation of metal sulfides is faster than sterile leaching at equal initial concentrations of ferric iron (Boogerd *et al.* 1991; Lizama and Suzuki 1991; Konishi *et al.* 1992). However, examination of the mineral oxidation rates in these experiments showed that the presented phenomena could also be explained by assuming an *indirect* mechanism (Boon 1996).

5.2.3 *Indirect* mechanism

Recent work has given strong evidence that bacterial oxidation of metal sulfides is most probably dominated by an *indirect* mechanism. Redox-stat experiments were used to simultaneously measure the oxidation rate of natural Ni_3S_2 (Crundwell 1987; Verbaan and Huberts 1988) or synthetic ZnS (Boon *et al.* 1998c) in the bacterial oxidation with *T. ferrooxidans* and in the (sterile) chemical oxidation at equal physical and chemical conditions. The redox potential in the solution is a measure for the ratio between ferric and ferrous iron species. In the sterile batch the redox potential and pH were kept equal to that in the batch culture with bacteria. In these experiments the role of a *direct* mechanism is negligible because the oxidation rate of the mineral is equal in the parallel batches.

The use of *L. ferrooxidans* has also played a major role in elucidating the mechanism of bacterial pyrite oxidation. For a long time *T. ferrooxidans* has been regarded as the most important organism in bacterial leaching. Nordstrom and Southam (1997) summarize which inorganic substances are utilized by different *Thiobacillus* and *Leptospirillum* species. From these data it is observed that all *Thiobacillus* species oxidize reduced sulfur compounds, whereas they are only able to utilize metal sulfides if they also have the capacity to oxidize ferrous iron (e.g. *T. ferrooxidans*). On the other hand *L. ferrooxidans* is able to oxidize ferrous iron and metal sulfides but not reduced sulfur compounds. Consequently, the capacity to utilize metal sulfides requires the capacity to oxidize ferrous iron instead of reduced sulfur compounds as has been assumed by authors who defended the *direct* mechanism.

Boon (1996) found that in pyrite oxidation with *L. ferrooxidans* the major part of the bacteria is not attached to the pyrite surface. It has been shown that at equal solution redox potential (i.e. equal concentrations of dissolved ferrous and ferric iron species) the oxygen consumption rate in batch cultures on pyrite is equal to that in the cell suspension without pyrite. Therefore, the only role of *L. ferrooxidans* in pyrite oxidation is the oxidation of ferrous iron. Sand *et al.* (1995) and Schippers *et al.* (1996) detected reduced sulfur compounds (e.g. thiosulfate) in the oxidation of pyrite with *L. ferrooxidans*. Based on these results they proposed that reduced sulfur compounds are produced as intermediates in the chemical oxidation of pyrite (Table 5.1). Thiosulfate is either chemically oxidized, or biologically if *Thiobacillus* species are present (Figure 5.3). This explains the higher bacterial growth yield of *T. ferrooxidans* on pyrite compared to its yield on ferrous iron. Norris (1983) showed a higher affinity for ferrous iron of *L. ferrooxidans* compared with *T. ferrooxidans*. Norris *et al.* (1987) and Boon *et al.* (1999b) showed that the pyrite oxidation rate is faster with *L. ferrooxidans* than with *T. ferrooxidans*. Therefore, the capacity of *Thiobacillus* species to oxidize reduced sulfur compounds does not increase the bacterial mineral oxidation rate.

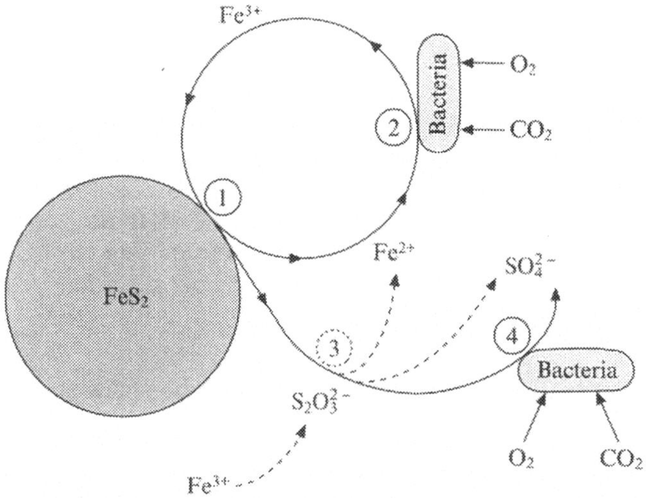

Figure 5.3. Refined *indirect* mechanism for the bacterial pyrite oxidation (Boon 1996). (1) Ferrous iron and thiosulfate are produced in the chemical oxidation of pyrite with ferric iron. (2) Ferrous iron is oxidized by *T. ferrooxidans* or *L. ferrooxidans*. (3) Thiosulfate is chemically oxidized with ferric iron. (4) Thiosulfate is oxidized by *T. ferrooxidans* but not *L. ferrooxidans*.

5.3 CHEMICAL OXIDATION OF SULFIDE MINERALS

In bacterial oxidation of sulfide minerals the chemical oxidation with ferric iron is a relevant sub-process. In an extended review of literature data, the chemical oxidation kinetics of pure sphalerite, chalcopyrite, and pyrite in acidic ferric sulfate solutions at 30 °C has been examined (Boon 1996). Galvanic interactions were excluded. Important conclusions are summarized in this section.

5.3.1 Stoichiometry

In their review Dutrizac and MacDonald (1974) predicted from thermodynamic data that mineral sulfides will completely oxidize under sterile conditions and in excess of ferric iron. However, elemental sulfur instead of sulfate is produced in the oxidation of metal sulfides (e.g. $CuFeS_2$, ZnS, CdS, CuS, etc.). In pyrite the sulfur moiety exists as S_2^{2-} instead of S^{2-}. Luther (1987) and Moses *et al.* (1987) concluded from theoretical studies of the pyrite oxidation mechanism that the consumption of oxygen is negligible. These authors also predicted that thiosulfate ($S_2O_3^{2-}$) is produced as an intermediate which is rapidly oxidized chemically to sulfate (Figure 5.3). Indeed, complete oxidation of pyrite to ferrous iron and sulfate was observed by most authors (McKibben and Barnes 1986; Mathews and Robins 1972; Boogerd *et al.* 1991). King and Perlmutter (1977) used 'coal' pyrite (marcasite) and found a ratio of 1:4.5 for FeS_2 : Fe^{3+}. Consequently, in bacterial oxidation of pyrite the chemical oxidation of intermediate thiosulfate is very rapid and is therefore only partly available to the bacteria. In bacterial oxidation of other sulfide minerals the chemical oxidation of elemental sulfur is relatively slow and is therefore available to bacteria. The mechanism of bacterial oxidation of elemental sulfur is not discussed in this chapter.

5.3.2 Mass transfer limitation at the mineral surface

The chemical oxidation reaction at the mineral surface may be inhibited by mass transfer limitation due to elemental sulfur or jarosite layers. Elgersma (1992) examined the kinetics of jarosite formation in acidic ferric sulfate media which can be described by:

$$M^+ + 3Fe^{3+} + 2SO_4^{2-} + 6H_2O \rightarrow MFe_3(SO_4)_2(OH)_6 + 6H^+$$

with M^+ = Na^+, K^+, NH_4^+ or H^+. Jarosites are usually produced at pH below 3, at higher pH ferric hydroxides and oxides are produced. Jarosite formation is promoted by the decrease of pH and increasing concentrations of M^+, Fe^{3+}, and SO_4^{2-}.

A decrease in the chemical reaction rate during a chemical batch experiment is often attributed to diffusion limitation. Whereas other process conditions might also cause a rate decrease, this argument needs to be supported by visual methods (e.g. X-ray techniques or scanning electron microscopy). A low activation energy or an apparent decrease of the activation energy indicates diffusion limitation. The reported activation energy of chemical sphalerite oxidation with ferric iron is between 50 and 80 kJ/mol. The value for chalcopyrite is between 50 and 85 kJ/mol and for pyrite between 60 and 95 kJ/mol. With the Arrhenius equation it is predicted that the reaction rate increases about three times with every ten degree increase in temperature (between 30 °C and 80 °C). Lowe (1970) estimated that between 25 and 68 °C the diffusion coefficient of ferric in sulfate solutions increases with a factor of 2.54 at ferric concentrations between 0.05 and 0.7 M. Accordingly, if the reaction rate is determined by the diffusion rate, applying the Arrhenius equation would yield an apparent activation energy of only 18 kJ/mol within this temperature range.

In the chemical oxidation of zinc sulfide elemental sulfur is produced. The measured activation energy is relatively large and most authors concluded that diffusion rate limitation due to S^0 formation does not occur. Although Rath *et al.* (1982) observed a decrease of the reaction rate during the experiment ("parabolic" kinetics), the measured activation energy (90 kJ/mol) is too high to explain the rate decrease with diffusion limitation. Zuo-Mei Jin *et al.* (1984) examined reaction residues at various levels of zinc extraction by SEM (scanning electron micrographs), energy dispersive X-ray analyses, and X-ray diffraction. These analyses showed increasing levels of sulfur at the surface of the particles. They concluded that the sulfur product layer is extremely porous and is therefore no diffusion barrier. Scanning electron micrographs reported by Rath *et al.* (1988) support these observations. These authors did not report jarosite formation during the chemical oxidation of zinc sulfide.

Several authors observed a decrease of the chemical oxidation rate of chalcopyrite during the course of a batch experiment and assumed that this parabolic kinetics was caused by diffusion limitation of ions through the sulfur product layer. However, the occurrence of diffusion limitation is contradictory with the high activation energy that was observed by those authors. Moreover, given the fact that chemical sphalerite oxidation is much faster than chalcopyrite oxidation, it is unlikely that the sulfur layer causes diffusion limitation in chalcopyrite oxidation. Lowe (1970) calculated that in the chemical oxidation of a chalcopyrite ore the aqueous diffusion of ferric sulfate ions through pores in the elemental sulfur is not rate determining. Linge (1976) also calculated that the diffusion rate through pores in a sulfur layer is large enough compared with the measured oxidation rate of chalcopyrite. This author observed parabolic kinetics and suggested that diffusion limitation through the solid state of elemental sulfur must be assumed. Unfortunately the porosity of sulfur layers on the chalcopyrite surface were not examined. Also, these authors did not examine jarosite formation. Boon (1996) rejected the hypothesis that jarosite formation possibly causes parabolic kinetics in chemical chalcopyrite oxidation. This author found strong arguments that parabolic kinetics is caused by the chemical reaction kinetics and not by diffusion limitation.

In chemical oxidation of pyrite neither formation of elemental sulfur- or jarosite layers nor rate decrease due to diffusion limitation, were reported.

5.3.3 Chemical oxidation kinetics

The chemical oxidation of mineral sulfides is a surface reaction. It is usually assumed that the chemical reaction rate is proportional to the chemical reaction surface area (e.g. BET surface area, Brunauer *et al.* 1938). Many authors apply a so called shrinking particle model. According to this model the conversion rate of the mineral, $d\xi/dt$ ($mol_{MeS}/mol_{MeS}/h$) is inversely proportional to the particle diameter. The chemical oxidation rate largely depends on the temperature which is expressed in the Arrhenius equation, and often increases with the concentration of ferric iron which is expressed in a reaction order, *n*. A typical example of the shrinking particle model in which the reaction rate is controlled by the chemical reaction is given by Rath *et al.* (1988):

$$1 - (1 - \xi)^{1/3} = k'' \frac{[Fe^{3+}]^{n}}{d_0} \exp\left(\frac{-E_{act}}{RT}\right) t \tag{5.1}$$

For ZnS $n = 0.62$, $k'' = 4.3$ m/s, $E_{act} = 58$ kJ/mol, and for $CuFeS_2$ $n = 0.38$, $k'' = 6.7 \times 10^3$ m/s, $E_{act} = 93$ kJ/mol. In this equation d_0 is the initial particle diameter, and t is the time during the course of the batch experiment. Most authors measured the chemical oxidation kinetics of sulfide minerals in batch experiments at pH between 0.5 and 2, T between 30 and 90 °C, and ferric iron approximately between 0.01 and 0.3 M.

In chemical oxidation of sphalerite Zuo-Mei Jin *et al.* (1984), Warren *et al.* (1985), Verbaan and Crundwell (1986), and Crundwell (1988) report that the increase of ferrous iron inhibits the chemical oxidation rate. The effect of pH is not reported. Crundwell (1988) and Palencia-Perez and Dutrizac (1991) report that higher chemical oxidation rates are observed at a higher iron content in sphalerite mineral.

In chemical oxidation of chalcopyrite most authors report a decrease of the oxidation rate during the batch experiment. This behaviour is called "parabolic" kinetics and is usually attributed to diffusion limitation due to elemental sulfur formation (see above). Saxena and Mandre (1992) proposed so called 'mixed control' kinetics to describe chemical reaction with mass transfer limitation. Munoz-Castillo (1977), Dutrizac (1978), and Hirato *et al.* (1986) reported that pH and ferrous iron do not influence the oxidation kinetics. Boon (1996) suggested that the chemical oxidation rate of pyrite and chalcopyrite increase at an increasing ferric to ferrous iron concentration ratio. Therefore, the parabolic kinetics in the chemical oxidation of chalcopyrite is caused by the increase of ferrous iron during the batch experiment. It is predicted that in batch experiments with pyrite the increase of ferrous iron decreases the chemical oxidation rate by a factor between 2 and 10 in the first few minutes (Boon and Heijnen 1998b). This explanation is in accordance with the observations of O'Malley and Liddell (1987) who reported a decrease of the $CuFeS_2$ oxidation rate at decreasing solution redox potential, and Hirato *et al.* (1987), who also measured a rate decrease at increasing ferrous iron concentration.

Most authors found that pyrite is completely oxidized to ferrous iron and sulfate. No mass transfer limitation to the surface was observed, and it was concluded that the reaction rate at the mineral surface is chemically controlled. The pyrite oxidation rate increases with increasing active ferric iron concentration below a ferric concentration of 0.1 M. An increase of the sulfate concentration causes a decrease of active ferric species and therefore inhibits the oxidation rate. Increasing concentrations of protons or ferrous iron inhibit the chemical oxidation rate as well.

5.4 BACTERIAL OXIDATION OF SULFIDE MINERALS

An extended review of literature data has examined the bacterial oxidation kinetics of pure sphalerite, chalcopyrite, and pyrite by *T. ferrooxidans* (Boon 1996). Important conclusions are summarized in the section below.

5.4.1 Gas-liquid mass transfer limitation

In bacterial oxidation of sulfide minerals transfer of oxygen and carbon dioxide from the gas to the liquid phase is required. Avecedo *et al.* (1987), Khinvasara and Agate (1987), Norris and Barr (1987), and LeRoux and Wakerley (1987) reported an increase in the measured bacterial oxidation rate when applying an experimental device with better oxygen- and carbon dioxide transfer capability was applied (e.g. an air-lift loop reactor gives better results than a pachuca tank or stirred tank reactor). It has been shown that many bio-oxidation experiments were carried out at carbon dioxide limited conditions (Boon and Heijnen 1998d).

5.4.2 Jarosites in bacterial oxidation

Several authors reported on the occurrence of jarosite in bio-oxidation experiments. Lazaroff *et al.* (1982) examined precipitate formation during the oxidation of ferrous iron in acid solution by resting suspensions of *T. ferrooxidans*. These precipitates consisted of crystalline jarosite and amorphous ferric hydroxy-sulfates, which however, differed in chemical composition and infra-red spectra from precipitates formed at similar abiotic conditions. These authors argued that bacteria promote the formation of jarosite because they lower the activation energy for the formation of ferric hydroxy-sulfate polymers. These polymers precipitate to either amorphous ferric hydroxy-sulfates or crystalline jarosite, depending on the ionic circumstances. Grishin *et al.* (1988) performed experiments in a packed bed bio-reactor with activated carbon particles as a carrier matrix for *T. ferrooxidans*. In this reactor ferrous sulfate was converted to ferric sulfate. Despite the low operating pH (1.35–1.5) precipitates were produced. X-ray diffraction indicated that the precipitates were well-ordered potassium jarosite. Toro *et al.* (1988) also reported precipitate formation in the biological oxidation of ferrous sulfate. Verbaan and Huberts (1988) carried out experiments with synthetic nickel sulfide (Ni_3S_2) in the presence and in the absence of *T. ferrooxidans* at controlled solution electrochemical potentials. They found that the dissolution of the mineral was inhibited in the presence of bacteria. Mineralogical inspection of the leach residues revealed that leaching in the presence of bacteria led to the deposition of a compact layer of material round most of the residue particles which was not present when the leach was conducted under sterile conditions. X-ray diffractometry, scanning electron microscope studies and electron micro-probe revealed the presence of a jarosite precipitate and a sulfur rich layer round the residue particles from the bacterial leach. Accordingly, the formation of jarosite does occur in bacterial oxidation experiments and may cause diffusion limitation of the reaction rate.

Bacterial oxidation is usually carried out at pH between 1.8 and 2.5 which is higher than pH in chemical oxidation experiments. Moreover, in bacterial oxidation of chalcopyrite and also bacterial oxidation of ferrous iron protons are consumed (Table 5.1). The pH will also rise in batch culture experiments if elemental sulfur accumulates. From the comparison of kinetic data (Boon 1996), it was shown that either significantly lower bacterial oxidation rates or a decrease of the oxidation rates during the batch culture experiments are reported by authors who did not apply pH control in bacterial chalcopyrite oxidation. Therefore pH control is required to prevent the formation of jarosites in bioleaching experiments.

5.4.3 Bacterial oxidation kinetics of ferrous iron and elemental sulfur

According to the *indirect* mechanism (Figure 5.2), the role of *T. ferrooxidans* in bacterial oxidation of sulfide minerals is to oxidize ferrous iron and reduced sulfur compounds produced in the chemical oxidation reaction, whereas the role of *L. ferrooxidans* is restricted to the oxidation of ferrous iron. Many authors have measured the ferrous iron oxidation kinetics of *T. ferrooxidans*, which has been reviewed elsewhere (Boon 1996). Most authors agree that the ferrous iron oxidation rate increases at increasing ferrous iron, whereas ferric iron inhibits the oxidation rate. An adapted Monod equation for competitive inhibition describes the kinetics. Boon (1996) and Scherpenzeel *et al.* (1998) showed that the same equation describes ferrous iron oxidation with *L. ferrooxidans*. However, *L. ferrooxidans* shows much higher oxidation rates at equal ferric and ferrous iron concentrations as compared with *T. ferrooxidans*, and this bacterial strain is able to achieve a much higher redox potential. It has been shown that the ferrous iron oxidation kinetics of *L. ferrooxidans* on ferrous iron measured in continuous cultures on ferrous iron (in the absence of pyrite) is equal to the ferrous iron oxidation kinetics in the presence of pyrite (Boon 1996).

Only limited research has been carried out to study the behaviour of *T. ferrooxidans* on the mixed substrate, Fe^{2+} and S^0. Espejo *et al.* (1988) measured bacterial oxidation rates on ferrous iron and on the mixed substrate in colorimetric experiments. They found that the ferrous iron oxidation rate is significantly less in the presence than in the absence of elemental sulfur. Wiertz (1993) came to the same results in respiration experiments. Boon *et al.* (1998c) have shown that in zinc sulfide oxidation due to its preference for elemental sulfur *T. ferrooxidans* can even lose its ability to oxidize ferrous iron

5.4.4 Bacterial oxidation kinetics of sulfide minerals

To examine the advantage of bioleaching, oxidation rates of pure sphalerite, chalcopyrite, and pyrite, in (sterile) chemical batch experiments and in batch culture experiments inoculated with *T. ferrooxidans* reported in the literature, were determined at 30 °C and similar initial particle diameters and ferric iron concentrations (Boon 1996). The mineral oxidation rate, d[MeS]/dt ($mol_{MeS}/l_{slurry}/s$), is proportional to the mineral surface area expressed as BET surface area (Brunauer *et al.* 1938) or as the surface area calculated from assuming spherical particles with an average diameter. Values of first-order kinetic constants ($k_{A,BET}$ in $mol/m^2_{BET}/s$; $k_{A,spher}$ in $mol/m^2_{spher}/s$; k_1 in $mol_{MeS}/mol_{MeS}/s$) were determined. The comparison of data at equal initial particle diameters is most reliable and allows the use of a first-order kinetic constant, k_1, which expresses a mineral specific oxidation rate proportional to the concentration of the mineral sulfide:

$$\frac{d\,[\mathrm{MeS}]}{dt} = -k_1\,[\mathrm{MeS}] \tag{5.2}$$

In Table 5.2 the average values of k_1 for chemical and bacterial oxidation are summarized. The conversion for the three different minerals in chemical and bacterial batch experiments is plotted in Figure 5.4A,B respectively. The lag-phase that usually occurs during the growth phase of bacteria is not plotted in Figure 5.4B. Chemical and bacterial oxidation rates of sphalerite are in the same range, which is in accordance with the *indirect* mechanism. It needs to be explained however, why bacterial chalcopyrite oxidation is approximately 5–10 times faster than (sterile) chemical oxidation, and in the case of pyrite this factor is 10–20 times.

The following equation presented by Mathews and Robins (1972) is an example for chemical pyrite oxidation kinetics measured in a batch experiment:

$$k_{A,BET} = \frac{1.75 \times 10^3}{[H^+]^{0.44}} \frac{[Fe^{3+}]}{[Fe^{3+}] + [Fe^{2+}]} \exp\left(-\frac{E_{act}}{RT}\right) \tag{5.3}$$

Table 5.2. Comparison of the maximum average chemical and bacterial (*T. ferrooxidans*) oxidation rate constants as calculated from the literature at 0.1 M Fe^{3+} (ferric sulfate medium), 30 °C, and initial particle size 38–45 μm

	Chemical oxidation		Bacterial oxidation	
Mineral	k_1 $*10^{-8}$ s^{-1}	Remarks	k_1 $*10^{-8}$ s^{-1}	Remarks
ZnS	200 (a)	at 0.01 M Fe^{3+}	10 to 80 (e)	Synthetic ZnS, Fe-free medium
	700 (b)	at 0.1 M Fe^{3+}	120 to 700 (f)	Iron containing sphalerite
$CuFeS_2$			4 to 23 (g)	No pH control
	5 to 8 (c)		40 to 110 (h)	pH control
FeS_2			20 to 50 (i)	No O_2 and/or CO_2 control
	8 to 10 (d)		120 to 200 (j)	O_2 control and/or CO_2 addition

(a) Zuo-Mei Jin *et al.* (1984); Warren *et al.* (1985); Rath *et al.* (1982, 1988); Verbaan and Crundwell (1986). (b) Su (1976); Bobeck and Su (1985); Zuo-Mei Jin *et al.* (1984); Warren *et al.* (1985); Rath *et al.* (1988); Verbaan and Crundwell (1986); Crundwell (1988); Palencia-Perez and Dutrizac (1991). (c) Haver and Wong (1971); Dutrizac and MacDonald (1971); Linge (1976); Munoz-Castillo (1977); Dutrizac (1978); Palmer *et al.* (1981); Dutrizac (1982); Hirato *et al.* (1986, 1987); Rath *et al.* (1988); O'Malley and Liddell (1987). (d) Mathews and Robins (1972); Zeng *et al.* (1986); McKibben and Barnes (1986); Kawakami *et al.* (1988); Boogerd *et al.* (1991). (e) Torma *et al.* (1970, 1972); Gormely *et al.* (1975); Torma and Sakaguchi (1978); Tributsch and Bennett (1981); Sanmugasunderam *et al.* (1985); Natarajan (1988). (f) Cwalina *et al.* (1988); Natarajan (1988); Konishi *et al.* (1992). (g) Yukawa *et al.* (1978); Blancharte-Zurita *et al.* (1986); Almendras *et al.* (1987); LeRoux and Wakerley (1987); Palencia-Perez *et al.* (1987); Cwalina *et al.* (1988); Natarajan (1988); Elzeky and Attia (1989); Sulka *et al.* (1990). (h) Kingma and Silver (1980); Blancharte-Zurita *et al.* (1987); Khinvasara and Agate (1987); Acevedo *et al.* (1987, 1989). (i) Atkins (1978); Hansford and Drossou (1987); Lawrence and Marchant (1987); Basaran and Tuovinen (1987); Norris *et al.* (1987); Cwalina *et al.* (1988); Elzeky and Attai (1989); Guay (1989 *et al.*); Olson (1989, 1991); Konishi *et al.* (1990); Baldi *et al.* (1992). (j) Pinches (1987) *et al.*; Norris and Barr (1987); Norris (1989); Boogerd *et al.* (1989).

In Equation 5.3 E_{act} = 86 kJ/mol. According to this equation a maximum oxidation rate will be observed if the ferrous iron concentration is negligible compared with the ferric concentration ($[Fe^{3+}]$ + $[Fe^{2+}]).[Fe^{3+}]$). Most authors proposed similar kinetic models in which the decrease of the chemical pyrite oxidation rate during batch experiments was described by using the actual ferrous iron concentration. However, very high oxidation rates at the start of sterile chemical batch experiments with pyrite were observed by several authors (Garrels and Thompson 1960; King and Perlmutter 1977; Wiersma and Rimstidt 1984; Kawakami *et al.* 1988).

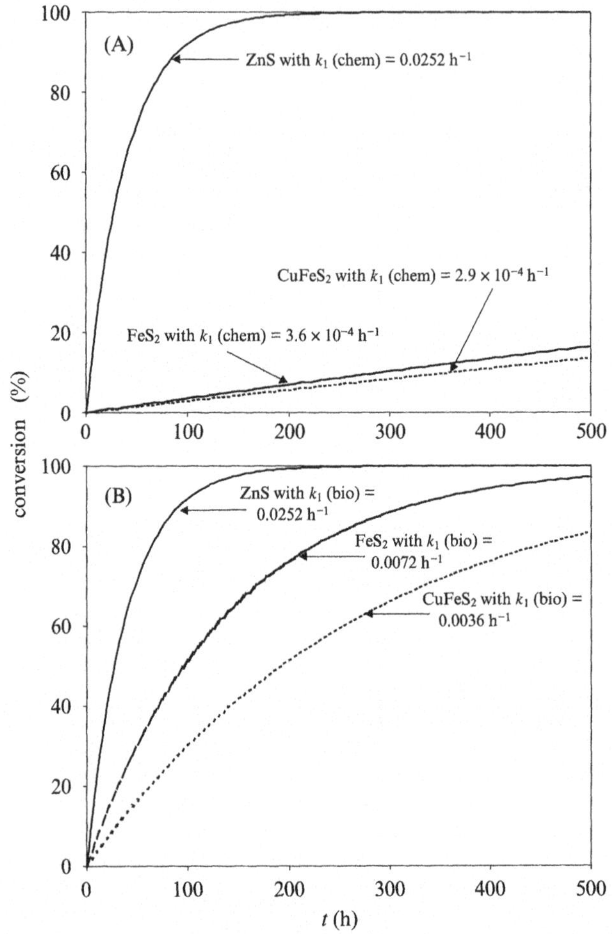

Figure 5.4. Conversion of sulfide minerals according to first order kinetics in the mineral concentration. (A) chemical and (B) bacterial batch experiments. The optimum values of k_1 summarized in Table 5.2 are used in these simulations. Conversion = $1 - \exp(-k_1.t)$.

These high initial rates are neglected in the proposed kinetic equations because the authors considered this phenomenon as anomalous behaviour, which was explained from assuming the occurrence of highly reactive sites at the pyrite. Unfortunately, the authors have not shown that in a repetition experiment using fresh medium and the same pyrite (of which the reactive sites have now been oxidized) the high rates in the initial phase did not occur. Boon (1996) observed that pyrite oxidation with *L. ferrooxidans* increases at an increasing redox potential (i.e. an increasing ratio of Fe^{3+} and Fe^{2+} species). Therefore, it is proposed that the chemical oxidation rate of chalcopyrite and pyrite increases at an increasing ratio of Fe^{3+}/Fe^{2+} species. Consequently, the reported high initial rates are caused by the high ferric to ferrous iron concentration ratios at the start of sterile chemical batch experiment; after several minutes the change of the redox potential is relatively slow (Boon and Heijnen 1998b).

Box 5.2. A kinetic model for pyrite oxidation with L. ferrooxidans

Boon and Heijnen (1998b) determined the chemical oxidation rate of pyrite in short term batch culture experiments with *Leptospirillum* bacteria and the step-wise addition of pyrite. They showed that the following empirical equation describes the chemical oxidation kinetics of pyrite with ferric iron:

$$\nu_{FeS_2} = \frac{\nu_{FeS_2,max}}{1 + B[Fe^{2+}]/[Fe^{3+}]} \tag{5.4}$$

In this equation $\nu_{FeS_2,max}$ and B are kinetic constants. ν_{FeS_2} is a pyrite specific oxidation rate ($molFeS_2/molFeS_2/h$). This equation is only applicable at a negligible decrease of the particle size. Accordingly, the production rate of ferrous iron in the chemical oxidation of pyrite, $r_{Fe^{2+},chem}$, is calculated from (see also stoichiometry for complete chemical pyrite oxidation in Table 5.1):

$$r_{Fe^{2+},chem} = 15\nu_{FeS_2}[FeS_2] \tag{5.5}$$

An adapted Monod equation for competitive ferric iron inhibition describes the biomass specific oxidation rate of ferrous iron, $q_{Fe^{2+}}$ ($molFe^{2+}$/C-mol/h), both by *T. ferrooxidans* (Boon *et al.* 1999a) and by *L. ferrooxidans* (Scherpenzeel *et al.* 1998). A simplified version of this equation is:

$$q_{Fe^{2+}} = q_{Fe^{2+},max} \Big/ \left(1 + \frac{K_s\,[Fe^{3+}]}{K_i\,[Fe^{2+}]}\right) \tag{5.6}$$

In this equation $q_{Fe^{2+},max}$ and K_s/K_i are kinetic constants. The ferrous iron oxidation rate is determined from the oxygen consumption in on-line off-gas analyses (Boon *et al.* 1998a). The biomass concentration, C_x, is expressed in moles organic carbon (C-mol/l). Accordingly, the bacterial oxidation rate of ferrous iron, $-r_{Fe^{2+},bio}$, is:

$$-r_{Fe^{2+},bio} = C_x q_{Fe^{2+}} \tag{5.7}$$

Thus, the ratio of total ferric and ferrous compounds in the solution is the rate determining variable in opposite directions for both bacterial ferrous iron oxidation and chemical pyrite oxidation. Figure 5.5 shows a simulation of the chemical ferrous iron production rate as a function of the ratio of the ferric to ferrous iron concentration for two different pyrite minerals. Also the simulated bacterial ferrous iron oxidation rate by *T. ferrooxidans* and *L. ferrooxidans* is shown in this graph. At the point of intersection where $r_{Fe2+}(\text{chem}) = -r_{Fe2+}(\text{bio})$ a pseudo-steady-state is obtained: the chemical production rate is equal to the bacterial consumption rate of ferrous iron. This graph shows that the oxidation rate of the less reactive pyrite by *T. ferrooxidans* is very low, whereas the pyrite oxidation rate in the presence of *L. ferrooxidans* is significantly higher. The pyrite specific oxidation rate at pseudo-steady-state, v_{FeS2}, increased at an increasing ratio between the biomass and pyrite concentration, $C_x/[FeS_2]$, which is in accordance with equations 5.4–5.7.

5.5 CHEMICAL VERSUS BACTERIAL OXIDATION: PRACTICAL CONSIDERATIONS

In bioleaching of sulfide minerals several sub-processes are involved, each of which might determine the overall metal solubilization rate in industrial processes. Design and optimization of these processes requires *mechanistic* kinetic models which account for the kinetics of relevant sub-processes and the effect of relevant process conditions (Boon *et al.* 1995). Relevant sub-processes are: (1) Oxygen and carbon dioxide transfer from aeration air to the slurry; (2) Formation of jarosites at the mineral surface; (3) Chemical oxidation of sulfide mineral with ferric iron; (4) Bacterial oxidation of ferrous iron and reduced sulfur compounds. In bio-oxidation of sulfide minerals bacteria maintain favourable conditions (i.e. high redox potential) for the chemical metal sulfide oxidation. This explains why the presence of bacteria can significantly enhance the chemical oxidation rate of metal sulfides compared with sterile oxidation at equal total dissolved iron

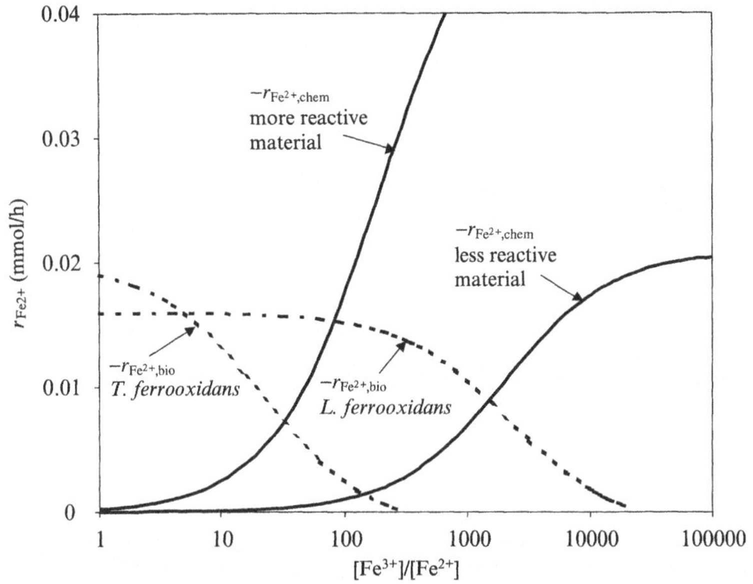

Figure 5.5. Simulation of chemical Fe^{2+} production rate, $r_{Fe^{2+},chem}$, and bacterial Fe^{2+} consumption rate, $-r_{Fe^{2+},bio}$, in bacterial oxidation of pyrite calculated at $C_x = 0.005$ C-mol/l, and 25 g/l FeS_2 (= 0.21 M). Kinetic constants for *T. ferrooxidans*: $q_{Fe^{2+},max}$ = 4.0 mol Fe^{2+}/C-mol/h, $K_s/K_i = 0.05$; *L. ferrooxidans*: $q_{Fe^{2+},max}$ = 3.2 molFe^{2+}/C-mol/h, $K_s/K_i = 0.0005$; Less reactive pyrite: $v_{Fe^{2+},max} = 0.10$ mol Fe^{2+}/mol FeS_2/h, $B = 2000$; More reactive pyrite: $v_{Fe^{2+},max} = 0.25$ mol Fe^{2+}/mol FeS_2/h, $B = 200$ (Boon *et al.* 1999b).

concentration, temperature, and particle diameter (e.g. a factor between 10 and 20 for pyrite). Therefore, bioleaching is particularly successful in recovery of gold from refractory pyrite, and copper from chalcopyrite.

5.6 REFERENCES

Acevedo F., Cacciuttolo M.A., Gentina J.C. (1987) Comparative performance of stirred and pachuca tanks in the bioleaching of a copper concentrate. p 385–394. In: *Biohydrometallurgy; Proc.Int.Symp. Warwick 1987*. Norris P.R., Kelly D.P. (eds.).

Acevedo F., Gentina J.C. (1989) Process engineering aspects of the bioleaching of copper ores. *Bioprocess Engineering* **4**, 223–229.

Almendras E., Arriagada F., Bustos S., Herrera L., Ruiz P., Vargas T. (1987) Surface transformation and electrochemical response of chalcopyrite in bacterial leaching process. p 259–271. In: *Biohydrometallurgy Proc.Int.Symp. Warwick 1987*. Norris P.R., Kelly D.P. (eds.).

Arkesteyn G.J.M.W. (1980) Pyrite oxidation by Thiobacillus ferrooxidans with special reference to the sulfur moiety of the mineral. p 39–53. In: Ph.D. thesis *Contribution of microorganisms to the oxidation of pyrite*. Wageningen, The Netherlands

Atkins A.S. (1978) Studies on the oxidation of sulfide minerals (pyrite) in the presence of bacteria. p 403–426. In: *Metallurgical Applications of bacterial leaching and related microbiological phenomena*. Murr L.E., Torma A.E., Brierley J.A. (eds.) Academic Press.

Baldi F., Clark T., Pollack S.S., Olson G.J. (1992) Leaching of pyrites of various reactivities by *Thiobacillus ferrooxidans*. *Appl.Environ.Microbiol.* **58**(6), 1853–1856.

Bärtels C.C. Charzitheodorou G., Rodriquez-Leiva M., Tributsch H. (1988) Novel technique for investigation and quantificantion of bacterial leaching by *Thiobacillus ferrooxidans*. *Biotechn.Bioeng.* **33**, 1196–1204.

Basaran A.H., Tuovinen O.H. (1987) Iron pyrite oxidation by *Thiobacillus ferrooxidans*: sulfur intermediates, soluble end products, and changes in biomass. *Coal Preparation* **5**, 39–55.

Blancharte-Zurita M.A., Branion R.M.R., Lawrence R.W. (1986) Particle size effects in the microbiological leaching of sulfide concentrates by *Thiobacillus ferrooxidans*. *Biotechn.Bioeng.* **28**, 751–755.

Blancharte-Zurita M.A., Branion R.M.R., Lawrence R.W. (1987) Microbiological leaching of chalcopyrite concentrates by *Thiobacillus ferrooxidans*; a comparative study of a conventional process and a catalyzed process. p 273–285. In: *Biohydrometallurgy Proc.Int.Symp. Warwick 1987*. Norris P.R., Kelly D.P. (eds.).

Bobeck G.E., Su H. (1985) The kinetics of dissolution of spalerite in ferric chloride solutions. *Metall.Trans.B.* **16B**, 413–424.

Boogerd F.C., Alphen van M.M.Q., Anrooij van W.J., Bruyn de J.C., Bos P., Kuenen J.G. (1989) The role of growth and maintenace in the oxidation of pyrite in batch culture by a moderately thermophilic, facultative chemolithoautotroph. p 735–761. In: *Proc.Int.Symp.Biohydrometallurgy*, Salley J., McCready G.L., Wichlacz P.L. (eds.), CANMET.

Boogerd F.C., Beemd C.v.d., Stoelwinder T., Bos P., Kuenen J.G. (1991) Relative contribution of biological and chemical reaction to the overall rate of pyrite oxidation at temperatures between 30^{0}C and 70^{0}C. *Biotechnol.Bioeng.* **38**, 109–115.

Boon M., Heijnen J.J., Hansford G.S. (1995) Recent developments in modelling bio-oxidation kinetics Part I: Measurement methods. p 41–61. In: *Minerals Bioprocessing II*, D.Holmes and R.Smith (Eds.) The Minerals, Metals and Materials Society, Warrendale, PA (1995) Proc. Eng. Found. Conf. Utah, USA, (July 1994).

Boon M. (1996) Theoretical and experimental methods in the modelling of bio-oxidation kinetics of sulfide mineral. PhD Thesis, October 1996, Technical University Delft, The Netherlands. ISBN 90-9009825-9.

Boon M., Luyben K.Ch.A.M., Heijnen J.J. (1998a) The use of on-line off-gas analyses and stoichiometry in the bio-oxidation kinetics of sulfide minerals. *Hydrometallurgy* **48**, 1–26.

Boon M., Heijnen J.J. (1998b) Chemical oxidation kinetics of pyrite in bioleaching processes. *Hydrometallurgy* **48**, 27–41.

Boon M., Snijder M., Hansford G.S., Heijnen J.J. (1998c) The oxidation kinetics of zinc sulfide with *Thiobacillus ferrooxidans*. *Hydrometallurgy* **48**, 171–186

Boon M., Heijnen J.J. (1998d) Gas-liquid mass transfer phenomena in bio-oxidation experiments of sulfide minerals: a review literature data. *Hydrometallurgy* **48**, 187–204.

Boon M., Ras C., Heijnen J.J. (1999a) The ferrous iron oxidation kinetics of *Thiobacillus ferrooxidans* in batch cultures. In press at *Appl. Microbiol. Biotechnol.*

Boon M., Brasser H.J., Hansford G.S., Heijnen J.J. (1999b) Comparison of the oxidation kinetics of different pyrites in the presence of *Thiobacillus ferrooxidans* or *Leptospirillum ferrooxidans*. In press at *Hydrometallurgy*.

Brierley C.L. (1978) Bacterial Leaching. *Critical Reviews in Microbiology*. Nov., 207–262.

Brunauer S.P.H., Emmett P.H., Teller E. (1938) Adsorption of gasses in multimolecular layers. J.Am.Chem.Soc. **60**, 309–319.

Corrans J.J., Harris B., Ralph B.J. (1972) Bacterial leaching: an introduction to its application and theory and a study on its mechanism of operation. *J. of South Afr. Inst. of Mining and Metallurgy*, March , 221–230.

Crundwell F.K. (1987) Kinetics and mechanism of the oxidative dissolution of a zinc sulfide concentrate in ferric sulfate solutions. *Hydrometallurgy* **19**, 227–242.

Crundwell, F.K. (1988) The role of charge-transfer mechanisms in the oxidative and non-oxidative dissolution of sphalerite. PhD thesis, Wits University, South Africa.

Cwalina B., Weglarz L., Dzierzewicz Z., Wilczok T. (1988) Dependence of effectiveness of leaching of metallic sulfides on enzymes involved in inorganic sulfur metabolism in *Thiobacillus ferrooxidans*. *Appl. Microbiol. Biotechnol.* **28**, 100–102.

Duncan D.W., Walden C.C., Trussell P.C. (1966) Biological leaching of mill products. CIM Bulletin **69**, 1075–1079.

Duncan D.W., Walden C.C., Trussell P.C. (1967) Recent advances in the microbiological leaching of sulfides, *Trans. Soc. Min. Eng., AIME* **238**, 122–128.

Dutrizac J.E., MacDonald R.J.C. (1971) The effect of sodium chloride on the dissolution of chalcopyrite under simulated dump leaching conditions. *Metall. Trans.* **2**, 2310–2312.

Dutrizac J.E., MacDonald R.J.C. (1974) Ferric ion as a leaching medium. *Minerals Sci. Eng.* **6**(2), 59–99.

Dutrizac J.E. (1978) The kinetics of dissolution of chalcopyrite in ferric ion media. *Metall. Trans. B.* **9B**, 431–439.

Dutrizac J.E. (1982) Ferric ion leaching of chalcopyrites from different localities. *Metall. Trans. B.* **13B**, 303–309.

Elgersma F. (1992) Integrated hydrometallurgical jarosite treatment. PhD Thesis Delft University of Technology, The Netherlands, ISBN 90-370-0073-8. p 95–111

Elzeky M., Attia Y.A. (1989) Effects of bacterial adaptation on bioleaching of ferrous sulfides. p 139–149. In: *Biotechnology in Minerals and Metal Processing*. Scheiner, Doyle, Kawatra (eds.) AIME.

Espejo R.T., Escobar B., Jedlicki E. (1988) Oxidation of ferrous iron and elemental sulfur by *Thiobacillus ferrooxidans*. *Appl.Environ.Microbiol.* July, 1694–1699.

Garrels R.M., Thompson M.E. (1960) Oxidation of pyrite by iron sulfate solutions. *Am.J.Sci.* **258A**, 57–67.

Gormely L.S., Duncan D.W., Branion R.M.R., Pinder K.L. (1975) Continuous culture of *Thiobacillus ferrooxidans* on a zinc sulfide concentrate. *Biotechnol.Bioeng.* **17**, 31–49.

Grishin S.I., Bigham J.M., Tuovinen O.H. (1988) Characterization of jarosite formed upon bacteral oxidation of ferrous sulfate in a packed-bed reactor. *Appl.Environ.Microbiol.* **54**(12), 3101–3106.

Guay R., Ghosh J., Torma A.E. (1989) Kinetics of microbiological production of ferric ion for heap and dump leaching. p 95–111. In: *Biotechnology in Minerals and Metal Processing.* Scheiner, Doyle, Kawatra (eds.) AIME.

Hansford G.S., Drossou M. (1987) A propagating pore model for the batch bioleach kinetics of refractory gold-bearing pyrite. p 345–358. In: *Biohydrometallurgy Proc.Int.Symp. Warwick 1987.* Norris P.R., Kelly D.P. (eds.).

Haver F.P., Wong M.M. (1971) Recovery of copper, iron, and sulfur from chalcopyrite concentrate using ferric chloride leach. *J.Metals* **23**, 25–29.

Hazeu W., Bijleveld W., Grotenhuis J.T.C., Kakes E., Kuenen J.G. (1986) Kinetics and energetics of reduced sulfur oxidation by chemostat cultures of *Thiobacillus ferrooxidans*. *Antonie v. Leeuwenhoek* **52**, 507–518.

Hazeu W., Schmedding D.J., Goddijn O., Bos P., Kuenen J.G. (1987) The importance of the sulfur-oxidizing capacity of *Thiobacillus ferrooxidans* during leaching of pyrite. *Proc.4th Europ.Congr.Biotechn.* Neysel O.M., Meer van der R.R., Luyben K.Ch.A.M. (eds.) **3**, 497–499, Elsevier Amsterdam.

Hirato T., Kinoshita M., Awakura Y., Majima H. (1986) The leaching of chalcopyrite with ferric chloride. *Metall.Trans.B.* **17B**, 19–28.

Hirato T., Majima H., Awakura Y. (1987) The leaching of chalcopyrite with ferric chloride. *Metall.Trans.B.* **18B**, 489–496.

Jyothi N., Brahmaprakash G.P., Natarajan K.A., Ramanand Rao G. (1988) Role of galvanic interactions in the bioleaching of mixed sulphides. p 111–115. In: Proc. 1st Inf. Conf. on Hydrometallurgy (ICHM'88) Beijng China 1988. Eds. Zheng Yulian Xu Jiazong, Int. Academic Publishers Pergamon Press Oxford.

Kawakami K., Sato J., Kusunoki K., Kusakabe K., Morooka S. (1988) Kinetic study of oxidation of pyrite slurry by ferric chloride. *Ind.Eng.Chem.Res.* **27**, 571–576.

Kelly D.P. *et al.* (1979) Metal extraction: the leaching of minerals. *Microbiol. Technology*, Current state, future prospects, eds. A.T. Bull, D.C. Ellwood, C. Ratledge, Cambridge 1979, 264–308.

Khinvasara N.J., Agate A.D. (1987) Bioreactor leaching of chalcopyrite concentrate from Mosaboni India. p 297–303. In: *Biohydrometallurgy Proc.Int.Symp. Warwick 1987.* Norris P.R., Kelly D.P. (eds.).

King W.E., Perlmutter D.D. (1977) Pyrite oxidation in aqueous ferric chloride. AIChE Journal **23** (5), 679–685.

Kingma J.G., Silver M. (1980) Growth of iron-oxidizing *Thiobacillus* in the presence of chalcopyrite and galena. *Appl.Environ.Microbiol.* March, 635–641.

Konishi Y., Asai S., Katoh H. (1990) Bacterial dissolution of pyrite by *Thiobacillus ferrooxidans*. *Bioproc.Eng.* **5**, 231–237.

Konishi Y., Kubo H., Asai S. (1992) Bioleaching of zinc sulfide concentrate by *Thiobacillus ferrooxidans*. *Biotechnol.Bioeng.* **39**, 66–74.

Lawrence R.W., Marchant P.B. (1987) Comparison of mesophilic and thermophilic oxidation systems for the treatment of refractory gold ores and concentrates. p 359–374. In: *Biohydrometallurgy Proc.Int.Symp. Warwick 1987.* Norris P.R., Kelly D.P. (eds.).

Lazaroff N., Sigal W., Wasserman A. (1982) Iron oxidation ond precipitation of ferric hydroxysulfates by resting *Thiobacillus ferrooxidans*. *Appl. Environ. Microbiol.* **43**(4), 924–938.

LeRoux N.W., Wakerley D.S. (1987) Leaching of chalcopyrite at 70^{0}C using *Sulfolobus*. p 305–317. In: *Biohydrometallurgy Proc.Int.Symp. Warwick 1987.* Norris P.R., Kelly D.P. (eds.).

Linge H.G. (1976) Reactivity comparison of australian chalcopyrite concentrates in acidified ferric solutions. *Hydrometallurgy* **2**, 219–233.

Lizama H.M., Suzuki I. (1991) Kinetics of sulfur and pyrite oxidation by *Thiobacillus thiooxidans*; competitive inhibition by increasing concentrations of cells. *Can.J.Microbiol.* **37**, 182–187.

Luther G.W. (1987) Pyrite oxidation and reduction: Molecular orbital theory considerations. *Geochim.Cosmochim.Act.* **51** (12), 3193–3199.

Lowe D.F. (1970) The kinetics of the dissolution reactions of copper and copper-iron-sulfide minerals using ferric-sulfate solutions. PhD thesis, Univ. of Arizona, Diss. Inf. Serv., Xerox Univ. microfilms Int. no. 70–18,175.

Mathews C.T., Robins R.G. (1972) The oxidation of iron disulfide by ferric sulfate. *Austral.Chem.Eng.* Aug, 21–25.

Mathews C.T., Robins R.G. (1972) The oxidation of aqueous ferrous sulfate solutions by molecular oxygen. *Proc.Aust.Inst.Met.* June, 242.

McKibben M.A., Barnes H.L. (1986) Oxidation of pyrite in low temperature acidic solutions: rate laws and surface textures. *Geochim.Cosmochim.Act.* **50**, 1509–1520.

Mehta A.P., Murr L.E. (1982) Kinetic study of sulfide leaching by galvanic interaction between chalcopyrite, pyrite and sphalerite in the presence of *Thiobacillus ferrooxidans* (30^{0}C) and a Thermophilic Microorganism (55^{0}C) *Biotechn.Bioeng.* **24**, 919–940.

Moses C.O., Nordsrom D.K. Herman J.S. Mills A.L. (1987) Aqueous pyrite oxidation by dissolved oxygen and by ferric iron. *Geochim.Cosmochim.Act.* **5**, 1561–1571.

Munoz-Castillo Pedro B. (1977) Reaction mechanism in the acid ferric-sulfate leaching of chalcopyrite. PhD thesis, Univ. of Utah, Diss. Inf. Serv., Xerox Univ. microfilms Int. no. 77–27,706.

Mustin C., Donato Ph de, Berthelin J. (1992) Quantification of the intergranular porosity formed in bioleaching of pyrite by *Thiobacillus ferrooxidans*. *Biotechnol.Bioeng.* **39**, 1121–1127.

Natarajan K.A. (1988) Electrochemical aspects of bioleaching multisulfide minerals. *Minerals Metallurg.Process.* May, 61–65.

Nordstrom D.K., Southam G. (1997) Geomicrobiology of sulfide mineral oxidation. Chap. 11 In: *Revs. in Microbiol.* Vol. 35. Geomicrobiology: interactions between microbes and minerals. Banfield J.F., Nealson K.H. (eds.).

Norris P.R. (1983) Iron and mineral oxidation with *Leptospirillum*-like bacteria. p 83–96. In: *Recent Progress in Biohydrometallurgy*. Rossi G., Torma A.E. (eds.) Calgliari Italy May 1983.

Norris P.R., Barr D.W. (1987) Bacterial oxidation of pyrite in high temperature reactors. p 532–536. In: *Biohydrometallurgy Proc.Int.Symp. Warwick 1987*. Norris P.R., Kelly D.P. (eds.).

Norris P.R., Barr D.W., Hinson D. (1987) Iron and mineral oxidation by acidophilic bacteria: affinities for iron and attachment to pyrite. p 43–59. In: *Biohydrometallurgy Proc.Int.Symp. Warwick 1987*. Norris P.R., Kelly D.P. (eds.).

Norris P.R. (1989) Factors affecting bacterial mineral oxidation: the example of carbon dioxide in the content of bacterial diversity. p 3–15. In: *Biohydrometallurgy Proc.Int.Symp. Wyoming*. Salley J., McCready R.G.L., Wichlacz P.L. (eds).

O'Malley M.L., Liddell K.C. (1987) Leaching of $CuFeS_2$by aqueous $FeCl_3$, HCl, NaCl: Effects of solution composition and limited oxidant. *Metall.Trans.B.* **18B**, 505–510.

Olson G.J. (1991) Rate of pyrite bioleaching by *Thiobacillus ferrooxidans*: Results of interlaboratory comparison. Appl.Environ.Microbiol. **57**(3), 642–644.

Olson G.J. (1989) Development of ore bioleaching standards. p 71–81. In: *Biohydrometallurgy Proc.Int.Symp. Wyoming*. Salley J., McCready R.G.L., Wichlacz P.L. (eds).

Palencia-Perez I., Garcia M.J., Alba J., Carranza F. (1987) Effects of pH and high density cultures on the silver catalyzed leaching of chalcopyrite. p 287–296. In: *Biohydrometallurgy Proc.Int.Symp. Warwick 1987*. Norris P.R., Kelly D.P. (eds.).

Palencia-Perez I., Dutrizac J.E. (1991) The effect of the iron content of sphalerite on its rate of dissolution in ferric sulfate and ferric chloride media. *Hydrometallurgy* **26**, 211–232.

Palmer B.R., Nebo C.O., Rau M.F., Fuerstenau M.C. (1981) Rate phenomena involved in the dissolution of chalcopyrite in chloride-bearing lixiviants. *Metall.Trans.B.* **12B**, 595–601.

Pinches A., Chapman J.T., Riele te W.A.M., Staden van M. (1987) The performance of bacterial leach reactors for the preoxidation of refractory gold bearing sulfide concentrates. p 329–344. In: *Biohydrometallurgy Proc.Int.Symp. Warwick 1987*. Norris P.R., Kelly D.P. (eds.) STL, Kew, Surrey, UK

Rath P.C., Paramguru R.K., Jena P.K. (1982) Kinetics of dissolution of zinc sulfide in aqueous ferric chloride solutions. Hydrometallurgy **6**, 219–225.

Rath P.C., Paramguru R.K., Jena P.K. (1988) Kinetics of dissolution of sulfide minerals in ferric chloride solution, 1: dissolution of galena, spalerite and chalcopyrite. *Trans.Instn.Min.Metall.* **97**, C150–158.

Rodriquez-Leiva M., Tributsch H. (1988) Morphology of bacterial leaching patterns by *Thiobacillus ferrooxidans* on synthetic pyrite. *Arch.Microbiol.* **149**, 401–405.

Sand W., Gerke T., Hallman R., Schippers A. (1995) Sulfur chemistry, biofilm, and the (in)direct attack mechanism - a critical evaluation of bacterial leaching. *Appl.Microbiol.Biotechnol.* **43**, 961–966.

Sanmugasunderam V., Branion R.M.R., Duncan D.W. (1985) A growth model for the continuous microbiological leaching of a zinc sulfide concentrate by

Thiobacillus ferrooxidans. *Biotechn.Bioeng.* **27**, 1173–1184.

Saxena N.N., Mandre N.R. (1992) Mixed control kinetics of copper dissolution for copper ore using ferric chloride. *Hydrometallurgy* **28** (1), 111–117.

Scherpenzeel D.A. van, Boon M., Ras C., Hansford G.S., Heijnen J.J. (1998) Kinetics of ferrous iron oxidation by *Leptospirillum* Bacteria in continuous cultures. *Biotechnol.Prog.* **14**, 425–433.

Schippers A., Josza P.G., Sand W. (1996) Sulfur chemistry in bacterial leaching of pyrite. *Appl.Microbiol.Biotechnol.* **62**, 3424–3432.

Su Huai. (1976) The kinetics on dissolution of sphalerite in ferric chloride solutions. In: *PhD thesis*, Univ. of Idaho, Diss. Inf. Serv., Xerox Univ. microfilms Int. no. 76–19,899.

Sulka L.B., Chaudhury G.R., Das R.P. (1990) Effect of silver ion on kinetics of biochemical leaching of chalcopyrite concentrate. Trans.Instn.Min.Metall. (Sect C. *Min.Process.Extr.Metall.*) **99**, C43–46.

Torma A.E., Walden C.C., Branion R.M.R. (1970) Microbiological leaching of zinc sulfide concentrate. *Biotechn.Bioeng.* **12**, 501–517.

Torma A.E., Walden C.C., Duncan D.W., Branion R.M.R. (1972) The effect of carbondioxide and particle surface on the microbiological leaching of a zinc sulfide concentrate. *Biotechn.Bioeng.* **14**, 777–786.

Torma A.E. (1977) The role of *Thiobacillus ferrooxidans* in hydrometallurgical processes. *Adv.Biochem.Eng.* **3**, 1–37.

Torma A.E., Sakaguchi H. (1978) Relation between the solubility product and the rate of metals sulfide oxidation by *Thiobacillus ferrooxidans*. *J.Ferment.Technol.* **56** (3), 173–178.

Torma A.E. and K. Bosecker (1982) Bacterial Leaching. Progress in *Indust.Microbiol.* **16**, 77–118.

Torma A.E. (1988) Leaching of metals. *Biotechnology*, ed. H.J. Rehm **6b**, 367–399.

Toro L., Paponetti B., Cantalini C. (1988) Precipitate formation in the oxidation of ferrous ions in the presence of *Thiobacillus ferrooxidans*. *Hydromettalurgy* **20**, 1–9.

Tributsch H., Bennett J.C. (1981) Semiconductor-electrochemical aspects of bacterial leaching. II. Survey of rate controlling sulfide properties. *J.Chem.Tech.Biotechnol.* **31**, 627–635.

Vargas T., Sanhueza A., Escobar B. (1993) Studies on the electrochemical mechanism of bacterial catalysis in pyrite dissolution. p 579–588. In: *Biohydrometallurgical Technologies Vol I. Proc.Int.Symp. Wyoming 1993.* Torma A.E., Wey J.E., Lakshmanan V.I. (eds.).

Verbaan B., Crundwell F.K. (1986) An electrochemical model for the leaching of a sphalerite (ZnS) concentrate. *Hydrometallurgy* **16**, 345–359.

Verbaan B., Huberts R. (1988) An electrochemical study of the bacterial leaching of synthetic Ni_3S_2. *Int.J.Min.Proc.* **24**, 185–202.

Warren G.W., Henein H., Zuo-Mei Jin (1985) Reaction mechanism for the ferric chloride leaching of sphalerite. *Metall.Trans.B.* **16B**, 715–724.

Wiersma C.L., Rimstidt J.D. (1984) Rates of reaction of pyrite ans marcasite with ferric iron at pH 2. *Geochim.Cosmochim.Act.* **48**, 85–92.

Wiertz J. (1993) Ferrous and sulfur oxidation by *Thiobacillus ferrooxidans*. p 463–471. In: *Biohydrometallurgical Technologies Vol.II. Proc.Int.Symp. Wyoming 1993*. Torma A.E., Apel M.L., Brierley C.L. (eds.)

Yukawa T., Vanselow D., Ralph B.J.; Rogers P.L. (1978) A kinetic model for the bacterial leaching of chalcopyrite ($CuFeS_2$) *J.Ferment.Technol.* **56** (1), 45–52.

Zeng Q.C., Allen C.C., Bautista R.G. (1986) Kinetic study of the oxidation of pyrite in aqueous ferric sulfate. *Ind.Eng.Proc.Des.Dev.* **25**, 308–317.

Zuo-Mei Jin, Warren G.W., Henein H. (1984) Reaction kinetics of the ferric chloride leaching of sphalerite - an experimental study. *Metall.Trans.B.* **15B**, 5–12.

6

Sulfur transformations during sewage transport

Thorkild Hvitved-Jacobsen and Per Halkjær Nielsen

6.1 INTRODUCTION

In this chapter we focus on the microbial sulfur cycle in sewers because of the potential problems of hydrogen sulfide formation, e.g. concrete and metal corrosion, health related aspects and odour. Such in-sewer problems were reported as early as over 50 years ago (Pomeroy and Bowlus 1946; Parker 1945).

During the 1960s and 1970s, several models were developed to predict the sulfide generation in pressure mains and gravity sewers. These models have been used for both sewer design and for implementation of appropriate control methods to prevent sulfide formation or its effects. It is not surprising that the main theoretical work and practical model development concerning sulfide formation and prevention in sewers took place in countries like the USA and Australia. These countries have extended sewer systems, transporting wastewater at elevated temperature (app. 20 - 30 °C). Also the UK contributed with important knowledge.

It is interesting to note that sulfide problems related to wastewater transport in sewers were not generally considered important in Denmark, a country with a temperate climate, until the mid 1980s. Traditionally, the Danish gravity sewers - and especially sewers in the combined sewered catchments - were designed with a relatively low dry weather flow compared with the capacity. Under such conditions, the re-aeration is sufficient to prevent the negative effects of sulfide, although its formation may take place within the sewer slimes. The establishment of the central wastewater treatment concept in Denmark changed this situation substantially, especially because of the many pressure mains established in the 1980s. Pumping stations and concrete gravity sewer systems located downstream from pressure mains with the anaerobic conditions of the wastewater were often totally corroded in several municipalities after 2 - 4 years of operation owing to sulfide problems. As a result, the first investigation in Denmark on sulfide formation was initiated by the National Environmental Protection Agency in 1986 (Miljøstyrelsen 1988). In addition, recent investigations have shown that considerable sulfide formation takes place in pressure mains during winter periods with wastewater temperatures around 6 - 8 °C (Nielsen *et al.* 1998). Furthermore, the negative impact of sulfide on floc stability in activated sludge treatment plants has been documented (Nielsen and Keiding 1998). These examples indicate the worldwide importance and the complexity of sulfide formation in sewers.

Evaluation and prediction of sulfide problems are complicated and susceptible to great variability. Therefore, there is a need to review existing knowledge and to develop new, reliable methodologies for prediction and control. A major aim of this chapter is therefore to give an overview of the sulfur-related processes in sewers, and methods to predict and prevent hydrogen sulfide formation.

6.2 A GENERAL OVERVIEW OF SULFIDE FORMATION AND EXCHANGE IN SEWERS

Hydrogen sulfide formation is a microbial process taking place under anaerobic conditions. When dissolved oxygen (DO) and nitrate in wastewater of sewer systems are depleted, sulfate-reducing bacteria will use sulfate as an electron acceptor in their use of wastewater organic matter (organic carbon) as substrate. These processes result in the formation of sulfide (hydrogen sulfide). Schematically, and without taking into account the formation of low molecular organics, the sulfide producing process is:

$$SO_4^{2-} + \text{ organic carbon} \rightarrow HCO_3^- \text{ (carbon dioxide)} + H_2S$$

Sulfide production in gravity sewers takes place mainly in slow-flowing (< 30 cm/s), large pipes with insufficient reaeration and at relatively high temperatures (> 15 - 20 °C). The presence of sulfide in bulk water of gravity sewers in Northern Europe is rare owing to the moderate temperatures even during summer periods and because of a tradition for design of relatively large pipes improving the reaeration. Although sulfide may be produced within the biofilms and the sewer sediments, aerobic conditions in the bulk water phase typically prevents occurrence of sulfide in the wastewater. In the warm climates, e.g. Central and Southern Europe, the Southern States of the USA and in Australia, sulfide problems under gravity sewer conditions are common (Meyer and Hall 1979; ASCE 1989). In pressure mains, where wastewater is pumped in filled pipes, sulfide may occur even at rather low temperatures when the residence time of the wastewater exceeds 0.5 - 2 hours (Hvitved-Jacobsen *et al.* 1995; Nielsen *et al.* 1998).

Sulfide build-up during wastewater transportation in a sewer is the major indicator for all types of sulfide problems. Sulfide concentrations of 0.5, 3 and 10 mgS/l may be considered as low, moderate and high, respectively, in terms of problems that are typically reported. In countries with long pressure mains or high temperatures, significantly higher concentrations than 10 mgS/l are reported (e.g. Thistlethwayte 1972; Pomeroy and Parkhurst 1977).

The sulfate reduction and corresponding sulfide production in sewer systems takes place in the biofilms and in the sewer sediments, although some activity may occur in the bulk water due to e.g. detached biofilm particles (USEPA 1974; Nielsen 1987; Kuhl and Jorgensen 1992; Schmitt and Seyfried 1992; Norsker *et al.* 1995). The problems in terms of corrosion, odour and health are primarily related to the release of hydrogen sulfide from the wastewater to the atmosphere.

6.3 FACTORS AFFECTING THE FORMATION OF SULFIDE

Anaerobic conditions, i.e. the absence of DO and nitrate, are required for sulfate reduction. Under such conditions, the most important factors determining sulfate reduction rates will be discussed. These factors are often included in empirical models. Different types of model are typically developed for gravity sewers and pressure mains.

6.3.1 Presence of sulfate

Sulfate is typically found in all types of wastewater in concentrations greater than 5 - 15 mgS/l, i.e. in concentrations that are not limiting for sulfide formation in relatively thin biofilms (Nielsen and Hvitved-Jacobsen 1988). In sewer sediments, however, where sulfate may penetrate into the deeper sediment layers, the potential for sulfate reduction may increase with increasing sulfate concentration in the bulk water phase. Under specific conditions, e.g. in the case of industrial wastewater, it is important that sulfur components (e.g. thiosulfate and sulfite) other than sulfate may act as sulfur sources for sulfate reducing bacteria (Nielsen 1991).

6.3.2 Quantity and quality of biodegradable organic matter

Biodegradable organic matter is available in wastewater as a substrate for sulfate reduction. However, in wastewater from, for example, food industries with a relatively high concentration of readily biodegradable organics preferred by the sulfate reducing bacteria, the sulfate reduction rate may be higher than in wastewater from households. However, also in domestic wastewater, COD may be high in certain areas owing to a shortage or reuse of water, leading to a higher potential for sulfide formation. Several specific organics, e.g. formate, lactate and ethanol, have been identified as particularly suitable substrates for sulfate reducing bacteria (Nielsen and Hvitved-Jacobsen 1988).

6.3.3 Temperature

The temperature dependency of sulfate reduction for sulfate reducing bacteria is high corresponding to a temperature coefficient of about 1.13 per degree Celsius, i.e. a change in the rate with a factor Q_{10} = 3.0 - 3.5 per 10 °C of temperature increase. Because diffusion of substrate into biofilms or sediments is typically limiting sulfide formation, the temperature coefficient is reduced to about 1.03 (Nielsen *et al.* 1998).

6.3.4 pH

Sulfate reducing bacteria mainly exist between pH 5.5 and 9. A significant inhibition of the sulfate reducing bacteria will, however, not take place below a pH of about 10.

6.3.5 Area:volume ratio in pressure mains

Sulfide is primarily produced in the biofilms. The corresponding water phase concentration of the sulfide therefore relates to the area:volume (*A:V*) ratio of a sewer pipe. Relatively low sulfide concentrations in wastewater from large diameter pipes therefore exist compared with small diameter pipes.

6.3.6 Flow velocity in pressure mains

The potential production of sulfide depends on the biofilm thickness. If the flow velocity in the pipe is 0.8 - 1 m/s, the corresponding biofilm is rather thin, typically 100 - 300 μm. However, high velocities also reduce the thickness of the diffusional boundary layer and thereby the resistance against transport of substrates and products across the biofilm:water interphase.

6.3.7 Anaerobic residence time in pressure mains

The anaerobic residence time of the wastewater during transport is a factor that affects the level of sulfide concentration in the wastewater. The residence time is determined by the magnitude of wastewater inflow compared with the volume (length and diameter) of the pipe. The level of sulfide formation in a given pipe is therefore subject to the diurnal variation of the inflowing wastewater and to the precipitation pattern in combined sewered catchments (Figure 6.1).

6.4 PREDICTION OF SULFIDE FORMATION IN PRESSURE SEWERS

Rather simple empirical equations (Table 6.1) have been developed for the prediction of the hydrogen sulfide formation rate in pressure pipes (in units of $gS/(m^2 \cdot h)$). Equations 2 - 5 imply that the sulfate concentration in the wastewater is high and not limiting for sulfide formation. This is typically the case for sulfate concentrations exceeding 5 - 15 mgS/l (Nielsen and Hvitved-Jacobsen 1988). Equations 1 and 2 correspond to the expected maximum formation rates under "optimal" conditions. Equations 4 and 5 take into consideration the quality of the wastewater in terms of its biodegradability.

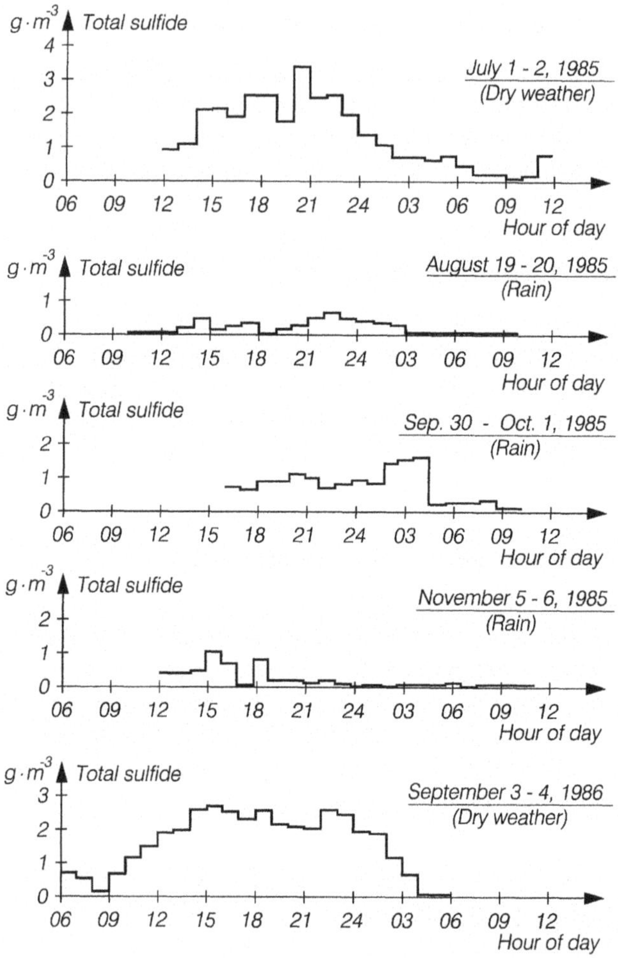

Figure 6.1. Diurnal variation in the sulfide concentration of a 4 km pressure main located in North Jutland, Denmark. The typical variation of the residence time during a dry weather period is between 6 and 14 hours corresponding to the daily wastewater flow. During a rain event, the residence time may be reduced to about 1 hour.

Table 6.1. Empirical equations to predict the sulfide production rate in pressure sewers in units of g S/(m^2 ·h)*

Equation	Reference
1. $r_a = 0.5\ 10^{-3}\ u\ BOD^{0.8}\ S_{SO4}{}^{0.4}1.139^{T-20}$	Thistlethwayte (1972)
2. $r_a = 0.228\ 10^{-3}\ COD\ 1.07^{T-20}$	Boon and Lister (1975)
3. $r_a = 1\ 10^{-3}\ BOD\ 1.07^{T-20}$	Pomeroy and Parkhurst (1977)
4. $r_a = k^{**}\ (COD_S\text{-}50)^{0.5}\ 1.07^{T-20}$	Hvitved-Jacobsen *et al.* (1988)
5. $r_a = a^{***}\ (COD_S\text{-}50)^{0.5}\ 1.03^{T-20}$	Nielsen *et al.* (1998)

* Can be expressed in units of g/(m^3·h) by division with the hydraulic radius (V/A)

** k = 0.0015 typical Danish domestic wastewater without industrial sewage
k = 0.003 wastewater from mixed domestic and industrial (foodstuff) sources
k = 0.006 wastewater mainly from foodstuff industries
see Figure 6.2, curve A, B and C

*** a = 0.001-0.002 typical Danish domestic wastewater without industrial sewage
a = 0.003-0.006 wastewater from mixed domestic and industrial (foodstuff) sources
a = 0.007-0.010 wastewater with biodegradable organic matter from mainly foodstuff industries

where: BOD = biochemical oxygen demand (mgO_2/l)
COD = chemical oxygen demand (mgO_2/l)
COD_s = soluble COD (mgO_2/l)
S_O = dissolved oxygen concentration (mgO_2/l)
S_{SO4} = sulfate concentration (mgS/l)
k and a = rate constants (-)
T = temperature (°C)
s = slope (m/m)
u = flow velocity (m/s)

6.5 PREDICTION OF SULFIDE FORMATION IN GRAVITY SEWERS

In addition to the DO consuming processes in wastewater and biofilm, the magnitude of reaeration determines whether sulfide problems in gravity sewers exist or not.

Typically, hydrogen sulfide does not exist in wastewaters from gravity sewers if the DO concentration is higher than 0.2 - 0.5 mg/l (USEPA 1974). If the DO concentration becomes lower, either because of a high DO consumption rate or because of reduced reaeration, equation 3 in Table 6.1 can be used if $r_a <$ about 10^{-3} gS/(m^2 · h) (USEPA 1974). Also, the other equations in Table 6.1 may result in a reasonable prediction of sulfide formation in gravity sewers at such low DO concentrations. However, in addition to sulfide formation in the water phase, release of H_2S into the sewer atmosphere and a possible oxidation of sulfide must be taken into account (Pomeroy and Parkhurst 1977; Tchobanoglous 1981; Wilmot *et al.* 1988).

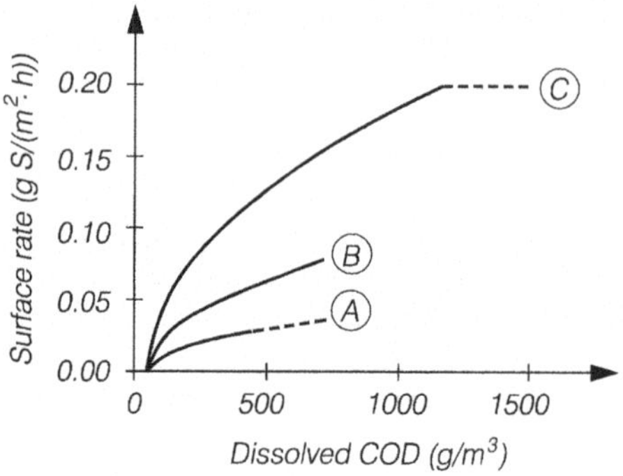

Figure 6.2. Hydrogen sulfide formation rates in pressure mains (Hvitved-Jacobsen *et al.* 1988).

A rather simple evaluation of sulfide problems in gravity sewers with a diameter smaller than about 0.6 m is based on the socalled *Z*-formula, (Pomeroy and Parkhurst 1977; ASCE and WPCF 1982):

$$Z = E\text{BOD}\,(s^{0.5}\,Q^{0.33})^{-1}\,(P/b)$$

where: EBOD = $\text{BOD}_5\,1.07^{T-20}$ (mg/l)
T = temperature (°C)
s = slope (-)
Q = flow (ft³/s, "cubic feet per second"), 1 m³ = 35,314 ft³
P = wet perimeter (m)
b = width at the water surface (m)

The size of the calculated *Z*-value determines the estimated magnitude of the sulfide problem (Table 6.2).

Table 6.2. Evaluation of sulfide problems in gravity sewers according to the Z-formula

Z-formula	Estimated magnitude of the sulfide problem
$Z<5{,}000$	Problems rather infrequent
$5{,}000<Z<10{,}000$	Risk of sulfide problems
$Z>10{,}000$	Risk of sulfide problems frequent

Table 6.3. Examples of sulfide problems in gravity sewers as estimated by the Z-formula; cf. text

Slope, s (%)	Velocity, u (m/s)	Flow, Q (l/s)	Load, PE	Z-value
0.03	0.32	45	19,000	19,370
0.1	0.58	82	35,000	8,740
0.3	1.00	141	61,000	4,220

As an example, a halffilled 0.6 m diameter gravity sewer, wastewater with EBOD = 250 mg/l, and slopes of s = 0.03, 0.1 and 0.3%, respectively, results in Z-values as shown in Table 6.3. Z-values from Table 6.3 indicate that significant sulfide problems may easily occur in trunk and intercepting gravity sewers with moderate slope.

6.6 AN INTEGRATED AEROBIC/ANAEROBIC APPROACH FOR PREDICTION OF SULFIDE FORMATION IN SEWERS

The empirical models for prediction of sulfide formation in pressure sewers shown in Table 6.1 all include the concentration of organic matter in the wastewater as a central parameter. Especially equations 4 and 5 described by Hvitved-Jacobsen *et al.* (1988) and Nielsen *et al.* (1998), respectively show the importance of biodegradable organics for sulfide formation by including the difference between dissolved COD and what is estimated as soluble and non-biodegradable. Furthermore, these two models have different levels of sulfide formation rates depending on the source of the wastewater. Laboratory investigations performed by Nielsen and Hvitved-Jacobsen (1988) also illustrate the importance of the organic substrate for the magnitude of the sulfate reduction. Based on field investigations, Nielsen *et al.* (1998) concluded that the sulfide production rates were strongly dependent on wastewater quality. However, they realized that there was no evidence to change dissolved COD in the model as an indicator of wastewater quality unless more fundamental wastewater processes were taken into account.

There are several reasons considering the microbial transformations of organic matter in sewers. The subsequent wastewater treatment and impacts on receiving waters from combined sewer overflows are important examples. An integrated aerobic and anaerobic process concept and a corresponding conceptual model were therefore developed for prediction of the heterotrophic carbon transformations in sewers in this respect. This concept and model have been further developed to include sulfur microbial transformations (Figure 6.3).

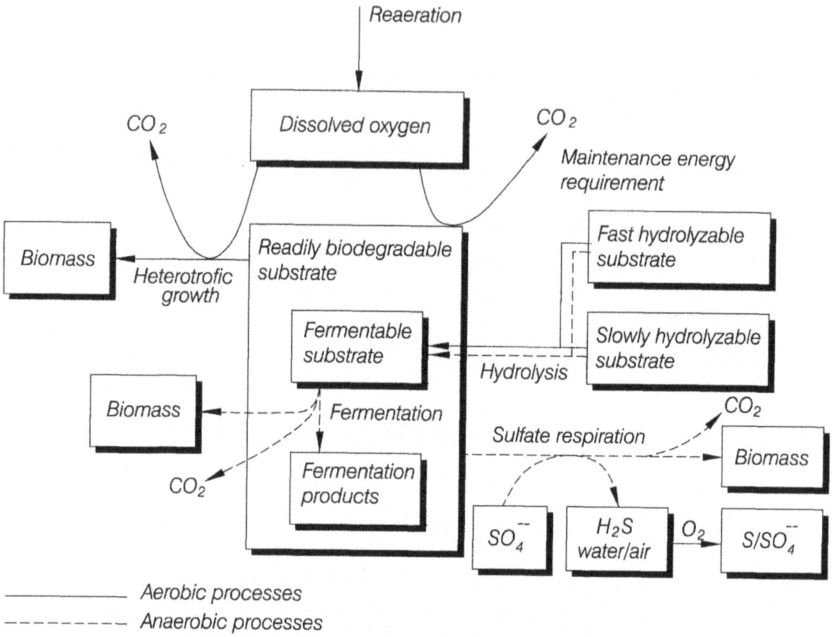

Figure 6.3. An integrated aerobic and anaerobic concept for transformations of organic matter and sulfur components of wastewater in sewers.

The concept depicted in Figure 6.3 and the corresponding model for carbon transformations were gradually developed and are described in Bjerre *et al.* (1995; 1998), Hvitved-Jacobsen *et al.* (1998a; 1998b), Tanaka *et al.* (1998) and Tanaka and Hvitved-Jacobsen (1998; 1999). The model integrates processes in sewers under aerobic and anaerobic conditions. In the water phase it includes growth of the heterotrofic biomass, growth and non-growth related oxygen consumption, hydrolysis and fermentation. Concerning the biofilm processes, e.g. sulfate respiration, relatively simple descriptions in terms of surface-based transformations are used. The improvement for prediction of sulfide formation compared to the empirical equations shown in Table 6.1 is that the wastewater quality in terms of its biodegradability of the organic matter is more directly expressed and included. Furthermore, sulfide formation can be predicted under changing aerobic/anaerobic conditions in both gravity sewers and pressure mains. The conceptual model is outlined in Tables 6.4 - 6.7. Further details are found in Hvitved-Jacobsen *et al.* (1998a; 1998b) and Tanaka and Hvitved-Jacobsen (1999).

Table 6.4. Characteristic values of COD components and dissolved oxygen at an upstream location of an intercepting sewer. Wastewater includes particulate (X) and soluble (S) fractions

Component	Description	Characteristic value	Unit
X_{Bw}	Heterotrofic active biomass in the water phase	20 - 100	$gCOD/m^3$
X_{Bf}	Heterotrofic active biomass in the biofilm	~ 10	$gCOD/m^2$
X_{S1}	Hydrolysible substrate, fast biodegradable	50 - 100	$gCOD/m^3$
X_{S2}	Hydrolysible substrate, slowly biodegradable*	300 - 450	$gCOD/m^3$
S_F	Fermentable substrate	0 - 40	$gCOD/m^3$
S_A	Fermentation products (i.e. VFAs)	0 - 20	$gCOD/m^3$
S_S	Readily biodegradable substrate (S_F+S_A)	0 - 40	$gCOD/m^3$
S_O	Dissolved oxygen	0 - 4	gO_2/m^3
COD	Total COD	about 600	$gCOD/m^3$

* Includes very slowly biodegradable and inert organic matter.

Equation g in Table 6.4, which was selected for the simulation of the sulfide production rate, originates from Nielsen *et al.* (1998). The dissolved COD used in this empirical model (Table 6.1, equation 5) as an indicator of the wastewater quality does not exist among the COD components of the sewer process model shown in Table 6.4. A substitute for the term (COD_s-50) in the original model (Table 6.1), interpreted as the biologically active COD components for sulfate production, must therefore be found. It is wellknown fact that sulfate reducing bacteria use readily biodegradable organic matter like alcohols, lactate, pyruvate and some aromatic substrates, but generally not directly higher carbohydrates (see Chapter 22). Based on studies in a pilot pressure sewer and field investigations, Tanaka *et al.* (1998) and Tanaka and Hvitved-Jacobsen (1999) found that, among different options, the term (S_s + X_{s1}) was an acceptable substitute for (COD_s-50) (Figure 6.4). Theoretically, experimentally, and from a modelling point of view, there are good reasons for this conceptual substitution.

The aerobic/anaerobic model concept presented for prediction of sulfide formation in sewers has further applications. The model is developed to simulate organic matter transformations and thereby the change in biodegradability during the wastewater transport. The model is therefore also a tool for prediction of quality changes of wastewater in sewers which

Table 6.5. Integrated aerobic and anaerobic process model concept for transformations of organic matter and sulfur components of wastewater in sewers. Symbols are defined in Table 6.5-6.7

	S_F	S_A	X_{S1}	X_{S2}	X_{Bw}	S_{H2S}	$-S_O$	Process rate
Aerobic growth in bulk water	$-1/Y_{Hw}$				1		$(1-Y_{Hw})/Y_{Hw}$	Eq. a
Aerobic growth in biofilm	$-1/Y_{Hf}$				1		$(1-Y_{Hf})/Y_{Hf}$	Eq. b
Maintenance energy requirement	−1						1	Eq. c
Aerobic hydrolysis, fast	1		−1					Eq. d, $n = 1$
Aerobic hydrolysis, slow	1			−1				Eq. d, $n = 2$
Anaerobic hydrolysis, fast	1		−1					Eq. e, $n = 1$
Anaerobic hydrolysis, slow	1			−1				Eq. e, $n = 2$
Fermentation	−1	1						Eq. f
Hydrogen sulfide production						1		Eq. g
Reaeration							−1	Eq. h

a: $\mu_H (S_F + S_A)/(K_S + (S_F + S_A))\ S_O/(K_O + S_O)\ X_{Bw}\ \alpha_w^{(T-20)}$

b: $k_{1/2}\ S_O^{0.5}\ Y_{Hf}/((1 - Y_{Hf})\ A/V\ (S_F + S_A)/(K_{Sf} + (S_F + S_A))\ \alpha_f^{(T-20)}$

c: $q_m\ S_O/(K_O + S_O)\ X_{Bw}\ \alpha_w^{(T-20)}$

d: $k_{hn}\ (X_{Sn}/X_{Bw})/(\ K_{Xn} + X_{Sn}/X_{Bw})\ S_O/(K_O + S_O)\ (X_{Bw} + \varepsilon X_{Bf}\ A/V)\ \alpha_w^{(T-20)}$

e: $\eta_{fe}\ k_{hn}\ (X_{Sn}/X_{Bw})/(\ K_{Xn} + X_{Sn}/X_{Bw})\ K_O/(S_O + K_O)\ (X_{Bw} + \varepsilon X_{Bf}\ A/V)\ \alpha_w^{(T-20)}$

f: $q_{fe}\ S_F/(K_{fe} + S_F)\ K_O/(S_O + K_O)\ (X_{Bw} + \varepsilon\ X_{Bf}\ A/V)\ \alpha_w^{(T-20)}$

g: $k_{H2S}\ 10^{-3}\ (S_F + S_A + X_{S1})^{0.5}\ \alpha_S^{(T-20)}\ 24\ A/V$

h: $K_L a\ \ (S_{OS} - S_O)$, where $K_L a = 0.86\ (1 + 0.20\ F^2)\ (s\ u)^{3/8}\ d_m^{-1}\ \alpha_r^{(T-20)}$

Table 6.6. Example of model parameters used in the sewer process model outlined in Table 6.4

Symbol	Definition	Value	Unit
μ_H	Maximum specific growth rate for heterotrofic biomass	6.7	d^{-1}
Y_{Hw}	Suspended biomass yield constant for heterotrofics	0.55	gCOD/gCOD
K_S	Saturation constant for readily biodegradable substrate	1.0	$gCOD/m^3$
K_O	Saturation constant for dissolved oxygen (DO)	0.05	gO_2/m^3
α_w	Temperature coefficient in the water phase	1.07	-
q_m	Maintenance energy requirement rate constant	1.0	d^{-1}
$k_{1/2}$	1/2 order rate constant	4	$gO_2^{0.5}m^{-0.5}d^{-1}$
Y_{Hf}	Biofilm yield constant for heterotrofic biomass	0.55	gCOD/gCOD
K_{Sf}	Saturation constant for readily biodegradable substrate	5	$gCOD/m^3$
ε	Efficiency constant for the biofilm biomass	0.15	-
α_f	Temperature coefficient in the biofilm	1.05	-
k_{h1}	Hydrolysis rate constant, fraction 1 (fast)	5*	d^{-1}
k_{h2}	Hydrolysis rate constant, fraction 2 (slow)	0.5*	d^{-1}
K_{X1}	Saturation constant for hydrolysis, fraction 1	1.5*	gCOD/gCOD
K_{X2}	Saturation constant for hydrolysis, fraction 2	0.5*	gCOD/gCOD
η_{fe}	Anaerobic hydrolysis reduction factor	0.14	-
q_{fe}	Maximum rate for fermentation	3	d^{-1}
K_{fe}	Saturation constant for fermentation	20	gCOD/gCOD
k_{H2S}	Hydrogen sulfide production rate constant	2 (3)	gS^{2-}/m^2
α_S	Temperature coefficient for hydrogen sulfide production	1.030	-

w: water phase; f: biofilm

* determined based on oxygen uptake rate (OUR) measurements on a wastewater sample.

Table 6.7. Reaeration, flow and system characteristics used in the sewer process model outlined in Table 6.4

Symbol	Definition	Value	Unit
K_{La}	Oxygen transfer coefficient		d^{-1}
T	Temperature		°C
S_{OS}	Dissolved oxygen saturation concentration at T °C		gO_2/m^3
F	Froude number = $u/(g\ d_m)^{0.5}$		-
u	Mean flow velocity		m/s
g	Acceleration due to gravity	9.81	m/s^2
s	Slope		m/m
d_m	Hydraulic mean depth		m
A/V	Ratio of biofilm area to bulk water volume		m^{-1}
α_r	Temperature coefficient for reaeration	1.024	-

Site specific parameters are not indicated with a value.

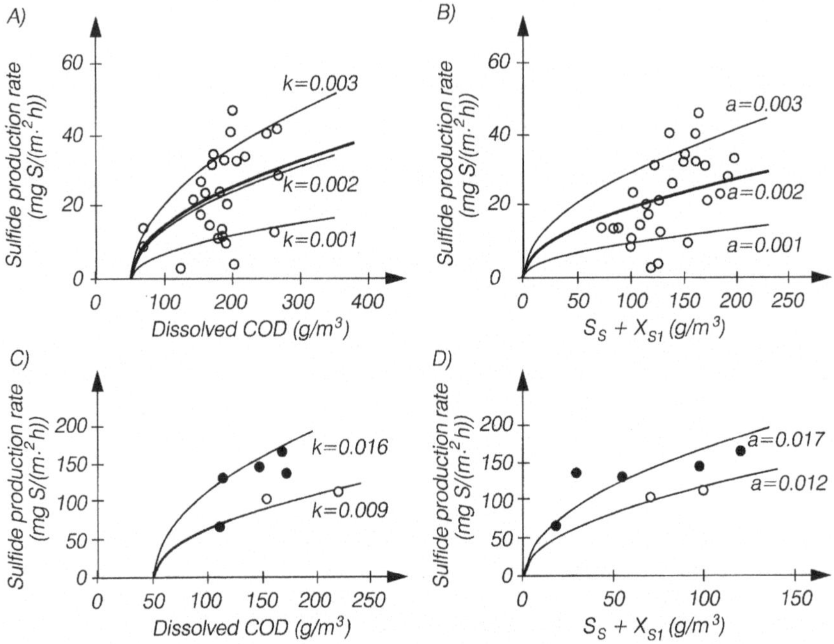

Figure 6.4. Sulfide production rates in sewers versus wastewater quality characteristics. The results originate from pilot plant studies (A and B) and field investigations (C and D) in Japan, Kawasaki town (o) and Oga city (•).

interacts with the successive treatment processes and the water quality processes taking place in watercourses during combined sewer overflows (Hvitved-Jacobsen and Vollertsen 1998).

6.7 EFFECTS OF SULFIDE FORMATION IN SEWERS

6.7.1 Release of hydrogen sulfide gas from wastewater to the atmosphere

Release of hydrogen sulfide gas from wastewater into the overlying atmosphere is of fundamental importance for effects of sulfide-like corrosion and odour problems. As long as the sulfide remains in the water phase, these effects do not occur.

It is important to note that the molecular form H_2S, and not the dissociated forms of sulfide, i.e. HS^- and S^{2-}, can be released from the water phase to the atmosphere. Increase of the wastewater pH will therefore at a

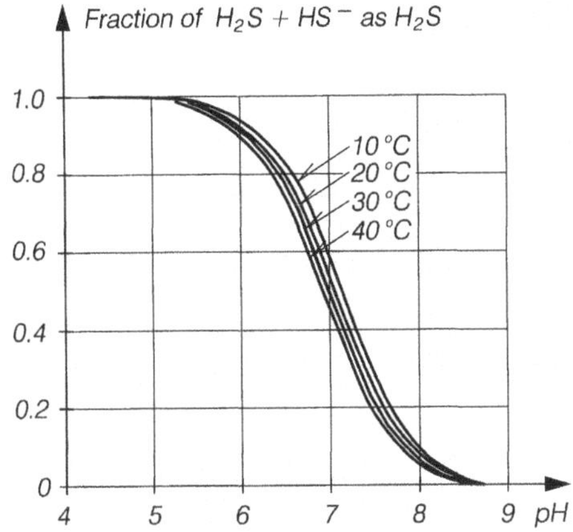

Figure 6.5. Equilibria for H_2S in aqueous solution (Melbourne and Metropolitan Board of Works 1989).

constant total sulfide concentration in the wastewater reduce the hydrogen sulfide concentration in the overlaying atmosphere.

Hydrogen sulfide (H_2S) is a weak acid, which dissociates as follows:

$$H_2S \rightleftharpoons H^+ + HS^- \rightleftharpoons 2H^+ + S^{2-}$$

In processes under sewer conditions only the equilibrium between H_2S and HS^- is of importance. Much S^{2-} will appear only if pH > about 12. The pH and temperature dependency of H_2S occurrence in a water phase is shown in Figure 6.5.

In a closed system, an equilibrium between the molecular form (H_2S) in the water phase and the overlaying atmosphere can be established. The theory is described by Henry's law, which in reality will depend on pH and temperature (Figure 6.6).

The curves shown in Figure 6.6 are essential for the evaluation of the potential risk for microbial corrosion and odour problems. Turbulence in the wastewater for establishment of the equilibrium shown in Figure 6.6 is important. Release of H_2S into the atmosphere at e.g. pumping stations and hydraulic jumps may be rather high and, therefore, cause a corresponding high corrosion. Matos and De Sousa (1992) have developed a model for prediction of hydrogen sulfide gas build-up in the atmosphere of a gravity sewer pipe.

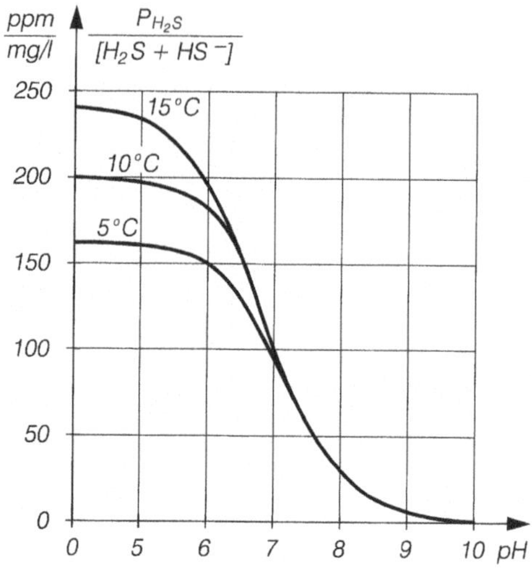

Figure 6.6. Partial pressure of H_2S measured in ppm on a volumetric basis in the atmosphere in equilibrium with a water phase of sulfide (H_2S and HS^-). The curves show the equilibrium H_2S-concentration in the atmosphere per unit concentration in the water phase.

6.7.2 Health aspects

Several health-related problems are potentially associated with the occurrence of hydrogen sulfide in the atmosphere of sewer systems. Hydrogen sulfide can be detected by its smell at very low concentrations down to about 0.1 ppm, and increasing health problems can be observed at increasing concentrations. Concerning the potential risks of hydrogen sulfide in sewer systems, Table 6.8 (from US National Research Council 1979) can be compared with Figure 6.6.

Table 6.8. Odour and health related effects of hydrogen sulfide in the atmosphere

Odour or human effect	Concentration in atmosphere (ppm)
Odour limit	0.1 - 0.2
Unpleasant smell	3 - 5
Recommended criterion per workday	10
Effect on eyes	50 - 100
Inactivation of smell	150 - 250
Serious water accumulation in lungs	300 - 500
Deadly impact on nervous system	500 - 1000
Immediate cessation of respiration	1000 - 2000

The density of hydrogen sulfide is slightly higher than that of the atmosphere (relative density 34/29). Therefore, hydrogen sulfide has a tendency to accumulate in, for example, pumping stations and manholes. As hydrogen sulfide typically will not be detected by its smell if the concentration > about 200 ppm, instruments or alarm systems for its monitoring must generally be used when working in sewer systems.

6.7.3 Concrete corrosion

The principles of the concrete corrosion process are depicted in Figure 6.7. More details about concrete corrosion are described in Chapter 22, whereas metal corrosion is described in Chapter 21. Those areas where corrosion may be especially noticeable are often close to where hydrogen sulfide is released, e.g. close to a permanent (anaerobic) water surface or turbulent points, or at points where moisture may be easily adsorbed. It is important to notice that corrosion does not take place under the water surface.

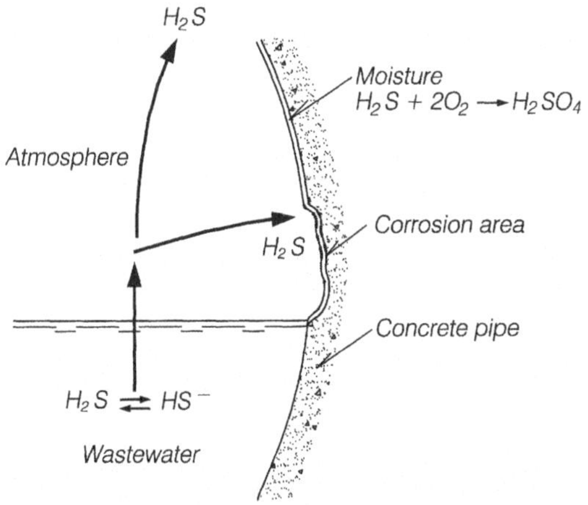

Figure 6.7. Principle for release of hydrogen sulfide and concrete corrosion.

When H_2S has been released into the atmosphere, the second step of the corrosion process is absorption of hydrogen sulfide at moist surfaces. After absorption the following microbial process carried out by aerobic bacteria of the *Thiobacillus* family may take place:

$$H_2S + 2O_2 \rightarrow H_2SO_4$$

Species of the *Thiobacillus* family have been observed to be active at solutions with up to about 7% of sulfuric acid (Milde *et al.* 1983). Although hydrogen sulfide may corrode, for example, metals by producing metal sulfides, the main corrosion problem is related to attacks on concrete surfaces caused by the sulfuric acid produced.

If the rate of sulfuric acid production is low, almost all acid produced on concrete surfaces will react with the cement in the concrete and only loosely bound, inert components are left. If the rate of sulfuric acid formation is high, part of the acid may be washed away before further reaction with the concrete takes place. Associated with acid corrosion is the formation of sulfate ions, which may cause sulfate attack on concrete surfaces. The principle of the concrete corrosion processes is given by the following equation:

$$\underset{}{H_2SO_4} + \underset{\text{(cement)}}{CaCO_3} \rightarrow H_2O + CO_2 + \underset{\text{(gypsum)}}{CaSO_4}$$

Based on this formula and the density of concrete, a simple formula for the corrosion rate c is (Melbourne and Metropolitan Board of Works 1989):

$$c = 11.4 f/A$$

where c = corrosion rate (mm/year).
f = rate of hydrogen sulfide absorption on a concrete surface (g/m^2h).
A = equivalent alkalinity of the concrete material in units of g$CaCO_3$ per g of concrete.

In cases of severe concrete corrosion in sewer systems, a corrosion rate of 1-5 mm/year may be observed.

6.8 METHODS FOR CONTROL OF SULFIDE FORMATION

A basic requirement is that a gravity sewer should be designed not to allow (serious) sulfide formation. A slope of the sewer pipe resulting in water velocities that prevent deposition of solids and enhance a high re-aeration is important.

In pressure mains, the risk for sulfide formation is much greater, and in many systems hydrogen sulfide control may be needed. Table 6.9 outlines methods that may be used. Principles for these control methods are

Table 6.9. Methods for control of sulfide formation in wastewater transport systems

General principle of the method	Specific measure
Prevention of sulfate reducing conditions	Addition to the wastewater of: - air - pure oxygen - nitrate
Prevention of adverse effects	Chemical precipitation of sulfides by: - iron (II) sulfate - iron (III) chloride
Methods aiming at specific effects on the biological system	- alkaline substances increasing pH - chlorine - hydrogen peroxide - ozone
Mechanical methods	- flushing - ball for detachment of biofilm

described in Melbourne and Metropolitan Board of Works (1989), ASCE (1982), ASCE (1989), USEPA (1974) and USEPA (1985). In the final selection of an appropriate corrosion abatement method, the potential negative effects, e.g. on the successive wastewater treatment, must also be considered. Further details are outlined in Chapter 22.

6.9 REFERENCES

ASCE, American Society of Civil engineers and WPCF, Water Pollution Control Federation (1982) Gravity sanitary sewer design and construction, ASCE manuals and reports on engineering practice no. **60** or WPCF manual of practice no.**FD-5**, pp. 275.

ASCE, American Society of Civil Engineers (1989) Sulfide in wastewater collection and treatment systems, ASCE manuals and reports on engineering practice no. **69**, pp. 324.

Bjerre, H.L., T. Hvitved-Jacobsen, B. Teichgräber and D. te Heesen (1995) Experimental procedures characterizing transformations of wastewater organic matter in the Emscher river, Germany. *Wat. Sci.Tech.* **31** (7), 201-212.

Bjerre, H.L., T. Hvitved-Jacobsen, B. Teichgräber and S. Schlegel (1998) Modelling of aerobic wastewater transformations under sewer conditions in the Emscher river, Germany, *Wat. Env. Res.* **70** (6), 1151-1160.

Boon, A.G. and A.R. Lister (1975) Formation of sulphide in rising main sewers and its prevention by injection of oxygen, *Prog. Wat. Tech.* **7** (2), 289-300.

Hvitved-Jacobsen, T., B. Jütte, P.H. Nielsen and N.Aa. Jensen (1988) Hydrogen sulphide control in municipal sewers. In: H.H. Hahn and R. Klute (eds.), Pretreatment in Chemical Water and Wastewater Treatment, Springer-Verlag, 239-247.

Hvitved-Jacobsen, T., K. Raunkjær and P.H. Nielsen (1995) Volatile fatty acids and sulfide in pressure mains. *Wat. Sci. Tech.* **31**(7), 169-179.

Hvitved-Jacobsen, T., J. Vollertsen and P.H. Nielsen (1998a) A process and model concept for microbial wastewater transformations in gravity sewers. *Wat. Sci. Tech.* **37** (1), 233-241.

Hvitved-Jacobsen, T., J. Vollertsen and N. Tanaka (1998b) Wastewater quality changes during transport in sewers - an integrated aerobic and anaerobic model concept for carbon and sulfur microbial transformations, *Wat. Sci. Tech.* **38** (10), 257-264 (read text pp. 249-256) or *Wat. Sci. Tech.* **39** (2), 242-249.

Hvitved-Jacobsen, T. and J. Vollertsen (1998) An intercepting sewer from Dortmund to Dinslaken, Germany, report submitted to the Emschergenossenschaft, Essen, Germany, pp. 35.

Kuhl, M. and B. B. Jorgensen (1992) Microsensor measurements of sulfate reduction and sulfide oxidation in compact microbial communities of aerobic biofilms. *Appl. Environ. Microbiol.* **58,** 1164-1174.

Matos, J.S. and E.R. de Sousa (1992) The forecasting of hydrogen sulphide gas build-up in sewerage collection systems, *Wat. Sci. Tech.* **26** (3-4), 915-922.

Melbourne and Metropolitan Board of Works (1989) Hydrogen sulphide control manual - septicity, corrosion and odour control in sewerage systems, Technological Standing Committee on Hydrogen Sulphide Corrosion in Sewerage Works, vol. 1 and 2.

Meyer, W.J. and G.H. Hall (1979) Prediction of sulfide generation and corrosion in concrete gravity sewers: A case study, J.B. Gilbert & Associates, A Division of Brown and Caldwell, Sacramento, Ca., USA.

Milde, K., W. Sand, W. Wolff and E. Bock (1983) Thiobacilli of the corroded concrete walls of the Hamburg sewer system, *J. Gen. Microbiol.* **129**, 1327-1333.

Miljøstyrelsen (1988) Hydrogen sulfide formation and control in pressure mains, The Danish Environmental Protection Agency, project report no. 96, pp. 109, (in Danish).

Nielsen, P.H. (1987) Biofilm dynamics and kinetics during high-rate sulfate reduction under anaerobic conditions. *Appl. Environ. Microbiol.* **53**, 27-32.

Nielsen, P.H. and T. Hvitved-Jacobsen (1988) Effect of sulfate and organic matter on the hydrogen sulfide formation in biofilms of filled sanitary sewers, *J. WPCF* **60**, 627-634.

Nielsen P.H. (1991) Sulfur sources for hydrogen sulfide production in biofilm from sewer systems. *Wat. Sci. Tech.* **23**, 1265-1274.

Nielsen, P.H. and K. Keiding (1998) Disintegration of activated sludge flocs in the presence of sulfide. *Wat. Res.* **32**, 313-320.

Nielsen, P.H., K. Raunkjaer and T. Hvitved-Jacobsen (1998) Sulfide production and wastewater quality in pressure mains. *Wat. Sci. Tech.* **37** (1), 97-104.

Norsker, N.H., P.H. Nielsen, and T. Hvitved-Jacobsen (1995) Influence of oxygen on biofilm growth and potential sulfate reduction in gravity sewer biofilm. *Wat. Sci. Tech.* **31** (7), 159-167.

Parker, C.D. (1945) Isolation of a species of bacterium associated with the corrosion of concrete exposed to atmospheres containing hydrogen sulphides, *Aust. J. Exptl. Biol. & Med. Sci.* **23**.

Pomeroy, R.D. and F.D. Bowlus (1946) Progress report on sulphide control research, *Sewage Works Journal* **18** (4).

Pomeroy, R.D. and J.D. Parkhurst (1977) The forecasting of sulfide buildup rates in sewers, *Prog. Wat. Techn.* **9** (3), 621-628.

Schmitt, F. and C.F. Seyfried (1992) Sulfate reduction in sewer sediments, *Wat. Sci. Tech.* **25** (8), 83-90.

Tanaka, N. and T. Hvitved-Jacobsen (1998) Transformations of wastewater organic matter in sewers under changing aerobic/anaerobic conditions, *Wat. Sci. Tech.* **37** (1), 105-113.

Tanaka, N., T. Hvitved-Jacobsen, T. Ochi and N. Sato (1998) Aerobic/anaerobic microbial wastewater transformations and reaeration in an air-injected pressure sewer, proceedings of the 71st Annual Water Environment Federation Conference & Exposition, WEFTEC'98, Orlando, Florida, USA, October 3-7, 1998, **2**, 853-864.

Tanaka, N. and T. Hvitved-Jacobsen (1999) Anaerobic transformations of wastewater organic matter under sewer conditions, accepted for the 8th International Conference on Urban Storm Drainage, Sydney, Australia, August 30 - September 3, 1999, pp. 8.

Tchobanoglous, G. (ed.) (1981) Wastewater Engineering: Collection and pumping of wastewater, Metcalf and Eddy, Inc., McGraw Hill Book Co., New York.

Thistlethwayte, D.K.B. (ed.) (1972) The control of sulfides in sewerage systems, Butterworth, Sidney, Australia.

US National Research Counsil, Division of Medical Sciences (1979) Hydrogen sulfide, report by Committee on Medical and Biological Effects of Environmental Pollutants, Subcommittee on H_2S.

USEPA (1974) Process design manual for sulfide control in sanitary sewerage systems, USEPA Technology Transfer, Washington, D.C., USA.

USEPA (1985) Odor and corrosion control in sanitary sewerage systems and treatment plants, USEPA 625/1-85/018, Washington, D.C., USA.

Wilmot, P.D., Cadee, K., Katinic, J.J. and Kavanagh, B.V. (1988) Kinetics of sulfide oxidation by dissolved oxygen. *J. WPCF* **60**, 1264-1270.

7

Biological treatment of organic sulfate-rich wastewaters

Piet N.L. Lens, Francisco Omil, Juan M. Lema and Look W. Hulshoff Pol

7.1 INTRODUCTION

Wastewaters containing organic matter and sulfate are generated by many industrial processes that use sulfuric acid (e.g. food and fermentation industry) or sulfate-rich feed stocks (e.g. sea food processing industry). Also the use of less oxidized sulfurous compounds in industrial processes, i.e. sulfide (tanneries, Kraft pulping), sulfite (sulfite pulping), thiosulfate (processing of photographs) or dithionite (pulp bleaching) results in the generation of sulfate-rich wastewaters. Besides these organic wastewaters, some sulfate-rich effluents contain hardly any organic matter. These are generated during the leaching of sulfur rich wastes (mine spoils, landfills) or during the scrubbing of sulfur containing off-gasses. Specific aspects of the treatment of these inorganic wastewaters are presented in Chapter 8.

Aqueous sulfate emissions can not be considered as a direct threat for the environment as sulfate is a chemically inert, non volatile and non toxic compound (Shin *et al.* 1995). However, there is concern about the direct application of biological anaerobic treatment of these wastewaters, because under anaerobic conditions, dissimilatory sulfate reducing bacteria (SRB) are enabled to use sulfate as a terminal electron acceptor for the degradation of organic compounds and hydrogen (Oude Elferink *et al.* 1994), which results in the generation of sulfide (Fig. 7.1). Therefore, the major problem associated with the anaerobic treatment of sulfate-rich wastewaters is the production of sulfide, which can be a toxic compound for the bacterial populations involved.

On the other hand, the occurrence of sulfate reduction in anaerobic reactors can serve as a biological method, together with a sulfide removal step, to remove sulfate (or other oxidized sulfur compounds) and SO_2 from sulfur containing waste streams and off-gasses. Table 7.1 lists some important disadvantages as well as potentials of sulfate reduction in anaerobic reactors.

7.2 ANAEROBIC TREATMENT OF SULFATE RICH WASTEWATERS

7.2.1 Competition between SRB and methane-producing bacteria

The presence of the sulfate ion in wastewaters considerably increases the complexity of the biodegradation routes (Widdel 1988). In the presence of sulfate, acidogenic, acetogenic (AB) and methanogenic (MB) bacteria compete with SRB for the available substrates (Fig. 7.1). The outcome of this competition is important, as it will determine to what extent sulfide and methane, the end-products of both anaerobic mineralization processes, will be produced. The mechanisms of sulfide toxicity as well as the factors which affect its production are addressed in Chapter 20.

The main intermediates in the anaerobic mineralisation of complex organic matter are hydrogen, acetate, propionate and butyrate (Fig. 7.1). Both from a thermodynamic (Table 7.2) and kinetic point of view, SRB should out-compete the methanogenic consortia during growth on these substrates. Thus, sulphate reduction will theoretically become the dominant process in anaerobic digesters treating sulfate-rich wastewaters.

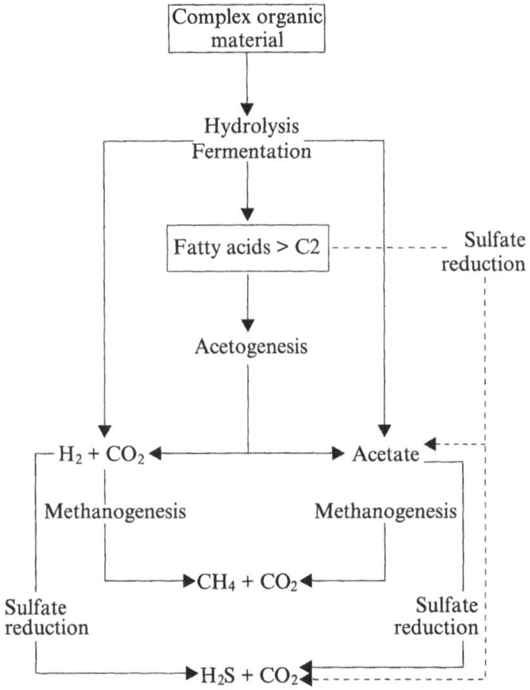

Figure 7.1. Substrate competition between sulfate reducing, methanogenic and acetogenic bacteria during the anaerobic digestion of organic matter.

Table 7.1. Effects of sulfide formation in anaerobic reactors

Disadvantages	Reference
Reduced COD-removal efficiency due to the presence of H_2S in the effluent	This chapter
Corrosion	Chapter 21
Accumulation of inert material in the sludge (e.g. metal sulfides)	Chapter 17
Less methane formation	Lettinga (1995)
Need for H_2S-removal from the biogas, Malodor	Chapter 13
Potential toxicity	Chapter 20
Deterioration of aerobic post treatment system (activated sludge bulking; excessive growth of phototrophs)	Chapter 19
Advantages	Reference
Biological sulfate removal	
- Organic wastewaters	This Chapter
- Inorganic wastewaters	Chapter 8
Degradation of xenobiotics	Chapter 16
Heavy metal removal	Chapter 17
Precipitated metal sulfides (e.g. FeS) form good precursors for granulation	Lettinga (1995)

Table 7.2. Stoichiometry of the anaerobic degradation of propionate, acetate and molecular hydrogen by SRB and MB

Reactions	$\Delta G°$
Propionate	
$CH_3CH_2COO^- + 3\ H_2O \rightarrow CH_3COO^- + HCO_3^- + H^+ + 3\ H_2$	+76.0
$CH_3CH_2COO^- + 0{,}75\ SO_4^{2-} \rightarrow CH_3COO^- + HCO_3^- + 0{,}75\ HS^- + 0{,}25\ H^+$	-37.7
$CH_3CH_2COO^- + 1{,}75\ SO_4^{2-} \rightarrow 3\ HCO_3^- + 1{,}75\ HS^- + 0{,}5\ H^+ + 0{,}25\ OH^-$	-88.9
Acetate	
$CH_3COO^- + H_2O \rightarrow CH_4 + HCO_3^-$	-31.0
$CH_3COO^- + SO_4^{2-} \rightarrow 2\ HCO_3^- + HS^-$	-47.6
Hydrogen	
$4\ H_2 + HCO_3^- + H^+ \rightarrow CH_4 + 3\ H_2O$	-32.7
$4\ H_2 + SO_4^{2-} + H^+ \rightarrow HS^- + 4\ H_2O$	-38.1

Data from Thauer *et al.* (1977) ($\Delta G°$ at 37°C in kJ/reaction).

Previous results have shown that, where there is no limitation of sulphate, hydrogen is completely consumed by SRB and propionate and butyrate are degraded faster by SRB than by the syntrophic consortia (Colleran *et al.* 1995). In contrast, for acetate, results have been reported where either MB or SRB were predominant.

Hydrogenotrophic SRB (HSRB) out-compete hydrogen utilizing MB (HMB) provided sufficient sulfate is present (Omil *et al.* 1996; Rinzema and Lettinga, 1988; Visser *et al.* 1993b). This corroborates with the process fundamentals, as HSRB gain more energy from the consumption of molecular hydrogen and have a higher substrate affinity than HMB, thus decreasing the hydrogen concentration below the threshold value of HMB (Oude Elferink *et al.* 1994). This explains the rapid inhibition of HMB when sulfate is fed to an anaerobic bioreactor.

The expected pre-dominance of acetate utilizing SRB (ASRB) over acetate utilising MB (AMB) in excess of sulfate has been confirmed in continuously stirred tank reactors and in the anaerobic contact process (Gupta *et al.* 1994; Middleton and Lawrence 1977). However, the outcome of the competition is less predictable in modern high-rate anaerobic reactors with sludge retention based on sludge immobilization. Several studies reported that acetate is completely converted into methane, even in excess of sulfate (Hoeks *et al.* 1984; Mulder 1984), while others report a predominance of ASRB (Omil *et al.* 1996). The relevance of this competition increases with a decrease in the chemical oxygen demand (COD):sulfate ratio (Fig. 7.2). Recently, two new ASRB were isolated from bioreactors (Oude Elferink *et al.* 1995, 1999). Their growth kinetic properties are only slightly better than those of *Methanosaeta* sp., the most

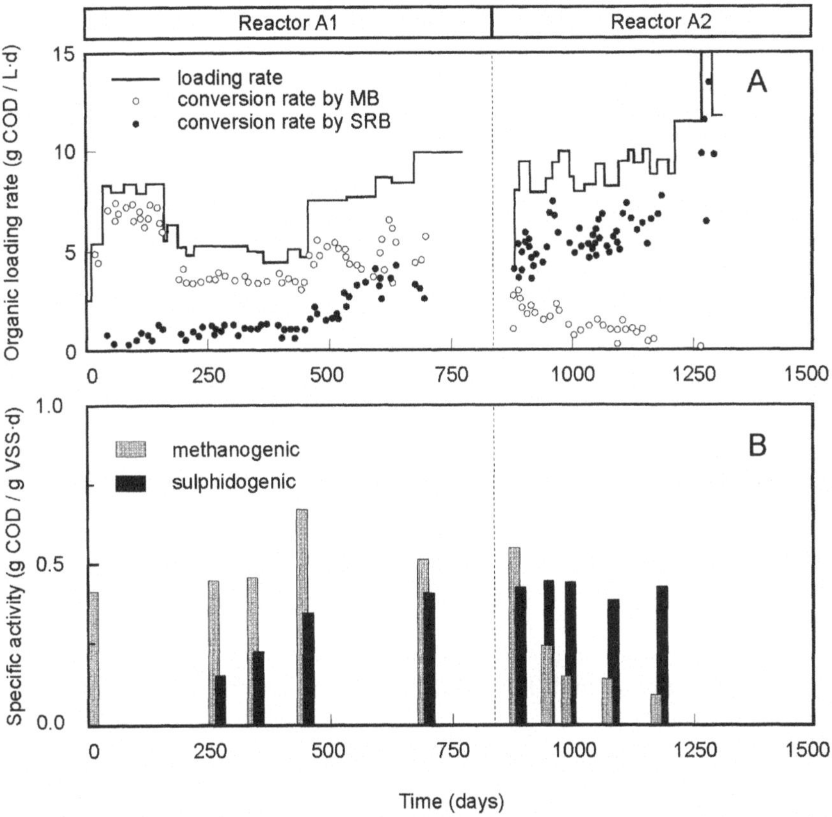

Figure 7.2. Evolution of the performance and sludge activity of two reactors treating acetate as the sole substrate. In this experiment, the COD:sulfate ratio was gradually decreased from 2 to 1 (reactor A1) and finally to 0.5 (reactor A2). (A) Organic loading rate and conversion rates by AMB and ASRB and (B) Specific methanogenic and sulfidogenic activities. (Adapted from: Omil *et al.* 1998)

abundant AMB in bioreactors (Oude Elferink *et al.* 1998). Besides growth kinetics, also many other factors influence the competition between ASRB and AMB (Table 7.3).

7.2.2 Process technology

In principle, all common anaerobic bioreactors can be applied for the treatment of sulfate containing wastewaters, provided proper measures to prevent the occurrence of high H_2S concentrations in the liquid or in the gas

Table 7.3. Factors determining the outcome of the competition between SRB and MB in high-rate anaerobic reactors (After: Hulshoff Pol *et al.* 1998)

Factor	Reference
Inoculum composition	
Type of seed sludge	McCartney and Oleszkiewicz (1991)
Bacterial composition	Harada *et al.* 1994; Omil *et al.* (1998)
Attachment properties of bacteria	Isa *et al.* (1986a, b)
Experimental run time	Harada *et al.* (1994); Omil *et al.* (1998)
Inoculation with new bacterial species	Omil *et al.* (1997a)
Influent composition	
Type of COD	Polprasert and Haas (1995)
Acetate concentration	Yoda *et al.* (1987)
Sulfate concentration	Overmeire *et al.* (1994)
Sulfide concentration	Omil *et al.* (1996)
Ca^{2+} and Mg^{2+} concentration	De Smul *et al.* (1999a)
Operational conditions	
pH, mixed liquor	Visser *et al.* (1996); De Smul *et al.* (1999a)
Temperature	Visser *et al.* (1992)

phase (Table 7.4), the precipitation of inorganic sulfides (leading to less active biomass) and low mass transfer efficiencies due to the lower biogas generation.

The sulfide generated during the anaerobic treatment will be distributed over the gas phase and the liquid phase according to the following expression:

$$[H_2S]_l = \alpha \cdot [H_2S]_g$$

in which $[H_2S]_l$ and $[H_2S]_g$ are respectively the concentrations of the H_2S in the liquid phase and the gas phase and α is a dimensionless distribution coefficient. In the liquid phase, the total dissolved sulfide is present as the unionized form (H_2S) and as HS^-. As the pK_a-value of this acid-base equilibrium is about 7, small pH-variations in the pH range 6 - 8 will significantly affect the free (unionized) H_2S concentration. At neutral pH values, free H_2S accounts to 50% of total dissolved sulfide, whereas at pH 8 it is only around 10%.

There are some particular aspects that should be considered when complete sulfate-reducing bioreactors are operated, as their relatively low gas production rates compared with fully methanogenic reactors, which can lead to a decrease of the mass transfer efficiency (Omil *et al.* 1996). Also, the possible precipitation of inorganic metal sulfides is another factor that could cause problems in some reactor configurations, such as the anaerobic filter design. Therefore, the selection of the reactor type has to be in

Table 7.4. Measures to reduce the reactor sulfide concentration, thus allowing the integration of methanogenesis and sulfate reduction

Measure	Reference
A. Dilution of the influent	
Non sulfate containing process water	Rinzema and Lettinga (1988)
Recycle of effluent after a sulfide removal step by:	
Sulfide stripping	Jensen and Webb (1995)
Sulfide precipitation	Särner (1990)
Biological sulfide oxidation to elemental sulfur	
Thiobacillus sp., oxygen	Buisman *et al.* (1990)
Thiobacillus denitrificans, nitrate	Gommers *et al.* (1988)
Chlorobium limicola, sunlight	Kim *et al.* (1993)
Chemical oxidation to elemental sulfur	
Ferric sulfate/silicone supported reactor	De Smul *et al.* (1999b)
B. Decrease of the unionized sulfide concentration	
Elevation of the reactor pH	Rinzema and Lettinga (1988)
Elevation of the reactor temperature	Rintala *et al.* (1991)
Precipitation of sulfide, e.g. with iron salts	McFarland and Jewell (1989)
Stripping of the reactor liquid using	
- high mixing degree inside the reactor	
- recirculation of biogas after scrubbing	Särner (1990)
- other stripping gas (e.g. N_2)	
C. Separation of sulfide production and methanogenesis	
Two stage anaerobic digestion	Rinzema and Lettinga (1988)
Multi step process	Sipma *et al.* (1999)
D. Selective inhibition of SRB	
Sulfate analogues (e.g. MoO_4^{2-})	Yadav and Archer (1989)
Transition elements (e.g. Co, Ni or Zn)	Clancy *et al.* (1992)
Antibiotics	Tanimoto *et al.* (1989)

accordance with the aim of the treatment, which can be either the removal of organic matter, sulfate, or the removal of both.

Considering the potential problems related to the occurrence of sulfate reduction in the anaerobic digestion process (Table 7.1), a complete suppression of the sulfate reduction and a complete conversion of the organic substrate into methane could be considered as the most optimal option. Therefore, attempts have been made to selectively suppress sulfate reduction by using specific inhibitors (Table 7.4). However, so far, no selective inhibitor of SRB has been found that can be used in full-scale anaerobic reactors. This implies that sulfate reduction can not be prevented in practice.

When wastewaters containing organic matter and sulfate are treated in an anaerobic bioreactor, organic matter will be removed both via sulfate reduction and methanogenesis. In practice, anaerobic treatment always

Table 7.5. Overview of reactor types, sulfate removal efficiencies and extent of sulfate reduction in anaerobic digestion of organic and sulfate-rich wastewaters (Adapted from: Lens *et al.* 1998a)

	Influent			SO_4^{2-} removal		
Reactor type	Type	COD (g/l)	SO_4^{2-} (g/l)	% SO_4^{2-} reduced	% COD_{SRB}	Reference
UASB	C_2	1.5-2.1	0.7-3.4	70	50-90	Visser *et al.* (1993b)
EGSB	$C_2/C_3/C_4$	0.5-2.5	1.2-4.6	27-68	59-97	Omil *et al.* (1996)
USSB	$C_2/C_3/C_4$	0.5-6.0	1.0-12.0	35	67-81	Lens *et al.* (1998b)
MUSB	C_2	0.2-0.4	0.1-0.2	40-80	100	Arora *et al.* (1995)
UASB	Ethanol	NR	0.84-5	80	NR	Kalyuzhnyi *et al.* (1997)
AF	Citric acid	25.8	3.4	93	18	Colleran *et al.* (1994)
BE	Glucose	0.4	0.7-3.0	35-55	NR	Watanabe *et al.* (1997)
HYBRID	Landfill leachate	19.6-42.0	5.9	> 90	NR	Nedwell and Reynolds (1996)
CAD	Sea food	10-60	0.6- 2.7	96.0	3-12	Omil *et al.* (1995)

Abbreviations: UASB upflow granular sludge bed reactor; EGSB expanded granular sludge bed reactor; USSB upflow staged sludge bed reactor; MUSB microaerophilic granular sludge bed reactor; AF anaerobic filter; HYBRID hybrid reactor; BE bio-electro reactor, CAD central activity digester, % COD_{SRB} percentage of COD used by SRB, C_2 acetate, C_3 propionate, C_4 butyrate, NR not reported.

proceeds successfully for wastewaters with COD:sulfate ratios exceeding 10, as for such wastewaters the H_2S concentration in the anaerobic reactor will never exceed the presumed critical value of 150 mg/l (Rinzema and Lettinga 1988). At COD:sulfate ratios lower than 10, process failures of anaerobic reactors have been reported, whereas in other cases the process proceeds successfully when precautions are taken to prevent sulfide toxicity (Table 7.4).

It is clear from different works that wastewaters containing organic matter and high sulfate concentrations can be successfully treated under anaerobic conditions, even at full scale. Table 7.5 summarizes some of the operational data obtained during the treatment of different types of wastewaters.

In fact, until recently, a COD:sulfate ratio higher than 10 was commonly regarded as a pre-requisite for a successful anaerobic treatment. Lower ratios were thought to be detrimental to methanogenesis because they produced excessive sulfide concentrations. However, Hilton and Archer (1988), Méndez *et al.* (1989) and Derycke and Pipyn (1990) reported successful anaerobic treatments at ratios of 8, 5 and 3, respectively. The anaerobic treatment of wastewaters from sea food processing industries can be considered as a case study in which the COD:sulfate ratio varies from 50 (fully methanogenic conditions) to values lower than 5 (Omil *et al.* 1995). Treating these effluents, in spite of the presence of high concentrations of salts (5-12 g Na^+/l) and ammonia (1.0 - 3.0 g N/l), high efficiencies in COD removal were obtained (70-90%) during a stable operation. The sulfate present in these effluents (0.6-2.7 g/l) was completely reduced by SRB, which accounted up to 11.6% of the total COD removed in the reactor. The strategy followed in this work consisted of the determination of the optimal pH to maintain the concentration of both the unionized H_2S and unionized NH_3 as low as possible, since the inhibitory effects of both compounds determine the overall efficiency (Fig. 7.3).

7.3 BIOLOGICAL SULFATE REMOVAL

Biological sulfate removal is a cost-effective alternative for costly and sometimes complex physico-chemical sulfate removal methods (Maree *et al.* 1991). Biological sulfate removal consists of two steps with (dissimilatory) sulfate reduction to sulfide as the first one. The sulfide produced in the first stage is then biologically oxidized to elemental sulfur (S^o). Since sulfur is a colloidal solid, it can be eliminated from the wastewater by gravity sedimentation, eventually after flocculant addition.

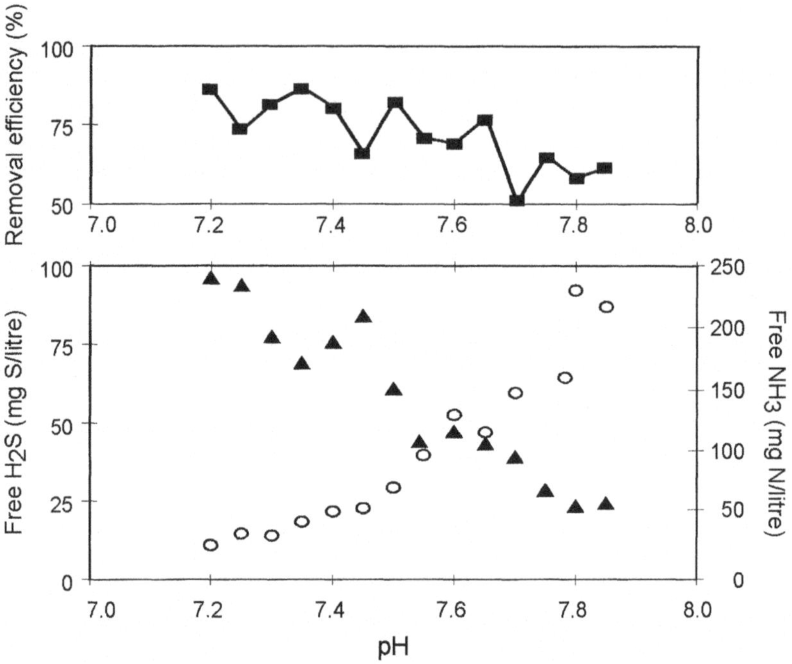

Figure 7.3. Average COD removal efficiencies (—■—), free hydrogen sulfide (▲) and free ammonia (o) concentrations vs. pH in a CAD digester during the treatment of sea food processing wastewaters (Adapted from: Omil *et al.* 1995).

To prevent sulfide inhibition, different process configurations can be proposed to integrate sulfate reduction, methanogenesis and sulfide removal in order to achieve the removal of both organic matter and sulfurous compounds (Fig. 7.4).

7.3.1 Sulfate reduction

In the sulfate-reducing stage, a complete reduction of sulfate to sulfide is desired. Channelling of reducing equivalents towards the SRB is enhanced by the ability of the SRB to effectively compete with other anaerobic bacteria for the available organic substrate and the sensitivity of the other bacteria for sulfide.

For wastewaters that contain no or insufficient electron donor and carbon source for a complete sulfate reduction, addition of an appropriate electron donor is required (Chapter 8). The selection of the electron donor depends on: (i) the costs of substrate required per unit of reduced sulfate;

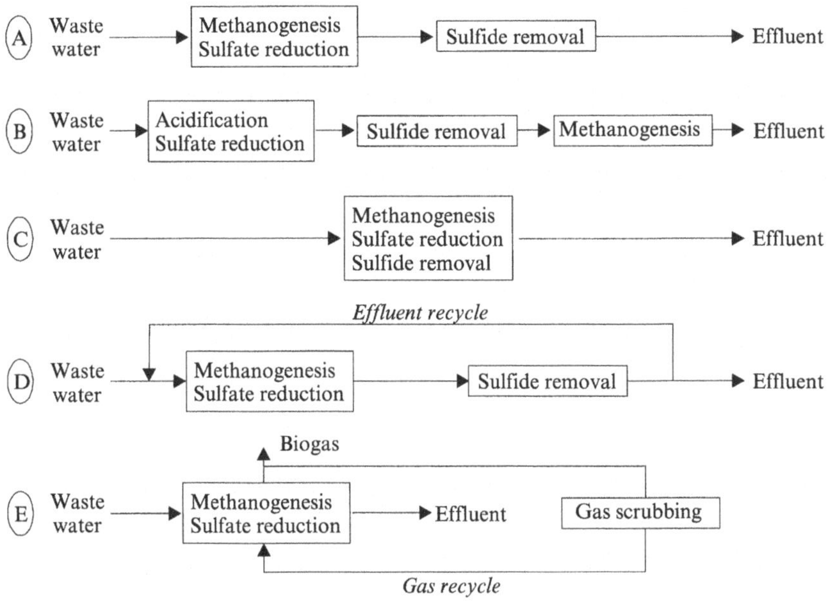

Figure 7.4. Process configurations integrating methanogenesis with sulfate reduction and sulfide removal.

and (ii) the biodegradability of this substrate. Based on the last criterion, simple organic compounds (ethanol, methanol) or synthesis gas (a mixture of H_2, CO and CO_2) are preferred above complex wastes (e.g. molasses).

Sulfate reduction was studied using laboratory UASB reactors at 35 °C with ethanol as the sole electron donor and carbon source (Kalyuzhnyi *et al.* 1997). High sulfate conversion efficiencies of up to 80% were achieved at sulfate loading rates (SLR) of up to 6 g SO_4^{2-}/l·d and influent sulfate concentrations of 0.84 - 5 g/l. SRB outcompeted MB for both the energy and carbon source in the system.

Under mesophilic conditions, hydrogenotrophic methanogenesis is minimal, as SRB out-compete other anaerobic bacteria more effectively with H_2 than with simple organic compounds as electron donor. Using a mixture of H_2 and CO_2 (80:20%), sulfate loading rates of 30 g SO_4^{2-}/l·d can be achieved at 30°C within 10 days of operation in gas-lift reactors (which provide good mass transfer rates) with pumice as carrier material (to immobilize SRB) when the free H_2S concentration is kept below 450 mg/l (Van Houten *et al.* 1994).

7.3.2 Sulfide oxidation

In the second step, sulfide can be removed from the liquid or H_2S enriched stripping gas using physico-chemical or biological techniques (Table 7.4). The relatively high energy requirements for stripping or the high chemical and disposal (chemical sludge like FeS or MnO_2) costs constitute important draw-backs of physico-chemical sulfide removal methods. Various biological sulfur removal methods exist (Table 7.4). One of these is the partial biological oxidation of sulfide to the solid S^o (Buisman *et al.* 1990). Under oxygen limited conditions, i.e. dissolved oxygen concentrations below 0.1 mg/l, S^o is the major end-product of the sulfide oxidation, whereas sulfate is mainly formed under sulfide-limiting conditions. S^o formation requires one quarter of the oxygen compared with complete oxidation and, consequently, a lower energy consumption for aeration. In reactors with low shear forces, i.e. when the reactor liquid is aerated in a spatially separated aeration unit, well-settling S^o particles (mean diameter 3 mm, settling velocity > 25 m/h for 90% of the sludge) are formed under autotrophic conditions (Janssen *et al.* 1997).

7.4 NOVEL DEVELOPMENTS AND PERSPECTIVES

7.4.1 Steering of the competition between SRB-MB

Methods that influence the outcome of the competition between SRB and MB would be useful to develop fully methanogenic or sulfate-reducing sludges, depending of the desired process application. Moreover, they can prevent potential process failures due to sulfide inhibition. Thus far however, adequate methods to steer the competition between ASRB and AMB that can be applied at full scale are not available (Lens *et al.* 1998a; Van Houten *et al.* 1997). The best way to steer a reactor towards the predominance of one population over the other involves the manipulation of the inoculum composition or the environmental conditions, as e.g. the reactor pH (Fig. 7.5). The MB:SRB ratio of a sludge can be manipulated by adding pure cultures of MB or SRB, or creating unfavorable conditions for the undesired population during short time intervals (Table 7.6). Successful methods that enhance the development of an SRB population after selective inhibition by MB include shocks of high sulfide concentrations (Omil *et al.* 1996) or high (65°C) temperatures (Visser *et al.* 1993a). However, more research is needed to develop methods that can be used for full-scale applications.

Table 7.6. Effect of changes in process conditions on the competition between SRB-MB: increase of the share of SRB related to the total COD removal (%). (Adapted from: Hulshoff Pol *et al.* 1998)

Measure	Increase (%)*	Reference
Manipulation of the influent composition		
Increase of acetate concentration	- 15	Omil *et al.* (1996)
Addition of iron (2 g/l)	0	Isa *et al.* (1986a, b)
Addition of transition elements	0	Clancy *et al.* (1992)
Availability of the electron donor	NR	Vroblesky *et al.* (1996)
Manipulation of the biomass composition		
Addition of *Desulforhabdus amnigenes*	0	Omil *et al.* (1997a)
Exposure to oxygen	35	Omil *et al.* (1997a)
Manipulation of operational conditions		
Alteration of pH	41	Omil *et al.* (1996)
Shock treatment		
Temperature decrease to 15°C	0	Omil *et al.* (1997b)
Temperature increase to 65°C	30	Visser *et al.* (1993b)
Manipulation of the reactor design		
EGSB (upflow velocity 4 - 6 m/h)	- 30	Omil *et al.* (1996)
Staged sludge bed (USSB) reactor	10	Lens *et al.* (1998b)
Baffled reactor	30	Lens *et al.* (1999)

* Negative value means an increase in the share of the MB to the total COD removal. NR: not reported.

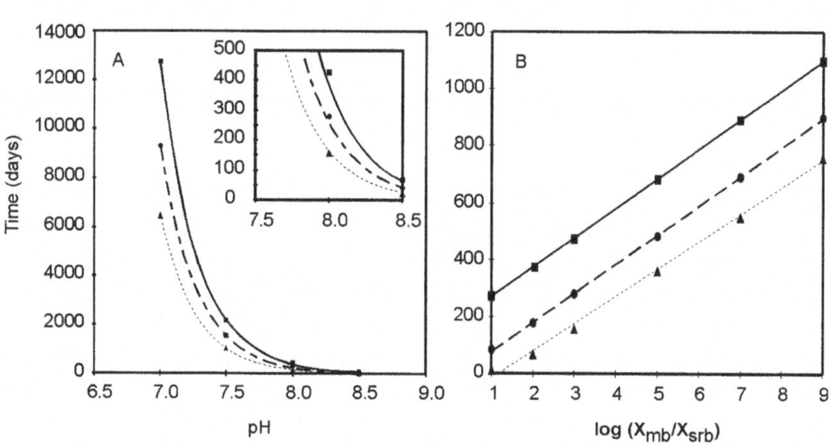

Figure 7.5. Simulation results for the time needed for ASRB to consume 10% (▲), 50% (●) and 90% (■) as a function of (A) pH and (B) sludge retention time (SRT) (Adapted from: Omil *et al.* 1998).

7.4.2 SRB based bioremediation techniques

SRB are present in anaerobic and even aerobic wastewater treatment sludges (Lens *et al.* 1995a). The development of treatment processes using their capacity to degrade a wide range of organic compounds (Widdel 1988) opens promising perspectives for environmental biotechnology. SRB do not require balanced growth with acetogens, which implies less sensitivity to organic overloads. Moreover, SRB are less sensitive to toxicants. Heavy metals are precipitated by sulfide, thus reducing their potential toxic effects (Chapter 17). SRB can metabolize organic inhibitors such as aromatics (toluene, ethylbenzene), alkanes, chlorinated compounds (chloroform) and long chain fatty acids (Widdel 1988). On the other hand, reactors based on organic matter removal by sulfate reduction lack one of the major advantages of methanogenic treatment: the recovery of methane from organic compounds.

7.4.2.1 Removal of organic matter

Anaerobic reactors in which organic matter is completely degraded by SRB might be an elegant alternative for methanogenic wastewater treatment. Sulfate-reducing Upflow Anaerobic Sludge Bed (UASB) reactors can be operated at sludge loading rates of 0.9 to 1.0 g COD per gram of volatile suspended solids per day (Visser *et al.* 1993b). Nedwell and Reynolds (1996) reported that sulfate-reducing and methanogenic hybrid reactors treating landfill leachates offered equal organic removal efficiencies at low organic loading rates (< 1 kg COD/m^3·d). However, the sulfate-reducing reactor was less effective in COD removal at higher loading rates. In contrast, mesophilic (30 °C) volatile fatty acid fed sulfate-reducing UASB reactors (liquid upward velocity 0.65 m/h) had a COD removal capacity comparable to methanogenic systems at volumetric loading rates of 7.5 kg COD/m^3·d (Alphenaar *et al.* 1993).

Sulfate-reducing granular sludge can be obtained by feeding a reactor, inoculated with methanogenic sludge, with wastewaters with a COD:sulfate ratio of 0.5. The population shift between MB and SRB proceeded relatively slow (about 100 days) and after the start-up phase, MB still consumed from 5% to 20% of the organic material (Omil *et al.* 1996). By treating the sludge with $CHCl_3$ (5 mg/l) for 1 - 5 days during the start-up, MB were completely eliminated in the inoculum, thus creating a sulfate-reducing sludge already from the start-up (Visser *et al.* 1993b).

Elimination of substrates by syntrophic associations in methanogenic reactors requires a low H_2 partial pressure. In contrast, their degradation by sulfate reducers does not depend on a low H_2 partial pressure (Table 7.2). Hence, high-rate sulfate-reducing removal systems for substrates as propionate and butyrate can be applied. The major drawback of sulfate-reducing reactor systems is their inefficient acetate removal, as this fatty acid generally accounts for most of the residual COD in their effluents (Lens *et al.* 1998b; Nedwell and Reynolds 1996; Omil *et al.* 1996). There are indications that ASRB have a poor affinity for sulfate. Hence, HSRB out-compete ASRB for the available sulfate. Optimization of the performance of sulfate-reducing reactors can be done by methods that integrate acetate scavenging processes, i.e. denitrification (Lens *et al.* 1999) or the use of staged reactors, which allows the development of primarily acetotrophic biomass in the last stages (Lens *et al.* 1998b).

7.4.2.2 Biodegradation of xenobiotics

The potential capacity of SRB to treat hardly biodegradable organic compounds, such as xenobiotics, is nowadays an important research subject. Many persistent xenobiotic compounds in the environment are nitro-substituted aromatics from pesticides, plastics, pharmaceuticals, industrial waste and military use of explosives. The anaerobic metabolic processes in the degradation of explosives and nitroaromatic compounds under sulfate-reducing conditions have been recently reviewed by Boopathy *et al.* (1998).

7.4.2.3 Heavy metal removal

Several sulfate-rich wastewaters, e.g. acid mine drainage, spoil leachates and landfill percolates often are also contaminated with heavy metals. Metal toxicity as well as the potential of sulfidogenic bioreactors to remove heavy metals from waste streams are presented in Chapter 17 of this book.

7.4.3. Micro-aerobic treatment of sulfate-rich wastewaters

Another way to utilize the capacities of SRB to degrade organic matter, is the use of reactors with combined aerobic/anaerobic conditions. These conditions are needed for a complete mineralization of certain xenobiotics (Field *et al.* 1995), where (facultative) aerobic bacteria utilize the degradation products of the SRB.

Micro-aerophilic conditions can be created in UASB reactors by dosing controlled amounts of oxygen to the sulfate-rich wastewaters. In hybrid reactors, the upper reactor part (with the carrier material) can be aerated and spatially separated from the anaerobic granular sludge section (Tilche *et al.* 1994). Alternatively, a single (Arora *et al.* 1995) or multiple

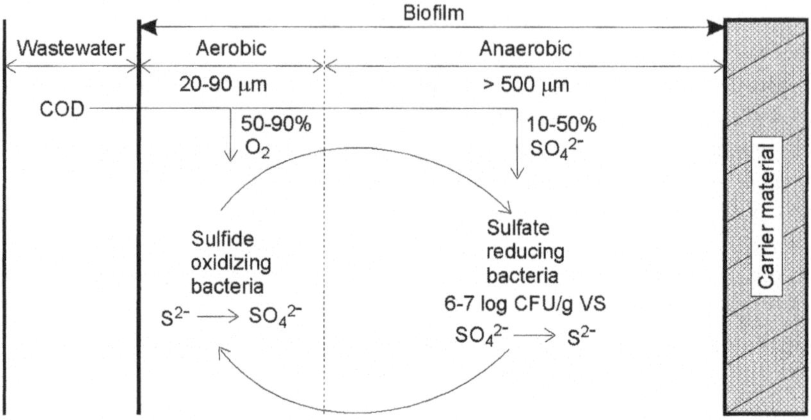

Figure 7.6. Schematic representation of the interaction between aerobic and anaerobic conditions in biofilm reactors (After: Lens *et al.* 1995a, b). CFU= colony forming units, VS = volatile solids.

(Takahashi and Kyosai 1991) sequence of aerated downflow and non aerated upflow (6 m/h) reactors can be applied. In the latter reactor type, a special type of granular sludge develops, composed of **SRB** in the inner anaerobic layers and sulfide-oxidizing (*Beggiatoa* sp.) bacteria in the oxygenated periphery of the granules (Takahashi and Kyosai 1991).

The close interaction between aerobic and anaerobic conditions also prevails in highly loaded aerobic fixed film reactors (Kuenen *et al.* 1986). Thus, a sulphur cycle at micro-scale can develop in biofilms growing on aerobic rotating biological contactors (Alleman *et al.* 1982), trickling filters (Fig. 7.6) or flow cells (Santegoeds *et al.* 1998). Stimulation of the activity of SRB by decreasing the COD:sulfate ratio of the wastewater from 20 to 0.5 resulted in a 50 % reduction of the waste sludge production (Lens *et al.* 1995b). Practical applications to reduce the waste sludge production of aerobic bioreactors are limited by the lack of sulfate removal by these aerobic fixed film reactors.

Besides oxygen, also nitrate can be used in microaerophilic systems (Lens *et al.* 1999). Nitrate is much more soluble compared to molecular oxygen. The nitrate is denitrified in a volatile fatty acid fed sulfidogenic reactor, although under certain conditions, nitrate is incompletely denitrified to nitrite or nitrous oxide. Besides denitrification, nitrate can also be converted into ammonium in the anaerobic biofilm layers. The latter process, called ammonification, is of no interest from an environmental engineering point of view. More research is needed to obtain full denitrification in nitrate based microaerobic reactors.

7.5 REFERENCES

Alleman J.E., Veil J.A. and Canaday J.T. (1982) Scanning electron microscope evaluation of rotating biological biofilm. *Wat. Res.* **16**, 543-550.

Alphenaar P.A., Visser A. and Lettinga G. (1993) The effect of liquid upward velocity and hydraulic retention time on granulation in UASB reactors treating waste water with a high sulphate content. *Biores. Technol.* **43**, 249-258

Arora S., Mino T. and Matsuo T. (1995) Mechanism and favourable operating conditions of organic substrate removal in microaerophilic upflow sludge bed reactor (MUSB). *Wat. Res.* **29**, 1895-1901

Boopathy R., Kulpa C.F., Manning J. (1998) Anaerobic biodegradation of explosives and related compounds by sulfate-reducing and methanogenic bacteria: a review. *Bioresource Technol.*, **63** (1), 81-89.

Buisman C.N.J., Geraats B.G., Ijspeert P. and Lettinga G. (1990) Optimization of sulphur production in a biotechnological sulphide-removing reactor. *Biotech. Bioeng.* **35**, 50-56.

Clancy P.B., Venkataraman N. and Lynd L.R. (1992) Biochemical inhibition of sulfate reduction in batch and continuous anaerobic digesters. *Wat. Sci. Tech.* **25**, 51-60.

Colleran E., Finnegan S. and O'Keeffe R.B. (1994) Anaerobic digestion of high sulphate-containing wastewater from the industrial production of citric acid. *Wat. Sci. Tech.* **30**, 263-273.

Colleran E., Finnegan S. and Lens P. (1995) Anaerobic treatment of sulphate-containing waste streams. *Antonie van Leeuwenhoek* **67**, 29-46.

De Smul A. and Verstraete W. (1999a) Retention of sulfate reducing bacteria in expanded-granular-sludge-blanket reactors. *Water Environ. Res.* **71**, 427–431.

De Smul A. and Verstraete W. (1999b) The phenomenology and the mathematical modeling of the silicone-supported chemical oxidation of aqueous sulfide to elemental sulfur with ferric sulfate. *J. Chem. Technol. Biotechnol.* **74**, 456–466.

Derycke, D. and Pipyn, P. (1990) Anaerobic digestion, ammonia stripping/recovery and (de)nitrification of a cytric acid factory effluent. *Meded. Fac. Landbouwwet. Rijksuniv. Gent*, **55** (4), 1481-1483.

Field J.A., Stams A.J.M., Kato M. and Schraa G. (1995) Enhanced biodegradation of aromatic pollutants in cocultures of anaerobic and aerobic consortia. *Antonie van Leeuwenhoek* **67**, 47-77.

Gommers P.J.F., Bijleveld W. and Kuenen J.G. (1988) Simultaneous sulfide and acetate oxidation in a denitrifying fluidized bed reactor - I. Start-up and reactor performance. *Wat. Res.* **22**, 1075-1083.

Gupta A., Flora J.R.V., Gupta M., Sayles G.D. and Suidan M.T. (1994) Methanogenesis and sulfate reduction in chemostats - I. kinetic studies and experiments. *Wat. Res.* **28**, 781-793.

Harada H., Uemura S. and Momonoi K. (1994) Interactions between sulphate-reducing bacteria and methane-producing bacteria in UASB reactors fed with low strength wastes containing different levels of sulfate. *Wat. Res.* **28**, 355-367.

Hilton M.G. and Archer D.B. (1988) Anaerobic digestion of a sulfate-rich molasses waste-water: inhibition of hydrogen sulfide production. *Biotechnol. Bioeng.* **31**, 885-888.

Hoeks F.W.J.M.M., Ten Hoopen H.J.G., Roels J.A. and Kuenen J.G. (1984)

Anaerobic treatment of acid water (methane production in a sulfate rich environment). *Progr. Ind. Microbiol.* **20**, 113-119.

Hulshoff Pol L., Lens P., Stams A.J.M. and Lettinga G. (1998) Anaerobic treatment of sulfate-rich wastewaters. *Biodegradation,* **9** (3/4), 213-224.

Isa Z., Grusenmeyer S. and Verstraete W. (1986a) Sulfate reduction relative to methane production in high-rate anaerobic digestion: Technical aspects. *Appl. Environ. Microbiol.* **51**, 572-579.

Isa Z., Grusenmeyer S. and Verstraete W. (1986b) Sulfate reduction relative to methane production in high-rate anaerobic digestion: Microbiological aspects. *Appl. Environ. Microbiol.* **51**, 580-587.

Janssen A.J.H., Ma S.C., Lens P. and Lettinga G. (1997) Performance of a sulphide oxidizing expanded bed reactor with a spatially separated aeration unit. *Biotech. Bioeng.* **53**, 32-40.

Jensen A.B. and Webb C. (1995) Treatment of H_2S-containing gases: a review of microbiological alternatives. *Enzyme Microb. Technol.* **17**, 2-10.

Kalyuzhnyi S.V., de Leon-Fragoso C., Rodriguez-Martinez, J. (1997) Biological sulfate reduction in an UASB reactor fed with ethanol as electron donor. *Mikrobiologiya,* **66** (5), 674-680.

Kim B.W., Kim E.H., Lee S.C. and Chang H.N. (1993) Model-based control of feed rate and illuminance in a photosynthetic fed-batch reactor for H_2S removal. *Bioprocess Eng.* **8**, 263-269.

Kuenen J.G., Jorgenson B.B. and Revsbech N.P. (1986) Oxygen microprofiles of trickling filter biofilms. *Wat. Res.* **20**, 1589-1598.

Lens P., Massone A., Rozzi A. and Verstraete W. (1995a) Effect of sulfate concentration and scraping on aerobic fixed film reactors. *Wat. Res.* **29**, 857-870.

Lens P., De Poorter M.P., Cronenberg C.C. and Verstraete W.H. (1995b) Sulfate reducing and methane producing bacteria in aerobic wastewater treatment. *Wat. Res.* **29**, 871-880.

Lens P., Visser A., Janssen A.J.H., Hulshoff Pol L.W. and Lettinga G. (1998a) Biotechnological treatment of sulfate rich wastewater. *Crit. Rev. Env. Sci. Technol.* **28**, 41-88.

Lens P., van den Bosch M., Hulshoff Pol L.W. and Lettinga G. (1998b) The use of staged sludge bed reactors for the treatment of sulfate rich wastewaters. *Wat. Res.* **32**, 1178–1192.

Lens P., Sipma J., Hulshof Pol L. and Lettinga G. (1999) Effect of staging and nitrate addition on sulfidogenic acetate removal. *Wat. Res.* **34**, 31–42.

Lettinga G. (1995) Anaerobic digestion and wastewater treatment systems. *Antonie van Leeuwenhoek* **67**, 3-28.

Maree J.P., Hulse G., Dods D. and Schutte C.E. (1991) Pilot plant studies on biological sulphate removal from industrial effluent. *Wat. Sci. Tech.* **23**, 1293-1300.

McCartney D.M. and Oleszkiewicz J.A. (1991) Sulfide inhibition of anaerobic degradation of lactate and acetate. *Wat. Res.* **25**, 203-209.

McFarland M.J. and Jewell W.J. (1989) In situ control of sulfide emission during thermophilic anaerobic digestion process. *Wat. Res.* **23**, 1571-1577.

Méndez R., ten-Brummeler E. and Hulshoff Pol, L.W. (1989) Start up of UASB reactors treating sucrose-containing substrates with a low COD/sulfate ratio. *Environ. Technol. Lett.* **10**, 83-90.

Middleton A.G. and Lawrence A.W. (1977) Kinetics of microbial sulfate reduction. *J. Wat. Pollut. Control Fed.* **229**, 1659-1670.

Mulder A. (1984) The effects of high sulfate concentrations on the methane fermentation of waste water. *Progr. Ind. Microbiol.* **20**, 133-143.

Nedwell D.B. and Reynolds P.J. (1996) Treatment of landfill leachate by methanogenic and sulphate-reducing digestion. *Wat. Res.* **30**, 21-28.

Omil F., Méndez R. and Lema J.M. (1995) Anaerobic treatment of saline wastewaters under high sulfide and ammonia content. *Biores. Technol.* **54**, 269-278.

Omil F., Lens P., Hulshoff Pol L. and Lettinga G. (1996) Effect of upward velocity and sulfide concentration on volatile fatty acid degradation in a sulphidogenic granular sludge reactor. *Process Biochem.* **31**, 699-710.

Omil F., Oude Elferink S.J.W.H., Lens P., Hulshoff Pol L. and Lettinga G. (1997a) Effect of the inoculation with Desulforhabdus amnigenus and pH or O_2 shocks on the competition between sulfate reducing and methanogenic bacteria in an acetate fed UASB reactor. *Biores. Technol.* **60**, 113-122.

Omil F., Bakker C.D., Hulshoff Pol L.W. and Lettinga G. (1997b) The effect of pH and low temperature shocks on the competition between sulphate reducing and methane producing bacteria in UASB reactors. *Environ. Technol.* **18**, 255-264.

Omil F., Lens P., Visser A., Hulshoff Pol L.W. and Lettinga G. (1998) Long term competition between sulfate reducing and methanogenic bacteria in UASB reactors treating volatile fatty acids. *Biotech. Bioeng.* **57**, 676-685.

Oude Elferink S.J.W.H., Visser A., Hulshoff Pol L.W. and Stams A.J.M. (1994) Sulfate reduction in methanogenic bioreactors. *FEMS Microbiol. Rev.* **15**, 119-136.

Oude Elferink S.J.W.H., Maas R.N., Harmsen H.J.M. and Stams A.J.M. (1995) Desulforhabdus amnigenus gen.nov. sp.nov., a sulfate reducer isolated from anaerobic granular sludge. *Arch. Microbiol.* **164**, 119-124.

Oude Elferink S.J.W.H., Luppens S.B.I., Marcelis C.L.M. and Stams A.J.M. (1998) Kinetics of acetate utilization by two sulfate reducers isolated from anaerobic granular sludge. *Appl Environ Microbiol.* **64**, 2301–2303.

Oude Elferink S.J.W.H., Akkermans-van Vliet W.M., Bogte J.J. and Stams A.J.M. (1999) *Desulfobacca acetoxidans* gen. nov. sp. nov., a novel acetate-degrading sulfate reducer isolated from sulfidogenic sludge. *Int. J. Syst. Bacteriol.* **49**, 345–350.

Overmeire A., Lens P. and Verstraete W. (1994) Mass transfer limitation of sulfate in methanogenic aggregates. *Biotech. Bioeng.* **44**, 387-391.

Polprasert C. and Haas C.N. (1995) Effect of sulfate on anaerobic processes fed with dual substrates. *Wat. Sci. Tech.* **31**, 101-107.

Rintala J., Sanz Martin J.L. and Lettinga G. (1991) Thermophilic anaerobic treatment of sulfate-rich pulp and paper integrate process water. *Wat. Sci. Tech.* **24**, 149-160.

Rinzema A. and Lettinga G. (1988) Anaerobic treatment of sulfate containing waste water. In *Biotreatment systems*, (ed. D.L. Wise), vol III, pp. 65-109, CRC Press Inc., Boca raton, USA.

Santegoeds C.M., Ferdelman T.G., Muyzer G. and de Beer D. (1998) Structural and functional dynamics of sulfate reducing bacterial populations in biofilms. *Appl. Environ. Microbiol.* **64**, 3731-3739.

Särner E. (1990) Removal of sulphate and sulphite in an anaerobic trickling (ANTRIC) filter. *Wat. Sci. Tech.* **22**, 395-404.

Shin H.S., Jung J.Y., Bae B.U. and Paik B.C. (1995) Phase-separated anaerobic toxicity assays for sulfate and sulfide. *Wat. Environ. Res.* **67**, 802-806.

Sipma J., Lens P.N.L., Vieira A., Miron Y., Van Lier J.B., Hulshoff Pol L.W. and Lettinga G. (1999) Thermofilic sulfate reduction in UASB reactors under acidifying conditons. *Process Biochem.* **35**, 509–522.

Takahashi M. and Kyosai S. (1991) Pilot plant study on microaerobic self-granulated sludge process (multi-stage reversing-flow bioreactor: MRB). *Wat. Sci. Tech.* **23**, 973-980.

Tanimoto Y., Tasaki M., Okamura K., Yamaguchi M. and Minami K. (1989) Screening growth inhibitors of sulfate-reducing bacteria and their effects on methane fermentation. *J. Ferment. Bioeng.* **68**, 353-359.

Thauer R.K., Jungermann K. and Decker K. (1977) Energy conservation in chemotrophic anaerobic bacteria. *Bacteriol. Rev.* **41**, 100-180.

Tilche A., Bortone G., Forner G., Indulti M., Stante L. and Tesini O. (1994) Combination of anaerobic digestion and denitrification in a hybrid upflow anaerobic filter integrated in a nutrient removal plant. *Wat. Sci. Tech.* **30**, 405-414.

Van Houten R.T., Hulshoff Pol L.W. and Lettinga G. (1994) Biological sulphate reduction using gas-lift reactors fed with hydrogen and carbon dioxide as energy and carbon source. *Biotech. Bioeng.* **44**, 586-594.

Van Houten R.T., Yun S.Y. and Lettinga G. (1997) Thermophilic sulphate and sulphite reduction in lab-scale gas-lift reactors using H_2 and CO_2 as energy and carbon source. *Biotech. Bioeng.* **55**, 807-814.

Visser A., Gao Y. and Lettinga G. (1992) The anaerobic treatment of a synthetic sulfate containing wastewater under thermophilic (55 °C) conditions. *Wat. Sci. Tech.* **25**, 193-202.

Visser A., Gao Y. and Lettinga G. (1993a) Effects of short-term temperature increases on the mesophilic anaerobic breakdown of sulfate containing synthetic wastewater. *Wat. Res.* **27**, 541-550.

Visser A., Beeksma I., van der Zee F., Stams A.J.M. and Lettinga G. (1993b) Anaerobic degradation of volatile fatty acids at different sulfate concentrations. *Appl. Microbiol. Biotechnol.* **40**, 549-556.

Visser A., Hulshoff Pol L.W. and Lettinga G. (1996) Competition of methanogenic and sulfidogenic bacteria. *Wat. Sci. Tech.* **33**, 99-110.

Vroblesky DA, Bradley PM and Chapelle FH (1996) Influence of electron donor on the minimum sulfate concentration required for sulfate reduction in a petroleum hydrocarbon-contaminated aquifer. *Environ. Sci. Technol.* **30** (4), 1377-1381.

Watanabe T., Imaoka H. and Kuroda M. (1997) Neutralisation and sulfate removal of acidic water containing high strength sulfate ion by using electrodes immobilized sulfate reducing bacteria on the surface. In *Proceedings of the 8th international conference on Anaerobic Digestion*, May 25-29, vol. 3, pp 397-400, Sendai, Japan.

Widdel F. (1988) Microbiology and ecology of sulfate-and sulfur reducing bacteria. In *Biology of Anaerobic Microorganisms* (ed. A.J.B. Zehnder), pp. 469-586, John Wiley and Sons, New York.

Yadav V.K. and Archer D.B. (1989) Sodium molybdate inhibits sulphate reduction in the anaerobic treatment of high sulphate molasses wastewater. *Appl. Microbiol. Biotechnol.* **31**, 103-106.

Yoda M., Kitagawa M. and Miyayi Y. (1987) Long term competition between sulfate-reducing and methane-producing bacteria in anaerobic biofilm. *Wat. Res.* **21**, 1547-1556.

8

Biological removal of sulfurous compounds from inorganic wastewaters

Barrie Johnson

8.1 INTRODUCTION

Sulfur-rich wastewaters that contain relatively little dissolved organic carbon may be generated as a result of several industrial processes: for example galvanic processes, scrubbing of flue gases at power plants and detoxification of metal-contaminated soils (Table 8.1). However, the mining of heavy metals and coal is, far and away, responsible for generating the bulk of this type of industrial wastewater (Banks *et al.* 1997: Geller *et al.* 1998). Metal- and sulfate-rich effluents produced at the sites of active and derelict mines (Table 8.2) are frequently referred to as "acid mine drainage" (AMD) or, alternatively, as "acid rock drainage". The pollution of water courses on a global scale as a result of AMD production is immense, and occurs in many western countries which have a long history in mining, and in developing countries where modern large-scale mining operations have been initiated. In the United Kingdom, for example, over 600 km of streams and rivers are estimated to be affected by AMD effluents from abandoned coal and metal mines.

8.2 SULFUR-RICH WASTEWATERS ASSOCIATED WITH MINING ACTIVITIES

There is a long history of sulfur-rich wastewater production resulting from mining activities. For example, the Rio Tinto in southern Spain has been associated with mining activities for over 5,000 years (Lopez-Archilla *et al.* 1995). Not all mining activities result in AMD production. Those which are chiefly implicated are coal and metal mining. The latter including the mining of base (e.g. copper, lead and zinc) and precious (e.g. gold) metals. Although AMD has the general characteristics of being rich in sulfate and dissolved metals, and generally of low pH, the chemical nature of these wastewaters can vary greatly on a regional basis and from site to site within a region (e.g. Table 8.2).

8.2.1 Origin of acid mine drainage

Acid mine drainage is formed by the dissolution of products resulting from the oxidation of sulfide minerals. Coal contains both inorganic sulfur (principally iron sulfides and sulfate) and organic sulfate (chiefly sulfide, thiophene, sulfoxide and sulfone). The relative proportions of these sulfurous compounds vary with different ranks of coal and degrees of oxidation. The content of sulfur in coals is generally between 1 and 10%. Many metals of commercial importance are chalcophilic; sulfides account for the most significant minerals and ores of, amongst others, copper, lead and zinc (Johnson 1995a). Metal ores frequently contain more than one type of sulfide mineral. Pyrite (FeS_2) is a major gangue mineral in these ores and, after separation (e.g. by froth flotation) from those minerals which have commercial value, often accounts for the chief reactive mineral in tailings deposits.

Sulfide minerals are stable in situations where both oxygen and water are excluded, such as an undisturbed ore body or coal seam. However, in moist aerobic environments these minerals tend to be unstable, and oxidize spontaneously to form (ultimately) sulfate, the free metal or metalloid (which may be subject to subsequent hydrolysis and precipitation) and, if the net metal:sulfur ratio is < 1, hydrogen ions. Several factors determine the rate at which sulfide minerals oxidize, such as the type of mineral and its surface area. In addition, certain lithotrophic ('rock eating') bacteria are known to accelerate the oxidation of sulfides by a factor of 10^4 - 10^6 and therefore render spontaneous chemical oxidation of these minerals of little importance. Much is now known of the biodiversity and nature of bacteria that oxidize mineral sulfides (e.g. Rawlings 1997; Johnson 1998), and a multi-billion dollar biotechnology industry ('biomining') has developed over the past decades which makes use of these microorganisms to extract metals from low-grade base metal ores and refractory gold ores (Rawlings 1997). In the case of gold, bioleaching is carried out with 'refractory' ores in which the fine-grain precious metal is in intimate contact with sulfide minerals (chiefly pyrite and arsenopyrite). Bacterial oxidation of the sulfides allows recovery of the gold by the conventional practice of cyanidation (Rawlings and Silver 1995).

The mechanism(s) by which bacteria actually solubilize these hard, dense minerals has been the subject of considerable debate (chapter 5). Current consensus is that ferric iron acts as the key mineral oxidant in both biological and chemical oxidation of sulfide minerals (Evangelou 1995; Schippers *et al.* 1996). The oxidation of pyrite by acidophilic bacteria that are either attached to the mineral or spacially removed from it (previously referred to as 'direct' and 'non-direct' attack) is outlined in Figure 8.1. The fate of the thiosulfate formed during pyrite oxidation is complex. This

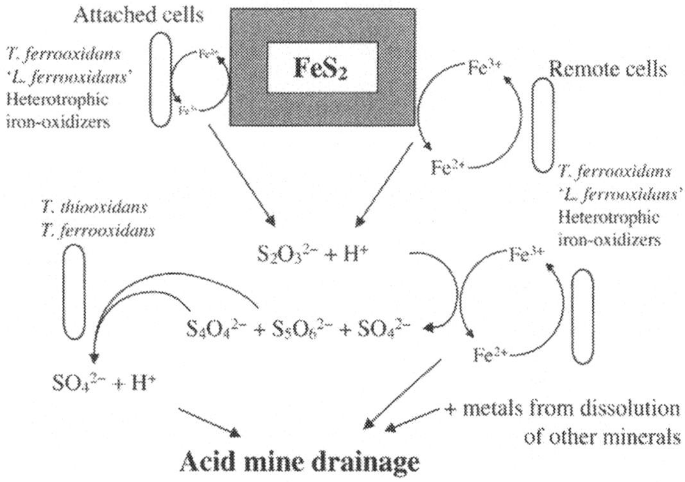

Figure 8.1. Role of acidophilic microorganisms in the genesis of acid mine drainage.

particular polythionate is highly unstable in even mildly acidic solutions, and in the presence of soluble ferric iron it is rapidly oxidized to tetrathionate, pentathionate and sulfate (Schippers *et al.* 1996). Whilst the concentrations of reduced inorganic sulfur compounds present in AMD are generally unknown, though Pichtel and Dick (1991) detected thiosulfate, trithionate and tetrathionate at concentrations of about 0.2-1 mmoles/kg spoil in pyritic coal spoils. These compounds are substrates for a number of acidophilic and neutrophilic chemolithotrophic bacteria (and some heterotrophic microorganisms), notably *Thiobacillus* spp., and are oxidized ultimately to sulfate (Figure 8.1).

Figure 8.2 illustrates the major locations of AMD production at a hypothetical abandoned deep mine where waste materials have been deposited in heaps. During the active life of the mine, the water table is maintained at artificially low levels by pumping water to the surface. However, when mines are abandoned, pumping generally ceases and water levels rebound, at highly variable rates. The humid aerated atmosphere in the drained mine provides favourable conditions for sulfide mineral

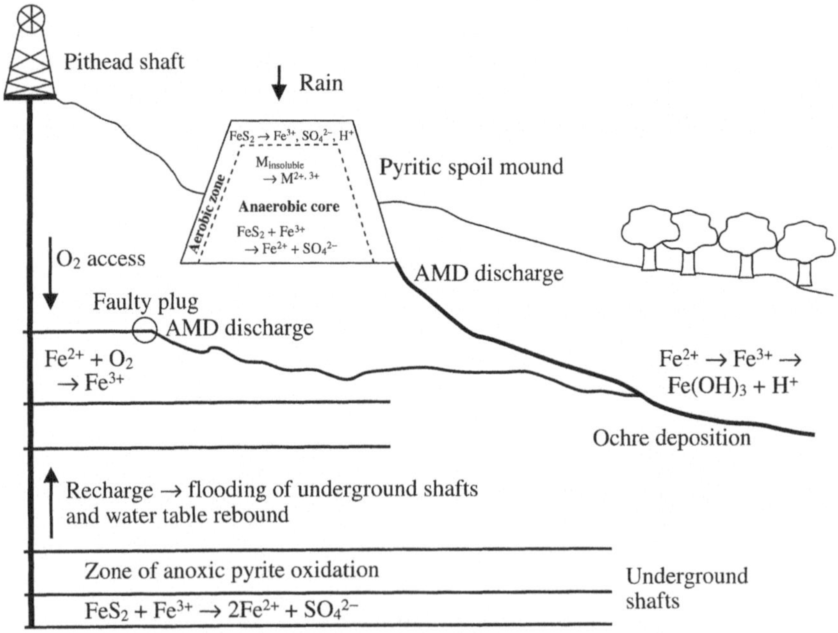

Figure 8.2. Acid mine drainage formation at a hypothetical derelict deep mine and associated spoil heap.

oxidation, and the up-welling waters cause iron, sulfate and other oxidation products to come into solution, so that the initial breakout of water tends to be the most acidic and polluting. Flooding mines limits oxygen diffusion rates and therefore sulfide oxidation, though this may be modified by fluctuating water tables. In the case of surface heaps and mounds, an aerated outer skin overlies an anaerobic core. In the aerobic region sulfide mineral oxidation may proceed rapidly. However, the oxidation of pyritic minerals can also occur within the anaerobic cores of mound, due to the anoxic attack of soluble ferric iron, contained in the waters percolating from the outer skin (equation 8.1). Similarly, drainage water from the mounds tend to be ferrous iron-rich at the point of discharge. Biological oxidation downstream changes the iron speciation, giving rise to the characteristic deposition of orange-brown 'ochre' (or 'yellow boy') in affected streams and rivers (Figure 8.2). Figure 8.3 shows the typical landscape of an abandoned mineral sulfide mine (Parys mountain, Anglesey, north Wales, once the world's major copper-producing mine) which has not been subject to restoration and remediation.

8.2.2 Chemical characteristics of AMD

The term 'acid mine drainage' is often used to describe all iron- and sulfate-rich wastewaters draining mine sites. Describing them as acidic is, however, not always appropriate, as these waters may have pH values close to neutral, particularly at the point of discharge (Clarke 1995). Those AMD waters which are at pH > 4.5 are considered to contain alkalinity, due to the presence of (predominantly) dissolved bicarbonate. The acidity of AMD derives from both hydrogen ions ('proton acidity') and dissolved metals ('mineral acidity'). The most important metals which contribute towards mineral acidity in AMD are iron (ferrous and ferric), aluminum and, usually to a lesser extent, manganese. The acidity of these dissolved metals is associated with their hydrolysis (preceded in the case of ferrous iron and manganese (II) by oxidation), as illustrated in equations 8.1 to 8.3:

$$Fe^{3+} + 3H_2O \rightarrow Fe(OH)_3 + 3H^+ \quad (8.1)$$

$$Al^{3+} + 3H_2O \rightarrow Al(OH)_3 + 3H^+ \quad (8.2)$$

$$Mn^{4+} + 2H_2O \rightarrow MnOOH + 3H^+ \quad (8.3)$$

Total acidity of freshly discharged AMD can be calculated from measuring the concentrations of these three metals. If this exceeds the bicarbonate alkalinity, the waters are assumed to be net acidic.

Figure 8.3. View of the former Parys copper mine on Anglesey, north Wales. The area remains mostly unvegetated after more than 100 years of being abandoned. A highly acidic (pH 2.5) iron- and sulfate-rich pool which has formed within one of the two main opencast voids can be seen.

The chemical composition of AMD is highly variable, as illustrated in Table 8.2. Elevated concentrations of iron and sulfate (compared to non-polluted surface waters) are their most characteristic feature, but other metals and metalloids may also be present at relatively high concentrations. The latter may also derive from the oxidation of sulfide minerals (e.g. copper from chalcopyrite; arsenic from arsenopyrite) or from accelerated weathering of non-sulfide minerals (e.g. aluminum from aluminosilicates). The solubility of most (cationic) metals (including some of the more toxic metals such as cadmium and lead) is far greater in acidic than in neutral and alkaline waters. The more acidic mine wastewater streams may therefore act as conduits for distributing metals to water courses (including groundwater) considerable distances away from their sites of origin.

Redox potential is a particularly useful physico-chemical index of AMD chemistry, since their *E*h values are dictated to one of two major redox systems that occur in this type of polluted water (Nordstrom *et al.* 1979;

Boulegue and Michard 1979). One is the ferrous/ferric couple and the other is the sulfide/sulfate couple. Both E^o values are affected by pH. In the case of the ferrous/ferric couple this is due to the very different solubilities of ferrous and ferric compounds above pH 2.5-3. The E^o value of the ferrous/ferric couple at pH 2 (where both ions are soluble) is +770 mV, whereas at pH ~7 the E^o of the ferrous carbonate/ferric hydroxide couple is +200 mV (Ehrenreich and Widdel 1994). For the sulfate/sulfide couple, the 'pH effect' results from the half-cell reaction (equation 8.3) which involves hydrogen ions:

$$SO_4^{2-} + 10H^+ + 8e^- \rightarrow H_2S + 4H_2O \qquad (8.4)$$

The E^o of this couple increases from -221 mV at pH 7 to +149 mV at pH 2. The large difference in E^o values is such that it may be readily inferred from measured *E*h values which of the two redox couples dictates the AMD redox potential in a particular situation. The two redox systems are mutually exclusive since ferric iron rapidly oxidizes sulfide. The relatively high redox potentials in surface waters at the Cae Coch and Parys mine sites imply the dominance of the ferrous/ferric couple. In contrast, the relatively low *E*h values of the discharge water at Ynysarwed (Table 8.2) and sediments in AMD streams are due to the sulfide/sulfate redox system.

Concentrations of dissolved inorganic carbon (DIC) in low pH (< 3.5) AMD waters tend to be very low (< 0.5 mg/L) due to the dissociation of carbonic acid and the low solubility of CO_2 in acidic waters. In acidic mining lakes, DIC concentrations in the hypolimnion may greatly exceed those in the epilimnion. Concentrations of dissolved organic carbon (DOC) also tend to be very small (generally < 10 mg/L), and are comparable with those in neutral/non-acidic oligotrophic waters. Some organic carbon may originate from the degradation of, for example, wooden pit props in underground mines. Primary production in AMD waters themselves may be mediated either by acid-tolerant phytoplankton and macrophytes, or by chemolithotrophic bacteria.

The dominant form of nitrogen in AMD tends to be ammonium, though significant concentrations of nitrate may occur where explosives are used. Levels of total nitrogen in AMD tend to be well in excess of total phosphorus. The latter is often considered to limit productivity in AMD streams and lakes (Nixdorf *et al.* 1998).

8.2.3 Impact of AMD on the biosphere

Acid mine drainage waters have a deleterious impact on most aquatic organisms, though these effects may be immediate or protracted. The major

Table 8.1. Examples of inorganic sulfate-containing wastewaters produced by industrial activities

Wastewater origin	Sulfate (g/l)	Sulfite (g/l)	Other pollutants	Reference
Industrial activity				
Chemical industry	0.2–50	0–5		Stucki *et al.* (1993)
Mining industry	0.1–20	–	Heavy metals	Banks *et al.* (1997)
Galvanic industry	0.2–50	0–25	Heavy metals	Tichy *et al.* (1998)
Flue gas scrubbing	1–2	1–2		Dijkman (1995)

effects are due to: (i) presence of toxic metals and metalloids; (ii) low pH values of the more severely acidic waters; (iii) precipitation of iron-rich ochre deposits on stream sediments, and (iv) osmotic stress. AMD-impacted water courses tend to be devoid of fish (due to mortality, and also because of loss of spawning gravel), have limited biodiversity in planktonic and benthic organisms, and display lower rates of primary production, compared to non-polluted surface waters.

Toxic metals and metalloids in AMD are more usually encountered in waters draining metal mines than in those draining coal sites (Table 8.1). In low pH AMD, proton acidity can have a direct toxic effect (e.g. to fish, causing damage to gills) or an indirect effect, by enhancing the solubility of metals. The deposition of ochre from ferruginous mine drainage (resulting from oxidation and hydrolysis of iron) can result in long stretches of down-stream waters being impacted. These iron-rich precipitates smother stream and river sediments, impeding oxygen diffusion and killing the majority of benthic organisms. This may have a dramatic effect on the food chain in impacted streams and rivers, with salmonid species being particularly sensitive to this type of pollution (Pentreath 1994). In faster-flowing waters, the ochre may remain in suspension rather than settle out. In that case, the turbidity of the water can have a major impact on light penetration, and thus negatively effect algal biomass and biodiversity (Robb 1994).

Biomass and biodiversity of phytoplankton in AMD lakes follow the same general trends as those described above (Nixdorf *et al.* 1998) though exceptions to this pattern have been observed. Some macrophytes may also establish in AMD waters. Of these, the most important in northern Europe is *Juncus bulbosus*, while *Phragmites australis* and *Typha latifolia* are also common (Pietsch 1998). Primary production by phytoplankton in acidic mine lakes in Germany was noted to be more limited by the availability of light, inorganic carbon and phosphorus rather than by the low pH *per se* (Nixdorf *et al.* 1998).

Most microorganisms are also severely damaged or killed upon exposure to AMD. Microbial communities that develop in acid-polluted streams are very different from those which are found in local non-polluted streams (Mills and Mallory 1987). However, it is now appreciated that the biodiversity of obligately acidophilic microorganisms active in highly acidic ($pH < 3$) metal-rich waters, such as AMD, is much greater than was previously recognised. It includes chemolithotrophic (iron- and/or sulfur-oxidizing) and heterotrophic bacteria, fungi and yeasts, protozoa and rotifera (Lopez-Archilla *et al.* 1995; Johnson 1998). Stable mixed communities of these microorganisms may establish in AMD and AMD-impacted waters; interactions between indigenous microbiota are akin to those which occur in non-polluted ecosystems (Johnson 1998).

8.3 PREVENTION, CONTAINMENT AND TREATMENT OF AMD

Various mechanical, chemical and biological approaches have been used to preclude the formation of AMD or, where this is not feasible, to remediate affected waters. 'Source control' measures are designed to prevent or limit AMD production, while 'migration control' is aimed at restricting the movement of contaminated waters.

8.3.1 Non-biological systems

Most source control approaches aim to limit AMD production by isolating potentially acid-forming materials from exposure to either oxygen or water. Flooding of underground mines is appropriate if they can be securely sealed, though this is often a more risk-laden strategy and is not feasible where underground passages are intricately connected, and detailed, accurate knowledge of mine workings are unavailable. Underwater storage of mine tailings may be effective, particularly when the tailings are covered with a layer of sediment to further limit oxygen penetration and reduce surface perturbations. Land-based waste heaps may be sealed with covers of clay, plastic liners or organic materials, both to reduce oxygen access and (in the case of plastic liners and clay skins) to limit water percolation.

Not all abandoned mines and mine spoils will produce acidic effluents, since some may be virtually devoid of acid-forming minerals, and others may contain sufficient quantities of neutralizing materials (carbonates etc.). Blending of potentially acid-generating with acid-consuming materials has been used to produce environmentally-benign composites (e.g. Mehling *et al.* 1997). A more recent technique is to form highly insoluble iron phosphates on the surfaces of oxidizing pyritic minerals by phosphate amendment (Evangelou 1995). Anionic surfactants have been demonstrated to inhibit bacteria which accelerate the oxidation of sulfides. However, the effectiveness of controlling AMD production by applying these chemicals tends to be highly variable (Loos *et al.* 1989).

The most widespread method used to remediate AMD is to aerate and add a neutralizing chemical, a process generally decsribed as 'active' treatment. Different neutralizing agents may be used, such as lime (calcium oxide), calcium carbonate and sodium hydroxide. The major objective is to accelerate the oxidation of ferrous iron (for which active aeration is also often necessary) and to precipitate iron and other metals as their hydroxides and/or carbonates. Some removal of sulfate (as gypsum) is achieved when calcium-containing neutralizing materials are used. Although active treatment can provide effective remediation of AMD, it has the disadvantages of high operating costs, and problems with disposal of the bulky sludge produced. The latter tends to be highly voluminous, typically containing only 2 - 4% (by weight) of solids. An alternative approach for adding alkalinity to AMD is the use of 'anoxic limestone drains' (ALDs). With these, waters containing little or no dissolved oxygen are fed through trenches or pits containing limestone. Iron oxidation has to be avoided with ALDs, as ferric iron deposits inactivate the limestone, a process known as 'armouring'. Anoxic limestone drains are often used in association with biological treatment of AMD, as described below.

8.3.2 Biological systems

There are a number of biological processes which generate alkalinity (or consume acidity) and which therefore have potential use in remediating AMD. These include photosynthesis, ammonification, and several reductive processes (methanogenesis, and iron and sulfate reduction). Of these, iron and sulfate reduction are obvious candidates for AMD bioremediation, given the chemical nature of AMD (see 8.2.2). Acid consumption resulting from the reduction of ferric iron is depends on whether the latter is present as soluble or mineral form. Whilst the reduction of soluble ferric iron does not generate alkalinity, the reduction of amorphous and soluble ferric iron minerals does (equation 8.5):

$$4Fe(OH)_3 + CH_2O \rightarrow 4Fe^{2+} + H_2CO_3 + 2H_2O + 8OH^- \quad (8.5)$$

The reduction of sulfate to sulfide consumes protons in acidic solutions, by transforming a strong acid (sulfuric) into a weak acid (hydrogen sulfide), as:

$$2H^+ + SO_4^{2-} + CH_2O \rightarrow H_2S + 2H_2CO_3 \qquad (8.6)$$

Both reactions require electron donors, which are generally organic ("CH_2O" in equations 8.5 and 8.6). An important additional feature of biological sulfate (as opposed to iron) reduction is that its product, sulfide, forms highly insoluble products with many toxic heavy metals that occur in AMD. The very low solubility products of many metal sulfides means that these metals are removed highly effectively in the presence of only trace amounts of soluble sulfide (Diaz *et al.* 1997).

8.3.2.1 Natural and constructed wetland ecosystems

An alternative approach to active treatment for remediating acidic metal-rich wastewaters which has become increasingly popular in the past two decades is to direct their flow through natural or constructed wetland ecosystems (Gazea *et al.* 1996). This 'passive' approach came about as a result of observations that outflows from sphagnum-dominated bogs displayed improved water chemistry compared with inflowing AMD. The increased pH of the outflow was accompanied by decreased concentrations of sulfate, iron and other metals. Wetlands are highly complex ecosystems, and modifications of water chemistry may be brought about by a number of mechanisms, including dilution, precipitation (by oxidative and reductive mechanisms), adsorption, and uptake by biomass. Where natural wetlands do not occur, artificial wetlands may be constructed if sufficient and suitable land area is available. This approach for bioremediating AMD has become particularly popular in the USA where many hundreds of wetland sites have been constructed to ameliorate acidic wastewaters from coal mining areas in Appalachia alone. Wetlands are also being used increasingly in Europe and other parts of the world for AMD remediation.

There are two main types of constructed wetland, which operate on radically opposed biogeochemical activities based on either oxidative or reductive processes. These may be used in tandem at a treatment site, often in association with anoxic limestone drains. The main objective of aerobic wetlands is to promote the oxidation of AMD, thereby causing metals (chiefly iron) to oxidize and precipitate. Other AMD constituents, such as arsenic, will co-precipitate with the ferric deposits. Aerobic wetlands are relatively shallow systems, within which acid- and metal-tolerant macrophytes are planted in soil or other stratum, to control the water flow.

Sphagnum mosses have proved to be somewhat sensitive to changes in AMD chemistry and accumulation of iron within constructed wetlands, and therefore cattails (*Typha latifolia* and *Typha orientalis*) are generally the preferred vegetation. The oxidation and precipitation of iron can result in significant production of proton acidity as AMD flows through aerobic wetlands, which slows down the rate of spontaneous (chemical) oxidation. This may have a negative impact on the performance of in-line compost (anaerobic) wetlands into which these waters flow.

Compost wetlands (or anaerobic organic substrate systems) function by actively generating alkalinity as a result of microbiological reductive activities. Dissolution of basic minerals within the compost mix will also contribute to acid neutralization. The reduction of both iron and sulfate are considered to be important in anaerobic wetlands. Since these processes are fuelled by organic electron donors, various organic materials, such as cow and horse manure, spent mushroom compost, peat, sawdust and wood chips, have been used as substrates in compost wetlands. In some systems, cattails and other emergent vegetation provide a continuous supply of carbon to the system. However, root penetration may introduce oxygen into the anaerobic zone and reduce the effectiveness of the reductive processes. In some systems, surface plants are excluded, and the anaerobic 'cell' is so constructed as to limit any oxygen diffusion. Microbial populations that develop in anaerobic wetlands are, at least in the initial phases, the endogenous populations of the composting materials. Microbial inocula tend not to be used. However, inflowing AMD contains its own microflora, including obligate and facultative acidophilic and acid-tolerant bacteria. It is conceivable that these bacteria also colonize the composting material and (in the more acidic zones) displace the initial populations.

Passive treatment of AMD has been one of the options evaluated at the site of the former Wheal Jane tin mine in Cornwall, UK (Figure 8.4). The mine closed in 1991 and, following the failure of an adit plug less than one year later, a catastrophic pollution event occurred in which an estimated 30,000 m^3 of acidic, metal-rich effluent flowed into the Carnon River and from there to the ecologically-sensitive Fal estuary within a period of 24 h (Lamb *et al.* 1998). Peak and mean metal concentrations in AMD from the Wheal Jane mine are given in Tables 8.2 and 8.4. While the main-stay of the treatment facility is a highly effective active process, a large-scale pilot plant passive system was installed to evaluate the most appropriate configuration for wetland bioremediation of this and similar drainage waters. The system which was designed used both constructed aerobic wetlands (shallow reed beds of *Typha, Phragmites* and *Scirpus*, planted in coarse-grain mineral

Figure 8.4. Constructed wetlands for treating AMD from the former Wheal Jane tin mine in Cornwall, south-west England. Aerobic, anaerobic (buried) wetlands and a rock filter system are shown, in sequence, from right to left.

tailings) and anaerobic cells (using mixtures of hay, sawdust and cattle manure, the latter providing both the priming substrates and the inoculum of sulfate reducing bacteria). The aerobic wetlands promoted the removal of iron (and arsenic) and the non-vegetated anaerobic cells the precipitation of copper, cadmium and zinc as metal sulfides. Manganese concentrations in Wheal Jane AMD were also significant, and rock filters were added as a final 'polishing' stage. The rock filters utilize algae to create high pH micro-environments where manganese oxidizes and precipitates as MnOOH Three variants of the basic aerobic-anaerobic wetland system were evaluated, which compared different pre-treatments of the raw AMD (Figure 8.5). Typical data from the performance of the anaerobic cell systems are shown in Table 8.3. In general, pre-treatment using anoxic limestone drainage has proved to be the most effective of the three variants, though all three systems have added further support for the efficacy of bioremediation of AMD using constructed wetlands. However, one of the problems highlighted by the Wheal Jane pilot plant system is the land area required to treat large volumes of AMD effectively. In order to achieve 95% compliance with water quality objectives, a land area far greater than that

Table 8.2. Physico-chemical data of AMD waters draining from various metal and coal mines

	Britannia Cu mine BC, Canada	Avoca Cu/Zn mine Ireland	Skorovas Cu/Zn mine Norway	Cae Coch Pyrite mine Wales	Parys Cu mine Wales	Wheal Jane Sn mine* England	Idaho Spr. Metal mines U.S.A.	Oatlands Coal spoil England	Ynysarwed Coal mine Wales
pH	2.4–3.6	2.7	2.36	2.2–2.5	2.2–2.8	2.8	2.9–3.0	5.5	6.3
E_h (mV)				+660–+690	+660–+780		+510–+710		+ 220
Conductivity (µS/cm)		8,010	5,922	1,500–1,800	650–1,500				1,450
SO_4^{2-} (mg/l)	200–1,950	10,579	7,291	1,300–2,700	1,300–2,700		1,660–1,750	146	2,000
Fe (mg/l)	0.4–55	1,031	2,284	1,356–2,261	150–550	5,070	33–46	287	300
Al (mg/l)	4–74			40–70	40–90	190		0.97	23
Mn (mg/l)				6–23	5–10	27	29–35	5.2	8
Zn (mg/l)	3–48	362	129	0.5–2	30–60	2130	8.5–10	0.05	
Cu (mg/l)	0.3–445	243	256	0.25–1.0	20–40	23	0.6–0.9	< 0.007	
Reference	Rowley *et al.* (1997)	Gray (1996)	Banks *et al.* (1997)	McGinness & Johnson (1993)	Walton & Johnson (1992)	Lamb *et al.* (1998)	Machemer & Wildeman (1992)	Banks *et al.* (1997)	Johnson (unpublished data)

*Peak discharge levels.

Table 8.3. Wheal Jane passive treatment plant: performance of the anaerobic cells (Data from: the Environment Agency, UK)

	Phase 1			Phase 2		
	Lime-free system	ALD system	Lime-dosed system	Lime-free system	ALD system	Lime-dosed system
Flow Rate (l/s)	0.2	0.1	0.1	0.2	0.1	0.1
pH change (inflow → outflow)	2.9/3.1 → 4.2/4.8	3.3/4.6 → 5.7/6.3	2.8 → 3.3/5.6	3.7 → 5.5	6.0 → 5.8	4.5/6.5 → 5.7/6.4
E_h change (mV) (inflow → outflow)	~+500 → +180	+235/+460 → −208/+75	~+500 → −93/+158	+305 → +164	+98 → +57	+50/+285 → −23/+126
SO_4^{2-} removal (%)	5–9	7 to 39	5 to 32	12	13	27 to 38
Fe removal (%)	net mobilization	−82 to 62	−42 to 81	33	42	52 to 81
Zn removal (%)	29 to 45	85 to >99.9	85 to >99.9	78	85	99 to >99.9
Cu removal (%)	>99.9	*	>99.9	>99.9	*	>99.9
Cd removal (%)	>99.9	*	>99.9	>99.9	*	>99.9

'Phase 1' data are from the earlier operation of the pilot plant, when water entering each of the anaerobic cells had flowed through the aerobic cell. 'Phase 2' data are results following modification of the pilot plant which allowed water to flow directly into the anaerobic cells after any pre-treatment (anoxic limestone drain (ALD) or lime-dosing, by-passing the aerobic cells. 'Net mobilization' and negative figures for iron removal indicate net release of ferrous iron as a result of flowing through the anaerobic cells.

* In the ALD system, most of the copper and cadmium was removed from the AMD before the anaerobic cell in the pre-ALD system.

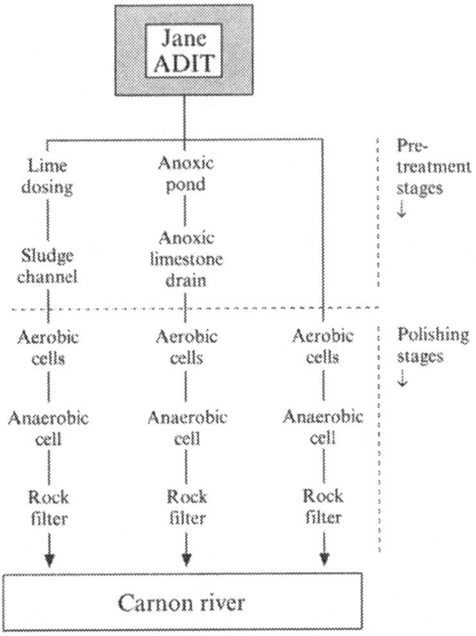

Figure 8.5. Layout of the pilot-scale passive treatment plant at the site of the former Wheal Jane tin mine, Cornwall UK (After: Lamb *et al.* 1998).

available at the site itself would be required to treat the AMD which discharges from the Wheal Jane mine at a rate of *ca.* 700 m^3/h.

The significance of sulfate reduction in wetlands impacted by AMD has been the subject of debate. Machemer and Wildeman (1992) produced evidence, from both field analyses and laboratory experiments, that net sulfate reduction occurred in a wetland site in Colorado used to remove heavy metals from an acidic (pH 2.9) waste stream (Idaho Springs; Table 8.2). The relative contributions of adsorption onto the organic materials (chiefly mushroom compost) and sulfide precipitation in removing heavy metals from the AMD were examined. These researchers concluded that, during the initial period of mine drainage flow into the wetland, adsorption was the more important process. However, latterly, as adsorption sites became saturated and sulfate reduction became more prevalent, this was superseded by sulfide precipitation. In contrast, Vile and Weider (1993) were unable to provide any evidence that net sulfate reduction occurred in five wetland sites in Kentucky that had been constructed using a variety of organic substrates, including mushroom compost. The reduction of ferric

Table 8.4. Bioremediation of acid mine drainage and related waters using SRB bioreactors: performance data

Site	Budelco Zn refinery		Kennecott Cu mine				Wheal Jane Sn mine			Britannia Cu mine	
			H_2S production		Cu recovery						
Parameter	Influent	Effluent	Influent	Effluent	Influent	Effluent	Influent	Thiopaq Effluent	Liming Effluent	Influent	Biosulfide Effluent
Flow (m^3/h)	300		0.2		5.5		1–1.5			3	
pH			2.5	8.5	2.6	2.2					
SO_4^{2-} (mg/l)	1,000	<200	30,000	<500				200	850		
Fe (mg/l)			675	<0.3	380	379	345	<0.1	0.2		
Zn (mg/l)	100	<0.05	65	<0.1	200	199	132	<0.05	0.03	16–17	0.01–0.4
Cu (mg/l)			60	0.1	180	<0.3	2	0.01	1.7	11–12	< 0.01
Cd (mg/l)	1	<0.001					0.08	<0.0005	0.004	0.09	< 0.01
Reference	Boonstra *et al.* (1999)		Boonstra *et al.* (1999)				Boonstra *et al.* (1999)			Rowley *et al.* (1997)	

oxyhydroxides was claimed to be the major process generating alkalinity in these wetlands. It was conceded that sulfate reduction was probably on-going at the sites, since gross sulfate reduction rates varying between 0 and 7.5 μmol g^{-1} dry substrate day^{-1} had earlier been recorded at these sites with the same organic substrates. In the later studies, the sulfides formed were considered to be oxidized abiotically by ferric iron, resulting in the observed absence of net sulfate reduction. Vile and Weider (1993) concluded that sulfate reduction may be an important mechanism for alkalinity generation (and iron retention) in newly-constructed wetlands, but with the subsequent accumulation of amorphous ferric compounds throughout the wetland substrate, biotic and abiotic reduction of iron becomes the dominant process.

8.3.2.2 Bioreactor systems

While it is generally accepted that biological processes (mostly driven by microorganisms) are the most significant involved in remediating AMD in natural and constructed wetlands, the fact that these ecosystems do not lend themselves to ready control and manipulation means that their performance is difficult to predict and is subject to seasonal and other variations. Various biotechnologies have developed from isolating 'useful' organisms from the environment and engineering conditions to promote their growth and activities in a controlled system. This approach has recently been applied to the mitigation of AMD and related waters. Two major biotechnology processes have been described: the *Biosulfide* and *Thiopaq* processes. In both, the central process is the generation of sulfide by neutrophilic sulfate reducing bacteria (SRB). These technologies have been applied as pilot plant, demonstration and full-scale treatments of acidic wastewaters from metal mines and related situations. The prime objective generally is to lower the levels of toxic metals to environmentally-acceptable concentrations. In addition, bioreactor systems offer the opportunity to lower sulfate concentrations often (to 25 - 500 mg/L), the selective precipitation of metals (which enables their recovery), and the production of waste sludges which, although susceptible to oxidation, are more compact and denser than equivalent metal hydroxide sludges.

The *Biosulfide* process (Rowley *et al.* 1997) essentially utilizes the biogenic production of H_2S by SRB to remove heavy metals from AMD waste streams. The biological and chemical systems of the process operate independently. Raw AMD entering the processing plant is dosed with biogenic H_2S. If the pH and sulfide concentrations are carefully controlled, this can result in the selective precipitation of a particular metal (e.g. copper in a waste stream containing a mixture of iron, zinc and copper). The waste stream can then proceed to a second precipitation tank where further (selective) metal precipitation (e.g. zinc) can be induced by controlling pH and flow rates. The biological and chemical circuits are designed to run independently. A major reason for this is to protect the SRB in the sulfide-generating bioreactor from the potentially debilitating effects of high acidity and metal concentrations in the AMD streams. Although it would be feasible, in theory, to feed AMD directly into the SRB system, the bioreactor would either have to be very large or else an excessively long retention time would be necessary, and both of these impositions would render the process non-viable. The sulfide-rich solution produced in the SRB reactor may be stored and used to treat peaks of AMD discharges. A pilot plant using the *Biosulfide* technology was established at the abandoned Britannia copper mine in British Columbia, Canada (Figure 8.6). The system involved two SRB bioreactors and used hydrogen gas as the bacterial energy source, and carbon dioxide gas as carbon source. The system was capable of effectively treating flow rates of up to 50 litres AMD/minute, and removal efficiencies of 99.9% copper, 97.8 - 99.9% zinc and 90 - 99% cadmium were routinely achieved (Table 8.4). The microbial consortium in the SRB bioreactors had originally been obtained from a mixture of sewage sludge and bog water and had been adapted to wide variations in feed water chemistry over a period of several years prior to commissioning of the pilot plant.

The *Thiopaq* process (de Vegt *et al.* 1997; Boonstra *et al.* 1999) utilizes

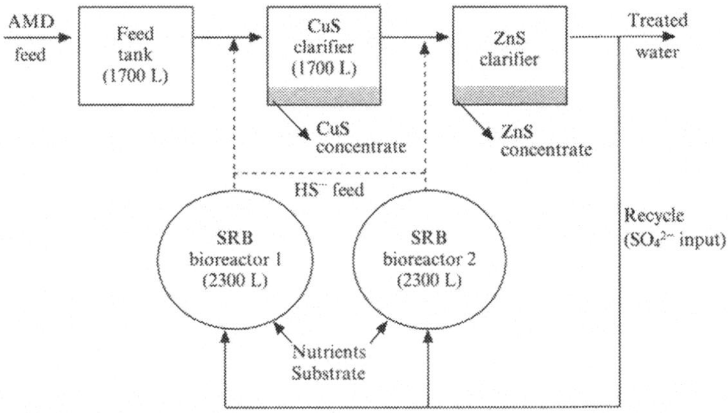

Figure 8.6. Flow diagram of the pilot-scale treatment of AMD at the Britannia copper mine, BC, Canada (After: Rowley *et al.* 1997).

two different biological processes, in contrast to the single bio-process in the *Biosulfide* technology. As before, SRB are used to convert sulfate to sulfide in the first stage in the operation (again generating net alkalinity and promoting the precipitation of chalcophilic metals in the wastewater). Excess sulfide is oxidized (under controlled redox potential) in a separate reactor to produce elemental sulfur. Two useful products may therefore be produced: metal sulfides which may be processed to recover the metal(s) concerned, and hydrophilic sulfur, which may be used to produce sulfuric acid or as an agrochemical by-product.

The technology has been demonstrated at a number of pilot scale operations (Table 8.4), and a full-scale plant has been in operation (since 1992) to remediate acidic metal-rich groundwater at the site of a zinc smelter (Budelco) in the Netherlands (Barnes *et al.* 1994; de Vegt *et al.* 1997; Boonstra *et al.* 1999). At Budelco, over 100 years of zinc refining has produced severe acid and metal contamination of groundwater and soils adjacent to the site, and there was concern about the pollution plume spreading to nearby aquifers which were sourced for drinking water. A hydrogeological containment system was designed, in which it was proposed that contaminated groundwater should be pumped to the surface, processed so that its quality fell within the limits set by the Dutch authorities, and then discharged into a nearby river. Various options were considered, and the decision was made, following extensive laboratory and pilot-scale testing, to install a biological treatment facility. The system installed utilized an upflow anaerobic sludge blanket (UASB) reactor for the sulfate reductions step, and a submerged fixed film reactor for the aerobic conversion of sulfide to elemental sulfur. Typical effluent concentrations of zinc, cadmium and sulfate in processed groundwater at Budelco are shown in Table 8.4.

Among the demonstration pilot plant operations that have shown the applicability of SRB bioreactors for remediating AMD are systems installed at the Kennecott Bingham Canyon copper mine (Utah, U.S.A.) and the Wheal Jane tin mine in Cornwall, U.K. A simplified flow diagram of the pilot plant at the Kennecott mine is shown in Figure 8.7. This system comprised a component in which the prime objective was to remove (and subsequently recover) copper from the AMD by contacting with biogenic hydrogen sulfide, and a second component in which sulfate reducing bacteria were used to lower the sulfate concentration of the AMD, and the hydrogen sulfide so produced was oxidized (as in the Budelco operation) to form elemental sulfur. Data from this pilot plant and a similar pilot plant operation set up at the Wheal Jane site are also shown in Table 8.4.

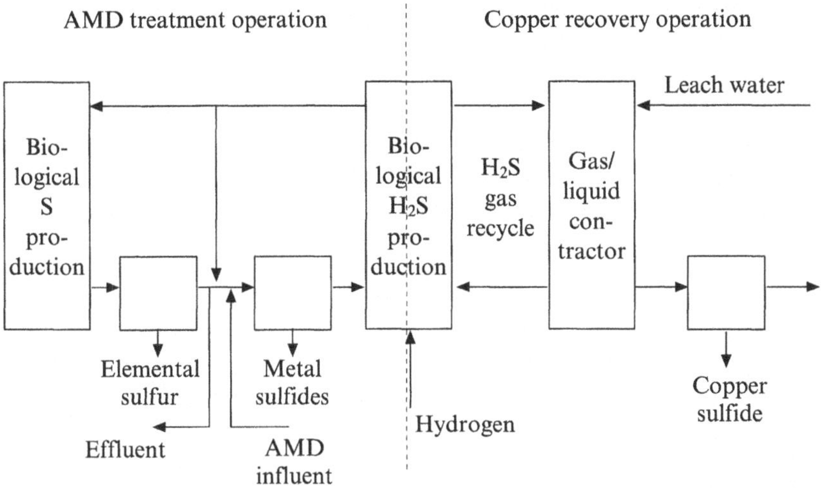

Figure 8.7. Flow diagram of the pilot-scale bioreactor at the Kennecott Utah copper mine for treatment of acid mine drainage and recovery of copper (After: Boonstra *et al.* 1999). Data of the leach water and AMD influent are given in Table 8.4.

Biological treatment of AMD and similar wastewaters has several major advantages over active (chemical) treatment. Operating costs of the latter are high, metal removal efficiencies tend to be relatively low, bulky sludges requiring dewatering and landfill are produced, and all potentially valuable metals which accumulate in the sludge are eventually lost. Biological treatment facilitates the recovery of chalcophilic metals (and of sulfur, when integrated with elemental sulfur production), the sludge produced is far denser (some 6 - 10 fold), and sulfate concentrations can be lowered to potable water levels. On the other hand, biological treatment requires fairly sophisticated and expensive engineering systems, and the process need to be fuelled. It has been estimated that, compared with the current active (lime) treatment operation employed at the Wheal Jane site, controlled biological treatment would result in higher quality effluents, reducing the annual discharge from the site by approximately > 600 kg iron, > 9,900 kg zinc, > 120 kg copper, 21 kg cadmium, 3,600 kg aluminium, 9,600 kg manganese and 3,895 ton sulfate (Boonstra *et al.* 1999).

8.3.3 Advantages and disadvantages of current AMD remediation technologies

Of the three types of technology currently used to remediate 'inorganic' wastewaters such as AMD, two make use of SRB to generate alkalinity, remove heavy metals and lower concentrations of sulfate. All three approaches have their limitations and detractions. Abiotic (lime) treatment is expensive to maintain, produces a bulky sludge and generally does not lower sulfate concentrations very effectively. Wetlands can be constructed only where there is sufficient suitable land area available. Their performance tends to be variable, and net sulfate reduction may not always be apparent, particularly in older systems. Advanced technology systems using SRB bioreactors have several advantages, including consistency of performance and the potential for recycling materials (metals and sulfur) from the wastewaters which have commercial value. Bioreactor plants, however, require significant start-up capital and current systems use pure-grade chemicals (ethanol or hydrogen) as electron donors for the bacteria. Alternatively, methanol may be utilized as electron donor by some SRB (Hard and Babel 1997). Alternatively, organic waste products, e.g. molasses (Maree and Strydon 1987), sewage digest (Kaufman *et al.* 1996) and potato skins (Castro *et al.* 1999), may act as suitable substrates for mixed cultures of SRB in bioreactors treating inorganic wastewaters.

8.4 BIOLOGICAL REMEDIATION OF INORGANIC WASTEWATERS: CONSTRAINTS AND OPPORTUNITIES

8.4.1 Evidence for *in situ* sulfate reduction in acidic wastewaters

The ubiquity and abundance of sulfate in AMD and related wastewaters would suggest that anaerobic respiration using this anion as an electron sink would be widespread in impacted stream and river sediments. However, two factors mitigate against this: (i) the fact that AMD generally contains very small concentrations of metabolisable organic materials (section 8.2.2.), and (ii) the well-documented sensitivity of characterized strains of SRB to even mild ($pH < 5.5$) acidity. There have, nevertheless, been periodic reports of active sulfate reduction in AMD-impacted ecosystems. Tuttle *et al.* (1969) noted that a pond which was located downstream of a porous dam (containing wood dust) through which an AMD stream flowed (pH 2.84) contained detectable numbers (ca. 876/ml) of SRB, though these only accounted for about 1% of the total heterotrophic bacterial population (enumerated via plate counts). The pH in the downstream pond was slightly greater (pH 3.38) and the sulfate concentration about 30% lower than the upstream AMD water. These researchers obtained mixed bacterial populations that reduced sulfate in the laboratory using sawdust as carbon and energy source at pH 3.0. However, none of the pure cultures of SRB they obtained were active below pH 5.5. Herlihy and Mills (1985) observed enhanced sulfate reduction in sediments in an arm of a freshwater lake that received AMD compared to another part of the lake that was not affected by the effluent. Gyure *et al.* (1990) detected reduction of $^{35}SO_4^{2-}$ in sediment slurries in a pH 3.8 strip mine lake, though the optimum pH for sulfate reduction in sediment samples was found to be pH 5. Johnson *et al.* (1993) described zones of iron and sulfate reduction in sediment (pH 2.8 - 4.4) in a stream (pH 2.3) draining a derelict copper mine. They also detected acid tolerant SRB in gelatinous 'acid streamers' (pH 2.3 - 2.4) within a subterranean stream in an abandoned pyrite mine. Incubation of surface ('aerobic') streamer growths under anaerobic conditions, amended with a suitable organic substrate (such as glycerol), resulted in an initial phase of iron reduction (marked by bleaching of the orange color of the AMD in which the streamers were immersed) followed by a phase of sulfate reduction (marked by the blackening of the cultures and the evolution of H_2S). Fortin *et al.* (1996) and Fortin and Beveridge (1997) noted that sulfate reduction occurred in acidic (pH 3 - 4) mine tailings and recovered SRB from these environments. However, these authors concluded that the SRB survived and were active within micro-environments within the tailings in which the pH was presumed to be higher and the redox potential lower that the measured bulk values of the sediments. SRB isolated from these mine tailings were unable to grow in low pH (< pH 5.5) media in the laboratory.

8.4.2 Physiological constraints on sulfate reduction processes

Hamilton (1998) described, in an excellent concise review, the physiological constraints on SRB and how these define their environmental impact. For example, sulfate is unique among electron acceptors used by microorganisms in its pre-requirement to be activated. The redox potential of the sulfate/bisulfite couple (E_h = -516 mV, at pH 7) is lower even than that of the hydrogen/proton couple (E_h = -414 mV, at pH 7). Sulfate activation, to form adenosine phosphosulfate, consumes two ATPs per sulfate activated. There is also the requirement for energy-dependent transport of sulfate (and co-transport with H^+ or Na^+) into the cell. Compared to aerobic microorganisms, and many anaerobes that use electron acceptors other than sulfate, SRB are severely disadvantaged with regard to the energy yield that is achievable from catabolism of organic substrates. Coupled to this is the fact that many SRB carry out incomplete oxidation of organic substrates, and excrete small molecular mass organic metabolites, such as acetate. Hamilton (1998) stressed, however, that standard redox potentials tend not to be identical with values experienced in actual metabolic situations. Under actual conditions, thermodynamic parameters might be modified by, for example, removal of reaction products in a linked series of reactions (i.e. syntrophism), resulting in dis-equilibrium.

Organic acids, such as lactate, have widespread use as carbon and energy sources among SRB and are frequently used in enrichment media for isolating these bacteria. However, acidophilic bacteria tend to be sensitive to organic acids, with concentrations of <1 mM often proving inhibitory or lethal. The major reason appears to be that many organic acids (depending on their pK_a values) exist predominantly as undissociated (and lipophilic) molecules in acidic waters such as AMD. As such, they are able to pass freely through bacterial cell membranes. The internal pH of acidophiles tends to be circum-neutral (Norris and Ingledew 1992), which causes the organic acids to dissociate and therefore dis-equilibrium between internal and external concentrations of the (undissociated) acid, and further influx. The resulting accumulation of protons within the bacterial cells can exceed the buffering capacity of cell cytoplasm and cause severe acidification. Postgate's medium 'B' (the standard version containing about 30 mM lactate) is used widely for isolating neutrophilic SRB. It is unlikely that an acidic variant of this medium would succeed in promoting growth of acidophilic SRB since, at low pH (e.g. 2-4), the dominant form in the medium would be lactic acid (pKa 3.86). For the same reasons, other organic acid-based media, and even the inclusion of some redox-poising chemicals (thioglycolic acid and ascorbic acid) are inappropriate for culturing SRB at low pH. This is not to say, however, that acidophilic or acid-tolerant SRB would not be able to utilize organic acids. *Acidiphilium* spp., obligately acidophilic aerobic heterotrophs, can generally grow on small molecular mass aliphatic acids when present in low (μM) concentrations. Higher (mM) concentrations are lethal to these bacteria (Johnson *et al.* unpublished data). In line with this, Gyure *et al.* (1990) found that 5 mM concentrations of organic acids inhibited sulfate reduction at pH 3.8, but concentrations of around 0.1 mM stimulated activity. Fortin *et al.* (1996) noted that SRB isolated from minerals tailings were able use 1 mM acetate and formate as electron donors in medium adjusted to pH 7.5, but not that poised at pH 5.5 and below.

Temperature constraints may also effect the feasibility of SRB bioreactors used for treating inorganic wastewaters. Daily and seasonal temperature fluctuations may effect the efficiency of biological sulfide production. In some areas, most notably in the high latitudes and at elevation, low temperatures (<10°C) may prevail, which will limit *in situ* SRB activity. However, psychrophilic SRB have been isolated (Isakson and Teske 1996) which have temperature optima at around 20°C and which are active at <10°C.

One other factor that has been observed to affect the growth and activity of SRB is the availability of suitable surfaces for bacterial attachment and colonization (e.g. Bass *et al.* 1996). Lyew and Sheppard (1997) noted that SRB utilized in passive treatment of AMD required the presence of a sediment, within which suitable micro-environmental conditions (pH, oxygen concentrations) could be established. Using column reactors containing different types and sizes of gravel, Lyew and Sheppard (1997) were able to demonstrate quantifiable relationships between SRB activity and physical parameters within the column beds.

8.4.3 Evidence for the existence of acidophilic and acid-tolerant SRB

Understanding of the microbial ecology of metal-rich acidic environments such as AMD has advanced considerably in recent years. It is known that a wide biodiversity of microorganisms may occur in extremely acidic (pH < 2 - 4) environments. These are mostly prokaryotic, though obligately acidophilic eukaryotes have also been described (Johnson 1998). However, little is known about anaerobic metabolism is these ecosystems. Some acidophilic bacteria are facultative anaerobes, and may use ferric iron as an electron acceptor in anoxic situations (Johnson 1998). No fermentative or methanogenic prokaryotes have yet been identified in AMD and related waters. Most SRB that have been isolated from these environments have turned out to be neutrophilic or only mildly tolerant of acidity (section 8.4.1). However, there have been some limited reports on the isolation of pure cultures of SRB which are either acid-tolerant or acidophilic. Johnson *et al.* (1993), using sediments from a stream draining a copper mine and fragments of 'acid streamers' (section 8.4.1) as inocula, found no SRB activity in acidified Postgate B medium. In contrast, positive enrichments were obtained by using a 10 mM glycerol/0.02% yeast extract (pH 3.5) medium. The mixed culture obtained reduced sulfate in glycerol medium poised initially between pH 2.9 and 7.0; no reduction was observed at pH 2.5 and below. A non-motile sporulating rod resembling *Desufotomaculum* spp. was observed in these cultures (Johnson *et al.* 1993). Hard and Babel (1997) isolated SRB from wastewater muds and a copper mine using methanol as electron donor. One of the seven isolates (a Gram-negative curved rod) obtained was able to grow at pH 4.0 while none of the others grew at pH <6. These researchers also noted that *Desulfovibrio salixigens*, obtained from a national culture collection, grew using lactate as electron donor in media at ≥ pH 5.0. When using methanol as electron donor, *D. salixigens* was able to grow at even lower pH (4.5). Sen and Johnson (1999) obtained SRB from two acidic mine sites in Wales and water draining a geothermal area on the island of Montserrat, West Indies. All of the isolates were Gram-positive spore-forming rods, and all grew in media maintained at pH 3.5 in laboratory-scale upflow fixed-bed bioreactors, using either ethanol or glycerol as electron donors. A mixed culture obtained from the Parys copper mine sites (north Wales) was observed to generate alkalinity in acidic iron-rich media between pH 1.7 and 6.0. Two pure cultures of acid-tolerant SRB, obtained from Montserrat and the Cae Coch mine (north Wales), were shown to be very similar phylogenetically, and both were tentatively identified as strains of *Desulfosporosinus orientis*.

8.4.4 Alternative approaches to remediating AMD using SRB

There have been several laboratory-scale demonstrations of AMD remediation using SRB technology. Wakao *et al.* (1979), following up earlier observations by Tuttle *et al.* (1968), also used a wood dust-based medium as a substrate for SRB. However, it was found that the addition of a more readily-degradable carbon source, such as peptone, lactate or pyruvate, was necessary to demonstrate sulfate reduction in AMD liquors. Problems associated with the recalcitrance of some organic substrates have also been reported by other researchers. Bechard *et al.* (1994) found that the rate of biodegradation of straw was too slow to sustain microbial remediation of simulated AMD. Some cellulosic substrates, such as alfalfa hay and timothy hay were found to be appropriate substrates in their experimental system. Whey has also been used with some success in a laboratory-scale AMD bioremediation process (Christensen *et al.* 1996).

A slightly modified approach has been suggested by Johnson (1995b) and recently demonstrated on the laboratory scale (Kolmert *et al.* 1999). This particular two-stage bioreactor utilizes, in sequence, acidophilic heterotrophic bacteria that reduce ferric iron and acid-tolerant/acidophilic SRB, respectively, in the first and second fixed-bed reactors. In both bioreactors, the bacteria are immobilised onto porous glass beads (Figure 8.8). The logic of this design is that any ferric iron contained in influent

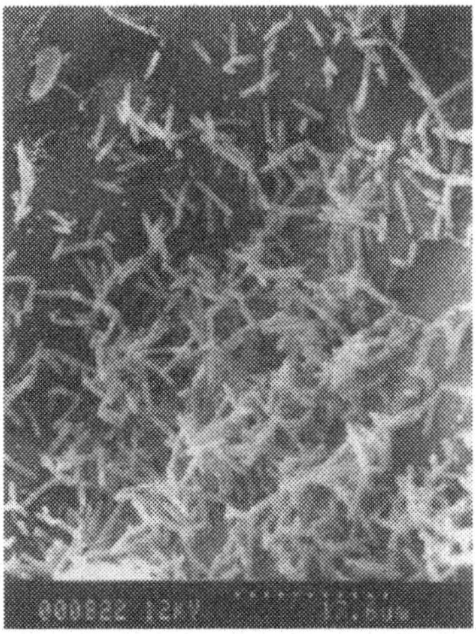

Figure 8.8. Biofilm of acidophilic SRB immobilised on porous glass beads in a fixed-bed bioreactor used for treating acid mine drainage.

AMD is reduced to ferrous before entering the second column, in order to eliminate abiotic sulfide oxidation (by ferric iron) in the latter. Both the redox potential and the dissolved oxygen concentration of the influent AMD are lowered in the first (iron reduction) bioreactor, since the iron-reducing acidophiles used (*Acidiphilium* spp.) are facultative anaerobic bacteria (Johnson 1998). In addition, the iron-reducing bioreactor provides suitable small molecular weight substrates for the SRB in the second column. These are produced, for example, during the biodegradation of cellulosic substrates. The iron-reducing *Acidiphilium* spp. are generally less fastidious in their range of substrates than are SRB. In view of the facts that (i) no alkaline material is added to the AMD at any stage, and (ii) iron reduction generates net acidity if the incoming ferric iron is soluble rather than in particulate form, the SRB in the second reactor (where there is net acid consumption and metal removal) need necessarily to be acidophilic or acid-tolerant. Bacteria described in section 8.4.3 are currently being evaluated in this system.

Future developments in AMD processing will doubtless involve SRB technology. There will be particular scope for developing low cost bioremediation systems, possibly utilizing readily-available organic waste materials (see 8.3.3) as direct or indirect carbon and energy sources for SRB, which have the potential for treating AMD waters of differing chemical composition, and which may be set up in areas where access is difficult and present options are not feasible. Coupled with this, advances in our knowledge of the biodiversity and physiology of SRB, including strains and species that are active in acidic liquors (pH <5) will further extend the possibilities for bioremediation of these polluting wastewaters.

8.5 REFERENCES

Banks, D., Younger, P.L., Arnesen, R-T., Iversen, E.R. and Banks, S.B. (1997) Mine-water chemistry: the good, the bad and the ugly. *Environ. Geol.* **32**, 157-174.

Barnes, L.J., Scheeren, P.J.M. and Buisman C.J.N. (1994) Microbial removal of heavy metals and sulfate from contaminated groundwaters. In *Emerging Technology for Bioremediation of Metals,* (eds. J.L. Means and R.E. Hinchee), pp.38-49, Lewis Publishers, Boca Raton, La.

Bass, C.J., Webb, J.S., Sanders, P.F. and Lappin-Scott, H.M. (1996) Influence of surfaces on sulphidogenic bacteria. *Biofouling* **10**, 95-109.

Bechard, G., Yamazaki, H., Gould, W.D. and Bedard, P. (1994) Use of cellulosic substrates for the microbial treatment of acid mine drainage. *J. Environ. Qual.* **23**, 111-116.

Boonstra, J., van Lier, R., Janssen, G., Dijkman, H. and Buisman, C.J.N. (1999) Biological treatment of acid mine drainage. In *Proceedings of the International Biohydrometallurgy Symposium*, Madrid, Spain (in press).

Boulegue, J. and Michard, G. (1979) Sulfur speciation and redox processes in reducing envir- onments. In *Chemical Modeling in Aqueous Systems* (ed. E.A. Jenne), ACS Symposium Series vol. 93, pp. 25-50, American Chemical Society, Washington, DC.

Castro, J.M., Wielinga, B.W., Gannon, J.E. and Moore, J.N. (1999) Stimulation of sulfate- reducing bacteria in lake water from a former open-pit mine through addition of organic wastes. *Wat. Environ. Res.* **71**, 218-223.

Christensen, B., Laake, M. and Lien, T. (1996) Treatment of acid mine water by sulfate-reducing bacteria; results from a bench scale experiment. *Water Res.* **7**, 1617-1624.

Clarke, L. B. (1995) *Coal Mining and Water Quality.* International Energy Agency Coal Research, London.

de Vegt, A.L., Krol, J. and Buisman, C. J. N. (1997) Biological sulfate removal and metal recovery from mine waters. In *International Biohydrometallurgy Symposium/Biomine 97 Conference Proceedings*, pp. 1-10, Australian Mineral Foundation, Glenside, South Australia.

Diaz, M.A., Monhemius, A.J. and Narayanan, A. (1997) Consecutive hydroxide-sulphide precipitation treatment of acid mine drainage. In *Proceedings of the Fourth International Conference on Acid Rock Drainage* vol. 3, pp. 1179-1194. Vancouver, British Columbia, June 1997.

Dijkman, H. (1995) Biological gas desulfurization. *Med. Fac. Landbouww. Univ. Gent* **60**, 2677-2684.

Ehrenreich, A. and Widdel, F. (1994) Anaerobic oxidation of ferrous iron by purple bacteria, a new type of phototrophic metabolism. *Appl. Environ. Microbiol.* **60**, 4517-4526.

Evangelou, V.P. (1995) *Pyrite Oxidation and its Control*, CRC Press, New York.

Fortin, D. and Beveridge, T.J. (1997) Microbial sulfate reduction within sulfidic mine tailings: formation of diagenetic Fe sulfides. *Geomicrobiol. J.* **14**, 1-21.

Fortin, D., Davis, B. and Beveridge T.J. (1996) Role of *Thiobacillus ferrooxidans* and sulfate- reducing bacteria in iron biocycling in oxic and acidic mine tailings. *FEMS Microbiol. Ecol.* **21**, 11-24.

Gazea, B., Adam, K. and Kontopoulos, A. (1996) A review of passive systems for the treatment of acid mine drainage. *Miner. Eng.* **9**, 23-42.

Geller, W., Klapper, H. and Salomons, W. (eds.) (1998) *Acidic Mining Lakes,* Springer-Verlag, Berlin.

Gray, N.F. (1996) Field assessment of acid mine drainage contamination in surface and ground water. *Environ. Geol.* **27**, 358-361.

Gyure, R.A., Konopka, A. Brooks, A. and Doemel, W. (1990) Microbial sulfate reduction in acidic (pH 3) strip-mine lakes. *FEMS Microbiol. Ecol.* **73**, 193-202.

Hamilton, W.A. (1998) Sulfate-reducing bacteria: physiology determines their environmental impact. *Geomicrobiol. J.* **15**, 19-28

Hard, B.C. and Babel, F.W. (1997) Bioremediation of acid mine water using facultatively methylotrophic metal-tolerant sulfate-reducing bacteria. *Microbiol. Res.* **152**, 65-73.

Herlihy, A.T. and Mills, A.L. (1985) Sulfate reduction in freshwater sediments receiving acid mine drainage. *Appl Environ. Microbiol.* **49**, 179-186.

Isaksen, M.F. and Teske, A. (1996) *Desulforhopalus vacuolatus* gen. nov., sp. nov., a new moderately psychrophilic sulphate-reducing bacteria with gas vacuoles isolated from a temperate estuary. *Arch. Microbiol.* **166**, 160-168.

Johnson, D.B. (1998) Biodiversity and ecology of acidophilic microorganisms. *FEMS Micro biol. Ecol.* **27**, 307-317.

Johnson, D.B. (1995a) The role of 'iron bacteria' in the biodegradation of minerals. *Biodeter. Abstr.* **9**, 1-7.

Johnson, D.B. (1995b) Acidophilic microbial communities: candidates for bioremediation of acidic mine effluents. *Int. Biodeter. Biodegr.* **35**, 41-58.

Johnson, D.B., Ghauri, M.A. and McGinness, M. (1993) Biogeochemical cycling of iron and sulphur in leaching environments. *FEMS Microbiol. Rev.* **11**, 63-70.

Kaufman, E.N., Little, M.H., and Selvaraj, P.T. (1996) Recycling of FGD gypsum to calcium carbonate and elemental sulfur using mixed sulfate-reducing bacteria with sewage digest as a carbon source. *J. Chem. Tech. Biotechnol.* **66**, 365-374.

Kolmert, A., Hallberg, K.B. and Johnson, D.B. (1999) Bioremediation of acid mine drainage by sulfate reducing bacteria in biofilm reactors. In: *5th International Symposium on In Situ and On-Site Bioremediation*, San Diego, California, April, 1999 (in press).

Lamb, H.M., Dodds-Smith, M. and Gusek, J. (1998) Development of a long-term strategy for the treatment of acid mine drainage at Wheal Jane. In *Acidic Mining Lakes* (eds. W. Geller, H. Klapper and W. Salomons), pp. 335-346. Acidic Mining Lakes. Springer-Verlag, Berlin.

Loos, M.A., Bosch, C., Maré, J., Immelman, E. and Sanderson, R.D. (1989) Evaluation of sodium lauryl sulfate, sodium benzoateand sorbic acid as inhibitors of acidification of South African coal waste. pp. 193-200. In: *Groundwater and Mining (Proceedings of the 5th Biennial symposium of the groundwater division of the Geological Survey of South Africa Rand berg, Transvaal)*, pp. 193-200, Geological Society of South Africa, Pretoria, South Africa.

Lopez-Archilla, A.I., Marin, I. and Amils, R. (1995) Microbial ecology of an acidic river: biotechnological applications. pp. 63-74. In *Biohydrometallurgical Processing* vol 2, (eds. T. Vargas, C.A. Jerez, J.V. Wiertz and H. Toledo), pp. 63-74, University of Chile, Santiago.

Lyew, D. and Sheppard, J.D. (1997) Effects of physical parameters of a gravel bed on the activity of sulphate-reducing bacteria in the presence of acid mine drainage. *J. Chem. Tech. Biotechnol.* **70**, 223-230.

Machemer, S.D. and Wildeman, T.R. (1992) Adsorption compared with sulfide precipitation as metal removal processes from acid mine drainage in a constructed wetland. *J. Contam. Hydrol.* **9**, 115-131.

Maree, J.P. and Strydon W.F. (1987) Biological sulphate removal from industrial effluent in an upflow packed bed reactor. *Water Res.* **21**, 141-146.

McGinness, S. and Johnson, D.B. (1993) Seasonal variations in the microbiology and chemistry of an acid mine drainage stream. *Sci. Tot. Environ.* **132**, 27-41.

Mehling, P.E., Day, S.J. and Sexsmith, K.S. (1997) Blending and layering waste rock to delay, mitigate or prevent acid generation: a case review study. In *Proceedings of the Fourth International Conference on Acid Rock Drainage* vol 2, pp. 953-970. Vancouver, British Columbia, June 1997.

Mills, A.L. and Mallory, L.M. (1987) The community structure of sessile heterotrophic bacteria stressed by acid mine drainage. *Microbiol. Ecol.* **14**, 219-232.

Nixdorf, B., Wollmann, K. and Deneke, R. (1998) Ecological potential for planktonic devel- opment in extremely acidic mining lakes in Lusatia. pp. 147-168. In: *Acidic Mining Lakes* (eds. W. Geller, H. Klapper and W. Salomons), pp. 147-168, Springer-Verlag, Berlin.

Nordstrom, D.K., Jenne, E.A. and Ball, J.W. (1979) Redox equilibria of iron in acid mine waters. In *Chemical Modeling in Aqueous Systems* ACS Symposium Series vol 93 (ed. E.A. Jenne), pp. 51-79, American Chemical Society, Washington D.C.

Norris, P.R. and Ingledew, W.J. (1992) Acidophilic bacteria: adaptations and applications. In *Molecular biology and biotechnology of extremophiles* (eds. R.A. Herbert and R.J. Sharp), pp. 115-142, Blackie, Glasgow.

Pentreath, R.J. (1994) The discharge of waters from active and abandoned mines. In *Issues in Environmental Science and Technology I - Mining and its Environmental Impact* (eds. R.E. Hestor and R.M. Harrison), pp. 121-131, Royal Society of Chemistry, Cambridge, UK.

Pichtel, J.R. and Dick, W.A. (1991) Sulfur,iron and solid phase transformations during the biological oxidation of pyritic mine spoil. *Soil Biol. Biochem.* **23**, 101-107.

Pietsch, W. (1998) Colonization and development of vegetation in mining lakes of the Lusatian lignite area in dependence on water genesis. In *Acidic Mining Lakes* (eds. W. Geller, H. Klapper, and W. Salomons), pp. 169-196. Springer-Verlag, Berlin.

Rawlings, D.E. (ed.) (1997) *Biomining: Theory, Microbes and Industrial Processes,* Springer- Verlag/Landes Bioscience, Georgetown, Texas.

Rawlings, D.E., and Silver, S. (1995) Mining with microbes. *Biotechnol.* **13**, 773-778.

Robb, G.A. (1994) Environmental consequences of coal mine closure. *Geograph. J.* **160**, 33- 40.

Rowley, M., Warkentin, D.D. and Sicotte, V. (1997) Site demonstration of the bisulphide process at the former Britannia mine. In *Proceedings of the Fourth International Conference on Acid Rock Drainage* vol. 4, pp. 1531-1548, Vancouver, British Columbia, June 1997.

Schippers, A., Jozsa, P-G., and Sand, W. (1996) Sulfur chemistry in bacterial leaching of pyrite. *Appl. Environ. Microbiol.* **62**, 3424-3431.

Sen, A.M. and Johnson, D.B. (1999) Acidophilic sulphate-reducing bacteria: candidates for bioremediation of acid mine drainage. In *Proceedings of the International Biohydrometallurgy Symposium, Madrid, Spain* (in press).

Stucki, G., Hanselmann, K.W. and Hurzeler, A. (1993) Biological sulfuric acid transformation: reactor design and process optimization. *Biotechnol. Bioeng.* **41**, 303-315.

Tichy, R., Grotenhuis, J.T.C., Bos, P. and Lens, P. (1998) Solid-state reduced sulfur compounds: environmental aspects and bioremediation. *Crit. Rev. Environ. Sci. Technol.* **28**, 1-40.

Tuttle J.H., Dugan P.R., and Randles C.I. (1969) Microbial dissimilatory sulfur cycle in acid mine water. *J Bacteriol.* **97**, 594-602

Vile, M.A. and Wieder, R.K. (1993) Alkalinity generation by Fe(III) reduction versus sulfate reduction in wetlands constructed for acid mine drainage treatment. *Wat. Air & Soil Pollut.* **69**, 425-441.

Wakao, N., Takahashi, T., Sakurai, Y. and Shiota, H. (1979) A treatment of acid mine water using sulfate-reducing bacteria. *J. Ferment. Technol.* **57**, 445-452.

Walton, K.C. and Johnson, D.B. (1992) Microbiological and chemical characteristics of an acidic stream draining a disused copper mine. *Environ. Pollut.* **76**, 169-175.

9

Anaerobic treatment of sulfate rich wastewaters: process modeling and control

Jean-Philippe Steyer, Nicolas Bernet, Piet N.L. Lens and Rene Moletta

9.1 INTRODUCTION

Anaerobic digestion has been largely applied to remove organic matter from high-strength wastewaters because of its relatively low sludge production and low energy needs compared with aerobic treatment. Sulfate-containing wastewaters can be treated in anaerobic reactors. However, a high sulfate content can cause major trouble to the process, mainly because of interactions between sulfate reducers and methanogens, but also because H_2S, the product of sulfate reduction, is a corrosive gas that can be toxic for microorganisms at elevated concentrations.

Several mathematical models of sulfate fed anaerobic reactors have been developed, mainly for continuous stirred tank reactor (CSTR) systems. They mainly focus on the interaction between sulfate-reducing bacteria and methanogenic bacteria while accounting for sulfide toxicity.

Sulfate fed anaerobic reactors can be very unstable, and implementing an efficient control system is essential to stabilize the process, especially in the presence of a fluctuating input pollution and/or a low chemical oxygen demand (COD)/sulfate ratio. For this purpose, sensors providing on-line data are required. In the case of sulfate rich anaerobic digestion, it would be useful to develop a control strategy taking into account sulfide production as a control parameter. This paper reviews the current state-of-the-art of these three aspects of process control (i.e., models, on-line sensors and control strategy).

9.2 MODELS

Dynamic models have been shown to be very powerful tools to improve monitoring and control of wastewater treatment plants in general. They can be used to analyze and to predict the performance of a plant under different operating conditions. See, for example, Queinnec *et al.* (1998) for a mathematical analysis of a nitrification process. Mathematical models can also help to understand the process from a global point-of-view and to train the process operators (Lawrence 1971; Iza *et al.* 1991; Mc Carthy and Mosey 1991). Last but not least, models can be also used directly in the control loop thus improving process operation.

Dynamic modeling of anaerobic digestion processes has been largely studied in the literature (see, for example, Andrews 1969; Mosey 1983; Denac *et al.* 1988; Heinzle *et al.* 1993; Husain 1998). A review of the kinetic parameters used in these models, such as the maximum specific substrate utilization, half-saturation constant, microbial growth yield and microorganism decay rate can be found in Pavlostathis and Giraldo-Gomez (1991).

9.2.1 Models at reactor level

Concerning more specifically sulfate-rich wastewater treatment, the first model of sulfate fed anaerobic reactors was developed by Gupta *et al.* (1994) for a chemostat. The main achievement of this model is a comprehensive description of the chemical subsystem where the complex chemistry involved was modeled by incorporating various buffer systems, acid-base and liquid-gas equilibria, ionic interactions and metal precipitation. The following important anaerobic digestion acid-base equilibria were included in the model: the carbonate, the phosphate, the ammonia and the sulfide ones. In addition, metal sulfide and carbonate precipitation was a key feature of this model. Liquid-gas equilibria (carbon dioxide, hydrogen sulfide, ammonia, methane, nitrogen and water vapour) have been given also special consideration to accurately predict the gas production rate and composition. On the contrary, the biological subsystem was represented in a simplified way - using the Monod equation for one, either methanogenic bacteria (MB) or sulfate reducing bacteria (SRB), or two (both MB and SRB) bacterial groups without pH modulation and without taking into account a sulfide inhibition:

$$\mu = \mu_m S/(K_S + S) \tag{9.1}$$

The model was calibrated by experimental data where methanogenic, sulfidogenic and mixed chemostats were fed by three different substrates (acetate, methanol and formate). Iron was added to precipitate the sulfide produced. The model was able to predict both steady state and transient batch spike experimental data fairly well for homogenic (methanogenic or sulfidogenic) environments. However, the description of mixed chemostats could be considered only on the qualitative level owing to the simplified conceptual structure of the biological subsystem in the model of Gupta *et al.* (1994).

A more complex model of sulfate fed CSTR was proposed by Vavilin *et al.* (1994). The modeling of the chemical subsystem was similar to Gupta *et al.* (1994). The representation of the biological subsystem included the Monod relationship (Eq. 9.1) for two bacterial groups (acetotrophic MB and SRB) with an inclusion of pH and sulfide inhibition:

$$\mu = \mu_m S \cdot F(pH) \cdot F(H_2S_f)/(K_S + S) \tag{9.2}$$

The inhibition terms were used in the following form (Vavilin *et al.* 1995):

$$F(I, K_2, K_{100}) = 1/(1 + (I/K_2)^{\ln 99/\ln(K_{100}/K_2)}) \tag{9.3}$$

where I = pH or H_2S_f

K_2 = concentration of I at which the growth rate is decreased twice

K_{100} = concentration of I at which the growth rate is decreased 100 times

The model of Vavilin *et al.* (1994) was calibrated on the experiments of Parkin *et al.* (1990) where anaerobic chemostats were operated at variation of COD (acetate):S ratios from 60:1 to 2:1. The simulations showed that, when the COD:S ratio was less than 10:1, both sulfate reduction and methanogenesis shut down, which agreed well with the experimental observations. The model revealed that free H_2S and pH inhibition were the main factors governing system failure. In addition, the former acted as a trigger stimulating a positive feedback loop between an increase in acetate and sulfate concentrations and a decrease in the pH level through microbial activity. Thus, an oscillating coexistence of MB and SRB with a period of 5-20 days under conditions close to system failure could be modeled. This phenomenon emphasizes the high complexity of sulfate fed anaerobic systems. The oscillating phenomenon was further investigated using a reduced model (Fomichev and Vavilin 1997). The latter authors found that the self-oscillation fed batch mechanisms disappeared at strong venting rates with an inert gas as well as at increased buffer capacity. Also, it was shown that, by decreasing the influent pH, one could shift the self-oscillation region to higher sulfate loads. Actually, oscillating coexistence is a form of competition.

The competition between sulfate reduction and methanogenesis for acetate in granular sludge was studied in detail by Omil *et al.* (1997, 1998) using a simple mathematical model based on the Monod kinetic parameters of acetate utilizing SRB and MB, without taking into account their sulfide and pH inhibition. The simulations confirmed the long term nature of the competition between these acetotrophs. It was shown that a high reactor pH (> 8), a short solid retention time (more than 150 days), and the presence of a substantial SRB population in the inoculum could considerably reduce the time required for acetate-utilizing SRB to outcompete MB.

All above-mentioned models were elaborated for sulfate fed reactors with a single substrate (acetate). Multiple substrate competition between sulfate reduction and methanogenesis has received a substantial attention in the dispersed plug-flow models for sulfate fed upflow anaerobic sludge blanket (UASB) (Kalyuzhnyi and Fedorovich 1997, 1998). There were no significant differences in this approach to describe the chemical subsystem in comparison with the models proposed by Gupta *et al.* (1994) and Vavilin *et al.* (1994). In contrast, the conceptual structure of the biological subsystem included seven bacterial groups (fermentative bacteria (FB), acetogenic bacteria (AB), two types of MB and three types of SRB) responsible for the microbial transformation of the multisubstrate influent consisting of sucrose, volatile fatty acids (VFA) and sulfate. The main features of the kinetic description can be summarized as the following:

1. The growth of each bacterial group proceeds according to Monod kinetics with simultaneous inhibition by undissociated H_2S. Instead of using true Monod kinetics, it is assumed that reaction kinetics for carbon dioxide was of zero order in its concentration because carbon dioxide is usually present in significant concentrations in anaerobic reactors. A dual substrate form of the Monod equation is postulated for SRB to account for their growth limitations under treatment of sulfate-deficient wastewaters.
2. The effect of pH on the growth rates is described by a bell-shape pH function normalized to give a value of 1.0 as the center value (Kalyuzhnyi 1997).
3. Undissociated H_2S inhibition proceeds according to first order inhibition kinetics for all bacteria. Thus, a specific growth rate equation for FB, AB and MB is expressed as:

$$\mu_j = \mu_{m,j} S_i (1 - H_2S_f / K_{I,j}) F(pH) / (K_{S,j} + S_i) \tag{9.4}$$

and for SRB bacteria:

$$\mu_j = \mu_{m,j} S_i [SO_4^{2-}](1 - H_2S_f / K_{I,j}) F(pH) / ((K_{S,j} + S_i)(K_n + SO_4^{2-})) \tag{9.5}$$

4. All reactions are effectively rate controlled, i.e. the effects of diffusion limitations of biomass aggregates are constant and incorporated into the kinetic term.

The developed structured model was calibrated and verified on the experimental study of Alphenaar *et al.* (1993) where the competition between sulfate reduction and methanogenesis in UASB reactors was investigated. The model was able to describe satisfactorily both the steady-state characteristics of reactor performance and the increase of the ratio of total COD converted by SRB relative to that converted by MB (main criterion) under the different hydraulic retention time (HRT). The model also predicted that MB and AB failed to compete for hydrogen and propionate but could effectively compete for acetate which agrees with experimental observations (Hoeks *et al.* 1984; Mulder 1984; Rinzema and Lettinga 1988; Alphenaar *et al.* 1993; van Houten *et al.* 1996; Omil *et al.* 1996).

Since many factors can influence the outcome of the competition between SRB and MB and not all of these influences can be thoroughly investigated experimentally, models can also be used to assess the impact of several process conditions on the outcome of the competition between SRB and MB (Kalyuzhnyi *et al.* 1998, Kalyuzhnyi and Fedorovich 1998; Omil *et al.* 1998). The modeling data indicated the following for sulfate fed UASB reactors: a decrease of the SO_4^{2-}:COD ratio has no influence under a SO_4^{2-}:COD ratio > 1, but a further decrease of this ratio leads to a decrease of the main criterion. The SRB/MB ratio in the seed sludge has a critical influence on the main criterion during the start-up period (first 2-3 months), but further continuation of the run leads to a progressive elimination of this influence. Sulfide resistance of SRB is very important for such systems, especially for acetate conversion.

9.2.2 Models at biofilm level

A unified model for the growth of the SRB *Desulfovibrio vulgaris* under different environmental conditions has been developed by Noguera *et al.* (1998). A combination of kinetic and thermodynamic mechanisms was used to model the general flow of electrons and the production and consumption of H_2 by the strain. The simulation results were in good agreement with experimental data. Moreover, the model can be used to interpret the behavior of these bacteria after changing culture conditions.

Since most anaerobic processes are not CSTR systems, it is necessary to develop models taking into account non-homogenous properties of the reactors (concentration gradients, biofilm structure, hydrodynamics, ...). The three-dimensional growth of multispecies anaerobic biofilms has been modeled using a two-species biofilm composed of SRB and MB (Noguera *et al.* 1999). The model predicts different biofilm structures in the presence and in the absence of sulfate. The role of mass transfer limitation of sulfate within UASB aggregates in the competition between sulfate reduction and methanogenesis was theoretically evaluated in the model of Overmeire *et al.* (1994). The steady-state microprofiles of sulfate inside a spherical granule were calculated using a reference set of diffusion and kinetic parameters for SRB obtained from literature. The effect of the parameters on mass transport limitation was tested by varying each reference value of the parameters with a factor of 3. The authors concluded from their analysis that sulfate limitation within UASB granules prevailed at sulfate concentrations in the bulk liquid below 0.2 g/l as well as in large aggregates (radius > 0.75 mm) and it could be a factor governing the competition between SRB and MB. The model calculations, however, did not allow a prediction of the ultimate outcome of this competition because they did not take into account mass transfer limitation of COD substrates though the importance of diffusion limitations for these substrates within biomass aggregates has been clearly demonstrated (Arcand *et al.* 1994; Guiot *et al.* 1992; Lens *et al.* 1993).

9.2.3 Models for process control

Kalyuzhni *et al.* (1998) recently developed a dispersed plug-flow model of sulfate fed UASB reactors. Concentration gradients on substrates, intermediates, products and bacteria inside the reactor as well as multiple-reaction stoichiometry and kinetics have been considered in this model. The four following blocks have been included: a hydrodynamic block describing liquid flow and transport and distribution of the different substrates and products; a kinetic block including biological phenomena in the system; a physico-chemical block for calculation of pH and a transfer block describing the mass transfer of gaseous compounds from the liquid to the gas phase. The model was validated on experimental data, both during process start-up with a non-sulfate-adapted sludge and on a mature sulfidogenic granular sludge reactor. This model could be easily adapted to describe sulfate-fed anaerobic fixed bed and fluidized bed reactors. See for example the models for anaerobic digestion using fluidized bed reactors developed by Buffière *et al.* (1995), Bonnet *et al.* (1997) and Buffière *et al.* (1998).

Finally, it is to be pointed out that other types of models can be very useful for the control of wastewater treatment processes. These models are usually simpler than those already presented (i.e., so as to be directly used into a controller) but they are only valid around a specific equilibrium point (i.e., a steady state of the process). As an example, Premier *et al.* (1999) compares black box linear models with the models obtained using artificial neural networks applied to a fluidized bed anaerobic digestion process (Artificial neural networks will be also later discussed in section 9.4.6).

9.3 SENSORS

The choice of appropriate on-line sensors is of crucial importance for controlling efficiently any kind of process. This is particularly true for biological processes since they are complex - due to their strong non-linearity and non-stationarity - and thus very difficult to control.

A few years ago, it was emphasized that "in general, sensors are the weakest part of the chain in real-time process control of wastewater treatment plants" (Harremoës *et al.* 1993). This belief has actually existed since the 1960s and, despite the sensor technology trying to catch up, the statement still exists nowadays. One of the reasons is the lack of appropriate and reliable sensor technology and the difficulty we have to act accurately on such processes. Additional criteria such as low cost or low maintenance efforts have also to be taken into account at the industrial scale. Thus, the engineer concerned with monitoring and control purposes of biological processes, and especially those with environmental issues, must handle an ill-defined process and he must keep in mind that, although in-cell reactions are critical, on-line information about them are usually not obtainable in the near future (Steyer 1991).

Concerning the anaerobic digestion process, a clear overview of the sensors available for monitoring anaerobic digestion processes can be found in (Vanrolleghem 1995). Among the available sensors, the choice of a measurement to be used for on-line control depends on several factors such as its sensitivity to variations and its response time to a disturbance. This is indeed one of the main criteria to select a sensor: it should react as fast as possible to any effect occurring on the process.

Based on these criteria, the hydrogen content is an interesting control parameter for the anaerobic digestion process since its concentration varies rapidly and more largely than any other parameter (Archer *et al.* 1986). Moreover, it plays an important role in the kinetics and stability of anaerobic digestion, particularly when the organic substrate is mainly composed of glucose (Pauss *et al.* 1990). The increase of hydrogen concentration in the gas phase indicates a process failure (Mathiot *et al.* 1992). Other parameters can of course be selected depending of the desired performance and objectives: rate of methane production (Andrews 1978), partial pressure of CH_4 (Ryhiner 1990), volatile acid concentrations (Slater *et al.* 1990), total organic carbon concentration (Alatiqi *et al.* 1990).

For sulfate rich wastewaters, sensors to probe sulfurous compounds are of utmost importance. An indirect method analyses the output gas composition as a measure of the overall activity of the reactor. This method is typically applied on lab-scale and consists of a combination of flow measurements in a set-up where one or more of the constituents is trapped in a washing bottle. The ratio of the flows before and after the bottle is representative of the gas composition. For instance, an alkaline washing bottle will trap all CO_2 and H_2S and will let CH_4 pass through (Vanrolleghem 1995).

The content of a component can also be monitored directly using more specific gas analysers. Typically, infrared absorption measurements are used to determine carbon dioxide and methane, whereas specific hydrogen analysers have been developed based on electrochemical cells (Mathiot et al. 1992; Pauss *et al.* 1993). Escoffier *et al.* (1992) trapped H_2S before the entrance of the biogas into the hydrogen monitor. Hydrogen sulfide measurements in the gas phase may be performed by monitoring the reaction of sulfide with a lead (Pb) strip. Subsequently, the black PbS that is produced is quantified by colorimetry. Meyer-Jens *et al.* (1995) used a silicone membrane probe connected to a quadrupole mass spectrometer as detector to measure on-line dissolved and gaseous hydrogen sulfide in a reactor.

Also, sensors for compounds in the liquid phase of a reactor can be used. A simple method for the determination of dissolved sulfide in colored complex media was developed by Percheron *et al.* (1996) using ion exchange chromatography. Its principle is based on the complete oxidation of sulfide into sulfate through the strong oxidant hydrogen peroxide. The difference in sulfate concentration, between a sample analyzed before and after H_2O_2 treatment, gives the total dissolved sulfide. To avoid autooxidation of H_2S, this oxidation has to be performed immediately after sampling, without bacterial separation. Sulfide from continuous and discontinuous digestion of sulfate rich wastewater were successfully assayed by this technique. A theoretical evaluation based on the Henry's law and the sulfide dissociation equilibrium led to a very good agreement with the analytical results.

Alternatively, the sulfide concentration can be measured using an ion selective electrode (Bandekar *et al.* 1995) or a redox electrode (Janssen *et al.* 1998). A big disadvantage of these ion selective electrodes is that they can be easily poisoned. To date, methods to measure the dissolved sulfate concentration are not available. Recently, ionophores selective for sulfate have been described (Nishizawa *et al.* 1998), which open perspectives for the development of ion selective electrodes for sulfate. Besides, titrimetric (Zancato *et al.* 1990) and spectrophotometric (Bandekar *et al.* 1995) methods have been developed to determine the sulfate concentration in aqueous solutions. However, application of these novel analytic techniques to mixed liquor of anaerobic digesters requires further research and development.

When no appropriate analytical solution can be selected, "software sensors" are of great help. These "sensors" can be defined as numerical algorithms that combine signals from available hardware sensors within a given mathematical framework (See Figure 9.1). They estimate an unknown parameter, taking advantage of the *a priori* knowledge available about a process (Bastin and Dochain 1990).

As an example of software sensors, an on-line alkalinity sensor has been

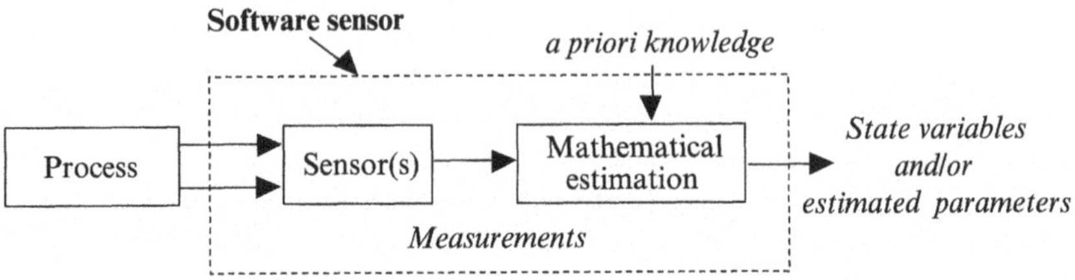

Figure 9.1. Schematic view of a software sensor.

developed at the "Laboratoire de Biotechnologie de l'Environnement" (LBE-INRA) for anaerobic digestion processes and is now commercially available. This sensor provides direct measurements of partial and total alkalinity and, using mathematical equations, it gives estimations of the bicarbonates and VFA concentrations in the reactor. Figure 9.2 shows the response of the sensor when the input liquid flow rate and the input COD concentration are manually changed. It clearly demonstrates the ability of the sensor to follow these fluctuations. In addition to be very accurate, this sensor requires very low maintenance effort, i.e. less than one manual operation per month, with one hardware apparatus providing four different measurements. Last but not least, this sensor has already been demonstrated to be very useful to monitor the anaerobic digestion reaction scheme.

9.4 CONTROL

The application of control systems for optimizing the functioning of biological wastewater treatment processes has been extensively studied in the last years (see, for example, the very interesting surveys by Agrawal and Lim 1984; Beck 1986; Bastin and Dochain 1990; Shimizu 1993; Andrews 1994; Bastin and Van Impe 1995). Indeed, the development and the improvement of high performance bioreactors, on-line facilities and application of automatic control have brought evidence that biological processes - when appropriate sensors are available - can be optimized and more efficient pollutant removal can be achieved when applying process control.

The lack of knowledge concerning anaerobic digestion processes in the past led to more or less serious breakdowns mainly due to the organic overload of various origins. They created some kind of suspicion towards this process and delayed its development at the industrial scale. Thus, the importance of implementing appropriate and efficient control systems for

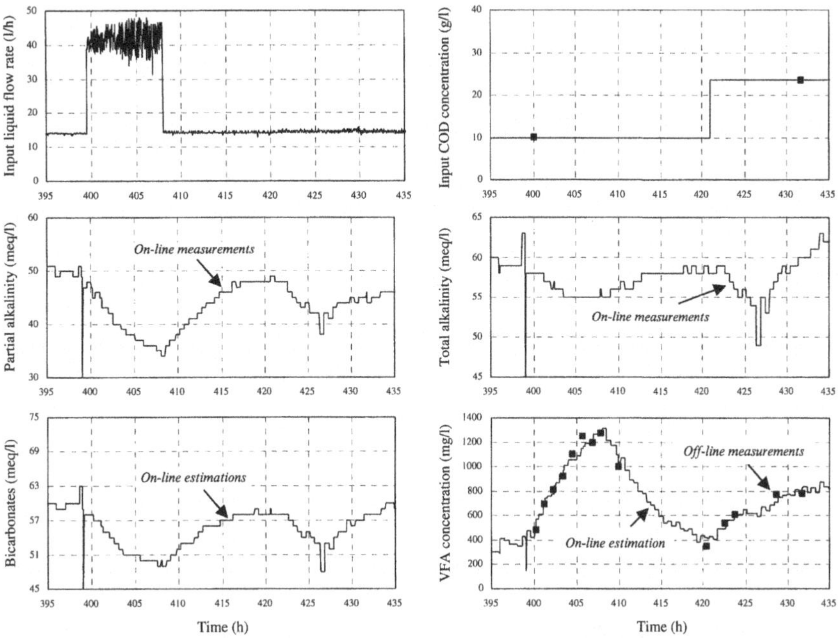

Figure 9.2. On-line alkalinity sensor combined with the mathematical estimation of bicarbonates and VFA concentration using the data of the sensor – the application of this sensor to an anaerobic digestion process treating wine distillery vinasses.

anaerobic digestion processes is of no doubt and several problems are to be handled by the controller: slow growth of methanogenic organisms, instability caused by toxic substrates or by overloading and, even though large progress has been made, insufficient knowledge of the process. In addition, anaerobic digestion is intrinsically a rather unstable process: variations of the input variables (hydraulic flow rate, influent organic load) may easily lead to a washout of the biomass from the reactor tank. This phenomenon takes place mainly under the form of acid accumulation in the reactor. It is therefore essential to implement controllers that are capable of stabilizing the process via a carefully designed control strategy.

Control algorithms applied to anaerobic digestion processes are mostly of the conventional proportional-integral-derivative (PID) type (Heinzle *et al.* 1993) and there are only few experimental results in the literature. When specifically concerned with the treatment of wastes containing sulfurous compounds, the number of experimental applications of control approaches is even more scarce. For example, McFarland and Jewell (1989) recommended pH adjustment for biogas sulfide control under conditions in which the influent sulfur level is well below sulfide inhibitory concentrations. They showed that control of gaseous sulfide levels through iron (3+) phosphate addition was an efficient control strategy with no adverse effects on the digester performance. Camp and Sublette (1992) used another system - a PC-based machine vision system - to continuously monitor changes in biomass concentration and to control the undesirable production of colloidal elemental sulfur (a reactor upset condition due to an excessive concentration of inhibitory sulfide substrate) in a bioreactor containing *Thiobacillus denitrificans*. A field of view of a video camera was established which contained regions of different background lighting. Mean values of the distribution of red, green and blue intensity components with corresponding regions of a digital image captured from the camera were used to monitor color changes, associated with changes in biomass concentration and to determine if the reactor is in an upset condition. They showed in particular that the ratio of red to blue intensity was an important parameter in detecting the formation of an elemental sulfur precipitant. Using a steeper motor-driven pressure regulator, process control was then performed by altering the hydrogen sulfide flow rate setpoint based on the vision system measurements. Finally, Janssen *et al.* (1998) showed that the formation of elemental sulfur from the biological oxidation of sulfide could be optimized by controlling the redox state of the reactor solution. They successfully applied a PI control of the redox potential so that nearly stoichiometrical amounts of oxygen are supplied to the system. In this way, the supplied oxygen suffices for oxidizing sulfide to sulfur, although about 10 % of sulfide is oxidized to sulfate.

Despite these very few studies specifically dedicated to the treatment of wastes containing sulfurous compounds, several control strategies will be briefly presented below. These studies were mainly developped for anaerobic (methanogenic) digestion processes but they will provide the reader with a view of potential controllers that are applied at the industrial scale or that could be applied in a near future to the treatment of wastes containing sulfurous compounds. A more detailed description and a comparative study of these approaches can be found in Steyer (1998).

9.4.1 The "disturbance monitoring" control

The control of wastewater treatment processes in general - and of processes for the treatment of waters polluted by sulfurous compounds in particular - can be made through a careful analysis of the disturbances affecting the main variables. The "disturbance monitoring" control strategy will thus first be presented. This approach was initially developed to control highly loaded anaerobic digestion processes (Steyer *et al.* 1999) and is not specific to wastewater containing sulfurous compounds. Nevertheless, because of its interest for sulfur treatment, the control approach will be briefly highlighted.

The purpose of the "disturbance monitoring" is to develop a control strategy that, compared to conventional controllers, faces the overloading problem all the way around. The basic idea of this control strategy is to add a disturbance on purpose on the input liquid flow rate. The response of only two simple parameters that were chosen because of their large use at the industrial scale (i.e., the pH and the output gas flow rate) is then analyzed to determine whether or not it is possible to increase the loading rate of the reactor. A negative effect of a positive disturbance (i.e., an increase of the input liquid flow rate) will be observed through a pH decrease and an increase of the output gas flow rate less high than it should be theoretically. In such a case, the reactor is assumed to be overloaded and the organic loading rate is automatically decreased. In case of no negative effect of the disturbance, the reactor is assumed to work safely (i.e., without being overloaded) and the loading rate is increased. Thus, the organic loading rate can be maintained as high as possible despite variations in the influent quality and the concentration of the treated effluent can be kept low and stable.

This control approach was successfully applied to an anaerobic 120 L fluidized bed reactor for the treatment of industrial wine distillery vinasses with sulfate concentrations varying between 50 and 200 mg/l. It was in particular possible to start-up the process automatically and to obtain a very high organic loading rate (i.e., from 0 to 100 $kg_{COD}.m^{-3}.d^{-1}$ in just five months) with the total organic carbon removal efficiency exceeding 70 % (Figure 9.3). Once the start-up was performed, this control approach was also able to prevent overloads of the process and to keep the process run steadily at very high organic loading rates (i.e., higher than 100 $kg_{COD}.m^{-3}.d^{-1}$).

9.4.2 Model Predictive Control

The Model Predictive Control (MPC) principle is largely used in the process industry (Qin and Badgwell 1997). In addition, as shown by the survey of Takatsu *et al.* (1997), MPC has a very large success being the most preferred control methodology by the end-users with 94% of satisfaction.

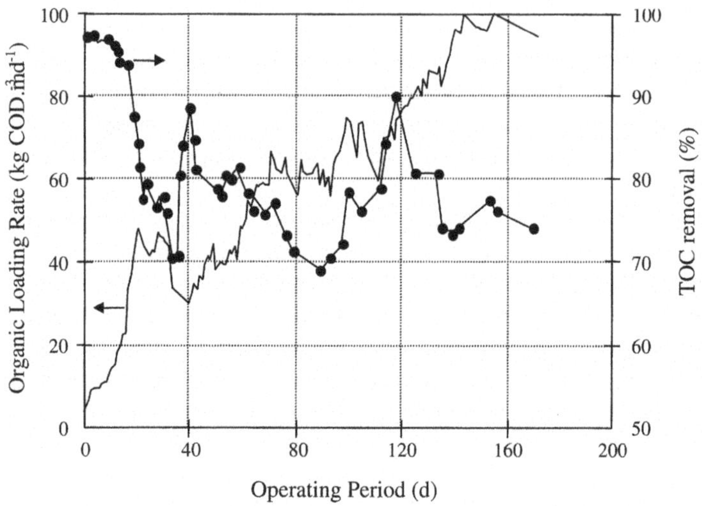

Figure 9.3. Organic loading rate and carbon removal efficiency during the automatic start-up of an anaerobic 120 L fluidized bed reactor using the "disturbance monitoring" control strategy (After: Steyer *et al.* 1999).

However, quite surprisingly, the MPC principle has not yet been implemented for anaerobic digestion processes but it can be easily guessed that this will change in a near future.

9.4.3 Adaptive Control

The key idea of the adaptive control design is to take advantage of what is well known about the dynamics of bioprocesses (basically, the reaction pathways and mass balances summarized in a dynamic mass balance model) while accounting for the model uncertainty (mainly the kinetics). Since the model is generally non-linear, the model-based control design might result in a linearizing control structure, in which the on-line estimation of the unknown variables (concentration of pollutant and its intermediates) and parameters (reaction rates and yield coefficients) are incorporated (Bastin and Dochain 1990).

Adaptive control schemes are also used for anaerobic digestion processes because of their good disturbance rejection ability (Renard *et al.* 1988; Dochain and Perrier 1993; Ben Youssef *et al.* 1995; Johnson *et al.* 1995). In such cases, the parameters of the controller are continuously adapted to follow - and sometimes to estimate - the changes in the operating conditions.

Adaptive linearizing controllers have also been designed, theoretically analyzed and experimentally validated on a pilot anaerobic digester for the control of the effluent COD concentration. For further details, see Dochain and Bastin (1984), Dochain (1986) and Renard *et al.* (1988). The industrial applicability of this control strategy may unfortunately still appear to be limited by the need of on-line measurements of COD or equivalents of the substrate concentration (i.e., Total Organic Carbon (TOC) or VFA analyzers). Therefore, there is a clear incentive to look for alternative substrate candidates. In this context, as an example, the use of hydrogen as a controlled variable has already been used (Dochain *et al.* 1991).

9.4.4 Robust control

As depicted before, one of the key issues to be addressed in controlling biological wastewater treatment processes is to reject the disturbances that can destabilize the reactor. Here, a clear distinction must be made between disturbance rejection and insensitivity to unmodeled phenomena or parameter variations. To tackle these two problems, robust control strategies appear as a very promising option and significant improvements over conventional proportional integral (PI) controllers are evidenced for wastewater treatment plants (Pohlmeier 1991; Steyer *et al.* 1995; Harmand *et al.* 1996).

As for adaptive control, robust control approaches strongly rely on mathematical models as those depicted in section 9.2. However, the main difference between robust and adaptive control schemes is that the later continuously adapt its parameters based on dynamic estimates of the process evolution whereas, in the former scheme, the uncertainty is explicitly accounted for in the building of the controller. The adaptive control scheme can thus achieve a better performance whereas the robust control scheme can handle more severe operating conditions.

9.4.5 Fuzzy control

As it has been already explained, anaerobic digestion is a very complex process, ill defined and difficult to control with classic PID methods. However, parallel with the imperfect knowledge of this process, there is often an important expertise obtained with the operation of pilot units. Since fuzzy logic naturally captures the acquired experience of human operators (see, for example, Driankov *et al.* 1983), it appears a powerful tool for controlling this type of process. In addition, fuzzy logic, first introduced by Zadeh (1965), can incorporate semi-quantitative information into simplified models and it can put subjective information into a form usable by computers.

In the general wastewater treatment field, one of the first attempts to accumulate rules for assisting the operator to understand the process was made for the Norwich Wastewater Treatment Plant in the UK (Tong *et al.* 1980). They used fuzzy reasoning techniques in the operation of the activated sludge process. Flanagan (1980) used techniques developed by Olsson and Andrews (1977) and fuzzy reasoning to estimate the organic loading and the biological activity in a wastewater treatment process. Another set of rules attempting to account for optimal performance conditions of activated sludge processes was developed by Joyce *et al.* (1974). A modification of this concept was used in the Koskinen's rule base (1989). A very interesting application of fuzzy logic to the control of an activated sludge process can also be found in Tsai *et al.* (1996).

Dealing more specifically with anaerobic digestion processes, we can mention the Marsili-Libelli's fuzzy control approach (Marsili-Libelli 1992). Using a multi-tank process, fuzzy logic was shown to be capable of reducing the maximum extent of COD overload by 50 % (Müller *et al.* 1995).

Fuzzy approximate reasoning was also used by Boscolo *et al.* (1993) for the management of a pilot scale anaerobic digester to improve the treatment of municipal solid waste. Estaben *et al.* (1997) used fuzzy control for the monitoring of the output gas flow rate in an anaerobic 15 L fluidized bed reactor. Steyer *et al.* (1997a) extended this approach to a hierarchical fuzzy controller to monitor additional variables (i.e., pH, temperature, ...) on a pilot-scale 120 L fluidized bed reactor for the anaerobic treatment of industrial wine distillery vinasses.

All these studies demonstrate the ability of a fuzzy controller to efficiently manage a wastewater treatment plant in general – and an anaerobic digestion process in particular. In addition, fuzzy logic has very interesting properties since human operator and/or control engineers generally find it difficult to decide upon the adequate parameters to use when dealing with a full-scale wastewater treatment plant. The fuzzy control theory can be used here to derive effective control strategies even when a suitable class of models or correct parameters are unknown. Last but not least, since a fuzzy controller can be easily tuned, this method can be extended to many different applications in the wastewater treatment field.

9.4.6 Artificial neural networks

As previously presented, several difficulties arise upon the implementation of control systems for wastewater treatment and for biological processes in general. Firstly, the behavior of biological processes, and therefore the models describing these systems (see section 9.2), is often highly nonlinear. These models are furthermore complex and their parameters are often hard to determine. The dynamics of biological processes also tend to vary in the course of time because of variations in metabolism and environmental parameters (e.g. Dochain 1998).

Application of artificial neural networks (ANNs) overcomes all these problems (Hunt *et al.* 1992). ANNs can indeed approximate any nonlinear mapping arbitrarily well (Funahashi 1989; Hornik *et al.* 1989) and their learning abilities together with their parallel structure make them fit for adaptive on-line application. It is thus clear that artificial neural networks have great promise for the modeling and control of biological processes in general (te Braake 1997; van Can 1997) and of anaerobic digestion processes in particular (Wilcox *et al.* 1995, Steyer 1998; Premier *et al.* 1999).

Various ANN based controller structures exist in the literature. See, for example, Thibault and Grandjean (1991) for a review of both the fundamentals of feedforward neural networks and their use in different control strategies. Concerning more specifically the use of ANNs in the control of biological processes, Bhat and McAvoy (1990) describe a model predictive control framework using an ANN model that provides n-step ahead predictions. They illustrate this approach on a general pH model using a CSTR with two input streams, one containing sodium hydroxide and the other acetic acid. Willis *et al.* (1991) also describe an ANN based Model Predictive Control and they present simulation results on a nonlinear exothermic CSTR. Their ANN model, providing a one-step ahead prediction, is trained on closed loop data and remains constant after convergence. In Willis *et al.* (1992), additional simulation results for a binary distillation column used to separate a 50-50 wt% methanol-water feed mixture are presented using basically the same controller. This ANN one-step ahead predictor is however used recursively to obtain n-step ahead predictions. An ANN based MPC used for one-step ahead prediction is also described by Ishida and Zhan (1995), presenting simulation results of a distributed parameter crystal growth process. Temeng *et al.* (1995) describe a MPC using a time lag recurrent ANN for n-step ahead predictions and they present an application to the control of an industrial vanadium pentoxide catalyst packed bed reactor where sulfur dioxide is aerobically

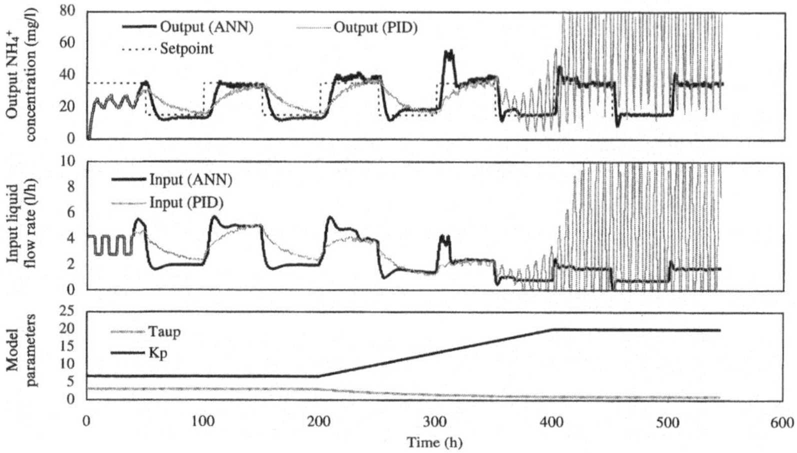

Figure 9.4. Simulation results for a nitrification process to illustrate the advantage of using an ANN based controller over a classical optimally tuned PI controller (the gain Kp and the time constant taup of the first order linear model of the process were dynamically changed – see figure at the bottom). For further details, refer to Devisscher *et al.* (1998).

converted to sulfur trioxide. Tan and Van Cauwenberghe (1996) describe a one-step ahead ANN based MPC and extend it to an *n*-step ahead MPC by using the ANN recursively. Stability analysis and simulation results for a general model of an exothermic first-order reaction in a CSTR are also presented.

Finally, Devisscher *et al.* (1998) detail an ANN Model Predictive Controller for stabilizing a fluidized bed nitrification process and both simulation and experimental results clearly highlight the interest of such an advanced control scheme over classical and simple PID controllers (see Figure 9.4). The same ANN control strategy was applied to regulate the output gas flow rate of an anaerobic fluidized bed reactor and similar results were experimentally obtained (Steyer 1998).

9.5 FAULT DETECTION AND ISOLATION

Despite the large interest of all the above mentioned control approaches, biological processes exhibit very specific behavior. Indeed, it has to be pointed out that the control needed is not the same as it would be usually understood by a control engineer. See for instance the survey paper by Andrews (1994) in which it is underlined that, in biological processes, things seem to work fairly well until some failure or fault occurs.

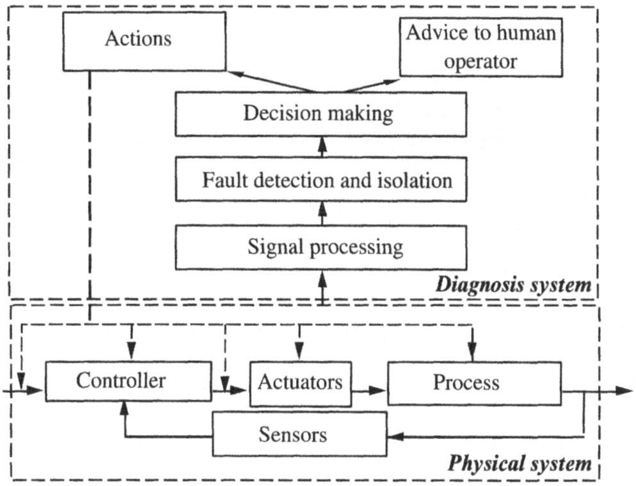

Figure 9.5. The different sub-tasks to be performed by a diagnosis system.

Hence, in addition to efficient control algorithms, there is an increasing need for Fault Detection and Isolation (FDI) systems. FDI systems are able to detect and diagnose any faults occurring on a process. This task can be divided in several sub-tasks (see Figure 9.5).

Automatic FDI systems use analytical or heuristic knowledge in order to detect as early as possible deviations from the normal operation that tend to degrade the overall system performance. Such malfunctions may occur in the sensors, the actuators, the components of the process (e.g., inhibition of the biomass in a biological reactor) or in the control system if the process is running in closed loop.

There has been a growing interest for using techniques such as artificial neural networks or fuzzy logic for fault detection purposes (Frank 1994; Isermann and Ballé 1997). Few demonstrations of the advantages of fuzzy-based FDI systems can be found in the literature (Ulieru and Isermann 1993; Montmain and Gentil 1993; Kiupel and Frank 1996; Giraud and Aubrun 1996). Other recent studies use this technique for sensor validation (Boudaoud and Masson 1998) or combine it with artificial neural networks (Steyer *et al.* 1997b).

A good FDI system for a biological process also requires inclusion of specific biological knowledge. Artificial Intelligence (AI) methods such as knowledge-based systems or fuzzy logic allow one to bring new insights into biological process control and to introduce a “biological dimension” (see

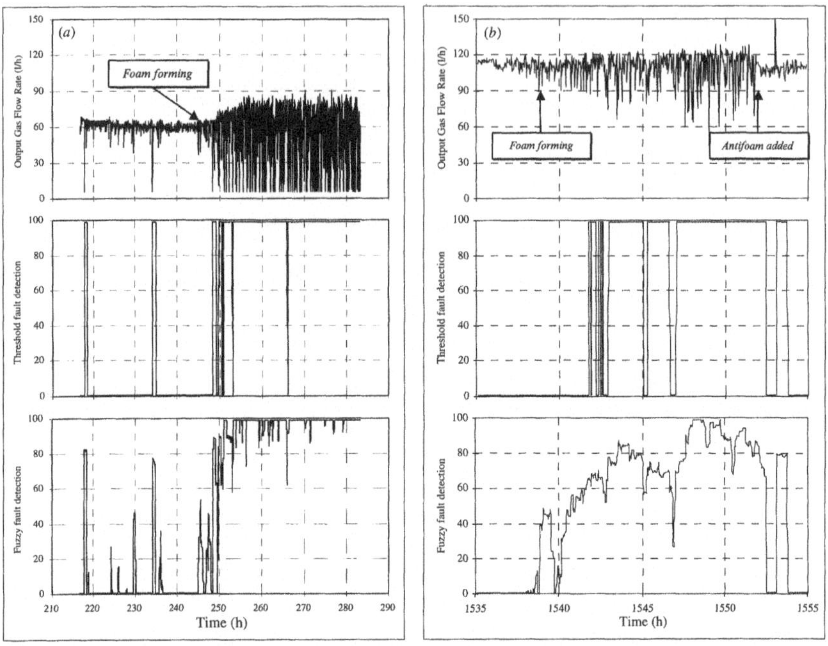

Figure 9.6. Foam forming in a 1 m^3 anaerobic fixed bed reactor (upper): Comparisons between fault detection using a threshold comparison (middle) and a fuzzy logic based (lower) approach.

for example Konstantinov and Yoshida 1992; Aynsley *et al.* 1993; Steyer *et al.* 1993; Siimes *et al.* 1995; Roca *et al.* 1996; Steyer *et al.* 1996).

As an example, Figure 9.6 presents comparative results between a threshold and a fuzzy logic FDI approach for the detection of foam formation based on the biogas flow rate measurements in a 1 m^3 anaerobic digestion upflow fixed bed reactor. In Figure 9.6.a, both detectors are optimized for this output gas flow rate data set and they show similar performances even if the fuzzy system detects the problem a little bit earlier. On the other hand, in Figure 9.6.b, the same detectors - with the same parameters as in Figure 9.6.a - were used to detect the foam formation in other operating conditions. The two fault detectors were thus not optimally tuned for this second data set and the threshold comparison shows a delay in the detection while the fuzzy logic based FDI approach gives a rapid and gradual increasing fault signal. These examples clearly demonstrate that the fuzzy logic based FDI approach is more robust and more effective than a threshold comparison algorithm. It can detect the faults earlier, and more

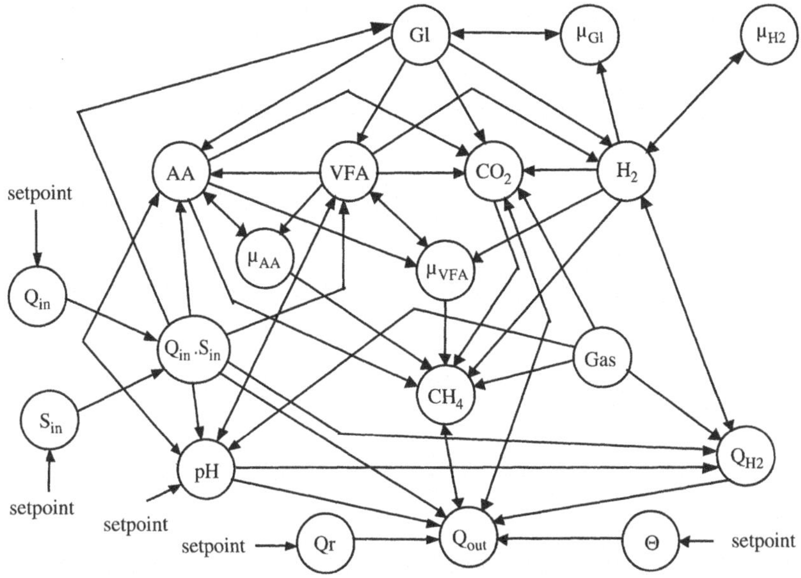

Figure 9.7. Qualitative causal graph of an anaerobic digestion process (Q_{in} = input liquid flow rate, S_{in} = COD concentration in the input, AA = acetic acid, VFA = volatile fatty acids, Gl = glucose, μ_X = specific growth rate of the X-consuming microorganisms (X = Gl, VFA, AA), Θ = temperature, Q_r = recirculation flow rate, Q_{out} = output liquid flow rate).

importantly, it provides the human operator with the degree of urgency of the problem (Genovesi *et al.* 1999a). Indeed, while the threshold comparison approach generates a binary fault detection signal of either 0 or 100 %, the fuzzy FDI approach provides a more accurate index between these limits: the larger the signal, the more significant is the fault.

All these studies were designed to cope with uncertain and incomplete knowledge. They allowed the coupling of quantitative information with the qualitative or symbolic expressions in order to reproduce the actions and decisions of an experienced process operator. Quantitative information is usually a mathematical model as those depicted in section 9.2 whereas qualitative knowledge can be introduced using a causal graph. As an example, Figure 9.7 gives a qualitative causal graph of an anaerobic digestion process (Steyer *et al.* 1997c).

Within the specific scope of the anaerobic digestion processes, an expert system has also been developed to control the dilution rate by managing four different control laws (Pullammanappallil *et al.* 1998). This expert system reproduced the decisions of a skilled human operator and prevented organic overloads and digester imbalance. Another paper by Genovesi *et al.* (1999b) details an FDI system for anaerobic digestion processes. This study concerns an object-oriented fuzzy logic based FDI approach capable of dealing with several disturbances that may affect anaerobic digestion processes. It allowed early detection of the problems: organic overloads as well as technical problems (i.e., pipe clogging, foam formation, sensor bias) or bad settings of local control loops. The results obtained demonstrate the ability of the fuzzy logic FDI scheme to handle human knowledge in a systematic way. It thus can be used as a powerful tool in industrial environments, i.e. wastewater treatment bioreactors.

9.6 REFERENCES

Agrawal, M. and Lim, H.C. (1984) Analysis of various control schemes for continuous bioreactors. *Adv. Biotech. Eng.* **16**(2), 61-90.

Alatiqi, I.M., Dadkhah, A.A. and Jabr, N.M. (1990) Dynamics and multivariable control analysis for anaerobic digestion. *Chem. Eng. J.* **43**, 81-89.

Alphenaar, P.A., Visser, A. and Lettinga, G. (1993) The effect of liquid upward velocity and hydraulic retention time on granulation in UASB reactors treating wastewater with a high sulphate content. *Biores. Technol.* **43**, 249-258.

Andrews, J.F. (1969) Dynamic Model of the Anaerobic Digestion Process. *J. Sanit. Eng. Div. Proc. ASCE* **95**(SA1), 95-116.

Andrews, J.F. (1978) The development of a dynamic model and control strategies for the anaerobic digestion process, *Mathematical Models in Water Pollution Control,* J. Wiley, Chichester, pp. 281.

Andrews, J.F. (1994) Dynamic control of wastewater treatment plants. *Environ. Sci. Technol.* **28**(9), 434A-439A.

Arcand, Y., Chavarie, C., and Guiot, S.R. (1994) Dynamic modelling of the population distribution in the anaerobic granular film. *Wat. Sci. Technol.* **30**(12), 63-73.

Archer, D.B., Hilton, M.G., Adams, P. and Wiecko, H. (1986) Hydrogen as a process control index in a pilot scale anaerobic digester. *Biotechnol. Lett.* **8**(3), 197-202.

Aynsley, M., Hofland, A.J., Morris, A.J., Montague, G.A. and Di Massimo, C. (1993) Artificial intelligence and the supervision of bioprocesses. In *Advance in Biochemical Engineering/Biotechnology*, Eds. A. Fiechter, Springer-Verlag, Heidelberg, Germany, **48**, 1-27.

Bandekar, J., Sethna, R. and Kirshner, M. (1995) Quantitative determination of sulfur oxide species in white liquor bt FT-IR. *Appl. Spectr.* **49**(11), 1577-1582.

Bastin, G. and Dochain, D. (1990) *On-line estimation and adaptive control of bioreactors*, Elsevier Science Publishing, Amsterdam.

Bastin, G. and Van Impe, J.F. (1995) Nonlinear and adaptive control in biotechnology: A Tutorial. *Eur. J. Control* **1**(1), 1-37.

Beck, B. (1986) Identification, estimation and control of biological wastewater treatment processes. *IEEE Proceedings* **33**(5), 254-264.

Ben Youssef, C., Roux, G. and Dahhou, B. (1995) Multivariable adaptive predictive control of non linear systems: application to a multistage wastewater treatment process. *ECC95*, Roma, Italy, 5-8 September 1995.

Bhat, N. and Mc Avoy, T.J. (1990) Use of neural networks for dynamic modeling and control of chemical process systems. *Computers Chem. Eng.* **14**, 573-583.

Bonnet, B., Dochain, D. and Steyer, J.P. (1997) Dynamical modeling of an anaerobic digestion fluidised bed reactor, *Wat. Sci. Technol.* **36**(5), 285-292.

Boscolo, A., Mangiavacchi, C., Drius, F., Rongione, F., Pavan, P. and Cecchi, F. (1993) Fuzzy Control of an anaerobic digester for the treatment of the organic fraction of municipal solid waste. *Wat. Sci. Technol.* **27**(2), 57-68.

Boudaoud, A.N. and Masson, M.H. (1998) An adaptive fuzzy diagnosis system for on-line sensor data validation. *3th IFAC Workshop on On-line Fault Detection and Supervision in the Chemical Process Industries,* Lyon, France.

Buffière, P., Steyer, J.P., Fonade, C. and Moletta, R. (1995) Comprehensive modeling of methanogenic biofilms in fluidised bed systems, *Biotech. Bioeng.* **48**(6), 725-736.

Buffière, P., Steyer, J.P., Fonade, C. and Moletta, R. (1998) Modeling and experiments on the influence of biofilm size and mass transfer in a fluidised bed reactor for anaerobic digestion, *Wat. Res.* **32**(3), 657-668.

Camp, C.E. and Sublette, K.L. (1992) Control of a *Thiobacillus denitrificans* bioreactor using a machine vision, *Biotech. Bioeng.* **39**(5), 529-538.

Denac, M., Miguel, A. and Dunn, I.J. (1988) Modeling dynamic experiments on the anaerobic degradation of molasses wastewater. *Biotech. Bioeng.* **31**, 1-10.

Devisscher, M., Steyer, J.P., Harmand, J., Bouvier, J.C., Bernet, N. and Moletta, R. (1998) Control of fluidised bed nitrification processes: comparison of an artificial neural network based control strategy with an industrial PI control. *Proc. of the INRA-SFM-CFRP European Conf. on New Advances in Biological Nitrogen and Phosphorus Removal for Municipal or Industrial Wastewaters,* 223-230, Narbonne, France, 12-14 Octobre 1998.

Dochain, D. and Bastin, G. (1984) Adaptive identification and control algorithms for nonlinear bacterial growth systems. *Automatica* **20**, 621-634.

Dochain, D. (1986) *On-line parameter estimation, adaptive state estimation and adaptive control of fermentation processes*. Ph.D. Thesis, Université Catholique de Louvain, Belgium.

Dochain, D., Perrier, M. and Pauss, A. (1991) Adaptive control of the hydrogen concentration in anaerobic digestion. *Ind. Eng. Chem. Res.* **30**, 129-136.

Dochain, D. and Perrier P. (1993) Control design for nonlinear wastewater treatment plant. *Wat. Sci. Technol.* **28**, 283-293.

Dochain, D. (1998). Modeling and control of wastewater treatment plants. *Proc. of Waste-Decision 98, Int. Workshop on Decision and Control in WWTPs, February 25-27, Narbonne, France (CD-Rom).*

Driankov, D., Hellendoorn, H. and Reinfrank, M. (1993) *An introduction to fuzzy control,* Springer-Verlag, Berlin, 316 pages.

Escoffier, Y., Ehlinger, F., Leyris, J.P., Coudert, J.P. and Moletta, R. (1992) On-line monitoring and control of an anaerobic fluidised bed digester. *Forum for Applied Biotechnology*", 24-25 Sept, Brugge, Belgium.

Estaben, M., Polit, M. and Steyer, J.P. (1997) Fuzzy control of anaerobic digestor. *Cont. Eng. Pract.* **5**(9), 1303-1310.

Flanagan, M.J. (1980) On the application of approximate reasoning to control activated sludge process. *Proc. Joint Automatic Control Conf.*, ASME, San Francisco, USA.

Fomichev, A.O. and Vavilin, V.A. (1997) The reduced model of self-oscillating dynamics in an anaerobic system with sulfate-reduction. *Ecol. Model.* **95**, 133-144.

Frank, P.M. (1994) Application of fuzzy logic to process supervision and fault diagnosis. *IFAC Symp. on Fault Detection Supervision and Safety for Technical Processes, SAFEPROCESS'94*, Espoo, Finland **2**, 531-538, 13-16 June 1994.

Funahashi, K. (1989) On the approximate realization of continuous mappings by neural networks. *Neural networks* **2**, 183-192.

Garcia, C.E., Prett, D.M. and Morari, M. (1989) Model predictive control: theory and practice; A survey. *Automatica* **25**, 335-348.

Genovesi, A., Harmand, J. and Steyer, J.P. (1999a) A fuzzy logic based diagnosis system for the on-line supervision of an anaerobic digestor pilot-plant, *Biochemical Engineering Journal*, accepted for publication.

Genovesi, A., Harmand, J. and Steyer, J.P. (1999b) Integrated fault detection and isolation: Application to a winery's wastewater treatment plant. *Applied Intelligence Journal*, accepted for publication.

Giraud, D. and Aubrun, C. (1996) A fuzzy fault diagnosis method applied to a steam circuit. *IEEE International Conference on Fuzzy Systems FUZZIEEE'96*, New Orleans, Louisiana, USA, vol. 3, pp. 1944-1950.

Gupta, A., Flora, J.R.V., Sayles, D.G. and Suidan, M.T. (1994) Methanogenesis and sulfate reduction in chemostats. II. Model development and verification. *Wat. Res.* **28**(4), 795-803.

Guiot, S.R., Pauss, A. and Costerton, J.W. (1992) A structured model of the anaerobic granule consortium. *Wat. Sci. Technol.* **25**(7), 1-10.

Harmand, J., Skelton, R.E. and Steyer, J.P. (1996) Disturbance accomodation control of a nitrification process with structural uncertainties. *IEEE-IMACS Int. Conf. CESA'96,* Lille, France, 9-12 July, pp. 70-75.

Harremoës, P., Capodaglio, A.G., Hellström, B.G., Henze, M., Jensen, K.N., Lynggaard-Jensen, A., Otterpohl, R. and Soeberg, H. (1993) Wastewater treatment plants under transient loading – Performance, modelling and control. *Wat. Sci. Technol.* **27**(12), 71-115.

Heinzle, E., Dunn, I.J. and Ryhiner, G.B. (1993) Modeling and control of anaerobic wastewater treatment. *Adv. Biochem. Eng. Biotech.* **48**, 79-114.

Hoeks, F.W., Ten Hoopen, H.J.G., Roels, J.A., and Kuenen, J.G. (1984) Anaerobic treatment of acid water (methane production in a sulphate rich environment). *Progr. Ind. Microbiol.* **20**, 113-119.

Hornik, K., Stinchcombe, M. and White, H. (1989) Multilayer feedforward networks are universal approximators. *Neural Networks* **2**, 359-366.

Hunt, K.J., Sbarbaro, D., Zbikowski, R. and Gawthrop, P.J. (1992) Neural networks for control systems - A survey. *Automatica* **28**, 1083-1112.

Husain, A. (1998) Mathematical models of the kinetic of anaerobic digestion – a selected review. *Biomass Bioen.* **14**(5/6), 561-571.

Isermann, R. and Ballé, P. (1997) Trends in the application of model-based fault detection and diagnosis of technical processes, *Contr. Eng. Pract.* **5**(5), 709-719.

Ishida, M. and Zhan, J. (1995). Neural model-predictive control of distributed parameter crystal growth process. *AIChE J.* **41**, 2333-2336.

Iza, J., Colleran, E., Paris, J.L. and Wu, W.M. (1991) Anaerobic treatment technology for municipal and industrial wastewaters - Summary paper. *Wat. Sci. Technol.* **24**, 1-16.

Janssen, A.J.H., Meijer, S., Bontsema, J. and Lettinga, G. (1998) Application of the redox potential for controling a sulfide oxidizing bioreactor. *Biotechnol. Bioeng.* **60**(2), 147-155.

Johnson, K.A., Wheatley, A.D. and Fell, C.J. (1995) An application of adaptive control algorithm for the anaerobic treatment of an industrial effluent. *Trans. IChemE* **73**, 203-211

Joyce, R.J., Ortman, C. and Zickefoose, C. (1974) How to optimize an activated sludge plant. *Wat. Sew. Works* **10**, 129.

Kalyuzhnyi, S.V. (1997) Batch anaerobic digestion of glucose and its mathematical modeling - 2. Description, verification and application of model. *Biores. Technol.* **59**(2-3), 249-258.

Kalyuzhnyi, S., and Fedorovich, V. (1997) Integrated mathematical model of UASB reactor for competition between sulphate reduction and methanogenesis. *Wat. Sci. Technol.* **36**(6-7), 201-208.

Kalyuzhnyi, S.V. and Fedorovich, V. (1998) Mathematical modelling of competition between sulphate reduction and methanogenesis in anaerobic reactors. *Biores Technol.* **65**(3), 227-242.

Kalyuzhnyi, S., Fedorovich, V., Lens, P., Hulshoff Pol, L. and Lettinga, G. (1998) Mathematical modelling as a tool to study population dynamics between sulfate reducing and methanogenic bacteria. *Biodegradation* **9**(3-4), 187-199.

Kiupel, N. and Frank, P.M. (1996) Fuzzy supervision for an anaerobic wastewater plant. *IEEE-IMACS Int. Conf. CESA'96,* Lille, France, vol. 1, pp. 362-367.

Konstantinov, K.B. and Yoshida, T. (1992) The way to adequate control of microbial processes passes via real-time knowledge-based supervision. *J. Biotechnol.* **24**, 33-51.

Koskinen, K. (1989) Expert systems as a top level controller of an activated sludge process. *Wat. Sci. Technol.* **21**, 1809.

Lawrence, A.W. (1971) Application of process kinetics to design of anaerobic processes. *Anaerobic Biological Treatment Processes,* F.G. Pohland (ed.) In *Advances in Chemistry Series*, vol. 105, pp. 163-189, American Chemical Society, Washington D.C.

Lens, P., de Beer, D., Cronenberg, C.C.H., Houwen, F.P., Ottengraf, S.P.P. and Verstraete, W. (1993) Heterogenous distribution of microbial activity in methanogenic aggregates: pH and glucose microprofiles. *Appl. Environ. Microbiol.* **59**, 3803-3815.

Mc Carthy, P.L. and Mosey, F.E. (1991) Modelling of anaerobic processes (a discussion of concepts). *Wat. Sci. Tech.* **24**, 17-33.

Mc Farland, M.J. and Jewell, W.J. (1989) In situ copntrol of sulfide emissions during the thermophilic (55C) anaerobic digestion process, *Wat. Res.* **23**(12), 1571-1578.

Marsili-Libelli, S. (1992) Fuzzy control of the anaerobic digestion process. *Int. Symp. On Anaerobic Digestion Solid Waste,* Venice, Italy, pp. 379-382.

Mathiot, S., Escoffier, Y., Ehlinger, F., Coudert, J.P., Leyris, J.P. and Moletta, R. (1992) Control parameter variations in an anaerobic fluidised bed reactor subjected to organic shockloads. *Wat. Sci. Technol.* **25**(7), 93-101.

Meyer-Jens, T., Matz, G. and Märkl, H. (1995) On-line measurement of dissolved and gaseous hydrogen sulphide in anaerobic biogas reactor. *Appl. Microbiol. Biotechnol.* **43**, 341-345.

Montmain, J. and Gentil, S. (1993) Decision-making in fault detection: A fuzzy approach. *Int. Conf. TOOLDIAG'93,* Toulouse, France.

Mosey, F. (1983) Mathematical modelling of the anaerobic digestion process: regulatory mechanisms for the formation of short-chain volatile acids from glucose. *Wat. Sci. Technol.* **15**, 209-232.

Mulder, A. (1984) The effect of high sulphate concentrations on the methane fermentation of waste water. *Progr. Ind. Microbiol.* **20**, 133-143

Müller, A., Marsili-Libelli, S., Aivasides, A. and Wandrey, C. (1995) Development of an integrated process control system for a multi-stage wastewater treatment plant - High level fuzzy control. *Forum for Applied Biotechnology FAB'95,* Gent, Belgium, pp. 2467-2473.

Nishizawa, S., Buhlmann, P., Xiao, K.P. and Umezawa, Y. (1998) Application of a bis-thiourea ionophore for an anion selective electrode with a remarkable sulfate selectivity. *Anal. Chim. Acta* **358**(1), 35-44.

Noguera, D.R., Brusseau, G.A., Rittman, B.E., and Stahl, D.A. (1998) A unified model describing the role of hydrogen in the growth of *Desulfovibrio vulgaris* under different environmental conditions. *Biotechnol. Bioeng.* **59**(6), 732-746.

Noguera, D.R., Pizarro, G., Stahl, D.A., Rittman, B.E. (1999) Simulation of multispecies biofilm development in three dimensions. *Wat. Sci. Technol.* **39**(7), 123-130.

Olsson G. and Andrews, J.F. (1977) The dissolved oxygen profile - A valuable tool for control of the activated sludge process. *Wat. Res.* **12**, 985.

Omil, F., Lens, P., Hulshoff Pol, L. and Lettinga, G. (1996) Effect of upward velocity and sulphide concentration on volatile fatty acid degradation in a sulphidogenic granular sludge reactor. *Proc. Biochem.* **31**(7), 699-710.

Omil, F., Lens, P., Hulshoff Pol, L. and Lettinga, G. (1997) Characterization of biomass from a sulfidogenic, volatile fatty acid-degrading granular sludge reactor. *Enzyme Microb. Technol.* **20**(3), 229-236.

Omil, F., Lens, P., Visser, A., Hulshoff Pol, L. and Lettinga, G. (1998) Long-term competition between sulfate reducing and methanogenic bacteria in UASB reactors treating volatile fatty acids. *Biotechnol. Bioeng.* **57**(6), 676-685.

Overmeire, A., Lens, P. and Verstraete, W. (1994) Mass transfer limitation of sufate in methanogenic aggregates. *Biotechnol. Bioeng.* **44**, 387-391.

Parkin, G.F., Lynch, N.A., Kuo, W.C. and van Keuren, E.L. (1990) Interaction between sulfate reducers and methanogens fed acetate and propionate. *Res. J. Wat. Pollut. Control Fed.* **62**(6), 780-788.

Pauss A., Beauchemin C., Samson R., Guiot S. (1990) Continuous measurement of dissolved H_2 in an anaerobic reactor using a commercial probe hydrogen/air fuel cell-based. *Biotechnol. Bioeng.* **35**, 492-501.

Pauss, A., Guiot, S., Dochain, D. and Perrier, M. (1993) Hydrogen as a key parameter for the monitoring and control of methane producing processes. In: *Med. Fac. Landbouww. Univ. Gent* **58**, 2011-2017.

Pavlostathis, S.G. and Giraldo-Gomez, E. (1991) Kinetics of anaerobic treatment. *Wat. Sci. Technol.* **24**(8), 35-59.

Percheron, G., Bernet, N. and Moletta, R. (1996) A new method for the determination of dissolved sulfide in strongly colored anaerobically treated effluents. *Bioproc. Eng.* **15**(6), 317-322.

Pohlmeier, J. (1991), Robust stability for a fixed bed reator using positive real systems. *European Control Conference, ECC'91,* Grenoble, France, pp. 22-27.

Pullammanappallil, P.C., Svoronos, S.A., Chynoweth, D.P. and Lyberatos, G. (1998) Expert system for control of anaerobic digesters. *Biotechnol. Bioeng.* **58**(1), 13-22.

Premier, G.C., Dinsdale, R., Guwy, A.J., Hawkes, F.R., Hawkes, D.L. and Wilcox, S.J. (1999) A comparison of the ability of black box and neural network models of ARX structure to represent a fluidized bed anaerobic digestion process, *Wat. Res.* **33**(4), 1027-1037.

Qin, S.J. and Badgwell, T.A. (1997) An overview of industrial model predictive control. in J.C. Kantor, C.E. Garcia and B. Carnahan, editors, *5th Int. Conf. on Chemical Process Control,* vol. 93 of AIChE Symposium Series N°316, pp. 232-256.

Queinnec, I., Harmand, J., Steyer, J.P. and Bernet, N. (1998) Mathematical analysis of a nitrification process. *Recent Res. Ferment. & Bioeng.* **1**, 153-163.

Renard, P., Dochain, D., Bastin, G., Naveau, H. and Nyns, E. (1988) Adaptive control of anaerobic digestion processes - A pilot-scale application. *Biotechnol. Bioeng.* **31**, 287-294.

Richalet, J., Rault, A., Testud, J.L. and Papon, J. (1978) Model predictive heuristic control: applications to industrial processes. *Automatica* **14**(5), 413-428.

Richalet, J. (1993) Industrial applications of model based predictive control. *Automatica* **29**, 1251-1274.

Rinzema, A. and Lettinga, G. (1988) Anaerobic treatment of sulphate containing wastewater. In: Wize DL (Ed), *Biotreatment Systems,* Vol.III, pp. 65-109. CRC Press Inc., Boca Raton.

Roca, E., Flores, J., Rodriguez, I., Cameselle, C., Nunez, M.J. and Lema, J.M. (1996) Knowledge-Based Control Applied to Fixed Bed Pulsed Bioreactor. *Bioproc. Eng.* **14**, 113-118.

Ryhiner, G. (1990) *Regelung und adaptive on-line Optimierung von ein- und zweistufigen anaeroben fliessbett-Reactoren* (in german) Diss. ETH, Zürich, Switzerland.

Shimizu, K. (1993) An overview on the control system design of bioreactors. In *Advance in Biochemical Engineering and Biotechnology,* Springer Verlag, Berlin.

Siimes, T., Linko, P., von Numers, C., Nakajima, M. and Endo, I. (1995) Real-time fuzzy-knowledge-based control of baker's yeast production. *Biotechnol. Bioeng.* **45**, 135-143.

Slater, W.R., Merigh, M., Ricker, N.L., Labib, F., Fergusson, J.F. and Benjamin, M.M. (1990) A microcomputer-based instrumentation system for anaerobic wastewater treatment processes. *Wat. Res.* **24**(1), 121-123.

Steyer, J.P. (1991) *Sur une approche qualitative des systèmes physiques – Aide en temps réel à la conduite de procédés biologiques* (in french), Ph.D. Thesis, Laboratoire d'Analyse et d'Architecture des Systèmes, LAAS-CNRS, Université Paul Sabatier, Toulouse, France.

Steyer, J.P., Queinnec, I. and Simoes, D. (1993) Biotech: a real-time application of artificial intelligence for fermentation processes. *Cont. Eng. Pract.* **1**(2), 315-321.

Steyer, J.P., Harmand, J., Bernet, N., Amouroux, M. and Moletta, R. (1995), Robust control as a solution for monitoring and control of biological nutrient removal processes. *Med. Fac. Landbouww. Univ. Gent* **60**, 2427-2434.

Steyer, J.P., Queinnec, I., Capit, F. and Pourciel, J-B. (1996) Qualitative rules as a way to handle the biological state of a fermentation process: an industrial application. *Journal Européen des Systèmes Automatisés RAIRO-APII* **30**(2/3), 381-398.

Steyer, J.P., Estaben, M. and Polit, M. (1997a) Fuzzy control of an anaerobic digestion process for the treatment of an industrial wastewater, *6th International Conference on Fuzzy Systems, FUZZ-IEEE'97,* **3**, 1245-1250, Barcelona, Spain, 1-5 July 1997.

Steyer J.P., Rolland D., Bouvier J.C. and Moletta, R. (1997b) Hybrid fuzzy neural network for diagnosis: Application to the anaerobic treatment of wine distillery wastewater in fluidised bed reactor. *Wat. Sci. Technol.* **36**, 209-217.

Steyer, J.P., Rolland, D., Jouanny, J-P. (1997c) Diagnosis of wastewater treatment processes, *Int. Conf. INRA-BayFORREST on the Treatment of Solid Wastes and Wastewater,* 28-29 April 1997, Narbonne, France, 183-192.

Steyer, J.P. (1998) *Modélisation, contrôle et diagnostic des procédés biologiques de dépollution* (in french), Habilitation à Diriger des Recherches, University of Perpignan, 192 pages.

Steyer, J.P., Buffière, P., Rolland, D. and Moletta, R. (1999) Advanced control of anaerobic digestion processes through disturbances monitoring. *Wat. Res.* **33**(9), 2059-2068.

Takatsu, H., Itoh, T., Araki, M. (1997) Future needs for control theory in industries – Report and topics of the control technology survey in Japanese Industry. *Int. Symp. on Advanced Control of Chemical Processes ADCHEM'97*, Banff, Canada, 7-11 June 1997, 263-268.

Tan, Y. and van Cauwenberghe, A. (1996) Nonlinear one-step-ahead control using neural networks: control strategy and stability design. *Automatica* **32**, 1701-1706.

te Braake, H. (1997) *Neural control of biotechnological processes.* Delft, Netherlands, Ph.D. thesis, Technical University of Delft.

Temeng, K.O., Schnelle, P.D. and Mc Avoy, T. (1995) Model predictive control of an industrial packed bed reactor using neural networks. *J. Proc. Cont.* **5**, 19-27.

Thibault, J. and Grandjean, B.P.A. (1991) Neural networks in process control - A survey. *Proc. Int. IFAC Conf. on Advanced Control of Chemical Processes,* Toulouse, France, pp. 251-259.

Tong, R.M., Beck, M.B. and Latten, A. (1980) Fuzzy control of the activated sludge wastewater treatment process. *Automatica* **16**, 695-704.

Tsai, Y.P., Ouyang, C.F., Wu, M.Y. and Chiang, W.L. (1996) Effluent suspended solid control of activated sludge process by fuzzy control approach. *Wat. Environ. Res.* **68**(6), 1045-1053.

Ulieru, M. and Isermann, R. (1993) Design of a fuzzy-logic based diagnostic model for technical processes. *Fuzzy Sets and Systems* **58**, 249-271.

van Can, E. (1997) *Efficient mathematical modeling for bioprocesses based on macroscopic balances and neural networks.* Delft, Netherlands, Ph.D. Thesis, Technical University of Delft.

van Houten, R.T., van der Spoel, H., van Aelst, A.C., Pol, L.H. and Lettinga, G.

(1996) Biological sulfate reduction using synthesis gas as energy and carbon source. *Biotechnol. Bioeng.* **50**(2), 136-144.

Vanrolleghem, P.A. (1995) Sensors for anaerobic digestion: An overview, *Int. Workshop on Monitoring and Control of Anaerobic Digestion Processes,* Narbonne, France, 6-7 December 1995, pp. 1-7.

Vavilin, V.A., Vasiliev, V.B., Rytov, S.V. and Ponomarev, A.V. (1994) Self-oscillating coexistence of methanogens and sulfate-reducers under hydrogen sulfide inhibition and the pH-regulating effect. *Biores. Technol.* **49**(2), 105-119.

Vavilin, V.A., Vasiliev, V.B., Rytov, S.V. and Ponomarev, A.V. (1995) Modelling ammonia and hydrogen sulfide inhibition in anaerobic digestion. *Wat. Res.* **29**(3), 827-835.

Wilcox, S.J., Hawkes, D.L., Hawkes, F.R. and Guwy, A.J. (1995) A neural network based on bicarbonate monitoring to control anaerobic digestion. *Wat. Res.* **29**(6), 1465-1470.

Willis, M.J., Di Massimo, C., Montague, G.A., Tham, M.T. and Morris, A.J. (1991) Artificial neural networks in process engineering. *IEE proceedings-D* **138**, 256-266.

Willis, M.J., Montague, G.A., Di Massimo, C., Tham, M.T. and Morris, J. (1992) Artificial neural networks in process estimation and control. *Automatica* **28**, 1181-1187.

Zadeh, L.A. (1965) Fuzzy sets, *Information and Control* **8**, 338-353.

Zancato, M., Pietrogrande, A. and Macca, C. (1990) Sequential determination of sulfur and chlorine in organic compounds by potentiometric titration of sulfate and chloride with an automatic titrator. *Annali di Chimica* **80**(9-10), 445-452.

10

Survey of H_2S and SO_2 removal processes

Jan A. Lagas

10.1 INTRODUCTION

Nowadays, there is a worldwide dependency on energy, especially that originating from fossil hydrocarbons. Coal was the dominant form in the past until the use of oil and natural gas increased significantly due to easier exploration, handling and transportation. Moreover, many forms of fluid hydrocarbons and natural gas also had a lower sulfur content than coal, and thus were environmentally more attractive. The intensive use of these fossil fuels still represented a considerable SO_2 emission into the atmosphere. Environmental problems caused by SO_2 emission such as acid rain have resulted worldwide in stringent emission regulations.

The demand to limit SO_2 emission has provided an incentive to develop or improve desulfurisation processes in order to remove the sulfur content in these fuels. This paper presents an overview of desulfurisation processes

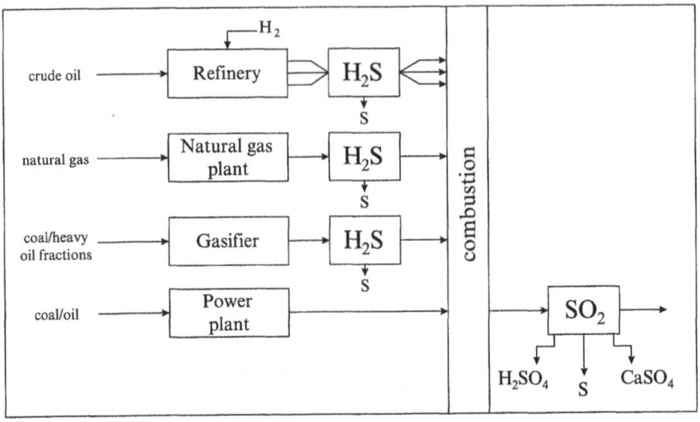

Figure 10.1. Desulfurisation of gases.

used in refineries, in purification of natural gas and syngas and for flue gas clean-up of power plants.

Desulfurisation of refinery gas, compared with natural gas and syngas, requires a different approach owing to the partial pressure of the acid gases and the more numerous impurities that may be present. All these processes desulfurise the fuel before combustion (Figure 10.1). Another way to prevent SO_2 emission is the desulfurisation of flue gas after combustion, known as Flue Gas Desulfurisation (FGD).

10.1.1 Refinery gas

The sulfur content of crude oils varies markedly, but is generally related to the API (American Petroleum Institute) gravity of the crudes. Very light crudes (API > 35) generally tend to be low in sulfur. In refineries several gas streams are treated mainly to produce fuel gas or to purify hydrogen-rich gas. These streams are typically scrubbed by alkanolamine solutions (chemical solvents) for the removal of H_2S. After purification the hydrogen-rich gas is normally recycled to units, such as catalytic reforming and hydroprocessing or the hydrogen is recovered from these streams by PSA (Pressure Swing Adsorption) or membrane separation.

10.1.2 Natural gas

The principal impurities in natural gas are H_2S and CO_2. Other impurities are COS (carbonyl sulfide), CS_2 (carbon disulfide), RSH (mercaptans) and

RSR (sulfides). Heavier hydrocarbons may also be present in significant quantities. The main objective worldwide for desulfurisation of natural gas is the production of sales gas with pipeline specification. Natural gas treatment takes place both directly at the well (mainly for transportation reasons) as well as in centralised gas desulfurisation plants. Normally, a hydrate inhibitor is added at the well to prevent hydrate formation in the transport pipe lines to the gas treating plant. In addition, condensate and heavier hydrocarbon separation may be necessary. Natural gas is treated to pipe line specification, which is less than 4 ppm by volume H_2S. The total sulfur specification of other sulfur species is often tolerated up to 100 ppm by volume and the CO_2 content should be less than 2 % by volume to meet the Wobbe index. For natural gas treatment, chemical solvents (alkanolamines) or physical solvents or a mixture of physical and chemical solvents (hybrid systems) are used to remove H_2S.

Selective absorption is becoming more important where a high CO_2 and low H_2S content are present with COS and mercaptans so as to produce a rich acid gas stream, which has a high H_2S and low CO_2 content and which contains all the removed COS and mercaptans. Such a specification of the acid gas requires additional purification steps and of course additional equipment. However, the need for reliable operation of the sulfur recovery plant (Claus plant) with Claus tail gas clean-up unit, as well as meeting emission regulations, are additional objectives.

10.1.3 Syngas

The first major attempt at precombustion desulfurisation was in the coal gas industry; several efficient and effective techniques for removal of sulfur compounds and other impurities from gasification processes were developed. Many of these techniques found application in the subsequent development of sour natural gas processing where large volumes of hydrogen sulfide had to be removed from the gas. Coal and residual oil fractions generally contain sulfur, nitrogen and metals such as iron, nickel and vanadium. Synthesis gas produced by gasification being mainly CO and H_2 will therefore contain H_2S, CO_2, COS, RSH, HCN, NH_3, $Fe(CO)_5$-iron carbonyl, $Ni(CO)_4$-nickel carbonyl and water.

Syngas after purification is used in combined cycle processes, where operation of a gas turbine is combined with generating high pressure steam (approx. 70 bar). Alternatively syngas after purification and CO shift conversion is used for the production of hydrogen, methanol or higher alcohols or to produce hydrocarbons by means of the Fischer Tropsch synthesis. For syngas purification in general physical solvents or physical/ chemical solvents are used.

10.1.4 Flue gas

Most of the FGD processes for clean-up of flue gas from power plants use an alkaline solution to absorb SO_2 chemically. There are more than 100 FGD processes that can be grouped in two categories: the regenerable and non-regenerable systems. In the non-generable systems, the SO_2 is permanently bound by the alkaline sorbent and has to be disposed of as a waste or sold as a by product such a gypsum. In the regenerable systems, the SO_2 is absorbed and during regeneration of the absorbent or adsorbent, the SO_2 is released and further processed to sulfuric acid, liquid SO_2 or elemental sulfur.

Most of the FGD units in operation are of the non-regenerable type using slaked lime ($Ca(OH)_2$) or limestone ($CaCO_3$) as sorbent. The new generation processes produce high quality gypsum ($CaSO_4$). Regenerable processes tend to be fairly complex and hence more costly. However, given the uncertainties about future waste management, these processes are likely to emerge. Most of the FGD processes are installed in the USA, Germany and Japan.

10.2 IMPURITIES IN GASES

In general, impurities present in refinery gas and natural gas are H_2S, CO_2, COS, CS_2, RSH and water. The variety and the various amounts of impurities will lead to problems during production and transportation, when not removed.

Syngas produced by gasification of residual oil fractions or coal may, besides the above components, contain HCN, NH_3 as well as Fe- and Ni-carbonyls.

10.2.1 Hydrogen sulfide

H_2S has to be removed for safety and corrosion reasons and to prevent air pollution by SO_2 when gas is burnt. It is well-known that H_2S is a very toxic gas, as are most of the other sulfur components. Hydrogen sulfide together with free water can cause stress corrosion and hydrogen-induced cracking.

10.2.2 Carbon dioxide

CO_2 can frequently be tolerated in refinery and natural gas. However, sometimes natural gas may contain significant amounts of CO_2 which have to be removed to meet the required Wobbe index. In syngas, CO_2 has to be removed down to a maximum value of about 500 bar partial pressure. For LNG (Liquefied Natural Gas) production, the CO_2 content should be less than 50 ppm by volume to prevent freezing of CO_2 during liquefaction.

Carbon dioxide together with free water causes pitting corrosion in carbon steel and low-alloy steels.

10.2.3 Total sulfur

Total sulfur in the form of COS, RSH, CS_2 and RSR are often part of the total sulfur specification. Natural gas, for example, should contain less than 100-150 mg S/Nm^3.

If the gas is used as a chemical feed stock, more stringent specifications have to be met, which can be less than 1 ppm by volume. If in coal gasification the gas is used as feed stock for a methanation unit, the total sulfur content should be less than 0.2 mg S/Nm^3 to prevent poisoning of the methanation catalyst.

Natural gas containing hydrogen sulfide may also contain elemental sulfur as vapour. Some gas fields, primarily in Canada, Germany and the US, contain such high quantities of sulfur that production wells and pipes plug up. Elemental sulfur in combination with free water is corrosive.

10.2.4 Hydrocarbons

If natural gas contains hydrocarbons higher than C_{2+}, the recovery of liquefied petroleum gas (LPG) and gas condensate becomes economically attractive. The desired Wobbe index could also require a reduction in the concentration of higher hydrocarbons.

Some pipeline gas specifications require that the hydrocarbon dew point should not exceed -2 °C at 80 bar. For LNG production a hydrocarbon dew point of -40 to -50 °C is usually specified.

10.2.5 Mercury

Natural gas can contain mercury in concentrations up to several milligrams per cubic metre, the bulk of which exists in elemental form. Separated liquid mercury causes mercury-induced corrosion in pipes and fittings, corrosion damage to aluminium heat exchangers in cryogenic plants, and damage to measuring and control valves containing non-ferrous metals by amalgam formation. Mercury must also be removed because of its toxicity. Normally mercury is removed to less than 10-20 ppb.

10.2.6 Ammonia

NH_3 is normally removed from syngas to less than 10 ppm by volme, before the gas is passed to the gas sweetening unit, to prevent accumulation of NH_3 in the overhead system of the regenerator. When the gas is used for methanation, the NH_3 concentration should be less than 0.01 ppm by volume.

10.2.7 Hydrogen cyanide

HCN reacts irreversibly with almost all solvents and for this reason it has to be removed upstream of the gas sweetening unit. The HCN concentration in the gas used for methanation should be less than 0.01 ppm by volume to prevent deactivation of the catalyst.

10.2.8 Iron- and nickel carbonyls

Depending on the operating pressure iron ($Fe(CO)_5$) and nickel ($Ni(CO)_4$) carbonyls are formed during synthesis gas production. The compounds lead to plugging problems in the gas treating and sulfur recovery unit.

When synthesis gas is used for generating electricity, these carbonyls have to be removed before the gas is passed to the gas turbine.

10.2.9 Water

Natural gas is normally dehydrated at the well, so that free water cannot condense in the pipeline and formation of solid hydrates is prevented.

The required water dew point of natural gas according to the pipeline gas specification is -5 °C at 80 bar. For very cold climates the water dew point can be specified down to -18 °C, whereas for hot climates the water dew point may be specified at +2 °C. For LNG production a water dew point of -150 to -160 °C is usually required.

10.3 PRODUCT GAS REQUIREMENTS

Product gas requirements depend on the use of the gas. In Table 10.1 some typical product gas specifications are given (Lagas 1988).

Table 10.1. Typical product gas specifications

		Typical product gas specification				
		Refinery fuel gas	Natural gas	Natural gas for LNG	Synthesis gas from coal [a]	Synthesis gas from coal [b]
H_2S	ppm by volume	100	3	3	3	0.05
CO_2	ppm by volume	[d]	(1–2%) [d]	50–100	[d]	50
Total S [c]	mg/Nm³	[d]	100–150	100–150	200	0.2
NH_3	ppm by volume				10	0.01
HCN	ppm by volume				3	0.01
H_2O dew point	°C		−5	−150–160		
HC dew point	°C		−2	−4–50	[d]	

[a] For production of fuel gas.
[b] For production of SNG (Synthetic Natural Gas), methanol and H_2.
[c] COS, CS_2, RSH and RSR.
[d] Normally no requirements.

10.4 APPLICATION OF VARIOUS GAS-TREATING PROCESSES

It seems obvious to desulfurise gases in one single purification step. If gases contain only H_2S in very low concentrations, the use of a solid absorbent that cannot be regenerated can be considered. Zinc oxide, for example, is converted to zinc sulfide which has to be disposed after saturation. Purification of large gas flows containing concentrations higher than a few ppm, will require high quantities of solid absorbent. It is expected that in the future such absorption processes will no longer be applied (Geus and Lagas 1993).

Another single purification step that may be considered is to combine the absorption and oxidation of H_2S to elemental sulfur. If separation of elemental sulfur out of a gas stream could be readily achieved, selective oxidation of H_2S to elemental sulfur would be the most attractive procedure. However, it is difficult to remove elemental sulfur completely out of a gas stream. Especially water and elemental sulfur can lead to severe plugging and corrosion problems. Normally in desulfurisation of gases the quantity of H_2S is large and moreover CO_2 and other sulfur components may be present. Therefore, the broadest application is gas purification in two separate process steps.

In the first step the acid gases are removed out of the main gas stream. In the second step, the sulfur components in the acid gas are converted to elemental sulfur. The procedures applied to purify refinery gas, natural gas or syngas can be classified according to the following schemes:

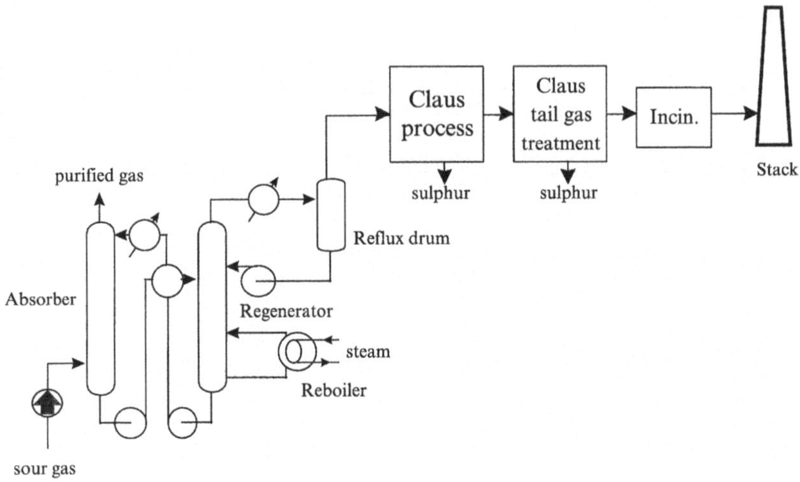

Figure 10.2. Gas treating and sulfur recovery process.

1. Separation of the acid gases by absorption into a solvent, desorption of the acid gases from the solvent, convertion of the sulfur compounds to elemental sulfur within a Claus process, removal of the sulfur compounds from the Claus tail gas, and finally incineration of the remaining sulfur compounds.
2. Separation of the sulfur compounds by absorption into a solvent, oxidation in the solvent to elemental sulfur and separation of the sulfur by filtration.

The H_2S removal processes will be discussed in section 10.5. There is much technical experience with processes using liquid absorption solvents and processing the separated sulfur compounds by means of the Claus process.

Also there is quite a commercial experience with Redox type processes using different kinds of H_2S absorption/oxidation solvents. Well known processes are Stretford, LO-Cat, and the newly developed SulFerox. The application of these processes for purification of gas will depend on the impurities and the product gas specification.

The selection of the solvent in a gas purification unit is most important, because solvent circulation rate determines equipment sizes, investment and operating costs.

The H_2S absorption process and the processing to elemental sulfur is schematically represented in Figure 10.2. Passing the sour gas through the

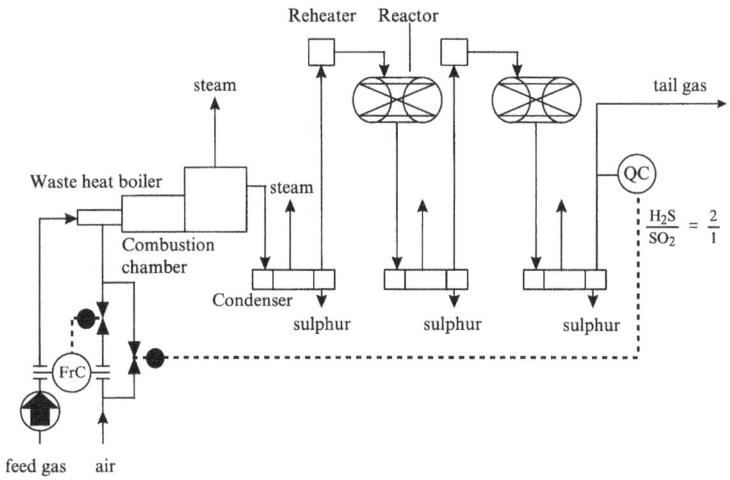

Figure 10.3. The Claus process.

absorber, in which the cold absorption solution is contacting the gas flow, leads to a purified gas flow. The loaded solvent is transported to the regenerator operating at a lower pressure, where the solvent is heated, and the H_2S is desorbed. The lean solvent is cooled and returned to the absorber. The acid gas leaving the top of the regenerator is fed to the Claus process, where the H_2S is converted to elemental sulfur by oxidation (see Equations 10.1 and 10.2). The Claus process consists of a thermal stage, followed by two or three catalytic reactor stages as shown in Figure 10.3.

In the thermal stage, one third of H_2S gas is burnt to SO_2 according to the reaction :

$$3\ H_2S + 1.5\ O_2 \rightarrow 2\ H_2S + SO_2 + H_2O \tag{10.1}$$

The remaining part of H_2S reacts subsequently with SO_2 to form S in the thermal and catalytic stages according to the reaction :

$$2\ H_2S + SO_2 \leftrightarrow 3/n\ S_n + 2\ H_2O \tag{10.2}$$

Reaction 10.2 is often referred to as the “Claus reaction”. Sulfur recoveries in conventional Claus plants vary from 90 to 96 % for a two-stage plant and 95 to 98 % for a three stage plant. The Claus tail gas is treated in a Claus tail gas clean-up unit increasing the sulfur recovery to more than 99.0 %. Finally the gas is passed to the incinerator, where the remaining H_2S is oxidised to SO_2 and the flue gas is emitted via the stack to the atmosphere.

10.5 SURVEY OF H_2S REMOVAL PROCESSES

Gas treating processes can be divided into five categories:

I. Chemical solvent processes
The alkanolamines were introduced industrially in 1928 and are generally used for gas sweetening in refineries and chemical plants. For the sweetening of natural gas and gases from coal gasification the amine solvents are not the favoured choice.

II. Physical solvent processes
The physical processes are mostly used for bulk CO_2 removal and selective H_2S absorption from natural gas, coal gas or synthesis gas treatment at high pressure with high concentrations of acid gas.

III. Physical/chemical solvent processes
Physical/chemical solvent processes use a mixture of a physical and a chemical solvent. The processes are often called hybrid systems. The processes are used for purification of natural gas or synthesis gases at a high pressure with high acid gas concentrations.

IV. Direct conversion processes
The direct conversion processes (or liquid oxidation processes) are mostly used for removal of H_2S and direct oxidation to sulfur from gases like coke-oven gas, coal gas and Claus tail gas. The processes, often called Redox processes, are best suited for low concentrations of H_2S and with a maximum sulfur production of about 20 t/d. The leading processes in this class are Stretford, LO-Cat and SulFerox.

V. Dry bed and other processes
Among the dry bed processes the iron sponge and molecular sieve processes are the most widely used for removal of small quantities of H_2S. The iron sponge or iron oxide process is the oldest treatment process and still in use today. For this paper processes like the Chemsweet process, making use of a slurry, and the gas permeation technology which uses membranes, were classified in this group.

Table 10.2 gives a survey of the characteristics of the five categories.

Table 10.2. Characteristics of chemical, physical and chemical/physical solvents, direct conversion and dry bed processes

	I. Chemical solvent	II. Physical solvent III. Physical/chemical solvent	IV. Direct conversion	V. Dry bed
H_2S removal principle	Chemical absorption	Physical absorption physical/chemical absorption	Chemical conversion	a. Chemical absorption b. Physical absorption
H_2S loading	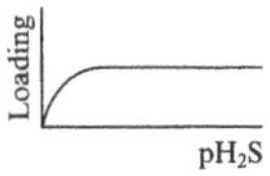	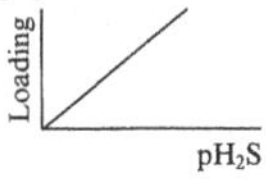	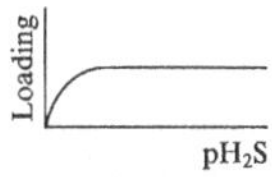	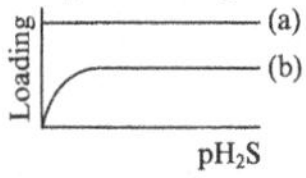
	Loading limited by stoichiometry	Loading proportional to partial pressure of H_2S	Loading limited by stoichiometry	Loading limited by a. stoichiometry b. surface area
Quantity of H_2S to be removed	Large	Very large	Small	Very small
Purity required	Moderate/high	High	Moderate/high	a. very high b. high
Desorption energy	High	Low	Moderate	a. not regenerable b. moderate
Typical application	General purpose continuous	Bulk removal continuous	Continuous	a. guard bed (batch) b. cyclic operation

The following Table 10.3 gives a survey of the commercial gas treating processes. The list is, however, not complete, as new processes are constantly being introduced and processes that are commercially unimportant have been left out (Burns and Maddox 1968).

Table 10.3. Commercial gas treating processes for H_2S removal

I. Chemical solvent processes		Molarity	Solvent
A.	**Alkanolamines**	**(typical)**	
	MEA	2.5 N	Mono-ethanolamine (= 15 % in water)
	DEA	2.5 N	Di-ethanolamine (= 26 %)
	SNEA-DEA	3 N	Di-ethanolamine (= 32 %)
	Amine Guard-UCAR	5 N	Mono-ethanolamine (= 30 %) with inhibitor
	Amine Guard-ST	5 N	Mono-ethanolamine (= 30 %) with inhibitor
	Amine Guard-ST	5 N	Di-ethanolamine (= 55 %) with inhibitor
	ADIP-DIPA	4 N	Di-isopropanolamine (= 54 %)
	ADIP-MDEA	4 N	Methyl-di-ethanolamine (= 48 %)
	Activated MDEA	4 N	Methyl-di-ethanolamine (= 48 %) with additives
	Ucarsol-HS	4 N	Methyl-di-ethanolamine (= 48 %)
	Gas/Spec	4 N	Methyl-di-ethanolamine (= 48 %)
	Textreat	4 N	Methyl-di-ethanolamine (= 48 %) with additives
	DGA	6 N	Di-glycolamine (= 63 %)
	Flexsorb-SE	3 N	Hindered amine
B.	**Hot Potassium Carbonate**		
	Catacarb		Potassium carbonate solution with catalyst
	Benfield		Potassium carbonate solution with activator
	Giammarco-Vetrocoke		Non-toxic organic-promoted hot potassium carbonate
	Alkacid-M		Potassium salt of methylaminopropionic acid
	Alkacid-DIK		Potassium salt of dimethylaminoacetic acid
	Ucarsol-CR		50 % aqueous solution
	Flexsorb-HP		Potassium carbonate solution with hindered amine

Table 10.3 (*cont.*)

II. Physical solvent processes	**Solvent**
Selexol	Di-methyl ether of polyethylene glycol (DMPEG)
Fluor solvent	Propylene carbonate
Purisol	N-methyl-2-pyrrolidone (NMP)
Rectisol	Methanol
Ifpexol	Methanol
Sepasolv MPE	Oligoethylene glycol methyl ispropyl ethers (mixed)
Cryofrac	Organic solvent

III. Physical/chemical solvent processes	**Solvent**
Sulfinol-D	DIPA with sulfolane
Sulfinol-M	MDEA with sulfolane
Amisol	MEA, DEA with methanol
Optisol	Organic solvent with amine
Selefining	Organic solvent with tertiary amine
Flexsorb-PS	Organic solvent with hindered amine

IV. Direct conversion processes	**Solvent**
Stretford	Sodium carbonate, sodium vanadate, anthraquinone disulfonic acid and traces chelated iron
Unisulf	Sodium carbonate/bicarbonate, vanadium, thiocyanate, carboxylate and aromatic sulfonate complexing agent
Sulfolin	Sodium hydroxide, sodium vanadate and organic catalyst
LO-Cat	Iron solution with two chelating (proprietary) agents
SulFerox	Iron solution with chelating (proprietary) agent
Hiperion	Chelated iron with naphthaquinone
Sulfint	Iron solution with chelating (EDTA based) agent
Bio-SR	Iron solution with bacteria which catalyse the re-oxidation of Fe^{2+} to Fe^{3+}

V. Dry-bed and other processes	**Medium**
Iron sponge	Iron oxide
Sulfatreat	Mixture of iron oxides (Fe_2O_3 and Fe_3O_4) with chemicals
Zinc oxide	Zinc oxide
Puraspec	Zinc oxide based material
Chemsweet	Slurry of zinc oxide in zinc acetate solution with dispersant
Sulfa-check	Sodium nitrite ($NaNO_2$) solution
Hondo HS-100	Potassium hydroxide nitrite solution
Sulfa-scrub	Hexahydrotriazine solution
Sulfa-Guard	Polyamine solution with chemicals
Sulfa-check 6138 / 6139	Alkylamine aldehyde solution
Sofnolime RG	Fixed bed of granular solids
Mol. Sieves	Molecular sieve
Activated carbon	Carbon
Gas permeation	Membranes
Shell Thiopaq	Biological desulfurisation / oxidation

10.5.1 Chemical solvent processes

10.5.1.1 Alkanolamines

In the alkanolamine processes, generally called Amine processes, the acid gas components react with the solvent and form a complex. The complex is

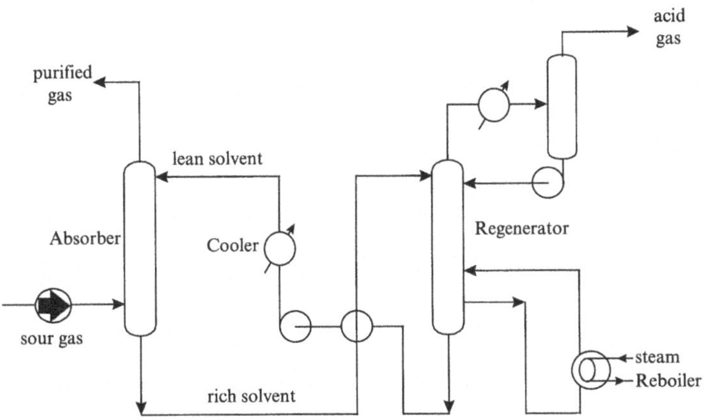

Figure 10.4. Schematic representation of the Amine process.

decomposed and acid gas is released when the temperature of the solvent is increased and pressure reduced. Aqueous solutions of alkanolamines (concentrations 15 to 50 %) are used to absorb H_2S and CO_2.

The process flow schemes for all chemical solvent processes are basically the same (Figure 10.4). Feed gas is contacted with the solvent in an absorber. The solvent with the absorbed acid gas components flows via a heat exchanger to the regenerator, in which the acid gases are desorbed from the solvent at reduced pressure with steam generated by reboiling. The acid gases leave the regenerator overhead system after condensation of the stripping steam. The stripped solvent is returned to the absorber via a heat exchanger and cooler.

Choice of amine solvents

An ideal solvent for the selective absorption of hydrogen sulfide (H_2S) is one that has a high absorption capacity for H_2S and a low absorption capacity for carbon dioxide (CO_2). Among the common amine solvents such a solvent is not available. Nevertheless, selective absorption of H_2S with an amine solvent can be achieved if use is made of the difference in absorption velocities of H_2S and CO_2.

The reactions taking place between primary (RNH_2), secondary (R_2NH) and tertiary (R_3N) amines and H_2S and CO_2 are as follows:

Primary and secondary amines:

(a) $H_2S + R_2NH \leftrightarrow R_2NH_2^+ + HS^-$ (instantaneous)

(b) $CO_2 + 2\ R_2NH \leftrightarrow R_2NH_2^+ + R_2NCOO^-$ (moderate)

(c) $CO_2 + R_2NH + H_2O \leftrightarrow R_2NH_2^+ + HCO_3^-$ (slow)

Tertiary amines:

(a)	$H_2S + R_3N \leftrightarrow R_3NH^+ + HS^-$	(instantaneous)
(b)	$CO_2 + R_3N$	(no reaction)
(c)	$CO_2 + R_3N + H_2O \leftrightarrow R_3NH^+ + HCO_3^-$	(slow)

Reaction (a) proceeds instantaneously for primary and secondary amines as well as for tertiary amines. In reaction (b), CO_2 reacts with primary and secondary amines at a moderate rate to form carbamate. With tertiary amines no carbamate is formed. Reaction (c) shows that for all types of amine solvent, CO_2 reacts slowly to bicarbonate upon further absorption.

Hydrogen sulfide reacts instantaneously with the amine (reaction (a)). Equilibrium is completed in the amine solvent. Absorption is limited by mass transfer of H_2S through the gas film (gas-film controlled) but the H_2S absorption rate is practically the same for all amines.

The absorption of carbon dioxide is determined by slow reactions (reactions (b) and (c)) and by diffusion in the liquid film. Both mechanisms slow down the absorption and are different for primary, secondary and tertiary amines.

To decide which amine solvents are most suitable for use in selective absorption systems, a comparison must be made between the best known primary amines (mono-ethanolamine - MEA - and di-ethylene glycolamine - DGA), secondary amines (di-ethanolamine - DEA - and di-isopropanolamine - DIPA), and tertiary amines (tri-ethanolamine - TEA - and methyl-di-ethanolamine - MDEA).

Primary amines

MEA and DGA are strong bases. For selective absorption MEA and DGA are not suitable, as CO_2 is absorbed at a high rate at the same time.

MEA can meet low acid gas specifications in the treated gas, but requires considerable amounts of energy (steam) for regeneration. COS degrades MEA and a reclaimer is required to eliminate the degradation products.

Secondary amines

CO_2 co-absorption in DEA is twice as high as for DIPA. Moreover, the allowable acid gas loading (mol acid gas / mol amine) is higher for DIPA than for DEA. Therefore the solvent circulation for DEA is higher than for DIPA. For this reason DEA is not a good selective solvent. DIPA, which has a CO_2 co-absorption percentage of 20-30 %, is a suitable solvent for selective absorption.

DEA and DIPA show a lower heat of reaction for absorption and desorption. Therefore less steam for regeneration is required compared with

Table 10.4. Heat of absorption and relative CO_2 absorption of alkanolamines

Solvent	Heat of reaction, kJ/kg		Relative CO_2 absorption, %
	H_2S	CO_2	
MEA	1905	1920	100
DEA	1190	1510	40
DIPA	1140	2180	20
MDEA	1050	1420	10

MEA. With DEA or DIPA, a reclaimer is normally not needed unless very high CO_2 levels are present in the gas. A certain selectivity can be obtained with DIPA.

Tertiary amines

The main difference compared with primary and secondary amines is that CO_2 does not react to form carbamate. Only a slow reaction to bicarbonate takes place. For this reason, tertiary amines have the highest selectivity for H_2S. TEA, however, is too weak a base to give acceptable H_2S loadings and thus can be eliminated. MDEA does not have this shortcoming and is becoming more and more attractive, especially if the CO_2 concentration is high. In such cases, the MDEA circulation rate is considerably lower than the DIPA circulation rate.

MDEA has a considerably better selectivity for H_2S. The heat of reaction is low and degradation products are not formed with COS or CO_2.

Proprietary MDEA based solvents with corrosion inhibitors and special additives allow for higher alkanolamine concentrations and higher loading or higher selectivity for H_2S and therefore lower circulation rates. Less corrosion problems and lower steam consumption can be obtained.

More higher selectivity of H_2S absorption can be met and considerable steam saving can be obtained by using sterically hindered amines such as Flexsorb. In Claus tailgas clean-up units, for example, the circulation rate in the gas treating section, by using Flexsorb SE, can be reduced by 50 % compared with generic MDEA.The relative rates of CO_2 absorption compared to MEA are summarised for a typical feed gas in Table 10.4 (Ullmann Volume A17 1991).

COS removal

Carbonyl sulfide (COS) is often present as an impurity in gas streams. It is partly removed (approximately 20 %) by DIPA; with MDEA there is no reaction. If COS must be removed, low temperature hydrolysis over a fixed catalyst bed ahead of the selective absorption unit can convert 90 %, or more, of the COS to H_2S and CO_2.

Table 10.5. Physical properties of amines and amine solvents

Amine	MEA	DGA	DEA	DIPA	MDEA	TEA
Properties of amine						
molecular weight	61.09	105.14	105.14	133.19	119.17	149.19
specific gravity	1.018 (20/20 °C)	1.055 (20/20 °C)	1.092 (30/20 °C)	0.989 (45/20 °C)	1.042 (20/20 °C)	1.126 (20/20 °C)
boiling point, °C @ 1 bar	171	221	decomp.	248	247	360
freezing point, °C	10.5	−9.5	28	42	−21	21
solubility in water, % (by mass) @ 20 °C	complete	complete	96.4	87	complete	complete
viscosity, cP	24.1 (20 °C)	26 (24 °C)	380 (30 °C)	198 (45 °C)	101 (20 °C)	1013 (20 °C)
Properties of amine solvent						
typical concentration, kmol/m^3	2.5	6	2	2	2	2
% solution (by mass)	15	63	21	27	24	30
boiling point, °C @ 1 bar	118	124	118	118	118	118
freezing point, °C	−5	−50	−5	−5	−6	−6
viscosity, cP @ 40 °C	1.0	6.5	1.3	1.06	1.06	1.06
vapour pressure at 40 °C, kPa	7.4	4.0	7.4	7.4	7.4	7.4

Note that a high degree of COS removal can be achieved using DIPA (ADIP process) for COS removal from liquid hydrocarbons.

Physical properties

The physical properties of MEA, DGA, DEA, DIPA, MDEA and TEA are given in Table 10.5 for comparison (Lagas 1982). The low freezing point of MDEA is interesting. It should be noted, however, that the freezing point of MDEA in aqueous solutions is raised sharply. DIPA and MDEA both have low vapour pressures, which allows absorption at low pressure without unacceptable solvent losses.

10.5.1.2 Hot potassium carbonate

The hot carbonate processes sometimes use single-stream, split-flow and two-stage operations for CO_2 removal as shown in Figures 10.5-10.7.

The selection of the processing scheme mainly depends on the required purified gas specification.

The process was originally developed using a solution of only potassium carbonate in water. In recent years, several additives and/or catalysts have been found that increase the amount of acid gas absorbed in the solution or reduce the heat required for regeneration (see Table 10.3).

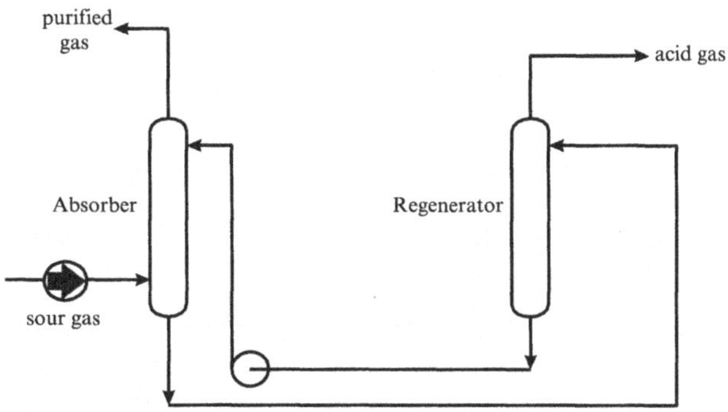

Figure 10.5. Single-stream operation for removal of CO_2 down to 2 % in purified gas.

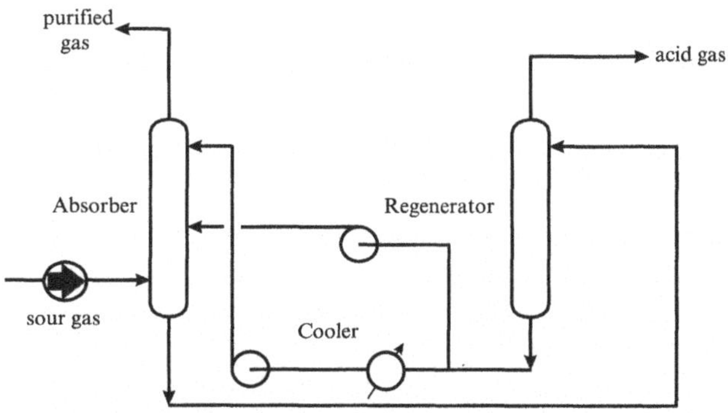

Figure 10.6. Split-flow operation for removal of CO_2 down to 1 % in purified gas.

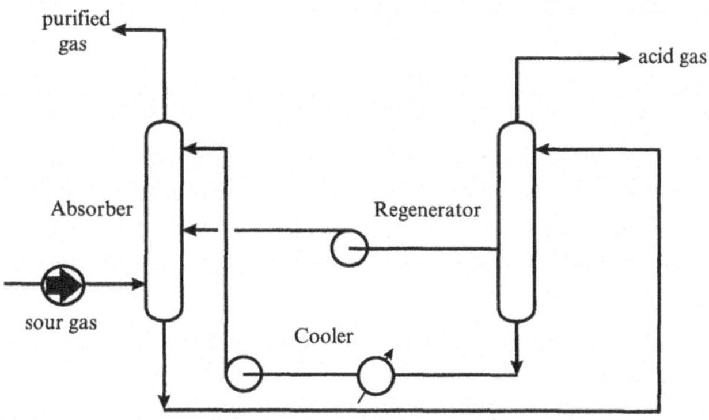

Figure 10.7. Two-stage operation for removal of CO_2 down to below 1 % in purified gas.

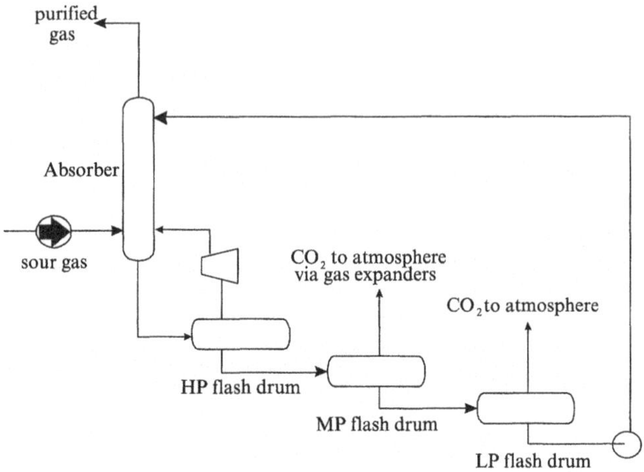

Figure 10.8. Schematic representation of CO_2 absorption unit using physical solvents.

10.5.2 Physical solvent processes

These processes use an organic solvent which absorbs the acid components as a function of their partial pressure. In general the higher the partial pressure and the lower the absorption temperature, the better the absorption. The process flow scheme for a physical solvent process depends on the type of acid gas components absorbed. For CO_2 absorption only, normally an absorber and 2 or 3 flash stages are sufficient (Figure 10.8).

For selective H_2S absorption, normally one or two flash stages are followed by a regenerator (Figure 10.9).

Lurgi in Germany offers two processes in which no aqueous solutions are involved, namely the Rectisol and the Purisol process. With the Rectisol process, methanol at about -40 °C is used. Prevention of loss of methanol is an important issue with the Rectisol process. The Purisol process employs N-methyl-2-pyrrolidon (Figure 10.9) as an absorbent. Allied Chemical Corporation has developed the Selexol process, in which the di-methylether of polyethyleneglycol is used as the solvent, which exhibits physical absorption of sulfur compounds.

Physical solvent processes are mostly used for bulk CO_2 removal and selective absorption of H_2S. Physical processes are characterised by high acid gas loadings, low circulation rates and low utility costs. In general they have two major disadvantages:

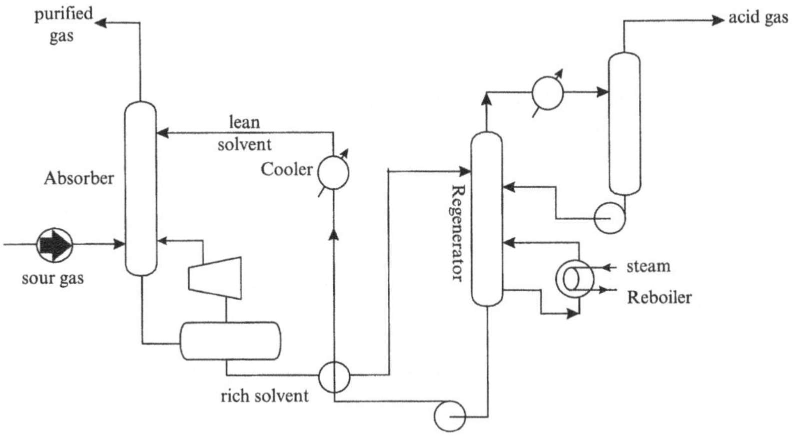

Figure 10.9. Flow diagram of selective H_2S absorption using physical solvents.

(a) Co-absorption of hydrocarbons (heavy hydrocarbons and aromatics are almost completely removed from the gas stream).
(b) Solvents are expensive.

When many heavy hydrocarbons are absorbed, an acid gas is produced, which can give problems in the Claus unit. Sometimes a charcoal adsorption or an acid gas enrichment unit is applied to improve the quality of the Claus feed gas.

10.5.3 Physical/chemical solvent processes

These processes use a mixture of a physical and chemical solvent and are called hybrid systems, because acid gases are both physically and chemically absorbed. As for physical solvent processes, the acid gas absorber is followed by a flash drum and a regenerator (Figure 10.9). Hybrid systems are used to remove H_2S / CO_2 and organic sulfur components like COS, CS_2 and mercaptans from gases like natural gas or synthesis gas at a high partial pressure.

One of the best known hybrid systems is the Shell Sulfinol process. In this system, the chemical component is either DIPA or MDEA, whereas the physical component is Sulfolane (Figure 10.10).

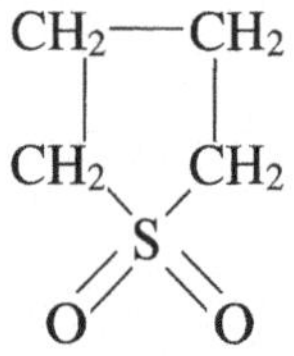

Figure 10.10. Molecular structure of Sulfolane (tetrahydrothiophene dioxide).

10.5.4 Direct conversion processes

These processes are based on the oxidation of H_2S to elemental sulfur by air with the aid of a compound that is easily oxidised by atmospheric oxygen and easily reduced by H_2S. The liquid oxidation processes can achieve a residual H_2S content in the treated gas of less than 1 ppm by volume with complete selectivity towards CO_2.

Absorption of hydrogen sulfide into solvents involves a chemical reaction. H_2S is removed from the gas stream by a circulating solution containing iron. The lean solution containing iron in the +3 valence state oxidises the H_2S to elemental sulfur. The iron is simultaneously reduced to the +2 valence state:

$$2Fe^{3+}L + H_2S \rightarrow 2Fe^{2+}L + S + 2H^+ \qquad (10.3)$$

The rich solution containing suspended sulfur and iron in the reduced form, is regenerated with air according to the following reaction and the sulfur is separated:

$$2Fe^{2+}L + \tfrac{1}{2}O_2 + 2H^+ \rightarrow 2Fe^{3+}L + H_2O \qquad (10.4)$$

Some processes such as SulFerox and LO-CAT use organic ligands or chelating agents, denoted by L in the above Equations (see Table 10.3). The chelate keeps the iron in solution and prevents the precipitation of $Fe(OH)_3$ and FeS. SulFerox differs from LO-CAT by a higher iron content in the solution ranging from 1 up to 4 % iron, compared to 0.05 to 0.2 % iron used in the LO-CAT process. H_2S conversion capacity in the SulFerox process is about 0.2 moles per litre solution and is about 10 times higher than for other processes. The absorption/oxidation processes are often indicated as REDOX processes (Figure 10.11).

The CO_2 solubility in these solutions is low and similar to water. Therefore a high selectivity for H_2S removal can be obtained. Sulfur removal efficiencies of 99.9 % or higher can be achieved (Quinlan 1991).

Redox processes are normally used for small scale H_2S removal up to

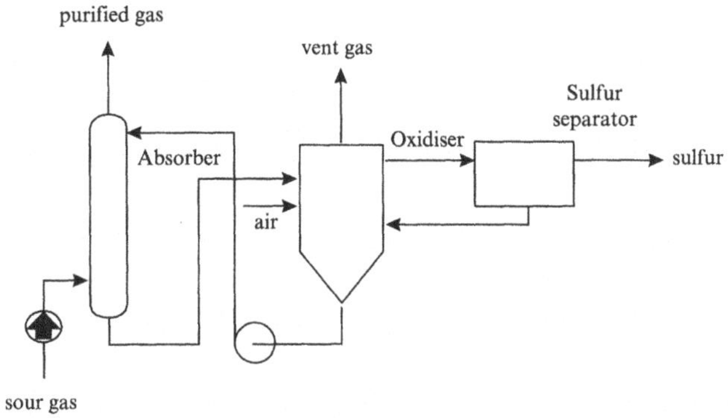

Figure 10.11. Process scheme of the Redox type processes for H_2S removal.

20 t/d of sulfur. In general, it is more attractive to apply a gas treating/Claus process with higher capacities.

In the LO-CAT process by-product reactions take place by slip of oxygen from the oxidiser to the absorber forming thiosulfate and sulfate salts. These reactions reduce the pH of the solution and thus the H_2S removal efficiency is reduced. Therefore, pH control by alkaline buffers, such as KOH or Na_2CO_3 is a need. In the SulFerox process the reaction with $Fe^{2+}L$ and O_2 is so fast that no O_2 slip to the absorber takes place. SulFerox operates at a lower pH (6-7.5) as compared with LO-CAT (pH 8 - 8.5). Impurities, such as mercaptans, COS and CS_2 are absorbed to a certain extent. Mercaptans are converted to disulfides and the latter are absorbed. If more than 5 % of the sulfur is coming from mercaptans, the sulfur quality will be poor. COS and CS_2 are removed to approximately 50 %.

Some hydrocarbons will be absorbed and end up in the produced sulfur and cause the sulfur product to darken. Aromatics such as benzene, toluene and xylene (BTX) will give an unpleasant odour to the sulfur product. In general it can be stated that the sulfur product from Redox processes is discoloured by carbon and remaining chemicals. The sulfur cake from the filter is often disposed off. Upgrading of the sulfur quality by washing, however, is possible. Melting and filtering the liquid sulfur can considerably improve the sulfur quality.

10.5.5 Dry bed and other processes

The dry bed processes use a fixed bed of solid material such as iron oxide, zinc oxide, molecular sieve (zeolites) or activated carbon. These dry bed processes can be divided into two categories:

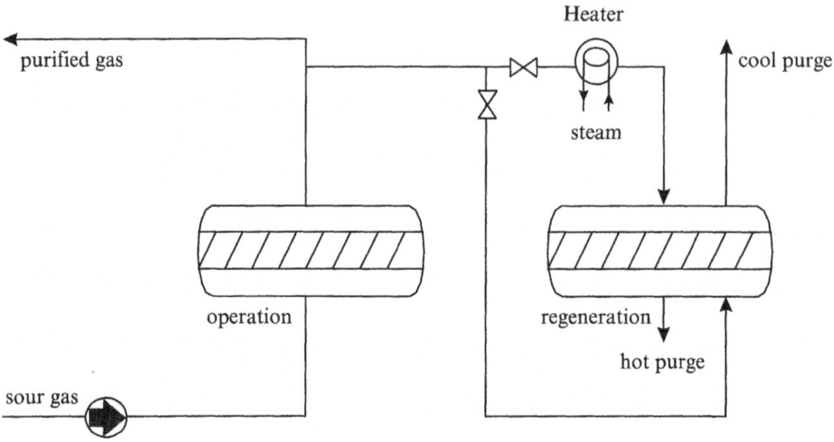

Figure 10.12. Process scheme using a molecular sieve to remove H_2S.

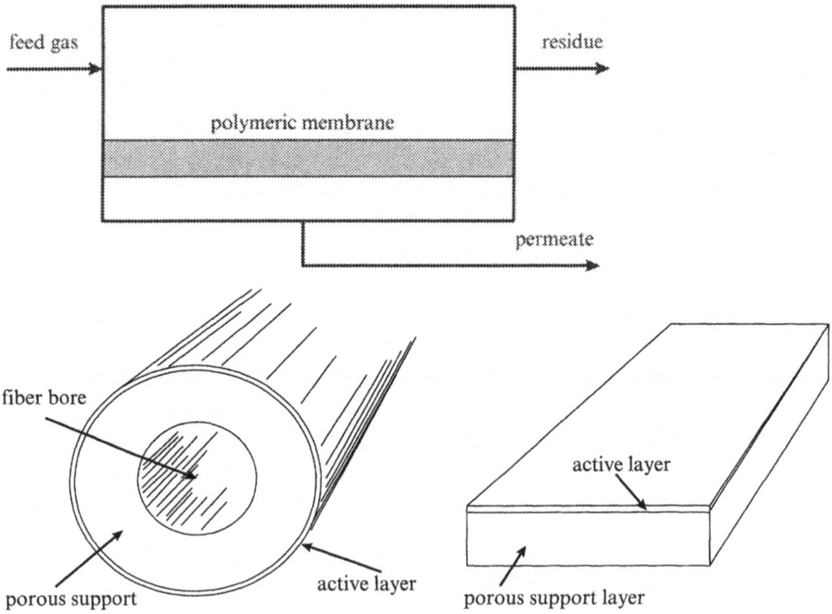

Figure 10.13. Principle of membrane separation of gas components.

- Chemical absorption (iron oxide, zinc oxide);
- Physical adsorption (molecular sieve, activated carbon).

Using physical adsorption the bed can be regenerated, which is not the case when chemical absorption is used. In the latter case, the bed has to be replaced. Using physical adsorption, the adsorbed acid components can be released during regeneration by a hot gas (Figure 10.12). With iron oxide and zinc oxide the H_2S reacts to iron and zinc sulfide, respectively.

A new development to purify gases is the use of membranes. Components such as CO_2, H_2S, H_2O and H_2 can be seletively separated. Membrane separation is based on the principle of gas molecules permeating through a membrane at different velocities. The differential pressure is the driving force.

The permeation of gases across polymeric membranes involves two steps, generally referred to as solution-diffusion. A solution of molecules of the permeable gas at high pressure dissolves into the surface of the membrane followed by diffusion of the permeable gas through the membrane to the permeate side and finally desorption at the low pressure surface. The permeability of each gas component is a function of the solubility and diffusion through the membrane and depends on the chemical composition and physical structure of the polymer.

The basic form of a membrane separation system is a vessel that is divided by a flat membrane into a high and a low pressure section (Figure 10.13). Membrane systems perform very effectively on high pressure streams of 20-100 bar. Feed gas entering the high pressure side selectively loses the fastest permeating components to the low pressure side.

For the gas permeation process, two main types of membrane configuration are applied (Figures 10.14 and 10.15).

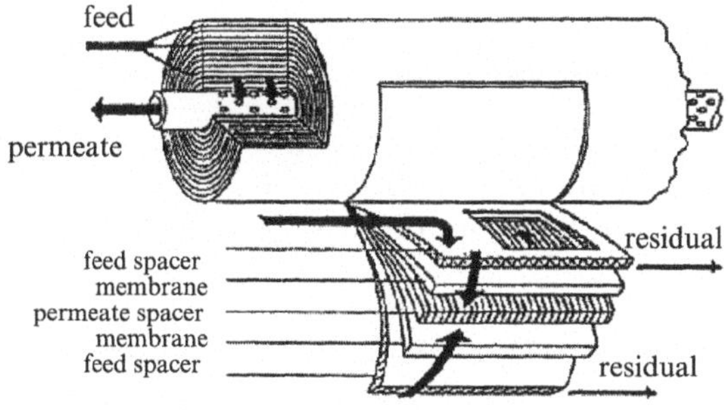

Figure 10.14. Spiral wound membrane.

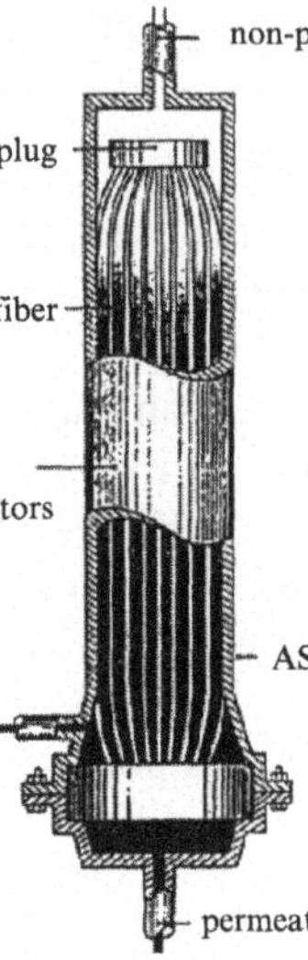

Figure 10.15. Hollow fibre membrane.

10.6 SURVEY OF SO_2 REMOVAL PROCESSES

Flue gas desulfurisation processes can be grouped into non-regenerable and regenerable processes. Most of the non-regenerable processes use wet absorption, spray drying and dry injection technologies. With wet scrubbing, a wide range of absorbents are used such as slaked lime, limestone, alkaline solutions, aqueous ammonia and sea water. The regenerable processes are more diversified and complex. Table 10.6 gives a survey of the commercial FGD processes (Cope and Klingspor1987).

10.6.1 Wet scrubbing processes

In the wet scrubbing processes, the flue gas is scrubbed with an alkaline slurry in a countercurrent open spray tower type absorber (Figure 10.16). A solution of slaked lime, limestone or slaked lime and alkaline fly ash can be used as a sorbent.

In the absorber, SO_2 reacts to an insoluble calcium compound which is separated and dewatered. Through the use of an oxidation step, gypsum is produced according to the overall reaction using limestone:

$$SO_2 + CaCO_3 + \frac{1}{2} O_2 + 2\ H_2O \rightarrow CaSO_4.2H_2O + CO_2 \qquad (10.5)$$

Table 10.6 Overview of different FGD processes

Non-regenerable processes	Sorbent	Product
Wet scrubbers	Lime/limestone	Gypsum or calcium sulfite / sulfate
	Lime/fly ash	Calcium sulfite / sulfate / fly ash
Spray-dry scrubbers	Lime	Calcium sulfite / sulfate
Dual-alkali	Primary: sodium hydroxide	
	Secondary: lime / limestone	Calcium sulfate / sulfite
Walther	Ammonia	Ammonium sulfate
Seawater	Primary: seawater	Waste sea water
	Secondary: lime	
Regenerable processes	**Sorbent**	**Product**
Bergbau - Forschung	Activated carbon	Concentrated SO_2
Wellman - Lord	Sodium sulfite	Concentrated SO_2 or elemental S
Linde - Solinox	Physical absorption solvent	Concentrated SO_2
Spray-dry scrubbers	Sodium carbonate	Elemental S
MgO process	Magnesium oxide	Concentrated SO_2

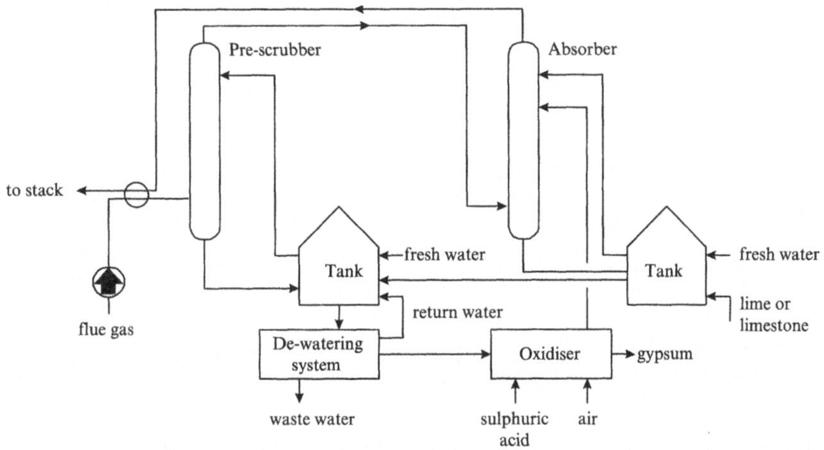

Figure 10.16. Process diagram of a FGD - wet scrubbing process.

Fly ash is separated in an electrostatic precipitator upstream from the spray tower.

In a pre-scrubber, the flue gas is cooled and humidified, and a significant portion of the halides (chloride and fluoride) is removed as well. Often, the pre-scrubber and the absorber section have separated liquid circulation systems to make the system work in a “double-loop” configuration to obtain a high quality of gypsum.

In some FGD processes, alkaline fly ash from the combustion process is used as a sorbent. In this case, the fly ash is separated in a venturi pre-

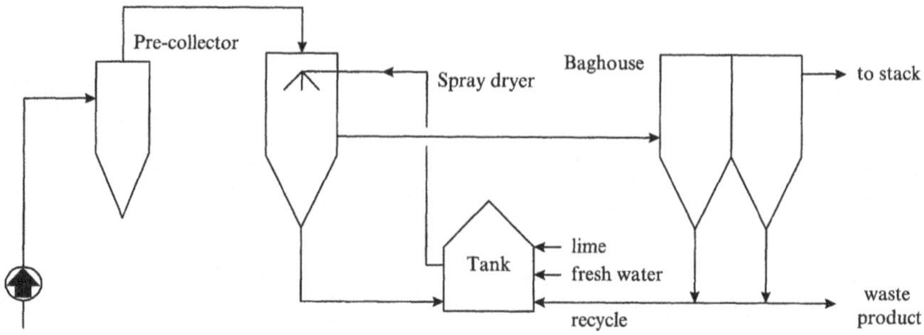

Figure 10.17. Process diagram of a FGD - spray dry scrubbing process.

scrubber and mixed with slaked lime to increase the efficiency for SO_2 absorption. This type of solvent can only be used for flue gas desulfurisation from low sulfur coal fired power plants.

Wet scrubbing processes are developed by many companies; among the best known are Mitsubishi Heavy Industries, Babcock-Hitachi, Babcock & Wilcox, Combustion Engineering, Saarberg-Hölter-Lurgi, Research Cottrell and Walter. Each of them constructed more than 15 plants.

10.6.2 Spray dry scrubbing processes

In the spray dry processes, the flue gas is contacted with a solution or slurry of slaked lime in a spray dryer. The solution or slurry is atomised by a rotary atomiser or through nozzles and is mixed intimately with the flue gas (Figure 10.17).

By evaporating water from the finely dispersed droplets, the flue gas is humidified to within 20 °C of the saturation temperature. As the solution or slurry is evaporated, salts are precipitated. This material is dried to less than a few percent free moisture. The solids and fly ash are entrained in the flue gas and carried from the spray drier to a separation/collecting unit. Often a baghouse is the selected separator because of its contribution of SO_2 removal from the flue gas.

The dry solid material from the spray drier and separator contains unreacted lime and is therefore commonly recycled to the solution or slurry feed tank. Recycling improves the efficiency of SO_2 removal and reduces the sorbent consumption.

The end product is a mix of calcium sulfite / sulfate and fly ash which is normally disposed in a landfill area. Some installations have facilities to upgrade this mix to a cement-like material. Some plants have fly ash pre-collectors to separate the fly ash from the waste product. An advantage of

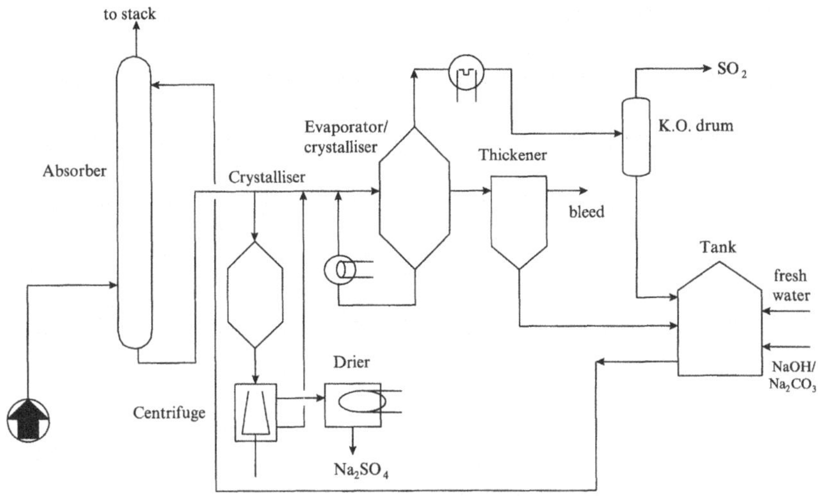

Figure 10.18. Flow diagram of the FGD-Wellman-Lord process.

the pre-collector is, that the FGD plant is protected against the abrasive effect of the fly ash. FGD waste free from fly ash and with a low chloride concentration can be oxidised to anhydrous calcium sulfate which can be used as a setting retarder in Portland cement. Also in this group of non-regenerable processes, many systems have been developed. Flaker Industri AB and Niro Atomiser are the most well-known.

10.6.3 Wellman-Lord

The Wellman-Lord process is the most common regenerable FGD process in operation today. Sulfur dioxide from the flue gas is absorbed in a sodium sulfite solution in a wet spray scrubber (Figure 10.18). The sulfite is converted to bisulfite according to:

$$SO_2 + Na_2SO_3 + H_2O \leftrightarrow 2\ NaHSO_3 \qquad (10.6)$$

The solution is regenerated by heat treatment, during which the bisulfite is decomposed in a evaporator/crystalliser unit, which is often called the heart of the regeneration system. In the evaporator, the sodium bisulfite is decomposed to sodium sulfite and SO_2 by the reverse of the above absorption reaction. The resulting vapour is condensed to remove the water and to achieve the desired SO_2 product quality.

The sodium sulfite precipitates out and builds a dense slurry of crystals in the evaporator, which is passed to a thickener. The overflow from the thickener is withdrawn as waste solution and is bleeded. The slurry of crystals from the underflow is diluted with condensate and fresh water and is recycled to the spray scrubber. To compensate for the losses in the bleed from the thickener, a solution of hydroxide or sodium carbonate is added to circulating solvent.

In a side reaction, part of the sulfite is oxidised to sulfate by residual oxygen and traces of sulfur trioxide present in the flue gas. To prevent build-up of the sulfate content, a part of the scrubbing solution is continuously withdrawn and is passed through a water-cooled crystalliser in which the sodium sulfate is precipitated. The sodium sulfate is separated from the solution in a centrifuge after which it is dried and discharged from the system for further processing.

The concentrated SO_2 may be used for sulfuric acid production or if a part is reduced to H_2S, elemental sulfur can be produced with the Claus process. Overall sulfur recoveries of 99.8 % have been reported (Link and Ponder 1977).

10.7 REFERENCES

Burns, M.D. and Maddox, R.N. (1968) Lease gas sweetening. *Oil and Gas Journal*, **65 - 66**, series of 15 articles.

Cope, D.R. and Klingspor, J.S. (1987) *FGD Handbook, Flue gas desulfurization systems*. IEA Coal Research, London, England.

Geus, J.W. and Lagas, J.A. (1993) Purification of synthesis gas. Paper presented at the Gas Processing Association, 10th Continental Meeting, Paris, France.

Lagas, J.A. (1982) Acid gas removal with selective amine processes. *Sulfur*, **159**.

Lagas, J.A. (1988) Selection of gas sweetening processes. Paper presented at the Sulfur Workshop Katwoude, The Netherlands.

Link, F.W. and Ponder, W.H. (1977) Status report on the Wellman-Lord/Allied Chemical fuel gas desulfurization plant at Northern Indiana Public Service Company's Dean H. Mitchell Station. Paper presented at the EPA Flue Gas Desulfurization Symposium, Hollywood, USA.

Ulmann's Encyclopedia of Industrial Chemistry (1991) Natural Gas, **A17**.

Quinlan, M.P. (1991) Characteristics and handling options for sulfur from liquid Redox processes. Paper presented at Sulfur '91 Conference, New Orleans, USA, Preprints, 41 - 65.

11

Novel biological processes for the removal of H_2S and SO_2 from gas streams

Albert J.H. Janssen, Henk Dijkman and Guido Janssen

11.1 INTRODUCTION

The increase in global population is inevitably associated with continued industrialization, urbanization and motorization. Increased pollution effects are recorded in industrial regions and on a global scale owing to an increase in industrial activities. Some of these environmental problems are related to the emission of sulfur dioxide (SO_2) and hydrogen sulfide (H_2S). The emission of sulfur dioxide greatly exceeds the emission of hydrogen sulfide because the latter is mostly burnt. When released into the atmosphere, SO_2 is a severe acidifying compound, causing acid rain. Emission of H_2S into the atmosphere is mainly a result of volcanic activities and evaporation from oceanic waters (Brimblecome et al. 1989). Problems due to H_2S pollution are encountered during the exploitation of H_2S containing resources, e.g. natural gas and crude oil, the use of H_2S in the production process such as tanneries, or when H_2S is an unwanted reaction product like in biogas.

In the atmosphere, H_2S causes acid rain owing to its reaction with ozone to sulfuric acid. Fossil fuel combustion accounts for approximately 90% of the global man-made emission of SO_2 (Brimblecome et al. 1989). Other major sources of SO_2 are the processing of ores, oil refining and sulfuric acid production. European sulfur emissions increased steadily from 1880 (interrupted only by World War II) up to a maximum of 60 million tonnes per year in 1980, followed by a steep decline. Since the 1970s in many industrialized countries various emission control strategies are in use, such as coal desulfurization, selection of fuels with a lower sulfur content, specialized combustion processes and waste gas treatment. These measures led to a yearly sulfur emission of about 30 million tonnes in 1995 (Brimblecome et al. 1989).

In this Chapter, two new biological processes for the desulfurization of respectively high pressure natural gas and the removal of SO_2 from gaseous streams are presented and data from pilot studies are shown. These processes are cheap and reliable alternatives for the conventional physico-chemical methods.

11.2 HIGH PRESSURE NATURAL GAS DESULFURISATION BY THE SHELL-THIOPAQ PROCESS

11.2.1 Introduction

In chapter 10, various well-established physico-chemical techniques for the removal of H_2S from sour, i.e. H_2S containing, gases are presented. These processes may be grouped into the categories listed in Table 11.1.

The application of the processes described in table 11.1 greatly depends on the amount of H_2S to be treated. In figure 11.1, it can be seen that at sulfur loads below 100 kg/day throw-away processes are in favour, whereas at sulfur loads between 0.1 and 15 tons/day liquid redox processes such as Lo-Cat or Sulferox are in use. At sulfur loads above 15 tons/day, amine-based processes are most common. The latter process concentrates the H_2S gas which is then sent to a Claus plant.

In liquid redox processes, H_2S is oxidised to elemental sulfur with iron(III) ions acting as oxidant. The formed iron(II) is subsequently

Table 11.1. Physico-chemical methods for H_2S removal (After: Jensen and Webb 1995)

Category	Example	Reagent	Products
Liquid phase chemical reaction	Amines	Alkanolamine	H_2S and CO_2
	Alkali salts	Potassium carbonate	H_2S and CO_2
Liquid phase physical absorption	Selexol	Dimethylether of polyethylene glycol	H_2S and CO_2
Dry bed	Iron sponge	Iron oxide	S°
	Molecular sieve	Crystalline alkali-metal aluminosilicates	S°
Direct conversion	Stretford	Sodium vanadate	S°
	Lo-Cat, Sulferox	Iron (III)complexes	S°
	Claus	H_2S and SO_2	S°
	Shell-Thiopaq process	Air-oxygen	S°

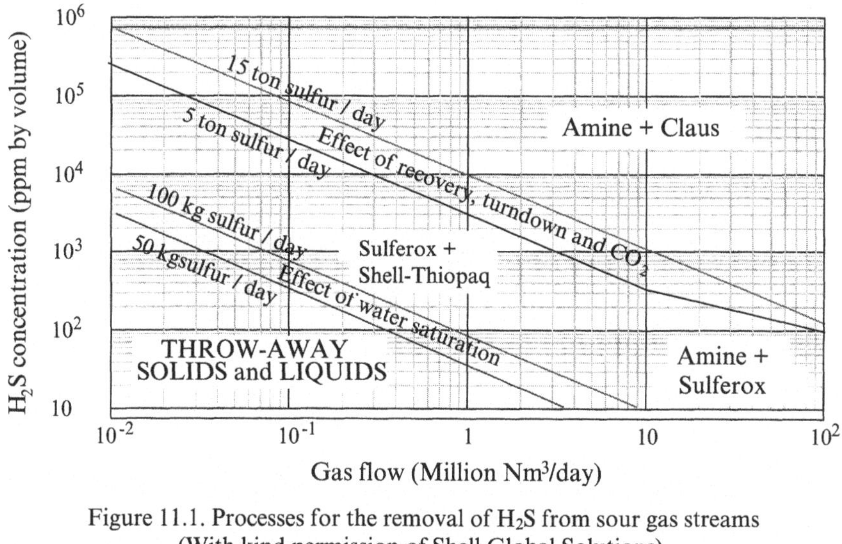

Figure 11.1. Processes for the removal of H_2S from sour gas streams (With kind permission of Shell Global Solutions).

oxidised. A major disadvantage of these processes is that the sulfur is formed in the absorber column which regularly leads to severe clogging problems. To readily re-oxidise the iron from the ferrous into the ferric form, the presence of certain complexing agents is required. These, however, are very often subject to chemical and biological decomposition resulting in severe chemical costs. A competing process for these liquid redox processes has been developed Paques Bio Systems B.V. and Shell Global Solutions.

The Shell-Thiopaq process relies on the biological oxidation of hydrogen sulfide to elemental sulfur by aerobic bacteria of the genus *Thiobacillus* (Janssen et al. 1998, 1999). A scanning electron micrograph of a sulfur excreting bacteria is shown in figure 11.2. It can be seen that sulfur droplets (globules) are deposited outside the bacterial cell. The droplets reach diameters of up to 1 μm but can grow into aggregates with a diameter up to 3 mm (Janssen et al. 1997, 1999). Due to the hydrophilic nature of the biologically produced sulfur particles severe clogging problems are avoided.

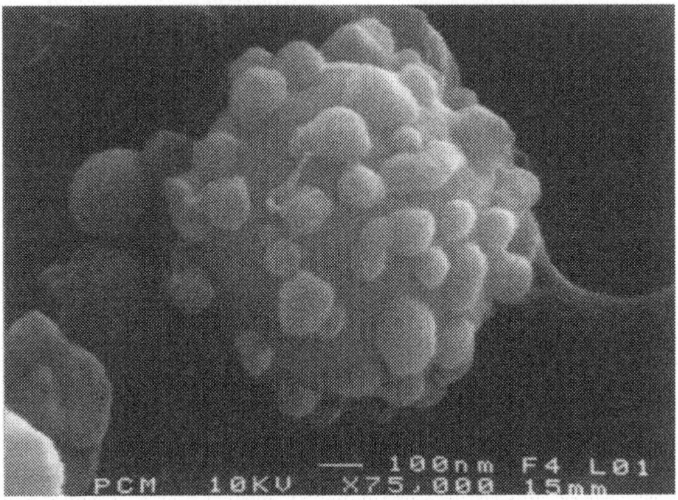

Figure 11.2. Scanning electron micrograph of the surface of a sulfur-excreting *Thiobacillus.* Sulfur excretion on the cell surface is visible.

The working principle of the Shell-Thiopaq process is based on the existing process for the removal of H_2S from biogas. Already in 1993, the first full-scale unit to remove H_2S from biogas was successfully taken into operation by Paques Bio Systems B.V. In a biogas stream (400 N m^3/h) generated from paper industry wastewater the H_2S concentration was reduced from 1-2 vol.% to less than 10 ppm. Currently, five of these units are in operation: one in The Netherlands, two in India, one in Germany and one in the USA.

A major difference between these existing processes and the Shell-Thiopaq process is the effect of the high pressure in the absorber column. Natural gas streams may occur at pressures up to 85 bar which means that the solubility of H_2S and CO_2 in the aqueous solvent is not comparable to the values that apply for the treatment of biogas. Also the effect of the pressure differences between the absorber coloumn and the bioreactor on the bacteria had to be studied. Besides the desulfurisation of high pressure natural gas streams is the Shell-Thiopaq process also suitable for the treatment of syngas streams and tail gas streams from Claus units and Selectox plants.

In the absorber H_2S gas is absorbed under high pressures (up to 60 bar). In the bioreactor, the dissolved sulfide is subsequently oxidised into elemental sulfur. The following reaction equations proceed:

Absorption and hydrolyses of H_2S:

$$H_2S_{(g)} + OH^- \rightarrow HS^- + H_2O \qquad (11.1)$$

Biological sulfur formation:

$$HS^- + \frac{1}{2} O_2 \rightarrow S^\circ + OH^- \qquad (11.2)$$

In case too much oxygen is supplied, the elemental sulfur particles can be further oxidized into sulfate:

$$S^\circ + 1\frac{1}{2} O_2 + H_2O \rightarrow SO_4^{2-} + 2\,H^+ \qquad (11.3)$$

The complete oxidation of sulfide into sulfate is almost completely prevented by controlling the redox-potential in the bioreactor (Janssen et al., 1998). From Reaction 11.2, it follows that the caustic which is used to absorb H_2S-gas is regenerated during the production of elemental sulfur.

11.2.2 Materials and methods

The Shell-Thiopaq pilot plant is located at the natural gas production facility of BEB Erdgas und Erdöl GmbH (Germany). It consists of two integrated parts, namely a high pressure absorption column for the absorption of H_2S and a bioreactor which is operated under atmospheric conditions. A schematic overview of the pilot plant is presented in Figure 11.3. The effluent from the bioreactor is recycled via a tilted plate settler to the absorption column to remove most the majority of the sulfur particles. The installation was completely automated so it could run without personal attendance.

The mixed gas stream which was used for the experiments was composed by mixing three separate gas streams. Gasstream 1 directly originates from the natural gas well. When this is desulfurized with a physical absorbent gasstream 2 is formed. Gasstream 3 is the completely desulfurised gas which

Table 11.2. Composition of gas streams used for the production of the 'mixed gas'

Mixed gas	H_2S (vol.%)	CO_2 (vol.%)	CH_4 (vol.%)	N_2 (vol.%)
Gas 1	9.6	8.5	76	5
Gas 2	450 ppm	6	90	3.8
Gas 3	<0.3 ppm	0.01	94	6

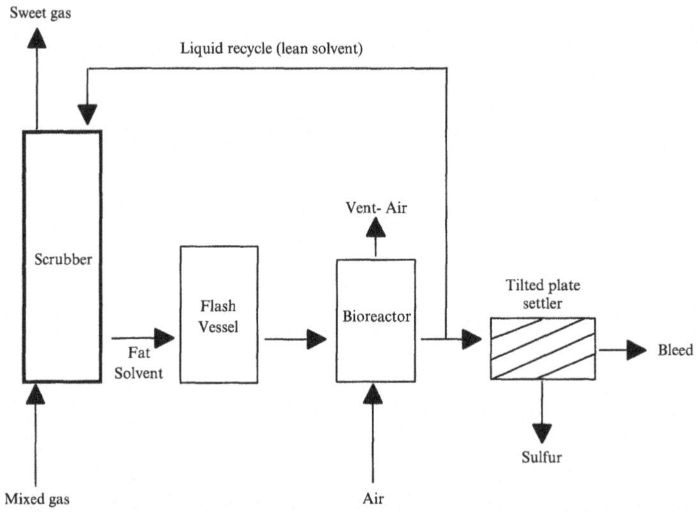

Figure 11.3. Block Process Diagram of the Shell-Thiopaq process for the treatment of high pressure natural gas.

is directed to the households. The composition of the gas streams is given in Table 11.2. The composition of the mixed gas stream depended on the required experimental conditions, such as high CO_2 or H_2S concentrations.

11.2.3 Results

11.2.3.1 Start-up of the system

Bacteria from a biogas treatment installation were used for the start-up of the bioreactor. Because this installation has been in operation since 1993 the bacteria are well adapted to the special environmental conditions, such as high pH values and high salt concentrations. The bacteria could be stored for more than 8 weeks in a closed vessel without any significant loss of activity. Accordingly, it is possible to use this bacterial population as a seed material for any installation in the world.

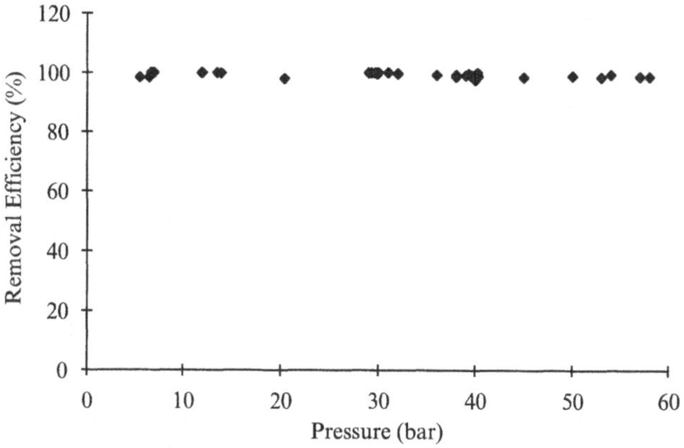

Figure 11.4. Effect of top pressure in absorber column on the activity of the biomass present in the bioreactor.

During the course of the experimental period (100 days) the top pressure in the absorber column varied between 5 and 60 bar. The suspended bacteria are exposed to these high pressures since they are recycled with the washing liquid from the bioreactor to the absorber column (Figure 11.4). From Figure 11.4 it becomes clear that the bacteria were not negatively affected by these high pressure alternations. All dissolved sulfide in the solvent was converted into elemental sulfur in the bioreactor.

Figure 11.5 shows the effect of varying sulfide concentrations in the influent on the removal capacity. Only the first day after start-up of the bioreactor sulfide could be measured in the liquid leaving the bioreactor. Afterwards, no sulfide could be detected in the effluent of the reactor set-up.

11.2.3.2 Process parameters

Absorption of carbon dioxide

Since an equilibrium situation exists between the CO_2 concentration in the gas-fase and the liquid fase almost no net absorption of CO_2 occurs. In the bioreactor, a small part of the dissolved CO_2 is removed due to aeration. This obviously leads to an in crease in pH which is compensated in the absorber column as a result of some CO_2 absorpotion. Figure 11.6 shows that this technology is highly selective for H_2S absorption.

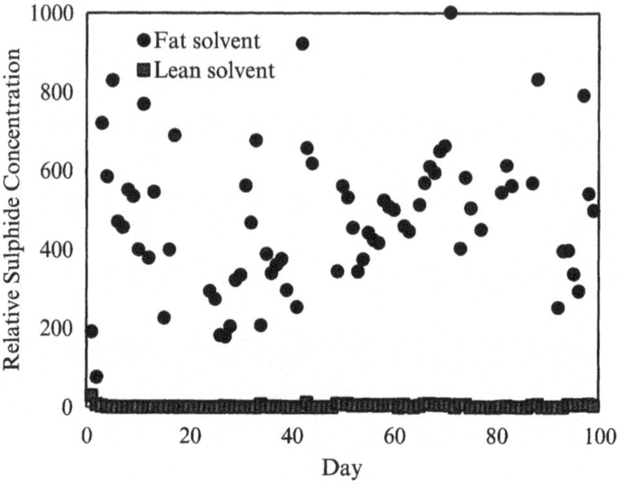

Figure 11.5. Effect of the relative sulfide concentration in the liquid from the absorber (fat solvent) on the sulfide concentration in the liquid leaving the bioreactor (lean solvent).

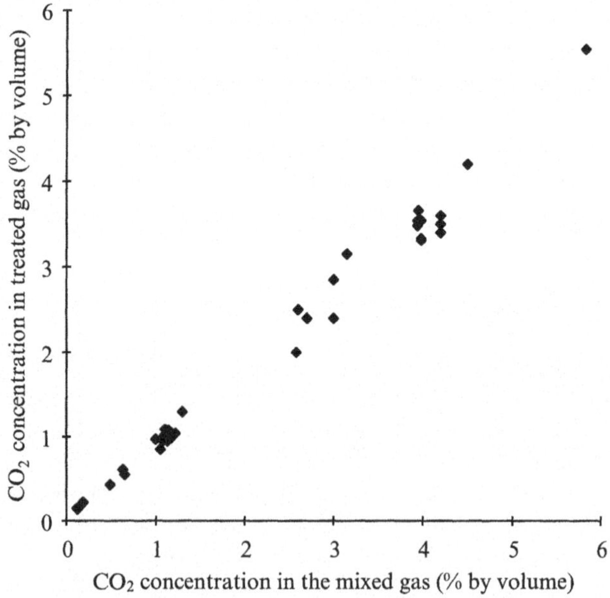

Figure 11.6. Carbon dioxide concentrations in the mixed gas and the treated gas.

Effect of elemental sulfur on sulfide loading capacity
The mechanisms for absorption of H_2S into the solvent are respectively (i) physical absorption (ii) chemical reaction with (bi)carbonate and hydroxide ions and (iii) chemical reaction with elemental sulfur. Physical absorption and reactions with (bi)carbonate and hydroxide will not be discussed here. The reaction of elemental sulfur with dissolved sulfide leads to the formation of polysulfides, according to (Chen and Gupta 1973) :

$$HS^- + 4\,S^\circ \rightarrow HS_5^- \tag{11.4}$$

Because of this reaction, the absorption capacity of the lean solvent is considerably increased. Future research will focus on the effect of the size distribution of the sulfur particles on the rate of polysulfide formation.

11.2.4 Process performance

The Shell-Thiopaq process for high pressure natural gas desulfurization was successfully demonstrated at the gas production facility of BEB Erdöl and Erdgas GmbH. The following process specifications were obtained:

- Deep H_2S removal (below 1 ppm) was proven at gasflows up to 350 N m^3/h.
- High degree of elemental sulfur formation in the bioreactor: less than 3% sulfate formation.
- Low chemical costs. Caustic and air-oxygen are used as the most important chemicals.
- Formation of hydrophilic 'bio-sulfur' with a purity of 95-98% (by mass). This can be upgraded to a high quality sulfur by application of a melting step.
- The raw bio-sulfur, i.e. without purification, is especially suitable for agricultural purposes (Tichý et al. 1994).
- No down time due to fouling problems were encountered.
- No problems due to corrosive chemicals.

11.3 BIOLOGICAL SO_2 REMOVAL FROM FLUE GAS

11.3.1 Introduction

With the introduction of ever-stricter environmental operating guidelines, capital expenditure restrictions and operational budget cutbacks in the area of flue gas treatment, a biological method of SO_2 removal has been developed. Biological treatment of SO_2 containing gas is a more cost effective technology in comparison to conventional sodium hydroxide scrubbing. Although the investment costs for sodium hydroxide scrubbing is lower, the operational costs are significantly higher. In addition, the end product of the biological installation is sulfur instead of sodium sulfate: no further wastewater treatment of the bleed from the biological installation is required since heavy metals are removed simultaneously.

The end product of the biological installation is elemental sulfur. The sulfur can either be produced as a sulfur cake (60% dry solids with a purity of 98% (by mass S^o) or as pure liquid sulfur. The sulfur cake can be used for the production of sulfuric acid at sulfuric acid plants which have facilities for burning waste-acid or slurries. It can also be used as soil amendment.

The general overall reaction that takes place when hydrogen is used as the reducing agent is as follows:

$$SO_2 + 3\,H_2 + \tfrac{1}{2}\,O_2 \rightarrow S^\circ + 3\,H_2O \qquad (11.5)$$

Instead of hydrogen gas, also ethanol (or methanol) can be used to reduce the sulfur dioxide into elemental sulfur:

$$SO_2 + \tfrac{1}{2}\,C_2H_5OH + \tfrac{1}{2}\,O_2 \rightarrow S^\circ + \tfrac{3}{2}\,H_2O + CO_2 \qquad (11.6)$$

An important feature from the above mentioned reaction equations is that the overall reaction is pH-neutral. This means that in contrast to conventional caustic scrubbing, hardly any alkalinity is consumed. Extensive pilot plant tests have been performed at the Elektriciteits Produktie Maatschappij Zuid-Nederland (EPZ, Geertruidenberg, The Netherlands) power station to develop this new biological process for the desulfurization of flue gas.

11.3.2 Process description

The process of removing sulfur dioxide consists of four main parts. These include an absorber, an anaerobic reactor, an aerobic reactor and a sulfur recovery step. Figure 11.7 schematically presents the process for sulfur dioxide removal. In Figure 11.8 a photograph of the pilot plant is shown.

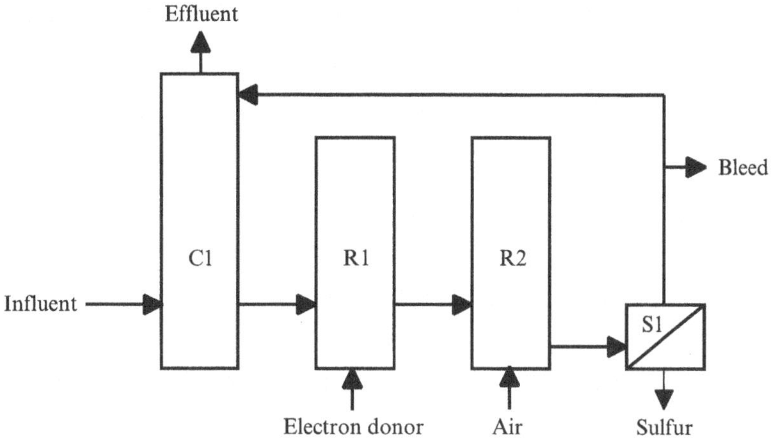

Figure 11.7. Schematic drawing of the pilot plant installation at the EPZ site, for the removal of sulfur dioxide; C1= absorber, R1=anaerobic reactor (V=5 m^3), R2= aerobic reactor (V=5 m^3), S1=settling unit for sulfur.

Figure 11.8. Pilot plant installation for the removal of SO_2 from a flue gasstream. The installation was located at the EPZ power station (Geertruidenberg, The Netherlands).

Table 11.3. Composition of the flue gas as used for the pilot-plant tests at the EPZ Power station

Compounds	Inlet
SO_2	2000 mg/Nm^3
H_2O	6.9 vol%
CO_2	12.9 vol%
N_2	75.2 vol%
O_2	5 vol%
HCl	250 mg/Nm^3
HF	25 mg/Nm^3
Fly Ash	75 mg/Nm^3

11.3.2.1 Absorber

The flue gases pass through an absorption tower where sulfur dioxide is removed. In this application, a vertical spray tower is used. The flue gases enter at the bottom of the tower and the washing water is sprayed downward from the top. The water is sprayed by means of nozzles that are designed to produce small droplets. In this way, a large surface is created for the absorption of the SO_2.

In the absorption unit, the physical chemical process of the absorption of sulfur dioxide into the water takes place. The SO_2 is absorbed and reacts with the water to form sulfites, according to:

$$SO_2 + H_2O \rightarrow HSO_3^- + H^+ \quad (11.7)$$

Under the condition applied in this process, sulfur will be mainly present in the form of bisulfite and sulfite. Depending on the amount of oxygen in the gas, up to 20% by volume, some of the sulfites (5-20%) are oxidized into sulfates.

The 2 MW demo plant treats about 6000 N m^3/h flue gas at a temperature of 120 °C. The composition of the flue gas is summarized in Table 11.3.

From figure 11.9, it becomes clear that a removal efficiency for SO_2 of 98% can be obtained.

11.3.2.2 Anaerobic reactor

The sulfite containing water from the absorber is directed to the first of two biological reactors. This is an anaerobic reactor where sulfite and sulfate are reduced by Sulfate Reducing Bacteria (SRB) to hydrogensulfide (HS^-). To facilitate the reduction of these compounds, an electron donor is required. The selection of this is determined by the costs of the added electron donor

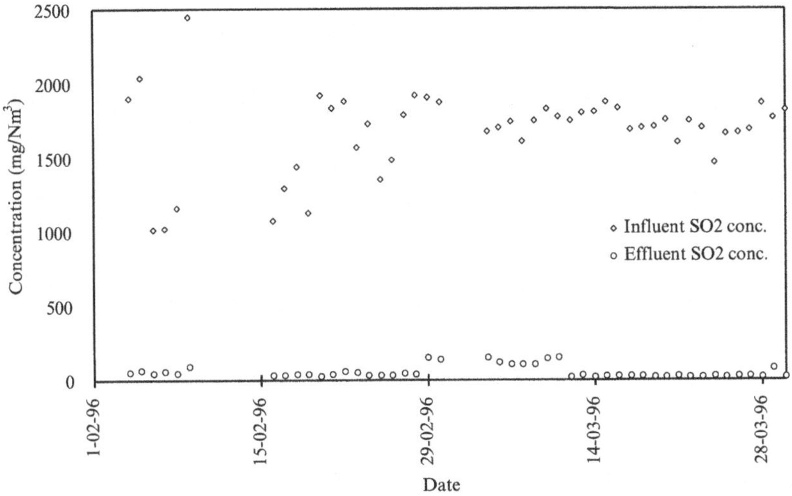

Figure 11.9. Influent and effluent SO_2 concentration of the pilot installation.

per unit sulfide produced and the absence of any remaining pollution. One should also consider that other micro-organisms, like methanogenic and acetogenic bacteria, are able to use these substrates. Therefore it is necessary to take the thermodynamics of the conversion reactions and the kinetic properties of the SRB and the competing organisms into account.

Possible candidates for use as an electron donor are ethanol, methanol (Weijma et al. 1999), hydrogen gas (van Houten et al. 1994) or syngas (Van Houten et al. 1996). The desulfurization of flue gases is mostly accompanied with the formation of 100 - 1500 kg of hydrogen sulfide per hour. At these loading rates, hydrogen gas is the most attractive electron donor. If hydrogen is not available, it is economically feasible to install a natural gas reformer to produce hydrogen and carbon dioxide. The formed carbon dioxide is needed to control the pH in the anaerobic reactor and as a carbon source for the micro-organisms. In the case where pure hydrogen gas is available on site, an additional CO_2 supply is required. The main reaction that occurs in the anaerobic reactor is the following.

$$HSO_3^- + 3\,H_2 \rightarrow HS^- + 3\,H_2O \qquad (11.8)$$

Van Houten et al. (1994) reported that *Desulfotomaculum* sp. species are the dominant organisms in the anaerobic reactor, when this is fed with hydrogen gas.

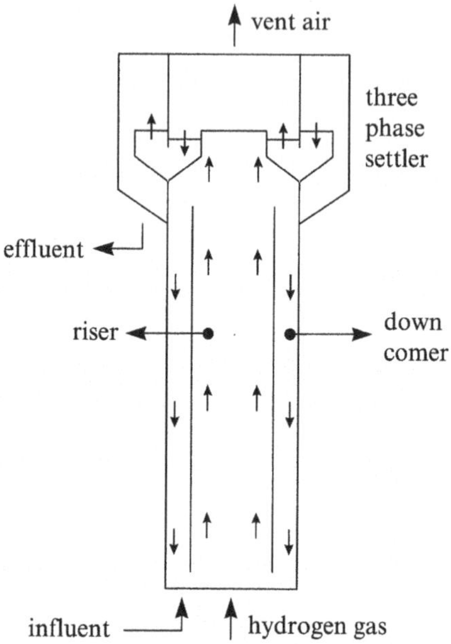

Figure 11.10. Schematic overview of a gaslift reactor for the reduction of sulfite and sulfate in the presence of hydrogen gas.

In the pilot plant experiments, a gas lift loop reactor was the preferred anaerobic reactor. This reactor type is characterized by a strong internal water recirculation. The volume of the reactor is designed with the optimal activity of the micro-organisms. This determines the residence time of the anaerobic reactor. The top of the reactor consists of a three-phase settler which separates the biomass from the effluent liquid and the effluent gas. For an efficient use of the hydrogen gas, a recirculation over the reactor is required. The effluent liquid will flow by gravity to the aerobic reactor. In figure 11.10 a schematic overview of a gaslift reactor is given.

11.3.2.3 Aerobic reactor

The aerobic reactor bacteria of the genus *Thiobacillus* that oxidize the formed sulfide into elemental sulfur. The conversion of sulfide into elemental sulfur has been described in section 11.2.

11.3.2.4 Products

The main product is elemental sulfur. This sulfur can either be produced as a sulfur cake (60% by mass with a purity of 95%) or as liquid sulfur with a higher purity. The sulfur cake can be used for the production of sulfuric acid at sulfuric acid plants with facilities for burning waste-acid or slurries. The liquid sulfur can be sold commercially.

Other products will differ with each case. Selectively recovered metal sulfides can often be used in the metal industry to recover the metal.

11.3.2.5 Applications

The biological sulfur dioxide removal is suitable for the following industrial situations:

- Sulfuric acid production plants
- Power stations
- Chemical and petro-chemical installations, especially FCC flue gases (Arena et al. 1998)
- Incineration of municipal sludge and industrial waste

The figures in this documentation are general values, approximate values or values which have been achieved by existing systems. They strongly depend on specific conditions. Thus, the values quoted do not apply universally. Paques Bio Systems B.V. emphatically reserves the right to interim revisions in technological and engineering matters.

11.4 REFERENCES

Arena, B.J., Robson, H.N., de Vegt, A.L. and Buisman, C.J.N. (1998) Biological treating of sulfur-containing waste streams. 1998 NPRA meeting, March 15-17, California.

Brimblecombe, P., Hammer, C., Rohde, H., Ryaboshapko, A., and Boutron, C.F. (1989) Human influence on the sulfur cycle. In: P. Brimblecombe and A. Yu Lein (Eds.) *Evolution of the global biogeochemical sulfur cycle*. Scope 39, pp. 77-121. Wiley, New York.

Buisman, C.J.N., IJspeert P., Hof, A., Janssen, A.J.H., Ten Hagen, R. and Lettinga, G. (1991) Kinetic parameters of a mixed culture oxidizing sulfide and sulfur with oxygen. *Biotechnol. Bioeng*. **38**, 813-820.

Chen, K.Y. and Gupta, S.K. (1973) Formation of polysulphides in aquous solution. *Environm. Letters* **4**(3), 187-200.

van Houten, R.T., Hulshoff Pol, L.W. and Lettinga, G. (1994) Biological sulfate reduction using gas-lift reactors fed with hydrogen and carbon dioxide as energy and carbon source. *Biotechnol. Bioeng*. **44**, 586-594.

van Houten, R.T., van der Spoel, H., van Aelst, C., Hulshoff Pol, L.W. and Lettinga, G. (1996) Biological sulfate reduction using synthesis gas as energy and carbon source. *Biotechnol. Bioeng*. **50**, 136-144.

Janssen, A.J.H., Ma, S.C., Lens, P. and Lettinga, G. (1997) Performance of a sulphide oxidizing bed reactor supplied with dissolved oxygen. *Biotechnol. Bioeng*. **53**, 32-40.

Janssen, A.J.H., Meijer, S., Bontsema, J. and Lettinga, G. (1998) Application of the redox potential for controling a sulfide oxidzing bioreactor. *Biotechnol. Bioeng.* **60**, 147-155.

Janssen, A.J.H., Lettinga, G. and A. De Keizer. (1999) Removal of hydrogen sulphide from wastewater and waste gases by biological conversion to elemental sulphur. Colloidal and interfacial aspects of biologically produced sulphur particles. Colloids Surfaces *A: Physicochemical and Engineering Aspects* **151**, 389-397.

Jensen, A.B. and Webb, C. (1995) Treatment of H_2S-containing gases: A review of microbiological alternatives. *Enzyme Microb. Technol.* **17**, 2-10.

Tichý, R, Janssen A., Grotenhuis, J.T.C., Lettinga, G. and Rulkens, W. (1994) Possibilities for using biologically-produced sulfur for cultivation of *thiobacilli* with respect to bioleaching processes. *Biores. Technol.* **48**, 221-227.

Weijma, J., Stams, A., Hulhoff Pol, L. and Lettinga, G. (1999) Thermophilic (65°C) methanogenesis and sulfate reduction with methanol in a high-rate anaerobic bioreactor. *Biotechnol. Bioeng.*, in press.

12

Biological treatment of gases polluted by volatile sulfur compounds

Veerle Herrygers, Herman Van Langenhove and Erik Smet

12.1 INTRODUCTION

Because of their bad smell and potential corrosive effect, the presence of volatile sulfur compounds (VSCs) in waste gases deserve special attention. These sulfur compounds mainly include hydrogen sulfide (H_2S), dimethyl sulfide (Me_2S), dimethyl disulfide (Me_2S_2), methanethiol (MeSH), carbon disulfide (CS_2) and carbonyl sulfide (COS). The most important characteristics of VSC are summarized in Table 12.1. Owing to their very low odour threshold value (ppb by volume range), VSCs contribute to odour pollution even upon the emission of very small amounts.

VSCs are mainly produced in those processes where organic matter is

Table 12.1. Properties of some VSCs (bp = boiling point, OT = odor threshold value, $H_{25°C}$ = air-water partition coefficient at 25 °C ((mol $m^{-3})_{air}$/(mol $m^{-3})_{water}$); MAK = maximum concentration value in workplace conditions)

Compound	bp[1] (°C)	OT[2,3] (ppbv)	$H_{25°C}$[4,5] (-)	MAK[2,6] (ppmv)	Odor quality[7,8,9]
H_2S	-60.7	8.5-1000	0.41	10	Rotten egg
MeSH	6.2	0.9-8.5	0.10	0.5	Decayed cabbage
Me_2S	37.3	0.6-40	0.07	20	Decayed vegetables
Me_2S_2	109.7	0.1-3.6	0.04	< 20	Putrification, foul
CS_2	46.2	9.6	0.65	10	Vegetable sulfide, aromatic
COS	-50.0	n.d.f.	1.94	n.d.f.	Pungent

Note: Superscripts refer to references: [1]Weast (1976); [2]De Zwart and Kuenen (1992); [3]De Vos *et al.* (1990); [4]Perry and Green (1987); [5]Przyjazny *et al.* (1983); [6]ACGIH (1991); [7]Verschueren (1983); [8]Bonnin *et al.* (1990); [9]Miller and Macauley (1988).

n.d.f. = no data found.

heated or decays under anaerobic conditions. Industrial applications of VSCs and their production during chemical processes are limited. In Table 12.2, some VSC-containing waste gases are given.

An overview of biotechnological abatement technologies to treat waste gases containing sulfur compounds is presented in this chapter. In the first part, the main biological waste gas treatment systems are described. In the next section, the use of these techniques for the abatement of sulfur compounds is reviewed. A distinction is made between the abatement of hydrogen sulfide and volatile organic sulfur compounds (VOSCs).

12.2 BIOTECHNOLOGICAL WASTE GAS TREATMENT METHODS

To date, three types of reactor are commonly used for biotechnological waste gas treatment: bioscrubbers, biotrickling filters and biofilters. Recently, two other types of bioreactors have been developed, mainly to treat waste gases containing poorly soluble pollutants with a high air-water partition coefficient: membrane bioreactors and two-phase bioreactors. Figure 12.1 overviews these different types of reactor concept. In all these types of bioreactor, two main processes take place: firstly, the pollutants are transferred to a liquid medium or to a biofilm, and secondly the pollutants are degraded by the microorganisms that are present in the liquid medium or in the biofilm.

Table 12.2. Examples of waste gases contaminated with odorous VSC compounds (After: Smet *et al.* 1998)

Sources	Odor compounds	Concentration (ppmv)	Reference
ANAEROBIC BIOPROCESSING OF ORGANIC WASTE			
Anaerobic brewery waste water treatment plant	H_2S	4	Gerards *et al.* (1995)
	Me_2S	30	
Production of compost to be used as mushroom cultivation substrate	H_2S, COS, MeSH, CS_2, Me_2S, Me_2S_2, Me_2S_3	0.024-0.840	Derikx *et al.* (1990)
Sludge composting facilities	Me_2S	10	Wilber and Murray (1990)
	Me_2S_2		
HIGH TEMPERATURE PROCESSES			
Rendering cookers	H_2S	<800	Chélu and Nominé (1984)
	MeSH	<200	
	H_2S, MeSH, CH_3CH_2SH, NH_3		Lecomte *et al.* (1995)
Thermal sludge treatment plants	Me_2S, Me_2S_2, thiophenes, pyrazines and indole		Schamp and Van Langenhove (1986)
INDUSTRIAL PROCESSES			
Viscose rayon manufacturing	CS_2, H_2S	20 - 1000	Revah *et al.* (1994)
Kraft paper production process	MeSH	94	Sivela (1980)
	Me_2S	16.6	
	Me_2S_2	21.7	

12.2.1 Bioscrubber

In a bioscrubber (Figure 12.1a), the pollutant is absorbed in an aqueous phase in an absorption tower. The aqueous phase containing the dissolved compounds is then treated in a separate activated sludge unit. The effluent of this unit is recirculated over the absorption tower, in a co- or countercurrent way to the gas stream. The pollutants are degraded by the microorganisms in the activated sludge unit (Kennes and Thalasso 1998). The scrubber compartment may be packed with synthetic packing materials, but also other designs have been proposed such as a spray column where the liquid is sprayed in very fine droplets (Cesário 1997).

Because the pollutants have to be absorbed in the water phase, this technology is only suitable for pollutants that are highly water soluble. As a general rule, bioscrubbers are only capable of removing compounds with an air-water partition coefficient lower than 0.01 ($(\text{mole m}^{-3})_{air}/(\text{mole m}^{-3})_{water}$) (Van Groenestijn and Hesselink 1993). The main advantage of this technology is the possibility to control the liquid medium composition (e.g. add appropriate nutrients) and to remove toxic or inhibiting products together with the water phase. The major drawback of this system, however, is the low scrubbing efficiency for compounds with a high air-water partition coefficient.

According to Diks (1992), typical values for the gas residence time and the superficial gas velocity in a bioscrubber are 1 to 20 s and 720 to 3600 m h^{-1}, respectively.

12.2.2 Biotrickling filter

In a biotrickling filter (Figure 12.1b), the waste gas is forced through a packed bed filled with a chemically inert carrier material. The latter can be colonized by microorganisms, similar to trickling filters in waste water treatment. The liquid medium, containing microbial nutrients, is circulated over the packed bed. The pollutants are first absorbed in the biofilm on the carrier material and are then degraded by the microorganisms in the biofilm. The liquid medium can be recirculated, co- or countercurrent to the gas stream. Thus, similar to the bioscrubber, the biotrickling filter allows optimal control of the biological process by the liquid medium composition and easy wash-out of reaction products.

Because of the presence of a recirculating water phase, the applicability of biotrickling filters is also limited to pollutants with a low air-water partition coefficient, but here this condition is less stringent than in a bioscrubber. Roughly, for biotrickling filters, the air-water partition

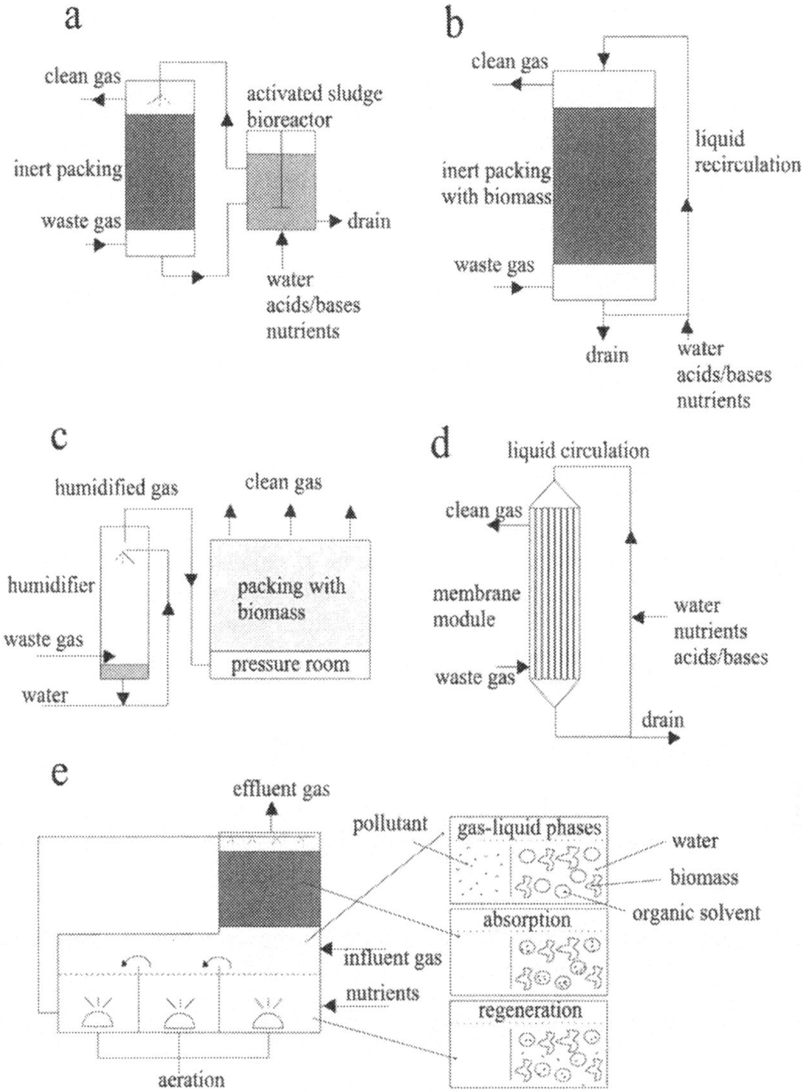

Figure 12.1. Process schemes of (a) a bioscrubber, (b) a biotrickling filter, (c) a biofilter, (d) a membrane bioreactor and (e) a two-phase bioreactor.

coefficient of the pollutants has to be less than 1 to allow a good removal of the gaseous compounds in the biotrickling filter (Van Groenestijn and Hesselink 1993).

Microbial carriers that are used in biotrickling filters are plastic or ceramic structured packings, unstructured celite, activated carbon or mixtures of different materials (Kennes and Thalasso 1998). The specific surface of these materials is normally quite low (100-400 m^2 m^{-3} reactor volume for rings and saddles of 2.54 cm). Upon start-up, the biotrickling filter should be inoculated with activated sludge or specific microbial strains.

A potential problem when using biotrickling filters is the excessive biofilm development on the carrier surface. It can thus partly or completely clog the reactor. This increases the pressure drop and leads to higher operation costs. The rate of clogging is mainly determined by the rate of carbon conversion achieved in the reactor (Okkerse *et al.* 1999). Methods have been developed to restrict clogging, including the limitation of biomass growth by control of the nutrient supply (nitrogen, phosphorus), regular filter-bed washing or limitation of the liquid supply (Kennes & Thalasso 1998).

On the other hand, the biofilm can completely slide off from the carrier material if the superficial liquid velocity is too high. This causes a temporary or permanent drop of the reactor performance. To prevent biofilm sliding off, Diks (1992) used a superficial liquid velocity lower than 15 m h^{-1} in his experiments for dichloromethane removal. The superficial gas velocity used under these conditions was 100-1000 m h^{-1}.

12.2.3 Biofilter

In a biofilter (Figure 12.1c), the gas to be treated is first humidified and then forced through a bed packed with an organic carrier material (compost, peat, bark or a mixture of these) on which the microorganisms are attached as a biofilm. Nutrients for the microorganisms are available in the organic carrier material, which eliminates the need for a circulating liquid phase. The carrier material must have a high surface area, which allows for a good biofilm development, while the void fraction must be high enough to prevent clogging of the biofilter and thus a too great pressure drop.

In this type of reactor, pollutants with an air-water partition coefficient up to 10 can be removed because both the residence time (30-60 s) and the specific gas/liquid surface area (300-1000 m^2 m^{-3}) in these biosystems are higher (Van Groenestijn and Hesselink 1993). Next to this, the apolar fraction in the organic carrier material promotes sorption and subsequent biodegradation of apolar compounds. In general, the organic carrier materials used have an endogeneous microbial community which can degrade the pollutants of interest. For specific compounds, however, the filter media can be inoculated with specialized microorganisms to enhance biodegradation.

The moisture content of the packing material is an important parameter in relation to microbial activity and operational stability. If the moisture content is too low, the degradation of the pollutants by the microorganisms will be inhibited. On the other hand, if the moisture content is too high, smaller voids will be filled with water. This lowers the availability of surface area for microbial colonization and can lead to anaerobic zones in the biofilter. An 'average' optimal moisture content for a biofilter is between 40 and 60 %. The moisture content can be controlled by prehumidification of the inlet air stream (Figure 12.1c) or by sprinkling water on the filter media.

In general, the biofilter media used have a neutral pH, which is in the optimal range (6-9) for the bacterial degradation. During the degradation of N, S and Cl containing pollutants, acidifying metabolites are formed (HNO_3, H_2SO_4 and HCl). At low pH (< 6), fungi may overgrow the bacteria which may affect the filter performance either negatively or positively (Cox *et al.* 1997; Van Groenestijn and Hesselink 1993). Contrary to the bioscrubber or biotrickling filter, control of the pH by the recirculating water is not possible in a biofilter, which is the main drawback of this method. A possible solution for this is to increase the buffer capacity of the biofilter packing, e.g. by mixing of limestone in the packing material (Smet *et al.* 1996b).

Acidification and drying out of the filter bed are two major causes of malfunctioning of biofilters. When using fungi instead of bacteria, tolerance to a low pH can be obtained. Cox *et al.* (1994) used a biofilter with fungi for the removal of styrene. At a volumetric load of about 0.7 kg styrene m^{-3} d^{-1}, removal efficiencies of 100% were obtained. The pH in the biofilter was 3. The minimal water content required for maximal biological activity was 40% on weight basis.

The superficial gas velocity in biofilters varies generally between 50 m h^{-1} and 150 m h^{-1} and the superficial gas residence time is 5-60 s (Diks 1992).

12.2.4 Membrane bioreactor

A membrane bioreactor (Figure 12.1d) contains a membrane that separates the gas phase and the liquid phase. Biomass grows on the "liquid" side of the membrane, pollutants and oxygen diffuse from the opposite side through the membrane (Reij 1998). In such a reactor, the advantages of both the biofilter and the biotrickling filter are combined. Indeed, there is no liquid phase between the pollutants and the biofilm as in the biofilter, which contributes to a high mass transfer to the biofilm. Control of the biofilm thickness, on the other hand, is possible via nutrient addition to and removal of toxic products from the liquid phase. Owing to the good mass transfer properties, pollutants with a high air-water partition coefficient (e.g. hexane: 74 $(mol\ m^{-3})_{air}\ (mol\ m^{-3})^{-1}{}_{water})$ can be treated with this type of reactor (Reiser *et al.* 1994).

Different types of membrane can be used: porous, dense and composite membranes. Porous membranes (e.g. polypropylene, polycarbonate) have pores with a diameters between 0.01 and 10 μm. These membranes have no selectivity towards different chemical compounds. A disadvantage of porous materials is the possibility of blockage of the pores by microorganisms (Reij 1998). Dense membranes (e.g. polydimethyl siloxane) have no pores and the mass transfer of the compound depends on its solubility and on the diffusion through the membrane. The mass transfer in these membranes is generally lower than in porous membranes. For a polypropylene porous membrane and a polydimethyl siloxane dense membrane, this factor varies between 7 (propane) and 1600 (propanol) for different kinds of organic compound (Reij 1998). Composite membranes have a thin layer of dense material coated on top of a porous membrane. This gives the advantage of a high mass transfer for the porous membrane and the thin layer of dense material located at the liquid site of the membrane avoids blockage of the pores by growth of microorganisms.

12.2.5 Two-phase bioreactor

The two-phase bioreactor is another method to treat pollutants with a high air-water partition coefficient. In this reactor concept, an organic solvent (e.g. silicone fluids and phthalates) is mixed into the water phase. The pollutants are first absorbed by the organic solvent, from which they are then slowly released into the water phase, where it is degraded by the microorganisms. The microorganisms grow in flocs in the water phase or in a biofilm. For a successful application, the organic solvent must be immiscible with the aqueous solution and it must also have a very high partition coefficient with respect to the pollutants.

Diks (1992) described a bioscrubber where the solvent is added in the water phase in concentrations up to 30% (Figure 12.1e). Budwill and Coleman (1997) described the addition of silicone oil to a biofilter. At an inlet pollutant concentration of 20-30 ppmv, the average removal efficiency

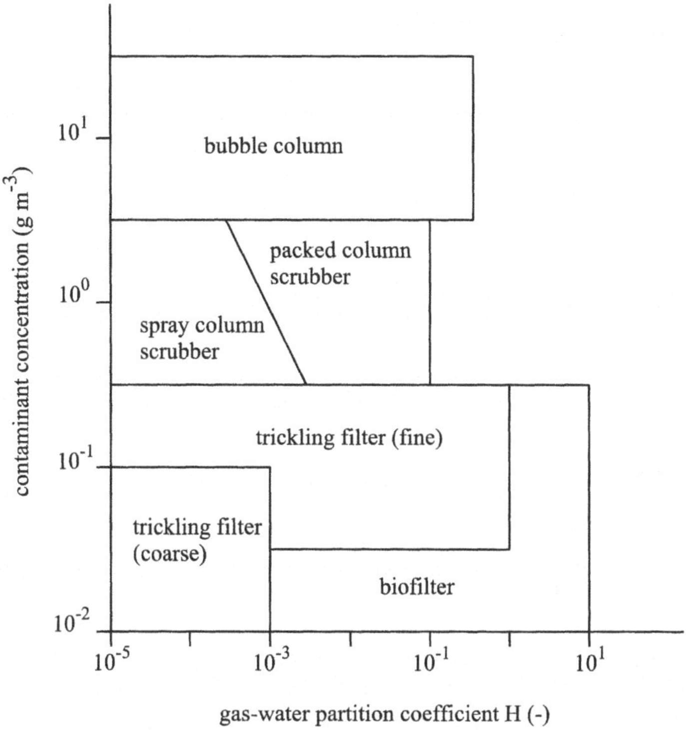

Figure 12.2. Application range of the conventional biotechniques (After: Kok 1992).

for hexane in the silicone-oil-treated biofilter was 60% in contrast to the untreated biofilter, which achieved an average removal efficiency of only 30%.

12.3 PRACTICAL ASPECTS OF REACTOR SELECTION FOR VSC TREATMENT

As was discussed in the previous paragraph, the air-water partition coefficient of the compound to be removed is the critical parameter in selecting a biological waste-gas cleaning system. As can be seen in Table 12.1, the air-water partition coefficient for VSC varies between 0.04 and 1.94. This suggests that mass transfer can be a problem for these compounds in 'wet' biosystems, like a bioscrubber or a biotrickling filter. This is also illustrated in Figure 12.2.

Therefore, of the three traditional biotechnological methods mentioned, mainly biofiltration is used in full-scale applications for the abatement of organic waste gases contaminated with VSC. Biotrickling filtration or bioscrubbing of VOSC and H_2S has thus far been poorly reported in the literature. As far as we know, the newer technologies (membrane bioreactor and two-phase bioreactor) have not yet been used for the abatement of sulfur compounds. Because of to the relatively high air-water partition coefficient of the VOSC, however, these two biosystems have good potential for the abatement of these compounds. This requires further research, such as the optimal membrane type to be used in a membrane bioreactor and the optimal solvent type and concentration to be used in the two-phase bioreactor.

No standard abatement technology is available for treating VOSC-contaminated waste gases since the optimal solution depends on several factors, such as the qualitative and quantitative gas composition, the flow rate and the temperature of the waste gas, the continuity of the gaseous emissions, the available space on the plant, the costs of the abatement technology and legislation. Sometimes a combination of chemical and biological treatment can be used. For instance, photochemical oxidation of hydrophobic pollutants as a pretreatment step for biological purification can improve the solubility in water as well as the biodegradability of these compounds, thus improving the elimination efficiencies (Cesário 1997). Whenever a waste gas treatment system is installed, pilot plant experiments should be performed to select for each specific problem the optimal (combination of) abatement technology(ies).

Biofiltration is generally considered to be the cheapest odour-abatement method. Diks (1992) reported investment costs of €2 to 5 $(m^3\ h^{-1})^{-1}$ and operation costs of €0.2 to 0.3 $(1000\ m^3)^{-1}$ for a biofilter. Typical power consumption rates for biofilters range from 1.8 to 2.5 kWh $(1000\ m^3)^{-1}$ (Leson and Winer 1991). For a bioscrubber, Diks (1992) reported investment costs of €5 to 20 $(m^3\ h^{-1})^{-1}$ and operational costs of €1 to 2 $(1000\ m^3)^{-1}$. For a biotrickling filter, operational costs of €2 to 5 $(1000\ m^3)^{-1}$ were reported (Diks 1992).

12.4 BIOTECHNOLOGICAL REMOVAL OF H_2S

In 1923, Bach discussed the basic concept of biofiltration for the control of H_2S emissions from sewage treatment plants (Leson and Winer 1991). Recent applications of biotechniques for H_2S removal, some of them applying inoculation with specialized microorganisms, are given in Table 12.3. This table shows that often *Thiobacillus* spp. are used for inoculation

Table 12.3. Removal capacities for H_2S (kg H_2S-S m^{-3} d^{-1}) in biotechnological systems with or without microbial inoculation

Reactor type[a]	Inoculation	Removal capacity (kg H_2S-S m^{-3} d^{-1})	Reference
BTF (polypropylene)	*T. thioparus* TK-m	0.56 (95%)	Tanji *et al.* (1989)
BF (peat)	Night soil sludge	0.57 (EC_{max})	Hirai *et al.* (1990)
BF (peat)	Night soil sludge	0.69 (EC_{max})	Zhang *et al.* (1991)
BF (peat)	*T. thioparus* DW44	0.56 (EC_{max})	Cho *et al.* (1991b)
BF (peat)	*Hyphomicrobium* I55	0.34 (EC_{max})	Zhang *et al.* (1991)
BF (peat)	*Thiobacillus* HA43	1.13 (EC_{max})	Cho *et al.* (1991c)
BF (peat)	*Xanthomonas* DY44	0.32 (EC_{max})	Cho *et al.* (1992b)
BTF (plastics etc.)	no	0.47-6.12 (85-98%)	Lanting and Shah (1992)
BF (dry act. sludge)	no	0.19-11.29 (100%)	Kowal *et al.* (1992)
BF (peat)	no	0.23 (99%)	Bonnin *et al.* (1994)
BF (mearl)	*Thiobacilli* sp.	1.36-1.58 (99%)	Bonnin *et al.* (1994)
BF (compost)	no	2.94 (EC_{max})	Yang and Allen (1994a)
BS	Activated sludge	0.19 (100%)	Brandy *et al.* (1995)
BF (gel beads)	*T. thioparus* CH11	0.52 (EC_{max})	Chung *et al.* (1996a)
BF (gel beads)	*Pseudomonas putida* CH11	0.46 (96%)	Chung *et al.* (1996b)
BTF (act. carbon)	*Thiobacillus* sp.	3.22 (90%)	Guey *et al.* (1995)
BF (dry waste water sludge)	no	2.9 (100%)	Degorce-Dumas *et al.* (1997)
BTF	yes	0.45 (EC_{max})	Melse and Kraakman (1998)

Note: The performance of the reactors is expressed as the maximum elimination capacity (EC_{max}) or as the removal efficiency (%) at the reported loading rate (After: Smet *et al.* 1998).

[a] BF = biofilter, BTF = biotrickling filter, BS = bioscrubber.

of the bioreactor. These are autotrophic bacteria which convert H_2S into elemental sulfur or sulfate, depending on the strain of *Thiobacillus* spp. used and the H_2S concentration in the reactor (Chung *et al.* 1996a). The formation of sulfuric acid causes a pH drop in the reactor.

However, Table 12.3 shows that high elimination capacities can also be obtained without specific inoculation, owing to the widespread occurrence of these *Thiobacillus* spp. The genus *Thiobacillus* includes both acidophobic bacteria which prefer a pH near 7 and acidophilic bacteria that grow at low pH values (Kasakura and Tatsukawa 1995). This allows an efficient H_2S-

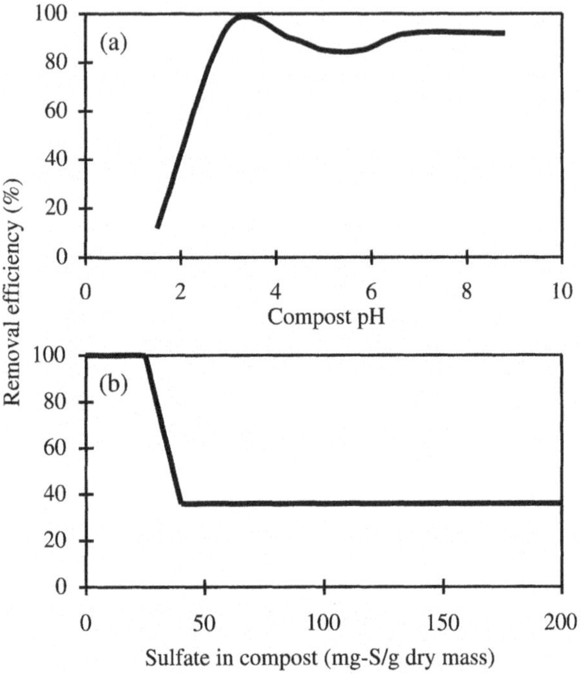

Figure 12.3. Effect of (a) compost pH and (b) compost sulfate concentration on H_2S removal efficiency. H_2S loading rate: 33.3 g H_2S-S m^{-3} h^{-1}, specific gas velocity: 26 m h^{-1} (After: Yang and Allen 1994a).

oxidation over a wide pH-range. Yang and Allen (1994 a,b) reported that the biodegradation of H_2S in a biofilter with compost as packing material strongly decreases at pH-values lower than 3 (Figure 12.3a) and at sulfate concentrations exceeding 25 mg SO_4^{2-}-S g^{-1} dry compost (Figure 12.3b). Washing the biofilter regularly, however, effectively mitigated these deficiencies and kept the compost material at an optimal moisture content of about 50%. In a bioscrubber, Brandy *et al.* (1995) observed an inhibition of the H_2S elimination efficiency when the sulfate concentration in the liquid exceeded 0.27 g SO_4^{2-}-S l^{-1} activated sludge, illustrating the need to drain part of the recirculation liquid. According to Williams and Miller (1992) the chemical oxidation of H_2S in the presence of organic matter, next to biological degradation, can contribute to the removal efficiency in biofilters. According to several authors (Hirai *et al.* 1990; Cho *et al.* 1991b; Zhang *et al.* 1991), the biofiltration of H_2S itself is unaffected by the presence of other volatile sulfur compounds such as Me_2S or MeSH.

Table 12.4. Organisms that can be applied in a H_2S treating biological waste gas treatment system

Group of organisms	Species	References
Autotrophic	*Thiobacillus* spp.	Cho *et al.* (1991c)
		Bonnin *et al.* (1994)
		Chung *et al.* (1996a)
Colourless sulfur bacteria	*Thiothrix* spp.	Lanting and Shah (1992)
	Beggiatoa spp.	
Phototrophs	*Chlorobium* spp.	Cork *et al.* (1983)
	Chromatium spp.	Jensen and Webb (1995)
	Ectothiorhodospira spp.	Then and Truper (1983)
Methylotrophs	*Hyphomicrobium* spp.	Zhang *et al.* (1991)
Cyanobacteria		Oren and Padan (1978)
Fungi	*Sporormia concretivora*	Cho *et al.* (1994)
Other heterotrophs	*Xanthomonas* spp.	Cho *et al.* (1992a)
	Pseudomonas putida CH11	Chung *et al.* (1996b)

New developments in biotechnological H_2S abatement mainly focus on the selection of better biofilter carrier materials and the use of specific inocula. Bonnin *et al.* (1994) reported high H_2S removal rates at very low superficial gas residence times (5-7 s) using mearl as a biofilter carrier material and sprinkling water at a water:gas ratio of 5×10^{-5} (Table 12.3). This sea mineral, consisting up to 82 % of $CaCO_3$, acts as pH buffering carrier material and a source of nutrients (CO_2 for the autotrophic *Thiobacillus* spp., nitrogen and micronutrients).

Besides *Thiobacillus* spp., numerous other H_2S converting microorganisms can be applied (Table 12.4). The degradation of H_2S by *Sporormia concretivora* and *Xanthomonas* spp. was speculated to be a detoxification process for these microorganisms. Since the end product of H_2S removal by *Xanthomonas* DY44 was polysulfide, a neutral pH was maintained in a biofilter inoculated with this culture (Cho *et al.* 1992a). As a drawback, organic compounds have to be supplied when applying this heterotrophic bacterium (e.g. glucose, maltose, acetate or yeast extract). To implement the other microorganisms mentioned for H_2S abatement, different process schemes have been established and investigated. However only very few of these processes have been proven on a large scale, and much optimization work still remains to be done (Jensen and Webb 1995).

12.5 BIOTECHNOLOGICAL REMOVAL OF ORGANIC SULFUR COMPOUNDS

12.5.1 Biodegradation of organic sulfur compounds

Most studies on the biotechnological removal of organic sulfur compounds have been performed using dimethyl sulfide, dimethyl disulfide, methanethiol or CS_2 as a model compound.

Numerous aerobic microorganisms which can degrade MeSH, Me_2S and Me_2S_2 were isolated (De Zwart and Kuenen 1992; Brennan *et al.* 1996; Smet *et al.* 1998). Among these, mainly methylotrophic *Hyphomicrobium* spp. and autotrophic *Thiobacillus* spp. have been used to inoculate biotechnological waste gas treatment systems. In Figure 12.4, a degradation pathway for dimethyl sulfide, dimethyl disulfide and methanethiol by *Hyphomicrobium* spp. or *Thiobacillus* spp. is given (Kelly and Smith 1990). The end-products of this pathway are CO_2 and H_2SO_4. The produced sulfuric acid again causes acidification of the bioreactor.

Methylated sulfur compounds can also be degraded anaerobically, e.g. by methanogens. This opens perspectives for the design of novel VSC removal techniques, i.e. biofilters containing an anaerobic zone or anaerobic bioreactors in combination with a scrubber. Methanethiol is an intermediate of methanogenesis and, next to CO_2 and H_2S, also methane is an end product, instead of CO_2. Anaerobic degradation of MeSH and Me_2S is less energetically favourable than aerobic conversion (Lomans *et al.* 1999).

Contrary to the methylated sulfides, CS_2 supports only autotrophic growth, narrowing the range of bacteria that will use these compounds as carbon and energy sources (Plas *et al.* 1993). According to Smith and Kelly (1988), CS_2 oxidation by *Thiobacillus thioparus* TK-m proceeds by hydrolytic cleavage into COS and H_2S, whereas the COS formed undergoes a similar hydrolysis to CO_2 and H_2S. In this biodegradation pathway, the oxidation of H_2S to sulfate is the only energy-yielding process.

12.5.2 Biofiltration of VOSC

12.5.2.1 Process performance

Contrary to the biological removal of H_2S and numerous volatile organic compounds, the removal efficiency for VOSC in biotechnological waste gas treatment systems is reported to be rather low and variable (Tanji *et al.* 1989; Phae & Shoda 1991; Kasakura & Tatsukawa 1995). In a full-scale

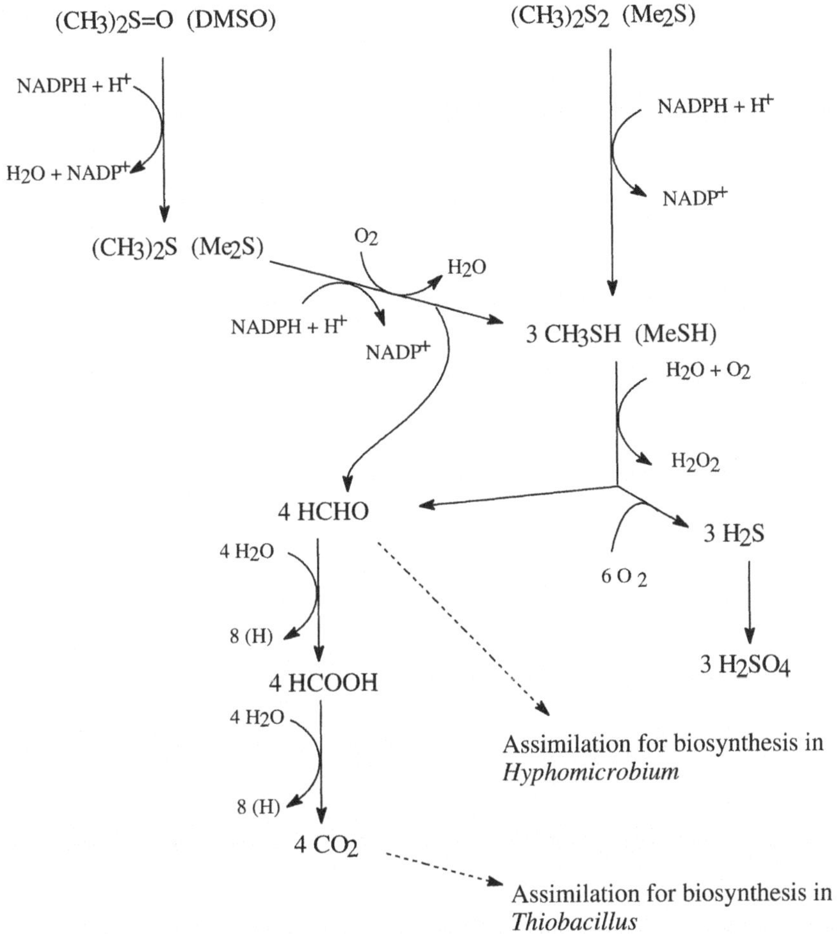

Figure 12.4. Metabolic pathway for Me_2S-, MeSH and Me_2S_2 degradation in *Thiobacillus* and *Hyphomicrobium* species (After: Kelly and Smith 1990).

biofilter at a rendering plant, Van Langenhove *et al.* (1992) reported fluctuating filtration efficiencies of 0-81% for methyl sulfides, whereas carbonyl compounds were very efficiently (> 98%) removed. According to Cho *et al.* (1991a), H_2S-oxidizing microorganisms are inherent in raw biofilters, contrary to the micro-organisms degrading VOSC. Smet *et al.* (1996a) observed very low elimination capacities (< 0.005 kg Me_2S-S m^{-3} d^{-1}) for Me_2S in a wood bark and compost lab-scale biofilter. To improve these treatment efficiencies, biotechnological waste gas treatment systems

Table 12.5. Removal capacities (kg Me_2S-S m^{-3} d^{-1}) of biotechniques for Me_2S

Reactor[a]	Inoculation	Removal capacity (kg Me_2S m^{-3} d^{-1})	Reference
BTF (polypropylene)	*T. thioparus* TK-m	0.12 (95%)	Tanji *et al.* (1989)
BF (peat)	night soil sludge	0.04 (EC_{max})	Hirai *et al.* (1990)
BF (peat)	*T. thioparus* DW44	0.05 (EC_{max})	Cho *et al.* (1991b)
BF (peat)	*Hyphomicrobium* I55	0.06 (EC_{max})	Zhang *et al.* (1991)
BTF (polyurethane)	*Hyphomicrobium* VS	0.17 (EC_{max})	Pol *et al.* (1994)
BF (bark)	*Hyphomicrobium* MS3	0.015 (EC_{max})	Smet *et al.* (1996a)
BF (compost)	*Hyphomicrobium* MS3	0.35 (EC_{max})	Smet *et al.* (1996a)
BF (compost/dolomite)	*Hyphomicrobium* MS3	0.86 (EC_{max})	Smet *et al.* (1999)

Note: The performance of the reactors is expressed as the maximum elimination capacity (EC_{max}) or as the removal efficiency (%) at the reported loading rate (After: Smet *et al.* 1998).

[a] BF = biofilter, BTF = biotrickling filter.

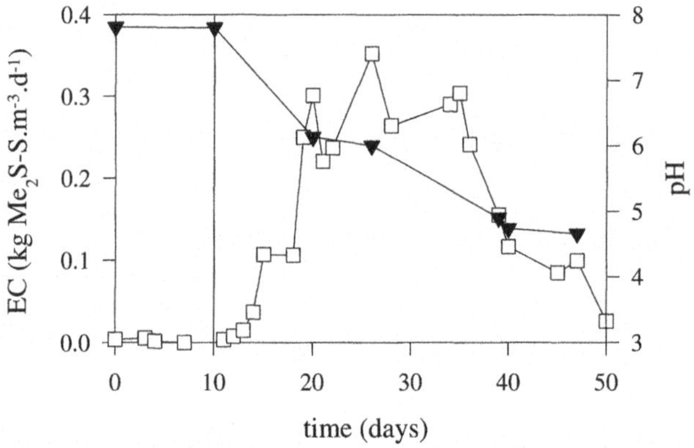

Figure 12.5. Elimination capacity (EC) of a compost biofilter for Me_2S (□) and pH of the compost material (▼) before (days 0-10) and after (days 10-50) inoculation with *Hyphomicrobium* MS3 (After: Smet *et al.* 1996a).

are inoculated with specialized aerobic micro-organisms. Table 12.5 overviews the Me_2S removal capacity in lab-scale biofilters and biotrickling filters, inoculated with some of these isolated micro-organisms.

In Figure 12.5, the effect of inoculating a biofilter with *Hyphomicrobium* MS3 on the elimination capacity of a biofilter for Me_2S is illustrated (Smet *et al.* 1996a).

12.5.2.2 *Long-term stability of the biofiltration process for VOSCs*

Figure 12.5 illustrates the decrease in performance of the *Hyphomicrobium* MS3-inoculated biofilter after a 30-day period due to acidification (accumulation of the metabolite H_2SO_4) in the biofilter. When the pH of the carrier material dropped below 5, the elimination capacity (EC) for Me_2S in the biofilter was less than 50% of its maximum value. Two solutions have been proposed against acidification: trickling tap water or a phosphate buffer over the biofilter or mixing the filter media with a buffer material.

Regeneration of the acidified biofilter by trickling tap water or a phosphate buffer solution over the reactor was not effective (Smet *et al.* 1996b). Since the protons produced in the biofilter displace the nutrient cations (e.g. Na^+, Ka^+, ...) from the cation exchange sites on the compost material, 95% of the SO_4^{2-} was leached as the corresponding sulfate salts and not as sulfuric acid. As a result, essential microbial nutrients were leached and the pH of the carrier material remained low.

Mixing limestone powder into the Me_2S-degrading compost biofilter was a successful approach to control the pH. A stoichiometric neutralization reaction (molar ratio $CaCO_3$:H_2SO_4 = 1.1) in the biofilter between the $CaCO_3$ added and the metabolite of the Me_2S degradation was observed. No sulfate toxicity was observed in the limestone-supplemented *Hyphomicrobium* MS3-inoculated biofilter, since the toxic sulfate concentration for *Hyphomicrobium* MS3 (= 10 - 13 g SO_4^{2-}-S l^{-1}) was about 15 times the maximum solubility of $CaSO_4$ (Smet *et al.* 1996b). For full-scale applications, however, mixing the carrier material with a neutralizing agent at regular time periods ($CaCO_3$ was consumed every 50 days upon addition of 23.9 kg $CaCO_3$ m^{-3}) is not practicable for controlling the acidification. Instead of mixing limestone powder in the biofilter, Smet *et al.* (1999) used a carbonate carrier material (dolomite) as filter medium for biofiltration of Me_2S. The grain size of the dolomite was between 1.40 and 4.76 mm. Although direct inoculation of the dolomite particles with *Hyphomicrobium* MS3 was not successful, start-up of Me_2S-degradation in this biofilter was observed when the dolomite particles were mixed with 33% by mass of *Hyphomicrobium* MS3-inoculated compost. Under optimal conditions, elimination capacities (EC) up to 0.86 kg Me_2S-S m^{-3} d^{-1} were obtained for the compost/dolomite biofilter (Table 12.5). No reduction in activity due to acidification was observed in the biofilter over a 235-day test period because of direct on-site neutralization of the sulfuric acid with the carbonate in the dolomite material. However, nutrient limitation resulted in a gradual decrease of the EC. The elimination capacity for Me_2S could be stabilized when nitrogen was dosed to the biofilter water supply at a Me_2S-C/NH_4Cl-N ratio of about 10.

12.5.2.3 Biofiltration of VOSCs in mixed waste gases

In practical situations, biofilters seldom have to treat pure gas streams with only one pollutant. Instead, mixtures of different organic and inorganic pollutants have to be treated. In comparison with waste gases containing only one compound, the complex chemical composition of the waste gas can have a negative or a positive effect, or no effect at all on the elimination capacity of the target compound.

Upon supplementation of 56-450 ppm by volume isobutyraldehyde as a second gaseous substrate next to Me_2S to a *Hyphomicrobium* MS3-inoculated biofilter, Smet *et al.* (1997) observed sequential degradation profiles of first isobutyraldehyde and subsequently Me_2S in physically separated sections of the biofilter. It was found that the inoculum first degraded the carbonyl compounds in the presence of both Me_2S and isobutyraldehyde. Complete purification of a waste gas stream containing both isobutyraldehyde and Me_2S should be performed using a sufficient high biofilter with different zones or a two-stage biofilter with isobutyraldehyde-degradation in the first non-inoculated and Me_2S-degradation in the second inoculated biofilter. The addition of toluene (± 50 ppmv) or limonene (20-100 ppmv) as a second substrate did not affect the Me_2S-degradation rate in the biofilter (Smet *et al.* 1997). Zhang *et al.* (1991) reported a decrease of the Me_2S-removal efficiency in a *Hyphomicrobium* I55-inoculated peat biofilter from 80% to 25% when treating a H_2S, Me_2S and MeSH gas stream due to a preferential degradation of the H_2S and MeSH. Also, Hirai *et al.* (1990) reported that the removal of VOSC can be inhibited by the presence of H_2S. In a peat biofilter inoculated with *Thiobacillus thioparus* DW44, however, the elimination capacity for Me_2S was inhibited by the presence of methanethiol, but accelerated by the presence of hydrogen sulfide (Cho *et al.* 1991b).

Following supplementation with a 20 ppm by volume NH_3-concentration for 24 h, Maes (1997) reported that the Me_2S degradation in a *Hyphomicrobium* MS3-inoculated compost/dolomite biofilter increased slightly from 0.139 kg m^{-3} d^{-1} to 0.152 kg m^{-3} d^{-1}. However, the addition of 150 ppm by volume NH_3 to this biofilter during a 6-day period resulted in a complete inhibition of the Me_2S-removal (Figure 12.6). Figure 12.6 shows that the pH of the carrier material strongly increased as a result of sorption of NH_3. About 10 days after this NH_3-supplementation, the biofilter recovered and even surpassed its initial Me_2S-removal rate. It was stated

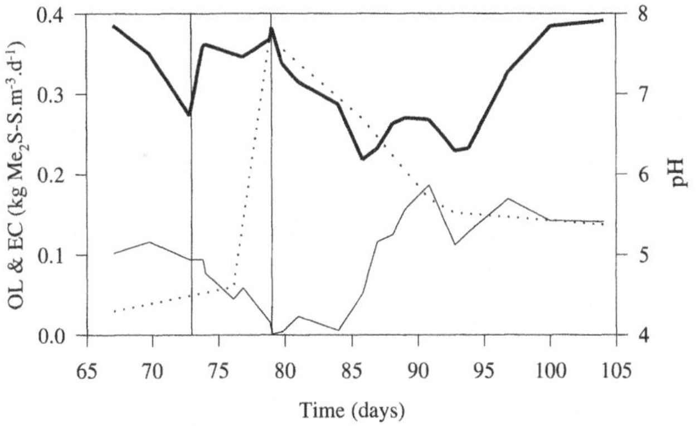

Figure 12.6. Organic loading rate (OL; bold line, left axis) and elimination capacity (EC; thin line, left axis) of a compost biofilter for Me_2S (kg Me_2S-S m^{-3} d^{-1}) and pH of the composting material (dashed line, right axis) before (day 65-73), during (days 73-79) and after (day 79-105) addition of 150 ppmv NH_3 during a 6-day period (After: Smet and Van Langenhove 1998).

that NH_3 is toxic for *Hyphomicrobium* MS3 at a concentration between 20 and 150 ppmby volume. At lower concentrations, however, the pH-neutralizing and the nutritional effect of NH_3 stimulates the Me_2S-degradation in the biofilter (Maes 1997). In a VOSC-loaded *Thiobacillus thioparus* DW44-inoculated peat biofilter, the presence of NH_3 (5-20 ppmv) in the waste gas resulted in a pH-correction because of its chemical reaction with the metabolite sulfuric acid (Cho *et al.* 1992b). Because of this, the pH decline in the peat was very slow, only from pH 7 to pH 6 in almost 200 days.

On the other hand, several authors (Bremner & Bundy 1974; Juliette *et al.* 1993) reported that VOSCs inhibit the nitrification of ammonia, since they are a substrate for the ammonia monooxygenase of *Nitrosomonas* spp. According to Bremner and Bundy (1974), CS_2 is a very effective inhibitor of nitrification in soils (95 % inhibition after five days at a concentration of 1 µg CS_2/g soil), while Juliette *et al.* (1993) reported Me_2S and Me_2S_2 to be weak nitrification inhibitors. Thus, VSC treating biofilters can have incomplete nitrogen elimination.

12.5.3 Biotrickling filtration of VOSCs

Kasakura & Tatsukawa (1995) reported high (>90%) elimination efficiencies for both H_2S and VOSC using a two-stage trickling filter with separate degradation of H_2S and the organic sulfur compounds. In the first stage, a low pH (2-3) allowed efficient H_2S biodegradation by acidophilic bacteria, while the organic sulfur compounds were removed in the second stage where the pH was controlled at a value of 6-7. The sulfate ions are washed out by the circulating water. The carrier medium consisted of polyvinyl alcohol (PVA) particles. Powdered activated carbon was melted into the surface of the PVA particles. The activated carbon can provide extra adsorption sites for the microorganisms. Moreover, the pollutants can be sorbed to the activated carbon before degradation by the microorganisms. The microorganisms were seeded by putting dried PVA particles into cartridges and immersing them in activated sludge. When the PVA particles were swollen, they were transferred to the biotrickling filter module. Lee & Shoda (1989) used activated carbon (AC) as the microbial carrier in a biotrickling filter. Upon inoculation with night soil sludge, they obtained complete removal of MeSH up to a load of 0.48 g MeSH-S kg^{-1} AC d^{-1}.

The application of CS_2-degrading *Thiobacillus* spp. in waste-gas treatment systems was illustrated by Berzaczy *et al.* (1988) and Revah *et al.* (1994), who obtained removal rates up to 0.7 and 2.5 kg CS_2 m^{-3} d^{-1}, respectively in inoculated biotrickling filters. Revah *et al.* (1994) controlled the pH in the biotrickling filter at 4-8 for optimal biological sulfur removal by adding NaOH. Melse and Kraakman (1998) also used a two-stage biotrickling filter (inert support material not specified) for the removal of H_2S and CS_2 from air. In this biotrickling filter, the pH is controlled by the amount of water. In the lower phase, more acids are produced and a lower amount of water has to be dosed for pH control. In this way, the total water supply can be reduced, while pH in both layers is appropriate. The maximum removal rate in this reactor was 0.45 kg H_2S-S m^{-3} d^{-1} and 1.1 kg CS_2-S m^{-3} d^{-1}.

12.6 REFERENCES

ACGIH (1991) Guide to occupational exposure values - 1991. *American Conference of Governmental Industrial Hygienists*, Ohio.

Berzaczy, L., Niedermayer, E., Kloimstein, L. and Windsperger, A. (1988) Biological exhaust gas purification in the rayon fiber manufacture. *Chem. Biochem. Eng. Q* **2**(4), 201-204.

Bonnin, C., Laborie, A. and Paillard, H. (1990) Odor nuisances created by sludge treatment: problems and solutions. *Wat. Sci. Tech.* **22**(12), 65-74.

Bonnin, C., Coriton, G. and Martin, G. (1994) Biodeodorization processes: from organic media filters to mineral beds. *VDI Berichte* **1104**, 217-230.

Brandy, J., Fanlo, J.L. and Le Cloirec, P. (1995) Deodorization of gaseous emissions by a bioscrubber. *Odours & VOC's J.* **1**(3), 192-197.

Bremner, J.M. and Bundy, L.G. (1974) Inhibition of nitrification in soils by volatile sulfur compounds. *Soil Biol. Biochem.* **6**(3), 161-165.

Brennan, B.M., Donlon, M. and Bolton, E. (1996) Peat biofiltration as an odour control technology for sulphur-based odours. *Journal of the Chartered Institution of Water and Environmental Management.* **10**(3), 190-198.

Budwill, K. and Coleman, R.N. (1997) Effect of silicone oil on biofiltration of *n*-hexane vapours. *Forum for Applied Biotechnology, Med. Fac. Landbouww. Univ. Gent* **62**(4B), 1521-1527.

Cesário, M.T. (1997) Water-immiscible solvents for the biological treatment of waste gases. Ph. D. thesis Wageningen Agricultural University, The Netherlands.

Chélu, G. and Nominé, M. (1984) Air pollution abatement in the rendery industry: investigation of various methods. In *Characterization and control of odoriferous pollutants in process industries.* pp. 397-405. (Vigneron, S., Hermia, J. and Chaouki, J. Eds.) Amsterdam, Elsevier Science B.V.

Cho, K.-S., Hirai, M. and Shoda, M. (1991a) Removal of dimethyl disulfide by the peat seeded with night soil sludge. *J. Ferment. Bioeng.* **71**(1), 44-49.

Cho, K.-S., Hirai, , M. and Shoda, M. (1991b) Degradation characteristics of hydrogen sulfide, methanethiol, dimethyl sulfide and dimethyl disulfide by *Thiobacillus thioparus* DW44 isolated from peat biofilter. *J. Ferment. Bioeng.* **71**(4), 289-291.

Cho, K.-S., Zhang, L., Hirai, M. and Shoda, M. (1991c) Removal characteristics of hydrogen sulphide and methanethiol by *Thiobacillus* sp. isolated from peat in biological deodorization. *J. Ferment. Bioeng.* **71**(6), 384-389.

Cho, K.-S., Hirai, M. and Shoda, M. (1992a) Degradation of hydrogen sulfide by *Xanthomonas* sp. Strain DY44 isolated from peat. *Appl. Environ. Microbiol.* **58**(4), 1183-1189.

Cho, K.-S., Hirai, M. and Shoda, M. (1992b) Enhanced removal efficiency of malodorous gases in a pilot-scale biofilter inoculated with *Thiobacillus thioparus* DW44. *J. Ferment. Bioeng.* **73**(1), 46-50.

Cho, K.-S., Ohtani, T. and Mori, T. (1994) Newly found fungus helps to corrode concrete pipes. *W.Q.I.* **3**, 27.

Chung, Y.-C., Huang, C. and Tseng, C.-P. (1996a) Operation optimization of *Thiobacillus thioparus* CH11 biofilter for hydrogen sulfide removal. *J. Biotech.* **52**(1), 31-38.

Chung, Y.-C., Huang, C. and Tseng, C.-P. (1996b) Biodegradation of hydrogen sulfide by a laboratory-scale immobilized *Pseudomonas putida* CH11 biofilter. *Biotechnology Progress* **12**(6), 773-778.

Cork, D.J., Garunas, R. and Sajjad, A. (1983) *Chlorobium limicola* forma *thiosulfatophilum*: biocatalyst in the production of sulfur and organic carbon from a gas stream containing H_2S and CO_2. *Appl. Environ. Microbiol.* **45**(3), 913-918.

Cox, H.H.J., Doddema, H.J. and Harder, W. (1994) Biological removal of styrene from waste gases using fungi on inert support materials. *VDI Berichte* **1104**, 289-299.

Cox, H.H.J., Moerman, R.E., van Baalen, S., van Heiningen, W.N.M., Doddema, H.J. and Harder, W. (1997) Performance of a styrene-degrading biofilter containing the yeast *Exophiala jeanselmei. Biotech. Bioeng.* **53**(3), 259-266.

De Vos, M., Patte, F., Rouault, J., Laffort, P. and Van Gemert, L.J. (1990) Standardized Human Olfactory Tresholds. IRL Press at Oxford University Press, Oxford.

De Zwart, J.M.M. and Kuenen, J.G. (1992) C_1-cycle of sulfur compounds. *Biodegradation* **3**(1), 57-59.

Degorce-Dumas, J.R., Kowal, S. and Le Cloirec, P. (1997) Microbiological oxidation of hydrogen sulphide in a biofilter. *Can. J. Microbiol.* **43**(3), 264-271.

Derikx, P.J.L., Simons, F.H.M., Op Den Camp, H.J.M., Van Der Drift, C., Van Griensven, L.J.L.D. and Vogels, G.D. (1990) Evolution of volatile sulfur compounds during laboratory-scale incubations and indoor preparation of compost used as a substrate in mushroom cultivation. *Appl. Environ. Microbiol.* **57**(2), 563-567.

Diks, R.M.M. (1992) The removal of dichloromethane from waste gases in a biological trickling filter. Ph. D. thesis. Technical University of Eindhoven, The Netherlands.

Gerards, R., Gevaert, D. and Vriens, L. (1995) Biotreatment of waste gases with full-scale Seghobioclean-systems. *Odours VOC's J.* **1**(2), 144-145.

Guey, C., Degorce-Dumas, J.R. and Le Cloirec, P. (1995) Hydrogen sulfide removal on biological activated carbon. *Odours VOC's J.* **1**(2), 136-137.

Hirai, M., Ohtake, M. and Shoda, M. (1990) Removal kinetics of hydrogen sulfide, methanethiol and dimethyl sulfide by peat biofilters. *J. Ferment. Bioeng.* **70**(5), 334-339.

Jensen, A.B. and Webb, C. (1995) Treatment of H_2S-containing gases: a review of microbiological alternatives. *Enzyme Microb. Technol.* **17**(1), 2-10.

Juliette, L.Y., Hyman, M.R. and Arp, D.J. (1993) Inhibition of ammonia oxidation in *Nitrosomonas europaea* by sulfur compounds: thioethers are oxidized to sulfoxides by ammonia monooxygenase. *Appl. Environ. Microbiol.* **59**(11), 3718-3727.

Kasakura, T. and Tatsukawa, K. (1995) On the scent of a good idea for odour removal. *W.Q.I.* **2**, 24-27.

Kelly, D.P. and Smith, N.A. (1990) Organic sulfur compounds in the environment. *Adv. Micr. Ecol.* **11**, 345-385.

Kennes, C. and Thalasso, F. (1998) Waste gas biotreatment technology. *J. Chem. Technol. Biotechn.* **72**(4), 303-319.

Kok, H.J.G. (1992) Bioscrubbing of air contaminated with high concentrations of hydrocarbons. In *Biotechniques for air pollution abatement and odour control policies*, 51, pp.77-82. (Dragt, A.J. and Van Ham, J., Eds.) Maastricht, Elsevier.

Kowal, S., Fanlo, J.L., Degroce-Dumas, J.R. and Le Cloirec, P. (1992) Deodorization with a biofilter using a consumable support: example of BSE process for hydrogen sulfide removal. *Poll. Atm.* **133**, 34-42.

Lanting , J. and Shah, A.S. (1992) Biological removal of hydrogen sulfide from biogas. In *46th Purdue Industrial Waste Conference Proceedings*. Chelsea, Lewis Publishers, Inc.

Lecomte, I., Martin, G., Laplanche, A., Lemasle, M. and Bloquel, M. (1995) Odour pollution in carcass processing plants. Identification of malodorous flux. *Odours VOC's J.* **1**(2), 39-40.

Lee, S.K. and Shoda, M. (1989) Biological deodorization using activated carbon fabric as carrier of microorganisms. *J. Ferment. Bioeng.*, **68**(6), 437-442.

Leson, G. and Winer, A.M. (1991) Biofiltration: an innovative air pollution control technology for VOC emissions. *J. Air & Waste Manage. Assoc.* **41**(8), 1045-1054.

Lomans, B.P., Op Den Camp, H.J.M., Pol, A. and Vogels, G.D. (1999) Anaerobic versus aerobic degradation of dimethyl sulfide and methanethiol in anoxic freshwater sediments. *Appl. Env. Microb.* **65**(2), 438-443.

Maes, K. (1997) Interactieve effecten bij de biofiltratie van vluchtige verbindingen voorkomend in afvalgassen van composteringsinstallaties. Thesis, Faculty of Agricultural and Applied Biological Sciences, University of Ghent, Belgium.

Melse, R.W. and Kraakman, N.J.R. (1998) Biological treatment of waste gas containing H_2S and CS_2 combined with production of concentrated NaOH and H_2SO_4. *Med. Fac. Landbouww. Univ. Gent* **63**(4B), 1841-1847.

Miller, F.C. and Macauley, B.J. (1988) Odours arising from mushroom composting: a review. *Austr. J. Exp. Agric.* **28**, 553-560.

Okkerse, W.J.H., Ottengraf, S.P.P., Diks, R.M.M., Osinga-Kuipers, B. and Jacobs, P. (1999) Long term performance of biotrickling filters removing a mixture of volatile organic compounds from an artificial waste gas: dichloromethane and methylmethacrylate. *Bioprocess Engineering* **20**(1), 49-57.

Oren, A. and Padan, E. (1978) Induction of anaerobic, photoautotrophic growth in the cyanobacterium *Oscillatoria limnetica. J. Bacteriol.* **133**(2), 558-563.

Perry, R.H. and Green, D. (1987) Perry's Chemical Engineers' Handbook, McGraw-Hill International Editions Chemical Engineering Series, Singapore.

Phae, C.-G. and Shoda, M. (1991) A new fungus which degrades hydrogen sulfide, methanethiol, dimethyl sulfide and dimethyl disulfide. *Biotech. Letters*, **13**(5), 375-380.

Plas, C., Wimmer, K., Holubar, P., Mattanovich, D., Danner, H., Jelinek, E., Harant, H. and Braun, R. (1993) Degradation of carbon disulphide by a *Thiobacillus* isolate. *Appl. Microbiol. Biotechnol.* **38**(6), 820-823.

Pol, A., Op den Camp, H.J.M., Mees, S.G.M., Kersten, A.S.H. and Van Der Drift, C. (1994) Isolation of a dimethylsulfide-utilising *Hyphomicrobium* species and its application in biofiltration of polluted air. *Biodegradation* **5**(2), 105-112.

Przyjazny, A., Janicki, W., Chrzanowski, W. and Staszewski, R. (1983) Headspace gas chromatographic determination of distribution coefficients of selected organo-sulphur compounds and their dependence on some parameters. *J. Chromatogr.* **280**, 249-260.

Reij, M. (1998) Membrane bioreactors for waste gas treatment. *J. Biotech.* **59**(3), 155-167.

Reiser, M., Fischer, K. and Engesser, K.H. (1994) Kombination aus Biowäscher- und Biomembranverfahren zur Reinigung von Abluft und hydrophilen und hydrofoben Inhaltsstoffen. *VDI Berichte* **1104**, 103-112.

Revah, S., Hinojosa, A. and Morales, V. (1994) Air biodesulfurization in process plants. Presented at *Bioremediation: the Tokyo 1994 workshop. Organisation for economic co-operation and development (OECD)*, Tokyo, November 1994.

Schamp, N. and Van Langenhove, H. (1986) Volatile organic compounds in air. In *Reviews in Environmental Toxicology* **2**, pp. 251-301. (Hodgson, E., Eds.) Amsterdam, Elsevier.

Sivela, S. (1980) Dimethyl sulfide as a growth substrate for an obligately chemolithotrophic *Thiobacillus*. In *Commentationes Physico-Mathematicae, Dissert.* **1**, 1-69. (Simons, L. Ed.) Helsinki, Societas Scientiarum Fennica.

Smet, E., Chasaya, G., Van Langenhove, H. and Verstraete, W. (1996a) The effect of inoculation and the type of carrier material used on the biofiltration of methyl sulphides. *Appl. Microbiol. Biotechnol.* **45**(1/2), 293-298.

Smet, E., Van Langenhove, H. and Verstraete, W. (1996b) Long term stability of a biofilter treating dimethyl sulphide. *Appl. Microbiol. Biotechnol.* **46**(2), 191-196.

Smet, E., Van Langenhove, H. and Verstraete, W. (1997) Isobutyraldehyde as a competitor of the dimethyl sulfide degrading activity in biofilters. *Biodegradation* **8**(1), 53-59.

Smet, E., Lens, P. and Van Langenhove, H. (1998) Treatment of waste gases contaminated with odorous sulfur compounds. *Crit. Rev. Environ. Sci. Technol.* **28**(1), 89-117.

Smet, E. and Van Langenhove, H. (1998) Abatement of volatile organic sulfur compounds in odorous emissions from the bio-industry. *Biodegradation*, **9**(3/4), 273-284.

Smet, E., Van Langenhove, H. and Philips, G. (1999) Dolomite as a new microbial carrier material for biofilters degrading dimethyl sulphide. Submitted to *Appl. Microbiol. Biotechnol.*

Smith, N.A. and Kelly, D.P. (1988) Oxidation of carbon disulphide as the sole source of energy for the autotrophic growth of *Thiobacillus thioparus* TK-m. *J. Gen. Microbiol.* **134**, 3041-3048.

Tanji, Y., Kanagawa, T. and Mikami, E. (1989) Removal of dimethyl sulfide, methyl mercaptan and hydrogen sulfide by immobilized *Thiobacillus thioparus* TK-m. *J. Ferment. Bioeng.* **67**(4), 280-285.

Then, J. and Truper, H.G. (1983) Sulfide oxidation in *Ectothiorhodospira abdelmalekii.* Evidence for the catalytic role of cytochrome c-551. *Arch. Microbiol.* **135**(4), 254-258.

Van Groenestijn, J.W. and Hesselink, P.G.M. (1993) Biotechniques for air pollution control. *Biodegradation* **4**(4), 283-301.

Van Langenhove, H., Bendinger, B., Oberthür, R. and Schamp, N. (1992) Organic sulfur compounds: persistent odourants in the biological treatment of complex waste gases. In *Biotechniques for air pollution abatement and odour control policies* **51**, pp.177-182. (Dragt, A.J. and Van Ham, J., Eds.) Maastricht, Elsevier.

Verschueren, K. (1983) Handbook of Environmental data on organic chemicals. Van Nostrand Reinhold Company Inc, New York

Weast, R.C. (1976) Handbook of Chemistry and Physics. CRC Press, Boca Raton, FL.

Wilber, C. and Murray, C. (1990) Odor source evaluation. *Biocycle* **31**(3), 68-72.

Williams, T.O. and Miller, F.C. (1992) Odor control using biofilters. *Biocycle* **33**(10), 72-77.

Yang, Y. and Allen, E.R. (1994a) Biofiltration control of hydrogen sulfide. 1. Design and operation parameters. *J. Air & Waste Manage. Assoc.* **44**(7), 863-868.

Yang, Y. and Allen, E.R. (1994b) Biofiltration control of hydrogen sulfide. 2. Kinetics, biofilter performance, and maintenance. *J. Air & Waste Manage. Assoc.* **44**(11), 1315-1321.

Zhang, L., Hirai, M. and Shoda, M. (1991) Removal characteristics of dimethyl sulfide, methanethiol and hydrogen sulfide by *Hyphomicrobium* sp. I55 isolated from peat biofilter. *J. Ferment. Bioeng.* **72**(5), 392-396.

13

Methods of odour measurement and assessment

Richard A. Fenner and Richard M. Stuetz

13.1 INTRODUCTION

Wastewater treatment has always produced unpleasant odours. However, the recent increase in the number of complaints about the release of odours from sewage and sludge treatment has forced water utilities to re-think their strategies for odour prevention and control. These growing environmental concerns about odour pollution have also led governments to consider odour policies that try to regulate odour emissions from wastewater treatment works (Toogood 1994). The recurrent release of unpleasant odours into the atmosphere can constitute a nuisance to a local population. An additional pressure on sewage treatment has resulted from the increasing urbanization surrounding a treatment works, as residential housing has been allowed to encroach within the greenfield buffer zones adjacent to local sewage works (Schulz and van Harreveld 1996; Vincent and Hobson 1998).

As well as the increasing exposure of more people to sewage odours, it is also thought that the increasing number of complaints has occurred because people are more willing to complain about local environmental issues (Schulz and van Harreveld 1996). Nowadays people are more aware of their rights and expect high standards from many of the recently privatized water utilities (Vincent and Hobson 1998). Consequently, odour control of existing treatment facilities and the process of developing new sites to meet obligations under EC water quality directives can incur a significant capital and operating cost for wastewater treatment operators (McGovern and Clarkson 1994; Toogood 1994). In addition, the increasing concerns about odours from environmental applications have also stimulated interest in the techniques used for odour measurement and assessment (Hobbs *et al.* 1995).

Sulfur based compounds are often the most dominant substances associated with odour emissions from wastewater treatment. The most important of these compounds is hydrogen sulfide (H_2S). Other non-sulfur compounds that are also associated with biogenic odours include nitrogens, fatty acids, alcohols and ketones. Table 13.1 shows some of the most common substances associated with odours from wastewater treatment.

To be able to investigate ways in which these problems can be solved, an objective means of odour measurement must first be quantified (Hobson 1995). Current methods used to measure odours rely either on the use of panels of odour assessors to determine human thresholds of detection (Misselbrook *et al.* 1997) or the use of surrogates such as H_2S for determining odour strengths (Hobson 1995). Odour measurements have two main areas of application in sewage odour prevention and control: (i) to be used in combination with dispersion modeling to evaluate actual or likely odour nuisance from existing, proposed or modified treatment works; and (ii) to specify the performance of odour abatement units (Peirson *et al.* 1998).

In this chapter we review both the measurement techniques and the assessment techniques that allow scientists and engineers to make intelligent decisions on the impact that odorous emissions can have on a local community. These techniques are also used to evaluate the performance of odour abatement technologies, which have become necessary to control and prevent the release of unpleasant odour into the environment.

13.2 METHODS OF ODOUR MEASUREMENT

There are two distinct methodologies for the determination of odours in the environment. One set of analytical methods is used in the field to allow the

Table 13.1. Compounds associated with odours from wastewater treatment (Vincent and Hobson 1998)

Substances	Compound	Odour description	Odour threshold (ppb)
Sulfurous	Hydrogen sulfide	Rotten eggs	0.5
	Methyl mercaptan	Decayed cabbage, garlic	0.0014-18
	Ethyl mercaptan	Decayed cabbage	0.02
	Sulfur dioxide	Pungent, acidic	
	Dimethyl sulfide	Decayed vegetables	0.12-0.4
	Dimethyl disulfide	Putrefaction	0.3-11
	Thiocresol	Skunk, rancid	
Nitrogenous	Ammonia	Sharp, pungent	130-15300
	Methylamine	Fishy, rotten	0.9-53
	Ethylamine	Ammonical	2,400
	Dimethylamine	Fish	23-80
	Pyridines	Disagreeable, irritating	
	Scatole	Faecal, repulsive	0.002-0.06
	Indole	Faecal, repulsive	1.4
Acids	Acetic	Vinegar	16
	Butyic	Rancid	0.09-20
	Valeric	Sweat	1.8-2630
Aldehydes and Ketones	Formaldehyde	Acrid, suffocating	370
	Acetaldehyde	Fruit, apple	0.005-2
	Butyraldehyde	Rancid, sweaty	4.6
	Acetone	Fruit, sweet	4,580
	Butanone	Green apple	270

capture of analytical information at the source of the odour emission, whereas the second method is laboratory based and requires samples to be captured in the field and brought to the laboratory for analysis. The type of sample methodology employed depends on the subsequent odour measurement that is going to be used to analyse the samples (Jarke 1998). Two types of sampling technique most commonly used for wastewater odours are capture techniques and concentration techniques.

Capture techniques trap the sample as it exists at the point of sampling, which involves collecting the gaseous emissions in flexible bags, steel canisters and glass bombs (Jarke 1998). The most common of these are flexible bags, usually made out of an inert and impermeable material such as Tedlar. The major disadvantage with this technique is that the gaseous samples can interact with the bag materials. However, the technique allows large sample volumes (up to 100 L) to be captured for multiple analysis. The best way to fill flexible bags is to use the *lung* method, whereby samples are drawn directly into the bags by evacuating a pressure vessel containing the bag. The concentration technique achieves the collection of gaseous molecules by passing the air through a medium (Jarke 1998). The main advantage of the concentration techniques over the capture technique is that the concentration of the molecules can be increased without affecting their relative concentration ratios. The most common application of this technique is for the analysis of odours by gas chromatography (GC), gas chromatography mass spectrometry (GC-MS) and ion-specific electrodes, as it greatly aids the concentration of very low concentrations of gaseous constituents. A variety of media are used to concentrate gaseous molecules, including porous polymers, charcoal, polyurethane foam, treated media, cryogenic techniques and impingers (Jarke 1998). The different types of field- and laboratory-based methods used to determine the odour strengths are discussed in the following sections.

13.2.1 Olfactometry

Olfactometry is used to quantify the strength or concentration of odours using a panel of human sniffers. Brennan (1994) has given a comprehensive review of the science of olfactometry. The strength or concentration of an odour is defined as the number of dilutions at which 50% of the panelists can detect no odour or fail to distinguish it from odour-free air (Misselbrook *et al.* 1997) and is more commonly termed the Threshold Odour Number (TON). Although the concentration is dimensionless the odour threshold is usually expressed as odour units per cubic metre of air (ou/m^3) (Frechen 1994).

The procedure used for determining odour concentration or TON involves odour samples being collected usually in inert flexible bags and being transported to an odour laboratory for analysis. The odour samples are then measured using a dynamic dilution olfactometer, whereby a range of diluted samples are presented to a panel of six or more odour assessors. Each olfactometer has two sniffing ports with odourless air being presented to the panelists through one port and diluted odorous air through the other. A range of at least five dilution steps (each differing by a factor of 2) are presented to the panelists in ascending concentrations, each panelist has to indicate which port contains the odorous air.

Although olfactometry relates most closely to what is been detected by odour complainants, it has several limitations. The measurements are time consuming, labor intensive, expensive and are subject to large variations between panelists (Bliss *et al.* 1996) and laboratories (Schulz and van Harreveld 1996). One of the main reasons for the inconsistency in odour measurements is the lack of standards and inadequate information between different tests (Schulz and van Harreveld 1996). Until recently no clear standardization was available for threshold olfactometry. However, a new European standard using dynamic dilution olfactometry has been drafted by the Committee European Normalisation (CEN) working group CEN/TC254/WG2 "Odour" (van Harreveld 1997). The document implies that repeatability (within-laboratory) and reproducibility (between-laboratory) of odour concentration measurements under reproducible conditions should not differ by more than a factor of 3 or 4 in 95 % of cases (Misselbrook *et al.* 1997). However, as odour laboratories are often remote from the wastewater treatment works and with the increasing need to assess odours on site, it is becoming necessary to be able to monitor odours continuously or be able to do spot measurements using portable equipment.

13.2.2 Hydrogen sulfide

Hydrogen sulfide is a common component of odours associated with wastewater. Therefore, determination of H_2S concentrations using chemical analyses is often used by sewage treatment operators as a marker for determining odour strengths (Vincent and Hobson 1998). It is regarded as easier, more reproducible and cheaper than olfactometry, when many odour concentrations are required for indicating sources of odours in terms of emission rates and estimating the impact of a wastewater treatment works on their surroundings, through the use of H_2S maps (Hobson 1995). H_2S has also been reported to be more sensitive than TON measurements for measuring threshold odour concentrations (Hobson 1997) and it can be monitored continuously or with spot measurements using portable equipment (Vincent and Hobson 1998).

A range of field instruments are available for the detection and/or measurement of H_2S by using relatively inexpensive techniques such as electrochemical detectors, chemical indicator tubes or by using H_2S analysers (based on gold-film resistance detectors) that can detect H_2S concentrations down to ppb concentrations. Since H_2S is not the only compound present in odours generated by wastewater, it is sometimes not the most representative marker for indicating odour strengths. Comparisons between H_2S concentrations and TON over a range of odour concentrations (125-781066 ou/m^3) have shown that clear relationships are not always present (Stuetz *et al.* 1999a). Additionally, H_2S would be an unsuitable marker for measuring odours from secondary aerobic treatment, sludge drying or when odour is caused by a specific industrial discharge (Vincent and Hobson 1998).

13.2.3 Other compounds

Sewage odours contain many odorous compounds (Table 13.1). The concentrations of these compounds can also be used to characterize an odour. However, it can often be very expensive and of limited benefit to try to identify many marker compounds, unless the odour is generated from a specific industrial discharge (Vincent and Hobson 1998). A variety of analytical instruments exist for the direct measurement of chemical compounds. These analytical techniques are well established and understood. It is nevertheless important to emphasize that the chemical characteristics of the individual compounds present in a gaseous mixture and their corresponding background odours can have an enormous impact on the accuracy of the analytical tool being used (Trzupek 1998).

The most common analytical technique used to identify and quantify odorous chemicals in wastewater treatment are gas chromatography systems, such as GC-MS (Devai and DeLaune 1999). This technique involves two steps: (i) separation of the chemicals through the use of chromatography columns, separation temperature(s) and carrier gas pressures; and (ii) detection using an appropriate detector (flame ionization, electron counter and flame photometric).

Other types of analytical system that can be used for direct measurement of odourous chemicals are: flame ionization detectors (FID), infrared analysers (IR) and fourier transform infrared (FTIR) analysers (Trzupek 1998). FID analysers are used for the continuous measurement of volatile organic compounds (VOCs) emissions in ambient air (using a hydrogen flame), whereas IR and FTIR take advantage of a compounds ability to absorb infrared light and therefore can be used either to measure individual compounds (using single wavelengths) or scan the infrared spectrum to measure different compounds simultaneously. However, a major constraint to the widespread application of these types of detection system in wastewater treatment is their high cost.

13.2.4 Electronic nose technology

The use of electronic nose technology for characterizing sewage and sludge odours is an alternative approach to traditional methods of odour measurement. Electronic nose systems are analytical instruments that can characterize an odour. This is achieved through the use of an array of different non-specific sensors. The result is an odour-specific response pattern that is analogous to the human olfactory system (Gardner and Bartlett 1994). A variety of sensor technologies have been used in electronic

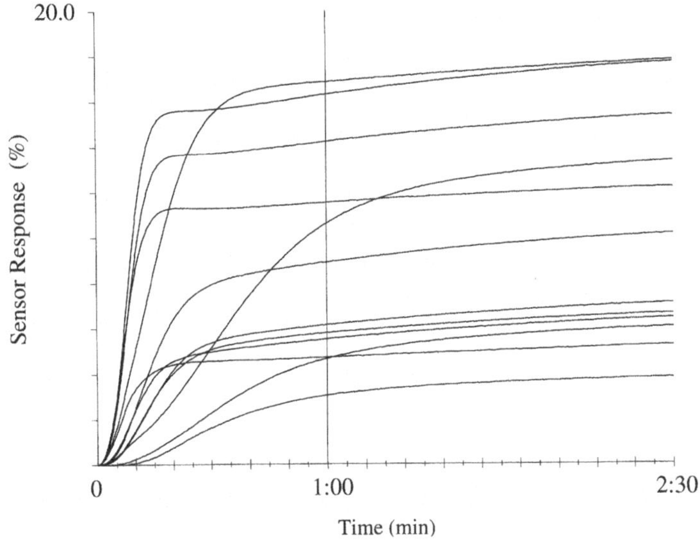

Figure 13.1. Response patterns of a 12-sensor polypyrrole array to sewage odour, showing the resultant odour profile over a 2.5 minute acquisition (Stuetz and Fenner 1998).

noses, the most common of these being metal oxide sensors (MOS), conducting polymers (CP), surface acoustic wave devices (SAW) and quartz crystal microbalances (Persaud and Travers 1997). The use of an array of non-specific sensors allows for responses from many thousands of chemical species, owing to broad selectivity of the sensor surfaces (Persaud *et al.* 1996). The relative responses between the sensors can then be used to produce a unique odour profile or fingerprint (Figure 13.1). These odour profiles can then be further analysed using pattern recognition techniques (such as multiple discriminant and principle component analysis) and/or neural network algorithms (Hodgins 1995; Persaud *et al.* 1996). The choice of analysis technique depends on the amount and nature of information available and the type of information required from the analysis, i.e. quantitative or qualitative.

Applications of electronic nose technology have evolved from the food and beverage industries (Hodgins and Simmonds 1995). Recently, studies have been reported on the assessment of a wide range of sample types, including the application to environmental odour problems (Stuetz and Fenner 1998). Hobbs *et al.* (1995) showed that an electronic nose (consisting of 20 conducting polymers) can discriminate between odours from livestock wastes (pig and chicken slurries), but the instrument was reported to have a low sensitivity compared with olfactometry measurements. Persaud *et al.* (1996) showed using a conducting polymer array that the response patterns of pig slurries are significantly different and that the sensor responses are proportional to the concentrations of various components in the pig slurry. Misselbrook *et al.* (1997) has also shown that when the output of averaged sensor responses is correlated against the odour concentration (using olfactometry) a reasonable fit can be obtained. Another significant feature of this work was the concentration of the odours being considered (100-1000 ou/m^3), which was considerably lower than what had previously been achieved.

More recent studies have shown using a 12-sensor conducting polymer array that when sensor responses from 10 sewage treatment works were correlated against a range odour concentrations (125-781,066 ou/m^3) no clear relation could be found (Stuetz *et al.* 1998). However, when only odour samples from a single works were considered a strong correlation was evident (Stuetz *et al.* 1999a). These observations show that odour profiles are site specific to individual treatment works and are most likely dependent on the composition of the wastewater. Similar relationships have also been found when the odour potential (Hobson 1995) of sewage liquors are compared with the sensor responses of quiescent sewage liquors (Stuetz *et al.* 1999a). A parallel investigation of the electronic nose responses to quiescent sewage samples has shown that the relationships between different odour responses are time dependent (Stuetz *et al.* 1999b).

These examples of environmental odour measurement demonstrate the potential application for electronic nose technology in terms of measuring sewage odours. However, several obstacles still remain in their application such as the need for the electronic nose responses to be calibrated against olfactometry measurements and the extend of the correlation between the sensor responses and odour concentrations (Stuetz *et al.* 1999a).

13.3 ODOUR ASSESSMENT AND PREDICTION TECHNIQUES

In addition to the need for a reliable and robust method of measuring the strength of odours released from wastewaters, procedures are also required that assess the impact of these odours on the surrounding area and the ability they have to cause nuisance. This will depend on the rate at which the odourous gases are released from their liquid sources, the location of these sources and the prevailing weather conditions. Nuisance may not occur unless residential areas are close to where the odour is generated and then the number of complainants will depend on the frequency, duration and time of day of any odour release. Vincent and Hobson (1998) have stated that as the nuisance limit of an unpleasant odour is typically 5-10 ou/ m^3, olfactometry cannot be used to track unpleasant odours in the environment down to the strength that people might find offensive. A range of assessment techniques have been developed which can be used to predict the impact of odourous discharges on local communities and also be used to evaluate odour control technologies and abatement measures. These techniques are reviewed in the following sections.

13.3.1 Odour potential

Odorous compounds are often generated through the on set of septicity in the sewerage network upstream of the wastewater treatment plant. Thus untreated wastewater may contain VOCs and other dissolved gases which may be only released from the liquid during points of turbulence created as the sewage flow passes through a sewage works. A fundamental question therefore is how to determine the inherent "smelliness" of a liquid sample and to assess its capability of releasing odour. An odour potential test was developed in the UK in the early 1980s (Toogood and Diaper 1986) and has been used extensively by the Water Research Centre (WRc) in determining how odour is transported in a sewage treatment system and as a means of estimating odour emission rates (Hobson 1995).

Odour potential is simply the odour associated with a liquid sample as measured by the odour strength of air (TON) after it has been bubbled through the sample in a controlled fashion using a standard apparatus (Figure 13.2). This value represents the maximum strength of odour, which the liquid could release if it passed through a treatment process of sufficiently high turbulence or agitation and not excessively aerated or vented (assuming no carry over of aerosols which might create a higher odour level).

Hobson (1995) suggests that the liquid sample should be at least as large as the volume of air to be passed and he describes a mobile rig (developed by the WRc to carry out measurements of odour potential) which uses a 100 L chamber. Sample liquid in this chamber is aerated while in total internal recycle mode to bring the headspace above the liquid to equilibrium. The unit is then switched to single pass mode and the emitted air is diluted 50:50 with ambient air (to reduce the risk of condensation in the sample bag) and 40 L of the mixture are collected in a Tedlar bag. This

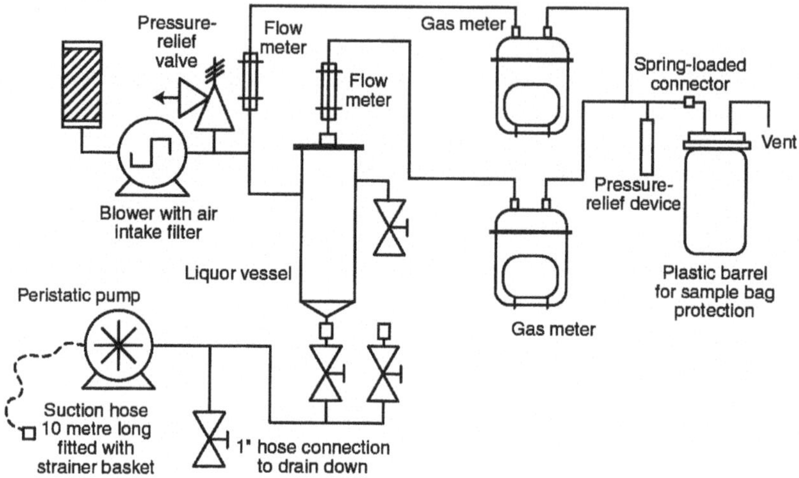

Figure 13.2. Diagrammatic arrangement of a mobile rig for measuring odour potential (Hobson 1995).

odour sample is then measured using conventional olfactometry and the TON value multiplied by 2 to produce the odour potential value. Some examples of odour potential values from a sewage works with activated sludge are shown in Table 13.2.

The odour potential test provides a greater understanding of the processes, which generate odour in wastewater treatment systems, although it may be less helpful in determining the impact of odour coming from different locations at a treatment works. Other techniques, including the use of odour mapping based on H_2S measurements, are more helpful to identify sources of odour release. However, odour potential measurement (based on olfactometry) can provide warning of possible odour nuisance that is not associated with H_2S. Hobson (1997) has suggested typical values of odour potentials that might be expected in a well-operated plant (Table 13.3).

13.3.2 Odour emission rates

The odour emission rate is an important concept which has practical application in the modeling of odour attenuation, as it is the fundamental parameter required by models of atmospheric dispersion, and it enables odour sources to be ranked in order of significance. The term is analogous to the mass emission rate, which is used to quantify pollution from a chimney. An odour emission rate is the product of odour concentration in

Table 13.2. Evolution of odour potential values and H_2S concentrations at different stages an activated sludge unit treating domestic sewage (Hobson 1995)

Source	Odour potential (ou/m^3)	H_2S concentrations
Influent	200-800	60 ppb
Feed to primary tank	3,000	20 ppb
Primary tank overflow	25,000	16 ppm
Final effluent	600	6 ppb
Press filtrate	120,000	170 ppm

Table 13.3. Expected odour potential values of raw and treated sewage (Hobson 1997)

Source	Expected odour potential values (ou/m^3)
Fresh crude sewage	Up to 5,000
Settled sewage	Up to 20,000 *
Activated sludge mixed liquor	2,000 (nitrifying); 5,000 (non-nitrifying)
Final effluent	Up to 2,000
Crude sludge and sludge liquors	Up to 5,000,000
Digested sludge and sludge liquors	Up to 200,000
Raw sewage plus return liquors	Up to 70,000
Significantly septic sewage	Around 100,000 (H_2S > 1mg/l)
Intensely septic sewage	Up to 1,000,000 (H_2S > 1 mg/l)

* Very much higher if tanks not effectively de-sludged or if flow include added liquors.

ou/m^3 and the volumetric flow rate in m^3/s. It therefore has units of ou/s and is a measure of the mass flow rate of odour from a source.

Hobson and Walsh (1990) point out the importance of realising that the highest odour emission rates on a site do not necessarily come from the points where the highest TONs or H_2S concentrations are measured, because the associated air flow rates are not known. For example, very high concentrations of odour may be detected within a sludge wet-well but the transfer of odour to the atmosphere may be quite low. Much lower levels of odour may be detected around a settlement tank or biological filter but because of the greater associated area or greater flow rate of process air these processes could have much higher odour emission rates. Problems arise when trying to determine odour emission rates for process tanks and other locations in a wastewater treatment plant because unlike point source discharges from a chimney, it is not easy to quantify a representative process air flow rate. Sewage-derived odours diffuse directly into the atmosphere making their odour emission rates more difficult to quantify.

Odour emission rates can be estimated for areas with an airflow (such as diffused air activated sludge plant or odour biofilters) using a passive hood which is placed over, or floated on, the source. The odour strength of the collected air is determined and the airflow rate is estimated from the process design or control unit. For area sources with no air flow, odour-free air can be blown into the hood by using a fan. The odour emission rate of the area covered by the hood is equal to the odour strength of the emitted air multiplied by the flow rate of air induced by the fan. The overall odour emission rate of the process is then calculated by scaling up from the area covered by the hood to the area of the process. The hood method developed in Holland by Belois and Anzion (1992) has been used in Germany by Frechen (1994) and in the UK by Hobson (1995). Vincent and Hobson (1998) point out several disadvantages of this method. Firstly the patterns of liquid flow in the flow can be disrupted and disturbed by the hood and secondly the airflow may impinge on water surfaces. The system should not be used for turbulent processes and cannot be used at points of interest such as weirs and flow distributors. If used over solid surfaces it may extract pore air, which would produce an unrealistically high measurement. Furthermore, it can be difficult to minimize contamination of the hood or connecting tubes and high or low pressures inside the hood need to be avoided. They suggest that such a method is neither accurate nor sensitive enough to measure the low odour-emission rates that are characteristic of a well run sewage treatment process.

An alternative micro-meteorological method uses measurements of odour strength and H_2S concentrations downwind of a source. An atmospheric dispersion model is then used to determine the emission rate by varying it until the predicted concentrations matches the measured odour or H_2S concentration profiles (Yang and Hobson 1998). This involves back calculating, by trial and error, the value of the source strength in the model, which gives rise to the observed odour strengths. The technique works best for strong sources. It is however, limited by problems of sensitivity for low odour concentrations from weak sources, which is typical for environmental odours, where it is easy to overestimate odour emission rates. In these circumstances a sensitive marker for odour, such as H_2S, should be used. The method requires an isolated source and a good estimate of the prevailing atmospheric turbulence.

Yang and Hobson (1998) have described an alternative approach using an environmental wind tunnel. Enclosed treatment process modules are placed inside a wind tunnel and operated normally whilst air is blown over their surfaces and the odour or H_2S emission rates are determined from the odour or H_2S strength and volumetric flow rate of air leaving the wind tunnel. This work, carried out by the WRc, has shown that emission rates from wastewater process tanks depend on wind speed, liquid flow rate through the process, tank surface area, length of weirs and filter distributors, height of drop over weirs or from a distributor pipe and the intensity of mixing, as well as the concentration of dissolved free H_2S for H_2S emission and odour potential for odour emission. Thames Water, UK, have used a portable wind tunnel to measure the rate of emission of odour from open sources (Scholes 1994). Filtered odour-free air was ducted over a known surface area of sources such as sludge cake stores. The airflow rate through the tunnel could be varied to simulate a range of wind speeds and odour concentration measurements were taken downwind of the source. Thus source strength per unit surface area could be directly measured against wind speed. Scholes (1994) also claims the technique allows a direct comparison of the impact of open sources to vented air streams and provides accurate emission rate data essential for reliable dispersion modeling predictions.

As the covering of treatment processes becomes a more common practice for wastewater treatment operators, it will be easier to gather site data on odour emission rates.

13.3.3 Odour emission capacity

This procedure has been proposed by Frechen (1998) and uses a substantially different methodology to that described above for the determination of the odour potential. The aim of determining the odour emission capacity (OEC) is to obtain knowledge about the amount of odourants in the liquid phase, which can possibly be emitted at liquid-gaseous interfaces. The key concept with OEC is that the potentially emittable odourants mass flow is related to the volume of the liquid rather than to the liquid surface and it is defined as the total mass of odourants (expressed as ou/m^3_{liquid}), which can be stripped from 1 m^3 of the liquid under standardized conditions (Frechen 1998). The liquid is aerated in a closed test reactor and samples of the off-gas are regularly taken for olfactometry analysis until the end of the test, which is defined as a "lower limit" when 100 ou/m^3 has been reached. It is not known during the execution of the test whether this limit has been reached and so the test is continued for sufficient time based on a direct judgement of the samples taken. With a degree of turbulence of 30 m^3_{air} /m^3_{liquid} hour (1/h) a test duration of up to 1h may be necessary. If the lower limit is not reached during the test time, then the time at which the odourants concentration would have fallen below 100 ou/m^3 has to be calculated by linear extrapolation on a semi-logarithmic scale. The OEC can then be calculated by integrating the area between the measurement values and the lower limit of 100 ou/m^3 (of odorous concentrations in ou/m^3 on a log scale) over the amount of air (plotted on a linear scale), and relating this to an amount of liquid of 1 m^3. Thus, the OEC is given in ou/m$^3_{liquid}$.

This technique allows the quality of liquids to be assessed concerning possible emissions and the results allow the appropriate design of countermeasures to avoid emissions. In addition, not only can the emission capacity of odourants be measured by evaluating TON using olfactometry, the emission capacity of specific compounds can also be determined using the same stripping principle. Thus, off-gas samples can be analysed for hydrogen sulfide and the results are expressed as H_2S Emission Capacity (H_2S EC) in mg H_2S/m$^3_{liquid}$.

Frechen (1998) presents the results when the test was used to assess the extent of odour emissions arising from the mixing of a domestic wastewater and a paper mill effluent. The effectiveness of dosing pure oxygen into a paper mill effluent before it was discharged to the main sewer has been evaluated. It was found that the H_2S EC significantly fell from 248 mg H_2S/m$^3_{liquid}$ without dosing to 32 mg H_2S/m$^3_{liquid}$ when dosing with oxygen (87% reduction), showing that H_2S can be stripped easily from the mixture. The corresponding OEC values were 41,000 ou/m$^3_{liquid}$ without dosing of oxygen and 10,000 ou/m$^3_{liquid}$ with dosing of oxygen (76% reduction). In this way dosing of oxygen was proved to be a suitable remediation measure. The data also indicate that H_2S was not the only relevant odourant, as in latter case the OEC would have dropped by the same percentage as the H_2S EC.

13.3.4 Odour mapping

A more direct approach to evaluating the distribution of odour across a site and its impact on adjacent communities is to use portable hand held equipment to measure H_2S concentrations. "Maps" of H_2S concentration down to parts per billion level can be produced for areas within and around a sewage treatment works by taking measurements at closely spaced points. A contour map of H_2S concentrations over the site can then be produced using suitable computer software containing surface fitting routines (Scholes 1994). These are very useful in helping to identify the position and extent of odour sources. This is done by focussing on those parts of the treatment process that fall within the contour of highest concentration. The direction in which the odour disperses can be found from movement in the contours and a reduction in concentration levels. An example of such a contour map is shown in Figure 13.3, whereas Figure 13.4 shows a projection profile of the spatial distribution of H_2S concentrations at a wastewater treatment plant before and after remedial work.

Background levels of H_2S in the UK are in the range 0.5–1.5 ppb, although this may increase near roads. Hobson (1995) has reported that levels in excess of 2 ppb, if clearly shown to be connected to a treatment works by the shape of the H_2S contours on the map, are generally associated with detectable odours. However there are limitations to the use of such maps based on individual marker compounds (such as H_2S) for an estimation of the general odour. Although they are useful in helping to identify the sources responsible for the odour release and to highlight the areas that might suffer the most impact, individual marker compounds are not so useful in identifying the causes of the odour emission. Hobson and Walsh (1990) have made the following comments on the limitations of the technique. If two processes fall in line with the wind direction, and the upwind process is producing significant hydrogen sulfide, it is not possible to say much about the downwind process. Ideally contour maps should be produced twice, under conditions when the wind directions are normal to each other. A further problem is that usually only ground-level measurements are made, which can be restrictive in cases where vertical dispersion is the more dominant component, leading to an odourous plume. Such a plume may be significant if it is swept back down to ground level at some distance from the source. It must also be appreciated that each contour map is specific to the atmospheric and wind conditions prevailing on the day of measurement and the operational activities being carried out. Lastly, although H_2S can be a useful surrogate for human perception of general sewage odour, it does not always correlate well with odours of specific origin, such as the characteristic smell of fresh sewage or the earthy smell characteristic of the activated sludge process. Equally H_2S may not be an appropriate marker compound to represent odours of industrial effluents.

13.3.5 Dispersion modelling

Odour dispersion modelling is frequently used to assess the impact on the surrounding area of key individual odour sources. It is particularly useful in predicting overall improvements, which can be achieved through the implementation of odour abatement measures. Dispersion models are commonly used to demonstrate the anticipated downwind areas potentially

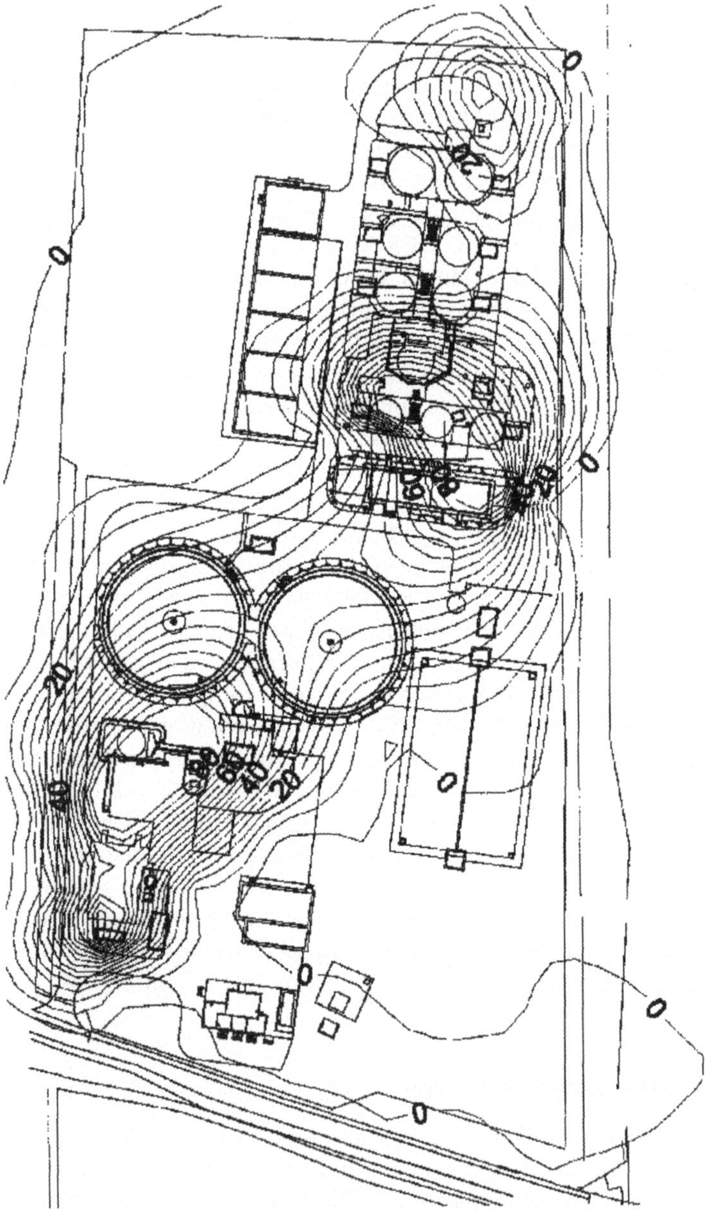

Figure 13.3. Example of an odour map (H_2S values as ppb) (Vincent and Hobson 1998).

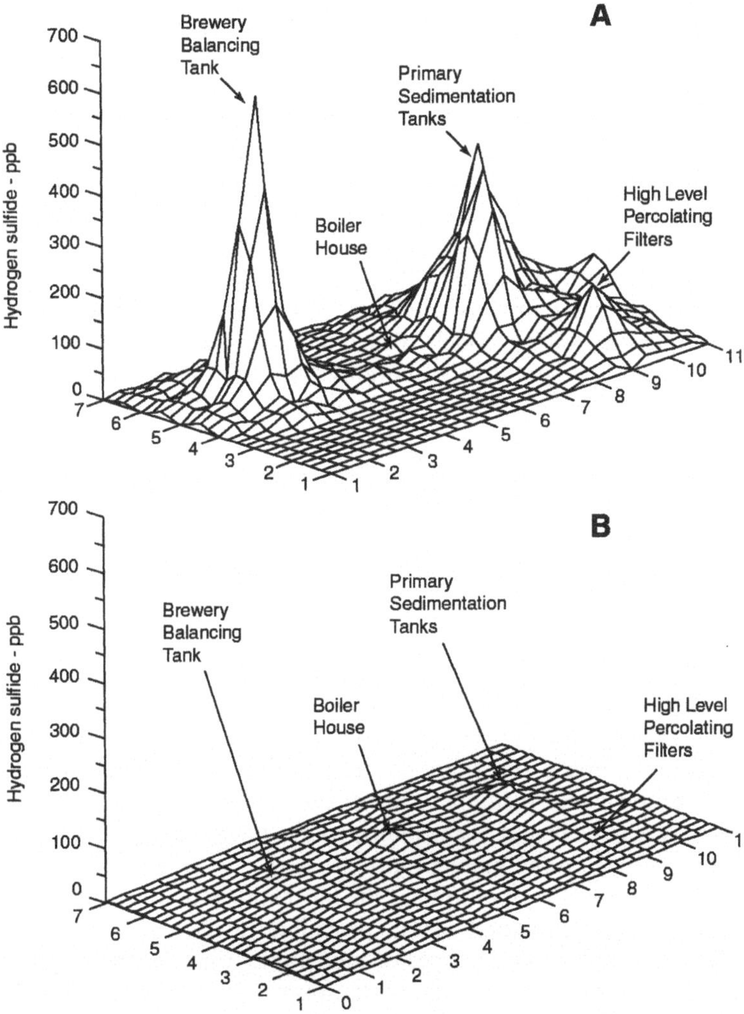

Figure 13.4. Odour map of a treatment works treating brewery and domestic wastewater before (A) and after (B) remedial work to reduce odorous emissions (Scholes 1994).

affected by odour from either planned works extensions, or new or yet to be built, sewage works. As such they have an important role to play in planning procedures and can help to demonstrate that local communities will not be affected by the installation of wastewater treatment facilities on new sites.

Once released, gaseous emissions will travel downwind from their source and be dispersed by the natural turbulence of the atmosphere. Thus the wind speed and direction, and the stability of the atmosphere are the main factors in determining the rate of dispersion of the emission. Atmospheric stability is categorized in classes from very stable (minimum dispersion) to very unstable (maximum dispersion). The prevailing stability class will depend on the temperature profile in the atmosphere (lapse rate) and can be determined from wind speed and cloud cover observations using a scheme originally developed by Pasquill (1971).

The most common models used for predicting odour dispersion are based on Gaussian plume models, which at their simplest deal with point source emissions. Under ideal conditions the dispersion process has the effect of producing a roughly cone shaped plume of air with the apex towards the point of emission. The concentration of odour C at any x, y, z location can be computed from the Gaussian equation (Kiely 1997), which allows for reflection of gaseous compounds from the ground surface when the plume comes in contact with the ground:

$$C_{(x,y,z)} = \frac{OER}{u}\frac{1}{2\pi\sigma_y\sigma_z}\exp\left[-\frac{1}{2}\left(\frac{y}{\sigma_y}\right)^2\right]\left\{\exp\left[-\frac{1}{2}\left(\frac{z-H}{\sigma_z}\right)^2\right]+\exp\left[-\frac{1}{2}\left(\frac{z+H}{\sigma_z}\right)^2\right]\right\}$$

for the co ordinate system: $x = 0$ at the point of emission, $y = 0$ at plume centre line and $z = 0$ at ground level. Where: OER = odour emission rate (ou/s); u = wind speed (m/s); σ_y and σ_z = diffusion coefficients (m), as functions of downwind distance x based on Pasquill's stability categories and H = the effective height of the plume source (actual height of the source modified by buoyancy and momentum).

Models based on this equation are the most widely used and give results consistent with experimental data, providing they are applied when conditions are suitable. These types of model may vary depending on how they are set up to deal with multiple sources, the treatment of local downwind topography, averaging periods for output data and the format of the model output. Several models are PC-based versions of the Gaussian dispersion model and are based on the "industrial source complex" such as the ISC3 developed by the US EPA. These are used as a regulatory tool for atmospheric emissions in many parts of the world.

Once a model has been selected and suitable calibration data prepared it is very important to carry out some form of model validation. This can be achieved by releasing single easily measured compounds and measuring their downwind concentrations and comparing these with predicted values.

For sewage odour this is not always easy as mass sample collection and measurement is not always possible owing to the cost. An alternative approach to validation therefore relies on comparing predicted odour nuisance with recorded odour complaints.

Care should be taken here to allow for the level of odour exposure that causes a population to complain. McGovern and Clarkson (1994) report that in recent exercises carried out by a UK water company predicted and actual complaints were found to produce a good correlation. Frechen (1994) describes field evaluations of dispersion calculations by using a field inspection programme carried out by a test person panel (duration normally one year) or questionnaires using the people in the nearby area. He notes some remarkable differences between these methods. Field inspection was found to be comparable to the results of dispersion calculations, whereas questionnaires will give results in terms of annoyance parameters of the people living in the respective area, which did not correspond to the actual odour dispersion.

Peirson *et al.* (1998) describe more recent non-Gaussian dispersion models such as UK-ADMS. This is similar to a Gaussian model as it calculates pollutant concentrations at x, y, z co-ordinates downwind from the source. However, this model takes a different approach in formulating the atmospheric boundary layer where a significant transfer of heat, momentum and moisture between the earth and atmosphere takes place. This layer is turbulent so that pollutants are dispersed and its depth varies according to wind speed, vertical temperature gradient and whether there is strong surface cooling. Upward dispersion is limited by an inversion above the layer. The model assumes a Gaussian shaped plume for neutral and stable conditions but allows a different characteristic to be employed for convective conditions. The ADMS has the capability of modelling building effects and wet and dry deposition, which is particularly important when considering ammonia emissions.

The advantage of using dispersion models rather than empirical relationships is that time-averaged concentrations can be calculated at any point down wind allowing the resulting odour concentrations to be assessed against agreed "nuisance" thresholds. However, one drawback is that time-averaged concentrations (ranging from 15 min to 1h) are used, whereas the scale of response of the human nose is within seconds. There is some uncertainty as to the ratio of the short-term peak concentrations sensed by the nose to, say, hourly mean values. Peirson *et al.* (1998) suggest a ratio of 10 may be a reasonable estimate. On the other hand, Vincent and Hobson (1998) suggest that the ratio of peak to average odour strengths is much nearer to 1 for large area sources. Hogstrom (1972) first pointed out that depending on variations of concentration it may often occur that the hourly mean is below the odour threshold and there are remarkable amounts of time within the hour where the concentration is above the limit value. Frechen (1994) reports on German practice which, to more correctly assess the extent of odour in ambient air, stipulates that 1h may be recognized as exceeding the limit value, if the limit value is exceeded during 10% of 1h, thus during 6 min. It is therefore assumed to be sufficient to multiply the hourly mean by a factor of 10.

13.3.6 "Nuisance" thresholds

Thresholds for odour nuisance are difficult to set objectively because human perception of a given odour can vary greatly between individuals and circumstances. Hassema (1987) has reported that olfactometric detection levels determined in comparison with totally odour-free air under laboratory conditions are defined as 1 ou/m^3 but under ambient conditions may be 3–5 times higher. The ability to recognize odour from a specific source or origin is often cited as about 5 times the detection threshold. It follows that the real detection threshold of an odour, under ambient conditions may be approximately 20 ou/m^3, which is 20 times the detection threshold assessed under laboratory conditions.

The reaction of an individual to an odour will depend on the environment in which it is experienced. For example, people living in towns and cities are more likely to be intolerant of sewage or agricultural odours compared with residents of small rural villages or isolated country properties. It has often been assumed that if there is no record of nuisance complaints then any odours are acceptably being dispersed. Peirson *et al.* (1998) have found very good correlations between the 5 ou/m^3 98th percentile odour concentration contour and complaints boundary for intensive livestock buildings. Whereas sewage works appear to be slightly less offensive to the public with complaints more likely to equate to the 10 ou/m^3 98th percentile level. The public may tolerate a low level of odour exposure without making formal complaints. The following guidelines have been used successfully in recent odour impact studies:

1. In suburban areas, average odour concentrations should not exceed 2.5 ou/m^3 for more than 2% of each hour.
2. In rural areas, average odour concentrations should not exceed 5 ou/m^3 for more than 2% of each hour (i.e. 175 hours a year).

The effect of an odour emission can be estimated using an equation derived from empirical data by the Warren Spring Laboratory (Department of Environment 1980), which relates the radius from an odour source in which complaints may be expected, to the flow rate and odour content of the gas:

$$\mathrm{OR} = (2.2E)^{0.6}$$

where OR is the odour complaint radius in metres and E is the odour emission rate calculated from multiplying the flow rate of air (m^3/s) by the odour concentration in ou/m^3. It should be noted that the accuracy of this approach is to give an "order of magnitude" estimate and that in practice the complaint radius can be up to only half this value.

13.3.7 Odour modelling

A wastewater treatment plant or sewerage system can be modelled using either odour emission formulae derived from wind tunnel experiments (see 13.3.2) or theoretical equations predicting sulfide formation and release. These can be correlated to site measurements of atmospheric and liquid-phase odour or H_2S concentrations. Once a model of a system has been developed, the effects of varying such factors as temperature, flow rate or odour load can be calculated, and the impact of any abatement measures assessed.

For example Boon (1994) presents a model for predicting the concentrations of sulfide (C^s) in a rising main sewer:

$$C^s = K_4 L_{COD} t_s \left(\frac{1 + K_5 d_1}{d_1}\right) 1.07^{(T-20)}$$

For domestic sewage the value of K_4 was found by Boon and Lister (1975) to be 0.00152 and K_5 to be 0.004 when L_{COD} was the average COD of the sewage (mg/l), t_s was the total retention period of the sewage in the main (minutes), d_1 was the internal diameter of the main (cm) and T was the sewage temperature (°C).

In the UK, WRc has developed a sewage treatment odour production model (STOP) by using an environmental wind tunnel. The model is based on the odour potential of the liquid being treated and the process design and operating parameters. It can be used to predict the odour emission rates from most commonly used processes for sewage treatment (Yang and Hobson 1998). The model consists of a set of formulae in which odour emission rates are expressed as functions of process parameters and the odour potential of the liquid flow. For example, the formula for the odour emission rate from a weir takes the following form:

$$OER = 7.16 \times 10^{-4}\, OP \times F_{weir} \times h \times K_{pH}$$

where: OER = Odour Emission rate per unit length of weir (ou/s/m); OP = odour potential of the liquid flowing over the weir (ou/m^3); F_{weir} = weir loading rate (m^2/h); h = height (m) of drops of liquid flow at weirs and K_{pH} = pH coefficient, takes a value of 1.17 at pH 7.

Similar formulae in which odour potential is replaced by free dissolved H_2S concentration are used to determine H_2S emission rates. The STOP model has been used in around 30 full-scale sewage works and in 8 of these, the odour emission rates estimated have been used as input to dispersion models.

Other types of models such as TOXCHEM attempt to quantify the volatilization of VOCs from industrial effluent treatment plants. These are based on the rate of volatilization of chemical compounds from water-based processes into the atmosphere. This rate is governed by the level of turbulence and/or interfacial area, the properties of the chemical compounds and the concentration gradients between the liquid and gas phases. In addition, such models have proved difficult for use in quantifying odour because they are costly and difficult to fully quantify the complex chemical mixture responsible for odour from the sewage. Furthermore, there is no reliable way to predict the odour impact from a sewage based on its chemical composition, because the required parameters are only known for a handful of the odorous compounds present. In addition, it is not possible to predict the odour strength of a mix of compounds, even when the threshold odour level of each compound is known.

13.4 REFERENCES

Belois, H. J. and Anzion, C. J. M. (1992) Measurement of emissions over surface areas using the hood method. In: *Biotechniques for Air Pollution Abatement and Odour Control Policies.* (eds. A.J. Draft and J van Ham), pp. 439-445, Elsevier, London.

Brennan, B. M. (1994) Recent advances in odour measurements. *Proc. 1st CIWEM Symposium-Odour Control and Prevention in the Water Industry, p.33-41.*

Boon, A.G. (1994) Septicity in sewers: causes, consequences and containment. *Proc. 1st CIWEM Symposium-Odour Control and Prevention in the Water Industry, p.1-21.*

Boon, A. G. and Lister, A. R. (1975) Formation of sulfide in a rising main sewer and its prevention by injection of oxygen. *Prog. Wat. Tech.* 7(2), 289-300.

Bliss, P. J., Schulz, T. J., Senger, T. and Kaye, R. B. (1996) Odour measurement-factors affecting olfactometry panel performance. *Wat. Sci. Tech.* **34**(3-4), 549-556.

Department of Environment (1980) *Odour control: a concise guide*. (eds. F. H. H Valentin and A. A. North), Warren Spring Laboratory, UK.

Devai, I. and DeLaune, R. D. (1999) Emission of reduced malodourous sulfur gases from wastewater treatment plants. *Water Environ. Res.* **71**(2), 203-208.

Frechen, F-B. (1994) Odour emissions of wastewater treatment plants-recent German experiences. *Wat. Sci. Technol.* **30**(4), 35-46.

Frechen, F-B. (1998) Odour Emission Capacity of Wastewaters-Standardisation of measurement method and application. *Wat. Sci. Technol.* **38**(3), 61-69.

Gardner, J. W. and Barlett, P. N. (1994) A brief history of electronic noses. *Sensors and Actuators B*, **18**, 211-220.

Hassema, H. (1987) Characterisation of exposure in odour annoyance situations. In: *Environmental Annoyance, Characterisation, Measurement and Control* (ed. H. S. Koelega), pp. 95-104, Elsevier, Amsterdam.

Hobbs, P. J., Misselbrock, T. M. and Pain, B. F. (1995) Assessment of odours from livestock wastes by a photoionization detector, an electronic nose, olfactometry and gas chromatography-mass spectrometry. *J. Agric. Engng. Res.* **60**, 137-144

Hobson, J. (1995) The odour potential: a new tool for odour management. *J. Chartered Inst. Water Envir. Mangt.* **9**(5), 458-463.

Hobson, J. (1997) Odour potential. *Water Qual. Internat.* (July/August), pp. 21-24.

Hobson, J. and Walsh, M. (1990) Application of Contour mapping of hydrogen sulphide levels to odour control at sewage treatment works. Foundation for Water Research Report Fr 0060, U.K.

Hodgins, D. (1995) The development of an electronic nose for industrial and environmental applications. *Sensors and Actuators B* **26-27**, 255-258.

Hodgins, D. and Simmonds, D. (1995) The electronic nose and its application to the manufacture of food products. *J. Auto. Chem.* **17**(5), 179-185.

Hogstrom, U. (1972) A method for prediction of odour frequencies from a point source. *Atmosph. Environ.* **6**, 103 – 121.

Jarke, F. H. (1998) Chemical Methods. In: *Odour and VOC Control Handbook* (ed. H. J. Rafson), McGraw-Hill, New York.

Kiely, G. (1997) *Environmental Engineering*. McGraw-Hill, London.

McGovern, J. E. and Clarkson, C. R. (1994) The development of Northumbrian Water Limited approach to odour abatement for wastewater facilities. *Proc.1st CIWEM Symposium-Odour Control and Prevention in the Water Industry, p. 45-73.*

Misselbrook, T. M., Hobbs, P. J. and Persaud, K. C. (1997) Use of an electronic nose to measure odour concentration following application of cattle slurry to glassland. *J. Agric. Engng. Res.* **66**, 213-220.

Pasquill, F. (1971) Atmospheric dispersion of pollution. *Qual. J. Royal Meteorolog. Soc.* **97**, 369.

Peirson, S., Buckland, A. and Barrie, I. (1998) The use of odour measurement in the control of odours. *Proc. 2nd CIWEM Symposium-Odour Control in Sewage Treatment*, p. 3.1-3.15.

Persaud, K. C., Khaffaf, S. M., Hobbs, P. J. and Sneath, R. W. (1996) Assessment of conducting polymer odour sensors for agricultural malodours measurements. *Chem. Sci.* **21**, 495-505.

Persaud, K. C. and Travers, P. J. (1997) Arrays of broad specificity films for sensing volatile chemicals. In: *Handbook of Biosensors and Electronic Noses.* (ed. E. Kress-Rogers), pp. 563-592, CRC Press, Boca Raton.

Scholes, P. (1994) Odour control in practice - an end to odour at two well known sites. *Proc. 1st CIWEM Symposium-Odour Control and Prevention in the Water Industry*, p. 137-150.

Schulz, T. J. and van Harreveld, A. P. (1996) International moves towards standardisation of odour measurements using olfactometry. *Water Sci. Tech.* **34**(3-4), 541-547.

Stuetz, R. M. and Fenner, R. A. (1998) Electronic nose technology: a new tool for odour management. *Water Qual. Internat.* (July/August), pp. 15-17.

Stuetz, R. M., Engin, G. and Fenner, R. A. (1998) Sewage odour measurements using a sensory panel and an electronic nose. *Wat. Sci. Technol.* **38**, 330-335.

Stuetz, R. M., Fenner, R. A. and Engin, G. (1999a) Assessment of odours from sewage treatment works by an electronic nose, H_2S analysis and olfactometry. *Water Res.* **33**(2), 452-461.

Stuetz, R. M., Fenner, R. A. and Engin, G. (1999b) Characterisation of wastewater using an electronic nose. *Water Res.* **33**(2), 442-452.

Toogood, S.J. and Diaper, J. (1986) Developments in the assessment of odours from sludges. In: *Odour Prevention and Control of Organic Sludge and Livestock Farming*, (eds. V.C. Nielsen, J.H. Voorburg and P. L'Hermite), pp. 131-145, Elsevier, London.

Toogood, S. J. (1994) Changing attitudes towards odours-consequences for the water industry. *Proc.1st CIWEM Symposium-Odour Control and Prevention in the Water Industry.*

Trzupek, R. (1998) Monitoring for volatile organic compounds. In: *Odour and VOC Control Handbook* (ed. H. J. Rafson), McGraw-Hill, New York.

van Harreveld, A. P. (1997) Europe moves towards standardisation on odour measurement. *Water Qual. Internat. (July/August)*, p. 24.

Vincent, A. and Hobson, J. (1998) Odour Control. *CIWEM Monographs on Best Practice No.2*, Terence Dalton, London.

Yang, G. and Hobson, J. (1998) Validation of the sewage treatment odour production (STOP) model. *Proc. 2nd CIWEM Symposim-Odour Control in Sewage Treatment*, p. 2.1-2.19.

14

Treatment of solid materials containing inorganic sulfur compounds

Richard Tichý

14.1 INORGANIC SULFUR TRANSFORMATIONS IN WATER / SOLID PHASE SYSTEMS

In the environment, different types of solid phase system can contain high concentrations of inorganic sulfurous compounds, e.g. ores, soils and sediments. The key criterion in the assessment of environmental risks of inorganic sulfurous compounds present in these materials is whether they remain bound to the solid phase or whether they are mobilized into the solution. Mobility of inorganic sulfur in natural water / solid biphasic systems is predominantly controlled by oxidative or reductive changes and precipitation constants of sulfurous compounds. In anaerobic environments, sulfur is encountered in its most reduced, i.e. sulfidic form.

Table 14.1. Comparison of the effects of oxidative and reductive conditions on sulfur behaviour in solid materials

Conditions	Dominant reactions	Environmental consequences
Reductive	Sulfate reduction Release of OH^-	Alkalization Immobilization of cations (e.g. metals) Accumulation of sulfur
Oxidative	Sulfide oxidation Release of H^+	Acidification Dissolution of alumo-silicates Leaching of cations Enhanced weathering

Owing to a poor solubility of most sulfidic compounds, sulfurous compounds are predominantly immobile in anaerobic environments. On the contrary, in aerobic environments, the most stable inorganic sulfur compound is sulfate, which is usually well soluble in water.

Reductive and oxidative changes of inorganic sulfurous compounds often involve an interaction with protons. Sulfur reduction is usually accompanied with alkalization of the environment, whereas sulfur oxidation often results in acidification. The general overview of the effects of reductive and oxidative changes of inorganic sulfur in the environment is summarized in Table 14.1.

14.1.1 Reductive processes

The most common reductive process affecting sulfur in the environment is microbial sulfate reduction. Sulfate reducing bacteria require an anaerobic environment (E_h<-150 mV) and a pH higher than 5-6. However, microbial sulfate reduction was encountered even in anoxic systems loaded with highly acidic leachates (pH<3). This is due to the strong alkalizing capacity of the sulfate reduction, which results in the formation of micro-regions within the soil or sediment profile with conditions favourable for sulfate reducing bacteria. Moreover, some recent studies indicate an existence of acid-tolerant sulfate reducing bacteria. Detailed information on this subject can be found in Chapter 8, section 8.4.

As can be seen from the reaction stoichiometry of sulfate reduction (van Houten *et al.* 1994), its main products are sulfide and OH^-.

$$8H_2 + 2SO_4^{2-} \rightarrow 6\ H_2S + HS^- + 5H_2O + 3OH^- \qquad (14.1)$$

Sulfide itself is a rather mobile species that can escape from a system in the form of H_2S-gas or soluble HS^- (Stumm and Morgan 1981). However,

Table 14.2. Solubility of selected metal sulfides (Ellwood *et al.* 1992)

Metal ion	Solubility product ($mol.L^{-1}$)
Hg^{2+}	4×10^{-54}
Cu^{2+}	8×10^{-45}
Cd^{2+}	5×10^{-29}
Zn^{2+}	10^{-20}
Mn^{2+}	1.4×10^{-14}
Fe^{2+}	10^{-20}

sulfide reacts with numerous naturally occurring chemicals, especially cationic metals. This leads to the formation of poorly soluble metal sulfides (see Table 14.2). In this way, high amounts of reduced sulfur species are accumulated in solid materials under reductive conditions (Ivanov 1983).

Via precipitation of sulfides, sulfur is removed from the geochemical cycling and buried in anaerobic soils or sediments. As stated above, these sulfur stocks are closely associated with stocks of cationic metals due to the formation of metal sulfides. Moreover, increased alkalinity of these systems is favourable for the formation of other (non-sulfidic) solid-phase metal stocks, e.g. hydroxides, carbonates, and for higher sorption of metals on the surface of the soil (Ivanov 1983; Salomons and Stigliani 1995).

14.1.2 Oxidative processes

Oxidation of reduced inorganic sulfur compounds in nature proceeds via spontaneous chemical as well as microbial oxidation processes (Ivanov 1983). Spontaneous chemical processes dominate the oxidation at neutral to alkaline pH. At acidic conditions, rates of spontaneous chemical oxidation decrease, whereas microbial oxidation rates increase. At a pH<4, the rate of microbial oxidation of inorganic sulfur compounds is by 5-7 orders of magnitude higher than the rate of spontaneous chemical processes (Smith *et al.* 1988; Evangelou and Zhang 1995). Details about the microbial sulfur oxidation are given in Chapters 3 and 5. Here, attention is given to the facts that are specifically important for solid materials.

Owing to the acidifying character of sulfur oxidation, appropriate microbes developed a strong tolerance to the extremely low pH. Some microbes even require a low pH for their growth (Karavaiko 1985). Among them, types with different pH-tolerance exist: some types are able to operate at pH 3-6, and are active during the initial phase of sulfur oxidation. Other bacteria are more active at pH<3, succeeding after disappearance of the less acid-tolerant microbes (Blais *et al.* 1993).

The most broadly known genus of the bacteria responsible for inorganic sulfurous compounds oxidation is *Thiobacillus*. Therefore, the mixed microbial communities that oxidize inorganic sulfurous compounds are called thiobacilli. However, a reader should bear in mind that thiobacilli is not a clearly defined set of bacterial species.

Basically, oxidation of inorganic reduced sulfur compounds requires the presence of oxygen. However, oxidation may proceed even under anoxic conditions. Spontaneous chemical oxidation can be driven by a strong oxidizing capacity of ferric (Fe^{3+}) iron:

$$MeS + 2Fe^{3+} \rightarrow 6\ Me^{2+} + 2Fe^{2+} + S^0 \tag{14.2}$$

Here, Me stands for a cationic divalent metal, e.g. Cd, Cu, Pb, Zn.

Microbial oxidation in anoxic conditions is possible due to the capacity of certain bacteria to utilize nitrate as an electron acceptor (Mackintosh 1978). Furthermore, *Thiobacillus ferrooxidans* is able to grow anaerobically with elemental sulfur as its electron donor and ferric iron as its terminal electron acceptor (Pronk *et al.* 1992).

A key factor in microbial oxidation of solid inorganic reduced sulfur compounds is an adhesion of bacteria to the surface of appropriate sulfurous minerals (Bryant *et al.* 1983; Karavaiko 1985; Solari *et al.* 1992; Devasia *et al.* 1993; Mirajkar *et al.* 1997; Porro *et al.* 1997). Therefore, microbial sulfur oxidation is influenced by the presence of surface active agents. Addition of wetting agents increases the oxidation rate and some wetting agents are even produced by the bacteria themselves (Cook 1964; Schaeffer and Umbreit 1963; Bryant *et al.* 1983). On the contrary, excessive amounts of wetting agents can inhibit the adhesion of bacteria, which results in a suppression of microbial oxidation (Ondruschka and Glombitza 1993).

At $pH<4$, the activity and growth of sulfur oxidizing microbes is also suppressed when organic acids are present in the system. Since these acids are not dissociated in acidic media, they are taken up by microbes as electroneutral molecules. However, they dissociate in the cytoplasm, which has usually a circumneutral pH. Acidification of cytoplasm brings higher requirements on energy supply from the cell and may eventually lead to a die-off of the bacteria (Pronk *et al.* 1990).

The direct effect of sulfur oxidation is its release into the environment, usually in the form of sulfate. Sulfate is a rather stable and intact molecule which is well soluble in water. Three exceptions from high solubility of sulfate compounds should be noticed here: calcium sulfate - gypsum, lead sulfate, and jarozites (Evans 1989). Formation of calcium and lead sulfates is illustrated in Box 14.1 by equilibrium simulations using ECOSAT ver. 4.4, a program for calculating the soil / water chemical equilibria

Box 14.1. Precipitation of calcium and lead sulfate as a function of pH

Figure 14.1 shows the equilibrium situations in a solution initially containing varying amounts of gypsum (calcium sulfate) and different pH. The concentrations of gypsum were selected between 12.5 mmol/L and 42.5 mmol/L (401.25 - 1364.25 mg/L of S). The simulations included also the gaseous phase with a constant atmospheric CO_2 partial pressure. Between pH 3 and 7, no influence of pH is observed. At pH lower than 3, gypsum is solubilized, whereas at pH>7, formation of calcium carbonate prevails.

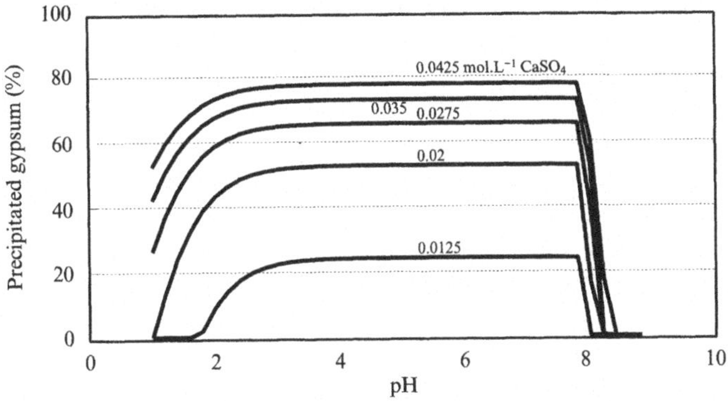

Figure 14.1.

Similar simulations were done to demonstrate the effects of pH on the solubility of lead sulfate. As can be seen from Figure 14.2, lead sulfate solubilization is also strongly pH dependent.

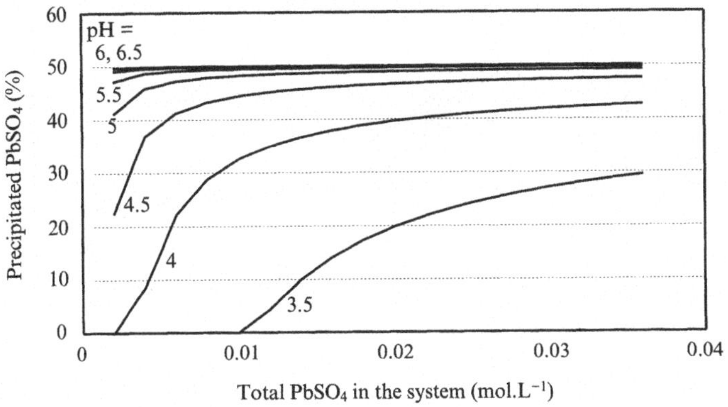

Figure 14.2.

(Wageningen Agricultural University, Department of Soil Chemistry and Plant Nutrition, The Netherlands), using stability and solubility constants from Lindsay (1979). Jarozites are formed at extremely low pH (<2). Jarozite is a basic ferric sulfate with the general formula $X_3Fe(SO_4)_2(OH)_6$, where X=K^+, Na^+, NH_4^+, H_3O^+ (Carlson *et al.* 1992; Karathanasis and Thompson 1995).

As described in Chapters 2 and 3, oxidation of sulfur usually results in aproduction of protons and thus acidification of a system. Increased acidity results in enhanced desorption of cations from solid materials and solubilization of metals precipitated as carbonates or hydroxides. This leads to the leaching of many cationic toxic metals (Förstner 1995).

Moreover, acidity induces a dissolution of mineral matrices, enhanced weathering, solubilization of soil alumo-silicates and leaching of aluminium (Stumm and Furrer 1987). This effect is pronounced at pH<4, as documented by Box 14.2. Leaching of aluminium from acidic soils is a world-recognized environmental problem associated with toxicity to aquatic life, adverse effects on soil biota and plant roots, and exhibiting direct toxicity on animals and humans (van Breemen 1973).

Box 14.2. Release of aluminium from the soil matrix as a function of pH

Figure 14.3 shows the solubilization of aluminium in a soil suspension at varying pH induced by sulfuric acid. Clay, silt and sandy soils were used at a solution/soil ratio of 0.415 kg/L (Tichý *et al.* 1996). Arrow indicates a threshold pH value below which a strong Al solubilization can be expected.

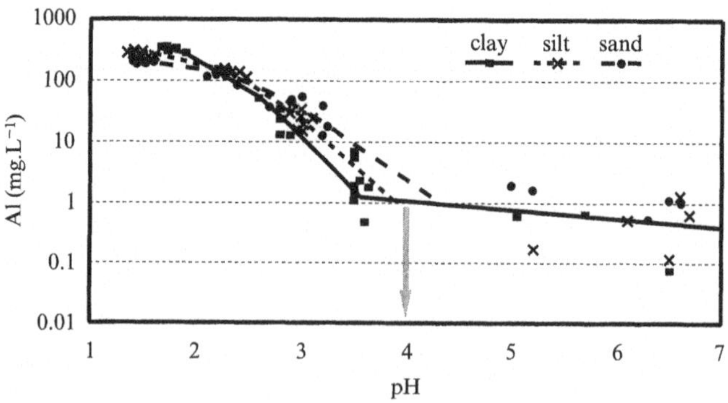

Figure 14.3.

14.1.3 Chemical time bombs

Stocks of reduced inorganic sulfur- and metal-containing compounds, that were built-up under anoxic conditions, are rather stable in oxygen-free conditions. However, when such a system is aerated, oxidative changes occur immediately, which results in a leaching of sulfate, acidity and toxic cationic metals. In popular literature, similar systems are called "chemical time bombs". They pertain in a landscape virtually unnoticed for long time periods without any direct effects on the environment. However, changes of certain chemical conditions, e.g. oxidation-reduction potential, trigger dramatic and unexpected deterioration of environmental quality (Förstner 1995; Salomons 1993; Salomons and Stigliani 1995).

Triggering of the chemical time bomb effect is shown on an example of freshwater sediment in Box 14.3. Here, the wetland sediment has been receiving acid mine drainage water (AMD) for decades, serving as an easy and cheap pollution preventive measure. Therefore, high amounts of sulfide and metals had accumulated. Nowadays, even a slight lowering of the water table results in a massive acidification and leaching of metals. Recently, repeated die-off of the fish population was recorded in the recipient creek and a concrete and steel construction of a watermill downstream was heavily damaged by the corrosion (Tichý *et al.* 1998b).

Box 14.3. Changes of chemical parameters of a sediment exposed to the air (Tichý et al. 1998a)

A wetland sediment loaded with acid mine drainage was exposed to aeration in an aqueous suspension. Figure 14.4 gives the time-courses of pH and oxidation-reduction potential (ORP) in this suspension.

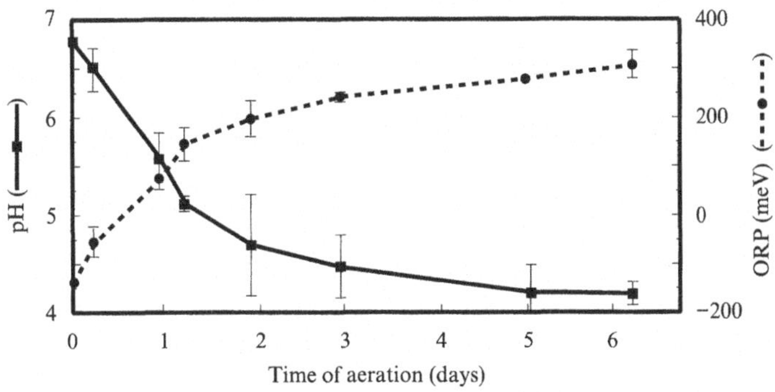

Figure 14.4.

Box 14.3 (*cont.*)

Parallel, sulfate, iron and divalent cationic metals (Cu, Zn) were released. Figure 14.5 shows the time-course of sulfate and iron release. The solubilization of other divalent cationic metals followed a monotonous increase with time (see the example of zinc).

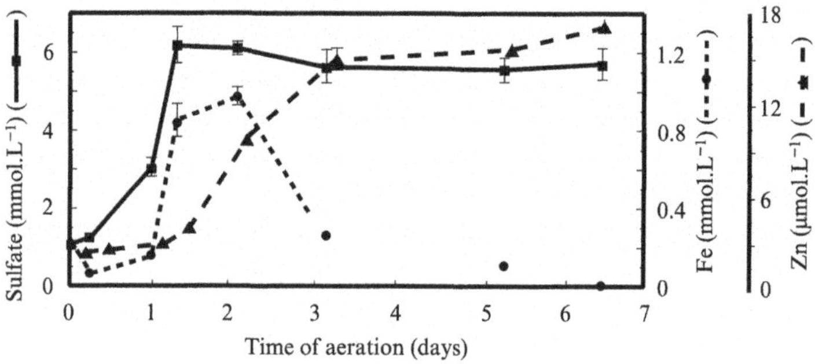

Figure 14.5.

14.2 SULFUR-CONTAINING MATERIALS WITH ENVIRONMENTAL CONCERN

Table 14.3 overviews different materials and systems threatened by the effects of sulfur transformations. The most important materials are described in detail below.

Table 14.3. Solid materials that are significantly affected with sulfur conversions and processes

Material	Examples	Trigger
Aqueous sediment	Freshwater sediments Marine sediments	Flooding Dredging Draught and temporary depression of the water table
Mineral soil containing reduced sulfur species	Spoil banks	Excavation and transport
	Mining waste Abandoned mines	Spontaneous seepage of groundwater
Periodically inundated soil	Acid sulfate soils in tropical regions Periodical swamps and wetlands	Periodical inundation and drying of the soil
Anaerobically digested organic waste	Sewage sludge	Transport and disposal of the waste
	Agricultural waste	Use of the waste as agricultural fertilizer

14.2.1 Sediments

Accumulation of substantial amounts of reduced sulfur is a well-known phenomenon encountered in all types of marine or freshwater sediments (Salomons and Stigliani 1995). When kept under the water level, sediments do not possess any significant risks for the environment. However, they can be exposed to oxygen when the water level decreases, when an incidental flood or storm-water event transports the sediment particles, or when the sediment is excavated (Tichý and Mejstřík 1996; Tichý *et al.* 1998a). The latter variant, i.e. the excavation of sediment, is often reinforced by the nautical requirements on rivers and in river estuaries, where the sedimentation is accelerated by increased salinity of the sea water.

Exposure of these sediments to aerobic conditions results in a sequence of oxidative reactions, leading to the decrease of pH, increase of sulfate levels in the water and the solubilization of most cationic metals (Salomons and Stigliani 1995). An illustrative example of these processes is given in Box 14.4.

14.2.2 Mining waste

Mining is a source of the most abundant waste that contains inorganic sulfurous compounds. Sulfur is here usually from geological origin and may be present in a whole range of different forms. Mining waste is not a homogenous category of material since it includes spoil, low-grade ores, mine-tailings, sedimentation lagoons, or even whole abandoned mining sites. These materials, when exposed to aeration, are oxidized, producing a significant pollution of effluents with acidity, sulfate, iron and other metals (Richards *et al.* 1993; Evangelou and Zhang 1995; Tichý and Mejstřík 1996). The effluents from abandoned mines, acid mine drainage (AMD), possess serious environmental concerns worldwide. AMD is usually produced at very high quantities and for very long times, sometimes decades or centuries. AMD is characterized by a high sulfate content (up to several grams per litre), sometimes extremely low pH (<4), and elevated levels of Fe and some other metals. More on the composition of AMD and its origin can be found in Chapter 8.

The character of AMD depends largely upon the amounts and character of the oxidized minerals, physical properties of the mineral surfaces, presence of electron acceptors in the water, and water flow velocity. The latter determines not only the dilution intensity, but also affects the overall chemical reactions in the system. This is illustrated on the example of pyrite oxidation: the oxidation of pyrite starts firstly with a reaction yielding

Box 14.4. Long-term chemical changes in a riverine sediment after its dredging and piling above the water level

Maass and Miehlich (1988) documented chemical changes in a pile of riverine sediment after its dredging from the river bottom and placement at a disposal site. The time course of pH-changes and leaching of main toxic metals (Cd and Zn) went through four different phases. The first phase was characterised by neutral pH and low concentrations of soluble Cd and Zn. Soon afterwards, oxidative processes started and the pH dropped down, resulting in accelerated metal solubilization. After this phase, the chemical conditions stabilized, reduced sulfur stocks were depleted, and natural buffering mechanisms allowed a significant pH-increase. The authors suggest that the fourth phase would appear in the future, characterised by a gradual pH-decrease and possible further release of metals (associated with long-term biochemical changes in the sediment, e.g. decrease of organic matter content).

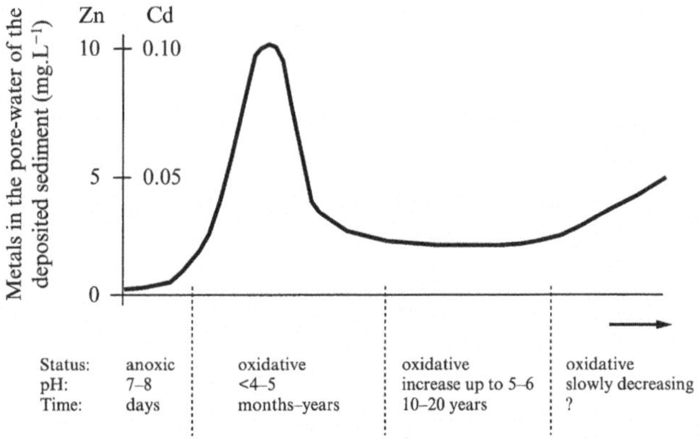

Figure 14.6.

sulfuric acid and ferrous iron (e.g. Equation 8.1). In stagnant conditions of pore-water (i.e. when the water in soil does not move), ferrous iron is oxidized at the same place, resulting in the formation of ferric precipitates and further acidification. However, movement of soil water may induce the leaching of ferrous iron out of the system. Ferrous iron may eventually enter a recipient stream or waterbody (Levy *et al.* 1997). This effect of distant oxidation leads to orange-red (ochre) cores on solid surfaces in recipient waterbodies and distant acidification of the recipient environments.

14.2.3 Acid sulfate soils and periodical swamps

Certain areas experience a periodical inundation, followed by a drying phase, e.g. inundation areas in estuaries of rivers, inundation by high flow of the sea or by some agricultural practices (e.g. rice production). When sufficient sulfate is present in the incoming water (e.g. seawater or brackish water), sulfate reduction can result in a periodic accumulation of reduced sulfur and *in situ* pyrite formation. This phenomenon is pronounced in tropical regions since the temperature substantially increases reaction rates (Bloomfield and Coulter 1973). After the water level drops down, pyrite is exposed to the air. Consequently, immediate acidification is observed *in-situ*. In some cases, effects of distant oxidation and acidification have also been reported, as described in Section 14.2.2.

Acid sulfate soils are sometimes called "cat-clays" and are found in tropical regions all over the world. Similar effects are known in periodical swamps and marshes even in moderate climate (Tichý and Mejstřík 1996; Tichý *et al.* 1998a, b).

14.2.4 Anaerobically digested sewage sludge

Surplus sludge from biological wastewater treatment is very often anaerobically digested. Owing to the high content of organic matter as well as biogenic and mineral sources of sulfur, sulfate reduction occurs during sludge digestion. As a consequence, anaerobically digested sludge usually contains elevated levels of various reduced sulfur compounds. When this material comes to contact with the air, oxidative processes will start. Environmental concerns here are associated primarily with the release of toxic metals, which impose legal limitations on the sludge manipulation and disposal (Couillard and Mercier 1992; Blais *et al.* 1993).

14.3 TREATMENT STRATEGIES

14.3.1 Prevention of sulfur pollution

Prevention of sulfur pollution is the primary measure to reduce risks associated with its accumulation in the environment (Ivanov 1983; Brimblecombe *et al.* 1989). This involves an improved efficiency of flue-gas treatment, desulfurization of fossil fuels, or application of clean technologies in ore processing and smelting (Tichý and Mejstřík 1996). More details about the desulfurization of fossil fuels and raw materials can be found in Chapters 4 and 5.

Systematic management of mines using the whole mine life-cycle can substantially reduce the production of AMD (Orava and Swider 1996; Gray 1997). Similarly, proper and sensitive management and policies in river basins can lead to a significant reduction of pollution, which consequently leads to lower risks associated with handling of riverine sediments (Salomons and Stigliani 1995). Preventive administrative measures are beyond the scope of this chapter.

14.3.2 Elimination of sulfurous compounds from solid materials

Elimination of inorganic sulfurous compounds from solid materials is usually a rather costly procedure. In practice, it is applied with a primary focus on the elimination of toxic metals accompanying the sulfur.

Any potential technique for the removal of sulfur from solid materials should meet the specific requirements imposed by the fact that sulfur is present in the solid state. Principally, the removal of solid contaminants from a solid material is possible using phase or molecular separation techniques (Rulkens *et al.* 1995, 1998).

14.3.2.1 Phase separation

Various techniques of gravity-separation and flotation are extensively applied for the enrichment of sulfur-containing non-ferrous ores (Spottiswood 1982; Richards *et al.* 1993; Rulkens *et al.* 1998). Phase separation is frequently used to treat metal-polluted sediments. To our knowledge, no case of phase separation is reported for the removal of solid sulfurous compounds from waste, soil or sediment.

14.3.2.2 Molecular separation

The molecular separation, i.e. the extraction of molecules containing the sulfur atom, implies the solubilization of the pollutants followed by its removal from the process liquor (Hinsenveld 1993). For the removal of metals, various extractants are used, e.g. mineral (HCl, H_2SO_4) or organic (lactate, formate) acids, complexing agents (EDTA, DTPA, NTA) or pure water. Because reduced inorganic sulfurous compounds are present mostly as solid particles, they need first to be dissolved. As mentioned above (see 14.1.2), inorganic sulfurous compounds are solubilized at oxidative conditions. Therefore, sulfur oxidation is a basic principle applied for the molecular separation of sulfurous compounds and associate toxic metals.

14.3.2.3 Bioleaching as a special case of molecular separation

In some cases, the oxidative process is dominated by microbial activities (growth of thiobacilli); then it is called "bioleaching" (Karavaiko 1985; Bruynesteyn 1989). Bioleaching is applied in practice especially in mining operations, e.g. for biohydrometallurgical mining, low-grade ore enrichment, etc. (Bruynesteyn 1989). The process has also been applied for the treatment of sewage sludge, soils or sediments (Iske and Glombitza 1988; Tyagi *et al.* 1991; Blais *et al.* 1993; Tichý *et al.* 1998a, b).

The aim of bioleaching is to achieve pH values that are low enough to solubilize a maximum of the toxic metals. In some cases, additional elemental sulfur or ferrous iron are added to the system as substrates for bacteria. Bioleaching has been tested in suspension reactors like Pachuca tanks, rolling reactors, or in propeller-agitated vessels (Tyagi *et al.* 1991). A less-intensive alternative for the removal of sulfur and associated pollution from highly voluminous wastes is its treatment on heaps at the place of permanent or temporary disposal. An enhancement of the heap leaching processes, e.g. by forced aeration or addition of hydrophilic sulfurous substrates, may be good alternatives to the expensive slurry leaching processes (Tichý *et al.* 1998a, b).

One of the few environmental applications of bioleaching has been used for the decontamination of anaerobically pre-treated sewage sludge and agricultural waste (Tyagi *et al.* 1991; Couillard and Mercier 1994). The application of anaerobic digestion for sludge containing heavy metals leads to qualitative changes in metal binding to the waste. Various bonds of metals to carboxylic or phenolic groups of organic matter, strong sorption to the cell walls, encapsulation within the vacuoles or other cellular components, chelation with complex organic molecules, and many other types of binding are replaced by sulfidic precipitates. Moreover, the structure of aerobically produced sludge is altered by the digestion, so that some forms of toxic metals can be more easily desorbed from the matrix. When the anaerobically digested sludge is aerated, the metals are solubilized in quantities that are never achieved without its anaerobic pre-treatment (Couillard and Mercier 1994).

This principle is at this moment one of few feasible alternatives to handle metal polluted sewage sludge and similar types of organic waste (Tyagi *et al.* 1993). An example of bioleaching of metals from sewage sludge is given in Box 14.5. Here, ferrous iron is used as a substrate for bacteria. Oxidation of ferrous iron leads to an acidification of the suspension, which results in solubilization of metals. Sulfides are oxidized and get into the process liquor as well.

Box 14.5. An example of bioleaching technology applied for the removal of sulfur and heavy metals from polluted anaerobically pre-treated sewage sludge (Couillard and Mercier 1994)

Anaerobically digested sludge is pre-acidified and pre-aerated in a mixing vessel. Afterwards, intensive aeration and agitation is performed either in a completely stirred or airlift reactor. After the solid/liquid separation, the cleaned sludge is disposed and wasted, whereas the excess process water undergoes neutralization and removal of pollutants.

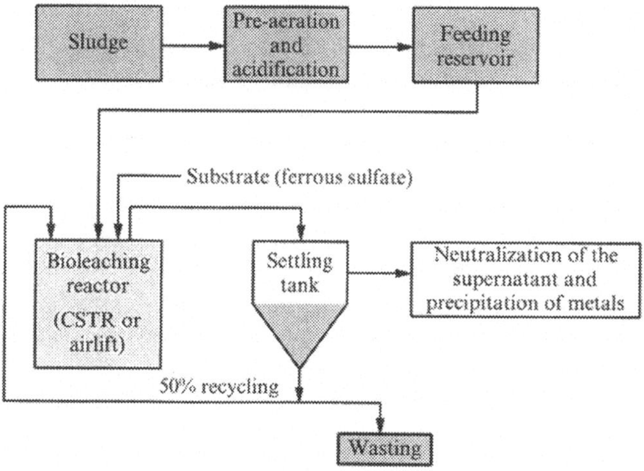

Figure 14.7.

14.3.3 Suppression of the sulfur oxidation

The other way to cope with the sulfur and associated pollution is to prevent its release into the environment. The text below summarizes the main strategies to prevent sulfur oxidation used in practice or proposed as feasible methods.

14.3.3.1 Inundation

For many systems, inundation can be a cost-effective and easy option. When capped with water, oxygen cannot penetrate to the sulfur compounds and the oxidation cannot start, as reported e.g. by Schuring *et al.* (1997) for sulfur-rich hard-coal tailings. However, this method can not always guarantee the full suppression of leaching, as bacterial leaching can proceed even in oxygen-free conditions. Moreover, spontaneous chemical solubilization can continue owing to the oxidative capacity of ferric iron (Karavaiko 1985; van Breemen 1988; Bruynesteyn 1989).

14.3.3.2 Pumping of seawater

Acidophilic thiobacilli are inhibited by sodium chloride at concentrations comparable to sea water (Cameron *et al.* 1984). Therefore, infiltration or pumping of sea water into the mine could inhibit microbial sulfide oxidation. On the other hand, some heavy-metal ions, like cadmium, form stable chloride complexes. Therefore, chloride addition might considerably enhance the leaching process instead of preventing it (Salomons 1993).

14.3.3.3 Hydrogeological isolation

Another strategy to suppress leaching is the hydrogeological isolation of the site. For this, isolation membranes or a mixture of clay and soil can be applied for preventing the vertical or horizontal water movements (Bell *et al.* 1995). Proper treatment and restoration of the surface can lead to a reduction of polluted outflows from mine tailings or spoil banks (Richards *et al.* 1993). Covering the deposit with soil and planting fast-growing trees is a generally applied method for the rehabilitation of large areas devastated by mining as well as for the covering of voluminous heaps of spoil worldwide (Tichý and Mejstřík 1996).

14.3.3.4 Coating and passivation

Leaching of sulfurous compounds can also be accomplished by using specific additives which alter the physical properties of the surface of solid inorganic sulfurous compounds. Addition of fly-ash and/or cement binding mixtures leads to an agglomeration of particles and a physical encapsulation of pyrite (Misra *et al.* 1996). Belzile *et al.* (1997) reported on a possible passivation of the pyrite surface with the use of acetyl acetone, humic acids, ammonium lignosulfonates, oxalic acid or sodium silicate. The efficiency of this coating was enhanced when the pyrite was exposed to a pre-oxidation, leading to the formation of a ferric oxide layer on its surface. Evangelou (1995) studied the possible micro-encapsulation of pyrite by a ferric phosphate layer. The coating involved a reaction mixture of H_2O_2 and phosphate. H_2O_2 oxidizes pyrite and the released Fe^{3+} precipitates as a ferric phosphate on the surface, thus preventing further pyrite oxidation. The coating was successful with solutions containing concentrations as low as 10^{-4} mol/L phosphate and 0.03 mol/L H_2O_2. Similar results were reported by Georgopoulou *et al.* (1996) and Nyavor *et al.* (1996).

14.3.3.5 *Use of surfactants*

Addition of surface active compounds (surfactants) can prevent the attachment of bacterial cells to the sulfur containing minerals. Thomson and Turney (1996) described the possible treatment of pyrite using various fatty-acid amines, resulting in a substantial reduction of the bioleaching process. Sasaki (1997) demonstrated a possible use of oxalic, tannic and fulvic acids to prevent bacterial attachment with the best results achieved with fulvic compounds. Takeuchi and Suzuki (1997) tested cell adhesion to pyrite in the presence of potassium phosphate. Their study showed a combined effect, since potassium phosphate both inhibited the adhesion of bacteria to pyrite and kept pH at values unfavourable for bacterial growth and activity. Addition of 20 mg/L of the tenzide sodium paraffin sulfonate (E30) to the percolate of a pilot ore dump successfully inhibited the growth and activity of thiobacilli (Ondrushka and Glombitza 1993). When the use of tenzides is considered, an estimation of the specific surface area of the sulfidic mineral is needed, because the applied tenzide concentrations should ensure full coverage of the mineral surface (Ondruschka and Glombitza 1993).

14.3.3.6 *Liming*

Adding lime to a system containing inorganic sulfurous compounds is another option to suppress the leaching. It leads to the neutralization of the acidic environment. Neutral pH is unfavourable for bioleaching (Evangelou 1995). Moreover, neutralization of acidic conditions suppresses leaching of toxic cationic metals. However, it is usually only a temporary measure, since the lime is continuously leached out from the system, and it is also rather expensive. Georgopoulou *et al.* (1996) even suggests that liming can be more expensive than the coating with ferric phosphate.

14.3.3.7 *Elimination of thiobacilli*

Thiobacilli can be inactivated by specific biocides. Suppression of leaching was demonstrated using the bactericides Fluorspar and Kathon (Ondruschka and Glombitza 1993). Fluorspar contains fluorides, which inhibit growth of *Thiobacillus* sp. at concentrations of minimum 400 mg F^-/L. The isothiazolone Kathon inhibits the activity of thiobacilli at concentrations of 300 mg/L. Application of both compounds to a pilot ore heap prevented leaching and no viable thiobacilli were found in the percolate (Ondruschka and Glombitza 1993). However, the use of biocides cannot be recommended for full-scale applications because of the large amount of biocides required (and associated costs) and the risk of adaptation of the endogenous bacterial population.

An inhibition of growth and activity of thiobacilli fed with pyrite was also demonstrated with activated carbon (Loi *et al.* 1993). This effect is likely caused by sequestering most of the cells from the suspension. Subsequently, microbes could not attach to the pyrite surface, and the oxidation could not initiate.

Padival *et al.* (1995) investigated the possible control of *Thiobacillus* sp. by means of microbial competition. This strategy uses excessive concentrations of nutrients to favour microbes that grow faster than thiobacilli. Experiments done with a mixture of thiobacilli and yeasts fed with glucose and thiosulfate have shown a substantial reduction of the activity of thiobacilli, and a significant reduction of the effluent sulfate concentrations. This method can be applied for prevention of corrosion of technological equipment. However, its large-scale application for treatment of overburdens or abandoned mines is not realistic.

14.3.4 Treatment of acid sulfate soils

In principle, the measures to decrease acidification of acid sulfate soils are similar to those described above (see 14.3.3). In practice, four basic strategies are applied, i.e. curtailing the pyrite oxidation, leaching of the reaction products, liming and self-neutralization at reductive conditions.

14.3.4.1 Curtailing the pyrite oxidation

As described in section 14.3.3, pyrite oxidation can be suppressed by limiting the supply of oxygen or by influencing the rate of one or more of the intermediate steps in oxidation, e.g. by decreasing the concentration of Fe^{3+}. However, in practical conditions, the only feasible prevention of oxygen entrance to the system is its permanent inundation (van Breemen 1975).

14.3.4.2 Leaching of reaction products

Another treatment strategy is to remove soluble or exchangeable acidity from the soil. Leaching with fresh water is efficient to remove free sulfuric acid and partly also the Fe and Al ions that had been solubilized by acidity. Leaching with salt or brackish water results also in the replacement of exchangeable Al by Na, Ca and Mg (van Mensvoort *et al.* 1991). The leaching is, however, inefficient when reducing conditions prevail in the soil profile. Therefore, efficient leaching can only be done with fully aerated soil, preferably a soil exposed to an intensive drying period before the leaching (van Mensvoort *et al.* 1991).

14.3.4.3 Liming of acid sulfate soils

Addition of lime or limestone can be applied to neutralize the acidity in acid sulfate soil. Typically, a dose of about 30 tons of $CaCO_3$ per 10000 m^2 and per 10 cm of soil is needed (van Breemen 1975). The liming is usually efficient only after the water-soluble acidity has been leached out. Otherwise, the huge amounts of lime required would make the process economically unfeasible. Excessive liming would also result in a strong salinization of the site that may become unfavourable for vegetation and wildlife.

14.3.4.4. Self-neutralizing in reductive conditions

Initiation of microbial sulfate reduction in the acid sulfate soils can efficiently remove acidity and stabilize the sulfate concentration. This makes the cultivation of e.g. rice on acid sulfate soils a desirable ameliorative measure, because it combines both the effect of inundation and stimulation of sulfate reductive processes. However, although the toxic effects of acidity and cationic metals (especially aluminium) are suppressed at anoxic conditions, they are replaced by the toxic effects of reduced Fe^{2+}, which is soluble in water. Park *et al.* (1971) showed, however, that Fe^{2+}-toxicity can be avoided by sulfate reduction. The produced sulfide reacts with free Fe^{2+} and forms pyrite.

14.3.5 Treatment of acid mine drainage

Mining waste, spoil or abandoned mines are very often hardly controllable and the production of mine drainage water cannot be avoided. In these cases, proper treatment of this water is necessary. The treatment aims at the removal of (1) acidity, (2) toxic metals, (3) sulfate, and (4) hardness of the water.

Numerous technologies were developed and are in use to treat AMD. However, many of them do not address all four above-mentioned pollutant groups in AMD. The two most important treatment principles in use are microbial sulfate reduction and the chemical neutralization.

14.3.5.1 Treating AMD by microbial sulfate reduction

Principles of using microbial sulfate reduction for AMD treatment is discussed in detail in Chapter 8. Sulfate reduction can be done in various technological configurations, e.g. suspension tanks (van Houten *et al.* 1994), anaerobic ponds (Rose *et al.* 1996), wetlands (Wildeman and Laudon 1989; Eger 1994; Tichý and Mejstřík 1996), or in a permeable reactive wall (Benner *et al.* 1997). In many installations, organic substrate is added to the system to increase the reaction rates. Application of sulfate reduction leads to the most complete treatment of AMD: it neutralizes acidity, removes toxic metals (by precipitation as metal sulfides), removes part of the sulfate from the water and via sulfides precipitation decreases the salts content in AMD (Eger 1994).

Among the outputs from sulfate reduction processing of AMD is a sludge containing toxic metals and sulfide. In suspension reactors, this sludge can be (relatively) easily removed from the system and treated as a hazardous material or recycled. However, in more extensive process configurations, e.g. wetland, this sludge remains at a treatment site. As a result, these sites accumulate high levels of reduced sulfur and toxic metals and therefore will become typical examples of chemical time bombs as described above. Owing to the pollution, such sites will require proper treatment and other pollution control measures (Gambrell 1994). In some wetlands, proper sanitation and regeneration of the site may far exceed the costs of AMD treatment (Gaydardjiev *et al.* 1996; Tichý *et al.* 1998b). Unfortunately, these items are often omitted from the cost breakdown, which falsely makes these types of AMD-treatment very economic and attractive (Kadlec 1995).

14.3.5.2 Chemical neutralization

The second major treatment principle for AMD is neutralization. Application of alkali results in neutralization of acidity and in precipitation of toxic cations. Neutralization is most typically performed using lime. The process is usually set up in a horizontally percolated series of drains or trenches filled with the limestone (Ziemkiewicz *et al.* 1997). A considerable negative effect of the lime application is usually an increasing hardness of the water.

Application of lime can result in a retention of sulfate as well. Sulfuric acid reacts with lime ($CaCO_3$), resulting in the formation of gypsum ($CaSO_4.2H_2O$). However, formation of gypsum depends on pH. At pH higher than 7 (which is typical for a system saturated with lime) and the presence of atmospheric CO_2, gypsum formation is thermodynamically unfavourable (see Box 14.1). This is overcome when limestone layers are isolated from the atmosphere (e.g. as anoxic lime drains, see Chapter 8, section 8.3.1.). The lack of free CO_2 in this system enables formation of gypsum even at high pH values. Moreover, the use of anoxic lime drains benefits from the prevalent existence of iron in its ferrous form. In this way, an undesired coating of lime particles with ferric precipitates (= armouring) is prevented.

14.3.5.3 Other principles of AMD treatment

Other AMD treatment principles reported in the literature are, e.g., possible electrochemical recovery of metals (Hatfield *et al.* 1996; Shelp *et al.* 1995, 1996), precipitation of toxic metals by apatite (Chen *et al.* 1997) or as phosphates produced by specific microbes (Roig *et al.* 1997) and removal of iron by induced ferrite production and subsequent magnetic separation. However, only limited practical experience with these methods is available and therefore, their feasibility and economy are unclear.

14.4 SULFUR IN SOLID MATERIALS: NOVEL ASPECTS OF ENVIRONMENTAL TECHNOLOGY

The preceding text demonstrated that sulfur in solid materials has numerous effects of environmental importance. However, the amount of sulfurous compounds present is usually not the key aspect of environmental quality or deterioration. Rather, the side-effects of the sulfur cycling are of concern, especially the coupling of sulfur transformations with acidification / alkalization of the environment and with the mobilization / retention of cationic toxic metals.

The negative side-effects like the appearance of chemical time bombs, self-acidifying soils, or AMD are evident. The treatment of sulfur pollution strongly depends on the extent of the pollution. Most of the techniques are intensive in respect to the use of chemicals, energy or equipment. However, they can only be applied for concentrated pollution in relatively small volumes of treated material (= point pollution). Sulfur is, however, mostly present in the form of diffuse pollution or as a pollution affecting large areas or volumes of the waste (AMD, spoil banks, mine tailings, etc.). Here, less intensive approaches should be preferred. Extensive treatment techniques require lower inputs of chemicals or energy. However, the treatment times are appropriately prolonged. Examples of such extensive measures can be the use of natural systems, preventive measures, and whole life management of mining sites.

Transformations of sulfur in soils, sediments or solid waste can also be exploited in a positive way. Coupling of the sulfur cycle with the mobility of cationic toxic metals can be employed to control heavy metal pollution. Techniques like sulfate reduction or bioleaching have been demonstrated for pollution control of highly polluted waste streams. Sulfur cycling can also offer numerous chances to control diffuse pollution of the whole landscapes. Amendment of sulfur at polluted sites may lead to metal mobilization, e.g. by desorption from the soil, removal from sludge, etc. and its deposition in other compartments of the environment. Application of wetlands downstream may be a next step of this landscape management: mobilized metals with sulfate will be entrapped in the wetland sediment and accumulated there to high concentrations. After the metal concentrations are built-up in the system, the sediment can be excavated and treated e.g. by bioleaching. In such a way, originally diffuse pollution can be converted into a concentrated and narrowly localized site, which can be periodically treated by intensive treatment systems. In the above described way, the whole microbial sulfur cycle may be employed as a tool to control heavy metal pollution in the landscape or in specific materials (see Box 14.6).

Box 14.6. Possible use of microbial sulfur cycle to control the heavy metal pollution (Tichý et al. 1998a)

Figure 14.8 shows a combination of bioleaching followed by a soil/solution separation, microbial sulfate reduction producing sulfide, neutralization of acidity and precipitation of metals, and a partial microbial sulfide oxidation leading to the formation of elemental sulfur, which can be reused.

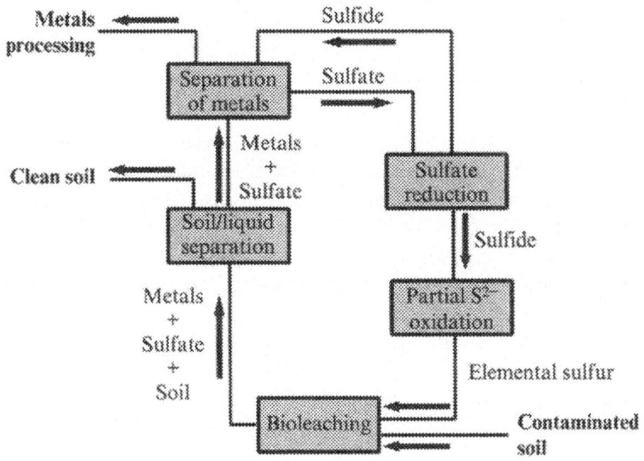

Figure 14.8.

14.5 REFERENCES

Bell, A. V., Riley, M. D., and Yanful, E. K. (1995) Evaluation of a composite soil cover to control acid waste rock pile drainage. *CIM Bull.* **88**(995), 41-46.

Belzile, N., Maki, S., Chen, Y. W., and Goldsack, D. (1997) Inhibition of pyrite oxidation by surface-treatment. *Sci. Tot. Environ.* **196**(2), 177-186.

Benner, S. G., Blowes, D. W., and Ptacek, C. J. (1997) Sulfate reduction in a permeable reactive wall for prevention of acid-mine drainage. *Abst. Papers Am. Chem. Soc.* **213**, 45-46.

Blais, J. F., Tyagi, R. D. and Auclair, J. C. (1993) Bioleaching of metals from sewage sludge: microorganisms and growth kinetics. *Wat. Res.* **27**, 101-110.

Bloomfield, C., and Coulter, J. K. (1973) Genesis and management of acid sulfate soils. *Adv. Agron.* **25**, 265-326.

Brimblecombe, P., Hammer, C., Rodhe, H., Ryaboshapko, A. and Boutron, C.F. (1989) Human influence on the sulfur cycle. In: Brimblecombe, P., Yu. Lein, A. (eds.): Evolution of the global biogeochemical cycle. SCOPE 39, John Wiley & Sons, Chichester, pp. 77-123.

Bruynesteyn, A. (1989) Mineral biotechnology. *J. Biotechnol.* **11**, 1-10.

Bryant, R. D., Costerton, J. W. and Laishley, E. J. (1983) The role of *Thiobacillus albertis* glycocalyx in the adhesion of cells to elemental sulfur. *Can. J. Microbiol.* **30**, 81-90.

Cameron, F., Jones, M. V. and Edwards, C. (1984) Effects of salinity on bacterial iron oxidation. *Current Microbiol.* **10**, 353-356.

Carlson, L., Lindstrom, E. B., Hallberg, K. B. and Tuovinen, O. H. (1992) Solid-phase products of bacterial oxidation of arsenical pyrite. *Appl. Environ. Microbiol.* **58**, 1046-1049.

Chen, X. B., Wright, J. V., Conca, J. L., and Peurrung, L. M. (1997) Evaluation of heavy-metal remediation using mineral apatite. *Wat. Air Soil Pollut.* **98**(1-2), 57-78.

Cook, T.M. (1964) Growth of *Thiobacillus thiooxidans* in shaken culture. *J. Bacteriol.* **88**, 620-623.

Couillard, D. and Mercier, G. (1992) Metallurgical residue for solubilization of metals from sewage sludge. *J. Environ. Engn.* **118**, 808-813.

Couillard, D. and Mercier, G. (1994) An economic evaluation of biological removal of heavy metals from wastewater sludge. *Wat. Environ. Res.* **66**, 32-39.

Devasia, P., Natarajan, K.A., Sathyanarayana, D.N. and Rao, G.R. (1993) Surface chemistry of *thiobacillus ferrooxidans* relevant to adhesion on mineral surfaces. *Appl. Environ. Microbiol.* **59**, 4051-4055.

Eger, P. (1994) Wetland treatment for trace metal removal from mine drainage: the importance of aerobic and anaerobic processes. *Wat. Sci. Technol.* **29**(4), 249-256.

Ellwood D.C., Hill M.J. and Watson J.H.P. (1992) Pollution control using microorganisms and magnetic separation. In: Fry J.C., Gadd G.M., Herbert R.A., Jones C.W., Watson-Craik I.A.: Microbial control of pollution. Proc. 48th Symp. Society for General Microbiology, Cardiff, U.K., Cambridge University Press, pp. 89-112.

Evangelou, V. P. (1995) Potential microencapsulation of pyrite by artificial inducement of ferric phosphate coatings. *J. Environ. Qual.* **24**(3), 535-542.

Evangelou, V.P. and Zhang, Y.L. (1995) Pyrite oxidation mechanisms and acid mine drainage prevention. *Crit. Rev. Env. Sci. Tech.* **25**, 141-199.

Evans, L.J. (1989) Chemistry of metal retention by soils. *Environ. Sci. Tech.* **23**, 1046-1056.

Förstner, U. (1995) Non-linear release of metals from aquatic sediments. In: W. Salomons, W.M. Stigliani (eds.) *Biogeodynamics of pollutants in soils and sediments*. Springer Verlag, Berlin, pp. 247-308.

Gambrell, R. P. (1994) Trace and Toxic Metals in Wetlands - A Review. *J. Environ. Qual.* **23**, 883-891.

Gaydardjiev S., M. Hadjichristova M., Tichý R. (1996): Opportunities for using two low-cost methods for treatment of metal bearing aqueous streams. *Minerals Engineering* **9**(9), pp. 947-964.

Georgopoulou, Z. J., Fytas, K., Soto, H., and Evangelou, B. (1996) Feasibility and cost of creating an iron-phosphate coating on pyrrhotite to prevent oxidation. *Environ. Geol.* **28**(2), 61-69.

Gray, N. F. (1997) Environmental impact and remediation of acid-mine drainage - a management problem. *Environ. Geol.* **30**(1-2), 62-71.

Hatfield, T. L., Kleven, T. L., and Pierce, D. T. (1996) Electrochemical remediation of metal-bearing wastewaters .1. Copper removal from simulated mine drainage waters. *J. Appl. Electrochem.* **26**(6), 567-574.

Hinsenveld, M. (1993) Physical/chemical extraction technologies. In: Demonstration of remedial action technologies for contaminated land and groundwater. Edited by Olfenbuttel, R.F. NATO/CCMS, Brussel.

Iske, U. and Glombitza, F. (1988) Moglichkeiten der nutzung biotechnologischer verfahren fur die gewinnung von metallen aus erzen, mineralen und abprodukten der chemischen industrie (Possible use of biotechnology for the extraction of metals from ores, minerals, and chemical waste) *Chem. Techn.* **40**(11), 464-468.

Ivanov, M.V. (1983) Major fluxes of the global biogeochemical cycle of sulfur. In: Ivanov, M.V., Freney, J.R.: The global biogeochemical sulfur cycle. SCOPE 19, John Wiley & Sons, Chichester, pp. 449-463.

Kadlec, R. H. (1995) Overview - Surface flow constructed wetlands. *Wat. Sci. Technol.* **32**(3), 1-12.

Karathanasis, A. D. and Thompson, Y. L. (1995) Mineralogy of iron precipitates in a constructed acid mine drainage wetland. *Soil Sci. Soc.* **59**, 1773-1781.

Karavaiko, G.I. (1985) Microbiological processes for the leaching of metals from ores. Centre of International Projects GKNT, Moscow.

Levy, D. B., Custis, K. H., Casey, W. H., and Rock, P. A. (1997) The aqueous geochemistry of the abandoned Spenceville copper pit, Nevada county, California. *J. Environ. Qual.* **26**(1), 233-243.

Lindsay, W.L. (1979) Chemical equilibria in soils. John Wiley & Sons, New York.

Loi, G., Mura, A., Passarini, N., Trois, P. and Rossi, G. (1993) Modification of oxidizing activity of *Thiobacillus ferroxidans* by some particulate solids. *Fuel* **72**, 1607-1611.

Maass, B. and Miehlich, G. (1988) Die Wirrkung des Redoxpotentials auf die Zusammnesetzung der Porenlösung in Hafenschlicks-feldern. *Mitt. Dtsch. Bodekunde Ges.* **56**, 289-294.

Mackintosh, M.E. (1978) Nitrogen fixation by *Thiobacillus ferrooxidans*. *J. Gen.*

Microbiol. **105**, 215-218.

Mirajkar, Y. R. K., Natarajan, K. A., and Somasundaran, P. (1997) Growth and attachment of thiobacillus-ferrooxidans during sulfide mineral leaching. *Int. J. Mineral Proces.* **50**(3), 203-210.

Misra, M., Yang, K., and Mehta, R. K. (1996) Application of fly-ash in the agglomeration of reactive mine tailings. *J. Haz. Mater.* **51**(1-3), 181-192.

Nyavor, K., Egiebor, N. O., and Fedorak, P. M. (1996) Suppression of microbial pyrite oxidation by fatty-acid amine treatment. *Sci. Tot. Environ.* **182**(1-3), 75-83.

Ondrushka, J. and Glombitza, F. (1993) Inhibition of natural microbiological leaching processes. In: Arendt F., Annokée G.J., Bosman R. and van der Brink W.J. (eds.) : Proc. Int. Conf. Contaminated Soil '93 May 1993, Berlin., Kluwer Academic Publishers, Dordrecht, pp. 1195-1196.

Orava, D. A., and Swider, R. C. (1996) Inhibiting acid-mine drainage throughout the mine life-cycle. *CIM Bulletin* **89**(999), 52-56.

Padival N.A., Weiss J.S., Arnold R.G. (1995) Control of *Thiobacillus* by means of microbial competition: implications for corrosion of concrete sewers. *Wat. Environ. Res.* **67**, 201-205.

Park, N. J., Park, J. S., and Kim, Y. S. (1971) Effect of lime on growth of rice and changes in pH, Eh, Fe2+ and Al in an acid sulfate soil. *J. Korean Soc. Soil Science and Fertilizer.* **4**, 167-175.

Porro, S., Ramirez, S., Reche, C., Curutchet, G., Alonsoromanowski, S., and Donati, E. (1997) Bacterial attachment - its role in bioleaching processes. *Process Biochem.* **32**(7), 573-578.

Pronk, J.T., Meulenberg, R., Hazeu, W., Bos, P. and Kuenen, J.G. (1990) Oxidation of reduced inorganic sulfur compounds by acidophilic *thiobacilli*. *FEMS Microbiol. Rev.* **75**, 293-306.

Pronk, J.T., de Bruyn, J.C., Bos, P. and Kuenen, J.G. (1992) Anaerobic growth of *Thiobacillus ferrooxidans*. *Appl. Environ. Microbiol.* **58**, 2227-2230.

Richards I.G., Palmer J.P. and Barratt P.A. (1993) The reclamation of former coal mines and steelworks. Studies in Environmental Science 56, Elsevier, Amsterdam, 720 pp.

Roig, M. G., Manzano, T., and Diaz, M. (1997) Biochemical process for the removal of uranium from acid-mine drainages. *Wat. Res.* **31**(8), 2073-2083.

Rose P.D., Maart B..A., Dunn K.M., Roswell R.A., Britz P. (1996) High rate algal oxidation ponding for the treatment of tannery effluents. *Wat. Sci. Technol.* **33**, 219-227.

Rulkens, W.H., Tichý, R. and Grotenhuis, J.T.C. (1995) Sites polluted with heavy metals: Current techniques for clean-up and desirable future developments. In: Heavy metals in the environment. Proc. Int. Conf., Hamburg, September 1995. Edited by Wilken, R.-D., Förstner, U. and Knöchel, A. CEP Consultants Ltd., Edinburgh, pp. 11-19.

Rulkens W.H., Tichý R., Grotenhuis J.T.C. (1998) Remediation of polluted soil and sediment: Perspectives and failures. *Wat. Sci. Technol.* **37**(8), 27-36.

Salomons, W.H. (1993) Non-linear and delayed responses of toxic chemicals in the environment. In: Arendt F., Annokée G.J., Bosman R. and van der Brink W.J. (eds.) : Proc. Int. Conf. Contaminated Soil '93 May 1993, Berlin, Kluwer Academic Publishers, Dordrecht, pp. 225-238.

Salomons W., Stigliani W.M. (1995) Biogeodynamics of pollutants in soils and sediments. Springer Verlag, Berlin.

Sasaki, K. (1997) A study on the dissolution of pyrite and its suppression in acidic environments. *Bunseki Kagaku* **46**(7), 611-612.

Schaeffer, W. I., and Umbreit, W. W. (1963) Phosphotidylinositol as a wetting agent in sulfur oxidation by Thiobacillus thiooxidans. *J. Bacteriol.* **85**, 492-493.

Schuring, J., Kolling, M., and Schulz, H. D. (1997) The potential formation of acid-mine drainage in pyrite-bearing hard-coal tailings under water-saturated conditions - an experimental approach. *Environ. Geol.* **31**(1-2), 59-65.

Shelp, G. S., Chesworth, W., and Spiers, G. (1995) The amelioration of acid-mine drainage by an in-situ electrochemical method. 1. Employing scrap iron as the sacrificial anode. *Appl. Geochem.* **10**(6), 705-713.

Shelp, G. S., Chesworth, W., and Spiers, G. (1996) The amelioration of acid-mine drainage by an in-situ electrochemical method. 2. Employing aluminum and zinc as sacrificial anodes. *Appl. Geochem.* **11**(3), 425-432.

Smith J.R., Luthy R.G. and Middleton A.C. (1988) Microbial ferrous iron oxidation in acidic solution. *J. Water Pollut. Control Fed.* **60**, 518-530.

Solari, J.A., Huerta, G., Escobar, B., Vargas, T., Badilla-Ohlbaum, R. and Rubio, J. (1992) Interfacial phenomena affecting the adhesion of *Thiobacillus ferrooxidans* to sulfide mineral surfaces. *Colloids and Surfaces* **69**, 159-166.

Spottiswood, E.G. (1982) Introduction to mineral processing. J. Wiley & Sons Inc., New York.

Stumm, W. and Furrer, G. (1987) The dissolution of oxides and aluminum silicates; Examples of surface-coordination-controlled kinetics. In: *Aquatic Surface Chemistry*, Stumm W. (ed.), John Wiley & Sons, New York, pp. 197-219.

Stumm, W. and Morgan, J.J. 1981. Aquatic chemistry. An introduction emphasizing chemical equilibria in natural waters. John Wiley & Sons, Inc., New York.

Takeuchi, T. L., and Suzuki, I. (1997) Cell hydrophobicity and sulfur adhesion of *Thiobacillus thiooxidans*. *Appl. Environ. Microbiol.* **63**(5), 2058-2061.

Thomson, B. M., and Turney, W. R. (1996) Minerals and mine drainage. *Wat. Environ. Res.* **68**(4), 542-546.

Tichý, R. and Mejstřík, V. (1996) Heavy metal contamination from open-pit coal-mining in Europe's Black Triangle and possible remediation. *Environ. Reviews* **4**, 321-341.

Tichý R., J.T.C. Grotenhuis, W.H. Rulkens, V. Nýdl (1996): Strategy for leaching of zinc from artificially contaminated soil. *Environ. Technol.* **17**, 1181-1182.

Tichý R., P. Lens, J.T.C. Grotenhuis, P. Bos (1998a) Solid-state reduced sulfur compounds: environmental aspects and bio-remediation. *Crit. Rev. Environ. Sci. Technol.* **28**(1), 1-40.

Tichý R., W.H. Rulkens, J.T.C. Grotenhuis, V. Nýdl, C. Cuypers, J. Fajtl (1998b) Bioleaching of metals from soils or sediments. *Wat. Sci. Technol.* **37**(8), 119-128.

Tyagi, R.D., Couillard, D. and Tran, F.T. (1991) Comparative study of bacterial leaching of metals from sewage sludge in continuous stirred tank and air-lift reactors. *Process Biochem.* **26**, 47-54.

Tyagi, R.D., Sreekrishnan, T.R., Campbell, P.G.C. and Blais, J.F. (1993) Kinetics of heavy metal bioleaching from sewage sludge. 2. Mathematical model. *Wat.*

Res. **27**, 1653-1661.

van Breemen, N. (1973) Dissolved aluminium in acid sulfate soils and in mine waters. *Soil Sci. Soc. Am. Proc.* **37**, 694-697.

van Breemen, N. (1975) Acidification and deacidification of coastal plain soils as a results of periodic flooding. *Soil Sci Soc. Amer. Proc.* **39**, 1153-1157.

van Breemen, N. (1988) Effects of seasonal redox processes involving iron on the chemistry of periodically reduced soils. *Iron in Soils and Clay Minerals*, D Reidel Publ Co, Dordrecht, 797-809.

van Houten R.T., Hulshoff Pol L.W. and Lettinga G. (1994) Biological sulfate reduction using gas-lift reactors fed with hydrogen and carbon dioxide as energy and carbon source. *Biotech. Bioeng.* **44**, 586-594.

van Mensvoort, M. E. F., Duang, V. N., and Schans, Van Der (1991) Improvement of acid sulphate soils by leaching with salt or brackish water. P.Deturck and F.N. Ponnamperuna (editors) Inst.Fundamental Studies, Kandy, 219-224.

Wildeman, T.R. and Laudon, L.S. (1989) Use of wetlands for treatment of environmental problems in mining: non-coal-mining applications. In: Hammer, D.A. (ed.), Constructed wetlands for wastewater treatment. Municipal, industrial and agricultural. Lewis Publishers, Chelsea, Michigan, pp. 221-231.

Ziemkiewicz, P. F., Skousen, J. G., Brant, D. L., Sterner, P. L. and Lovett, R. J. (1997) Acid-mine drainage treatment with armored limestone in open limestone channels. *J. Environ. Qual.* **26**(4), 1017-1024.

15

Agricultural aspects of sulfur

Wilfried H.O. Ernst

15.1 INTRODUCTION

Sulfur is a major nutrient which ranks fifth or sixth in quantity of macronutrients taken up by plants, comparable to or surpassing the demand for phosphorus. The attention paid to its impact on the performance of higher plants, however, is low compared with the enormous emphasis given to carbon dioxide, nitrogen and phosphorus (Cram 1990). Plants as part of the biosphere can only exist by their capability to integrate sources from the three abiotic spheres, i.e. the atmosphere, the hydrosphere and the pedosphere. Sulfur together with carbon and nitrogen can be derived from all three spheres owing to their different valence states, i.e. chemical speciation. Sulfur can be taken up by plant leaves from the atmosphere as very reduced (COS, CS_2 and H_2S) up to highly oxidized (SO_2) compounds (De Kok *et al.* 1997; Godzik and Krupa 1982; Protoschill-Krebs and Kesselmeier 1992). However, most of the sulfur is taken up by plant roots as water-soluble sulfate (Scheme 15.1).

Before playing its role in plant metabolism, sulfate has to be activated by ATP sulfurylase, transferred by adenosine 5′-phosphosulfate (APS) sulfotransferase or adenosine 3′-phosphate 5′-phosphosulfate (PAPS) to an S-S-carrier and finally reduced by thiosulfate reductase or sulfite reductase (Giovanelli 1990) for the biosynthesis of the amino acid cysteine as a central metabolite in a plant's sulfur metabolism. Methionine, the other S-containing essential amino acid, is formed from cysteine. Many S-containing peptides and proteins, and secondary S-based metabolites, are synthesized from cysteine and methionine. The nutritional quality of legume seeds and cereals strongly depends on the presence of sulfur-rich proteins (Bohlmann 1993; Müntz *et al.* 1997). From PAPS, sulfur is incorporated in sulfolipids which guarantee the integrity and functioning of biomembranes. General ecophysiological aspects of sulfur have been reviewed elsewhere (Ernst 1990, 1993, 1997).

Another important aspect of sulfur is its role in secondary plant compounds, such as glucosinolates (Schnug 1990) and alliins (Mazelis 1993) which govern the odour and flavour of many agricultural products such as cabbage, chives, garlic, onions and radish. Some of the secondary compounds play a role in plant protection against pests: glucosinolates protect cruciferous plant species against herbivores and pathogenic fungi, but hamper also the symbiosis with arbuscular mycorrhizal fungi (Cole 1997; Macfarlane Smith *et al.* 1991; Mithen *et al.* 1987; Schreiner and Koide 1993). Carbon disulfide released from the breakdown of L-djenkolic acid in Mimosaceae and thiophenes deliberated from Asteraceae are effective nematicides (Daimon and Mii 1995; Piluk *et al.* 1998). Many plants can store and detoxify a surplus of sulfur in vacuoles as flavonoid sulfates, phenol sulfates and cholin-sulfates (Ernst 1997) or can volatilize sulfur as COS, DMS, H_2S, and many other organic sulfur compounds.

This overview will elaborate how the various metabolic sulfur pathways determine the sulfur demand of agricultural crops and how the knowledge of sulfur metabolism can be implemented in solving agricultural problems in various environmental settings.

15.2 SULFUR DEMANDS OF AND RELEASE BY AGRICULTURAL CROPS

The primary pathways of sulfur to sulfolipids and sulfo-amino acids are present in all plant species (Scheme 15.1). The various pathways of secondary sulfur compounds, however, are restricted to taxa from the rank of order, like the glucosinolates in Capparidales, up to single species, like

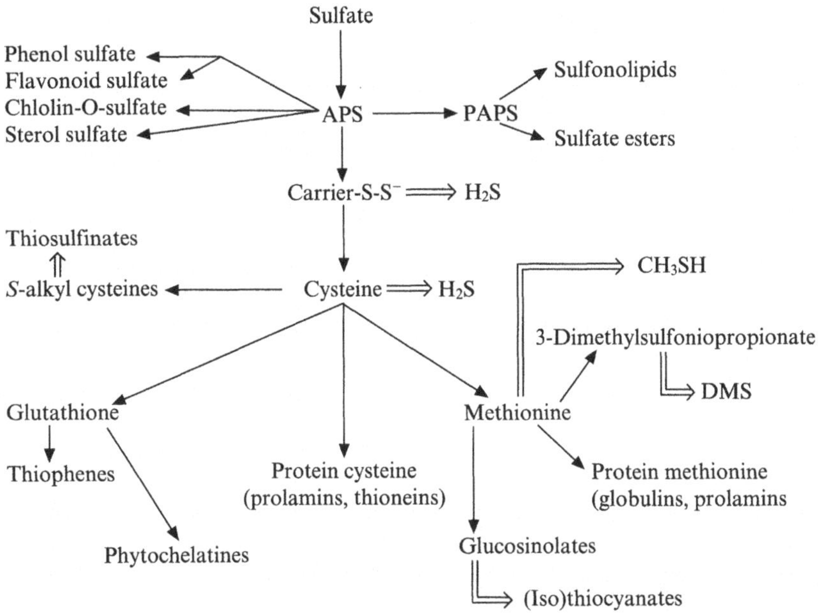

Scheme 15.1. Pathways of sulfur assimilation in higher plants (Modified after Ernst (1998a) and accomplished with data from Bohlmann (1993), Giovanelli (1990), Grill *et al.* (1989), Hanson and Gage (1991), Hanson *et al.* (1994), Jacobs *et al.* (1994), Mazelis (1993), Müntz *et al.* (1997). ⇒ indicates emission of sulfur compounds by plants.

the *p*-mentha-8-thiol in the aroma of grapefruit (*Citrus decuman*, Rutaceae; Simonsen 1947). Both primary and secondary sulfur metabolism determine the sulfur demand of the various plant species and the potential of the species to cope with the sulfur supply of their environment.

15.2.1 Species-specific demands

The sulfur demand of agricultural crops varies by a factor of five and more, from less than 10 kg S ha^{-1} (1 ha = 10^4 m^2) to 80 kg S ha^{-1} (Table 15.1). Wild herbs and grasses have the lowest sulfur demand varying from below 10 up to 20 kg S ha^{-1} yr^{-1}, the latter if applied as gypsum to wheat (McGrath *et al.* 1996). The sulfur supply to cereals has a certain impact on yield, but a more important one on the breadmaking quality of flour. A high concentration of thiol and disulfide groups of the S-rich storage proteins in the endosperm, i.e. the gliadins, enhances the visco-elasticity of the dough and increases the

Table 15.1. Annual sulfur extraction from soil by plants

Plant group/species	S-extraction (kg S ha^{-1})	Reference
Herbs	<10	
Wild grasses	10-30	Jolivet (1993)
Avena sativa	10	Withers *et al.* (1995)
Triticum aestivum	>15-20	McGrath *et al.* (1996)
Saccharum officinarum	24	Stanford and Jordan (1966)
Allium species	>15	
Beta vulgaris	30	Schachtschabel *et al.* (1992)
Gossypium hirsutum	30	Stanford and Jordan (1966)
Legumes	15-50	Rendig (1956); Jones (1964)
Helianthus annuus	20-45	
Cruciferous weeds	10-20	Ernst (1999)
Cruciferous crops	20-30	Jolivet (1993)
Brassica napus	45-100	Schachtschabel *et al.* (1992) Zhao *et al.* (1993a)

loaf volume of the bread (Haneklaus *et al.* 1992). In the case of barley the sulfur supply influences not only the content of the sulfur storage protein hordein (Shewry *et al.* 1993), but also the malting process (Gayler *et al.* 1997).

The sulfur demand of onions (*Allium cepa*) and other *Allium* species is determined by the concentration of S-alkylcysteine sulfoxides, the alliins, which make up 80% of the total sulfur in these plants. Alliins determine the odour and flavour of these vegetables. The content of the various alliins increases with increasing sulfur fertilization and with the storage behaviour of plant species and cultivars (Randle *et al.* 1994). During plant development there is a high dynamic of the flavour precursor (Lancaster *et al.* 1986). As in many plant-defence substances the substrate and the cleaving enzyme are separated in different cell compartments: the alliins are stored in the cytosol and the enzyme alliinase in the vacuoles (Lancaster and Collin 1981). After cellular disruption the alliinase cleaves alliin, yielding thiosulfanates and pyruvate. It may be expected that the expression of the odours and flavours will be regulated by different genes, determining the amount of substrate and enzyme, as is known for cyanogenic glucosides (Kakes 1989). Genetic manipulation of the amounts of substrate and enzyme opens new perspectives in influencing the flavour quality of crops.

In sunflower (*Helianthus annuus*) and in legumes the sulfur demand is strongly related to the synthesis of S-rich storage proteins in seeds (Blagrove *et al.* 1976; Kortt *et al.* 1991; Naito *et al.* 1995). Another reason for the enhanced demand for sulfur by legumes may be their good autochthonic supply with nitrogen owing to the symbiosis with nitrogen-fixing *Rhizobium*

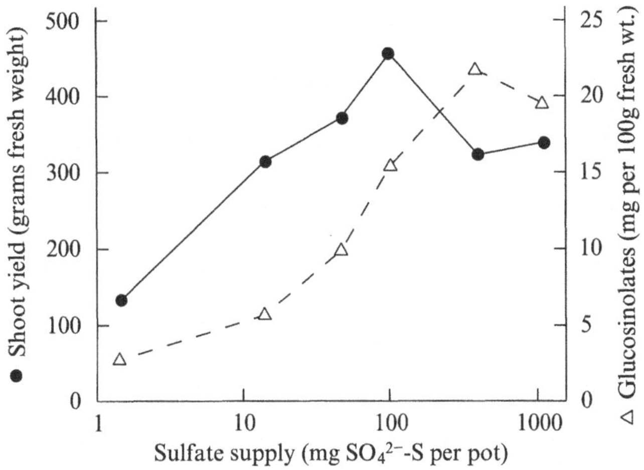

Figure 15.1. The response of shoot yield and glucosinolate content of seeds in *Raphanus oleiferus* (After: Marquard *et al.* 1968).

and *Bradyrhizobium* bacteria. The increase of the sulfur amino acid content in legume seeds by transgenic plants (Saalbach *et al.* 1995) will not only help to improve the nutritive value of legumes, but may also increase the sulfur requirement of legumes.

The high S-demand of cruciferous crops is closely related to the content in glucosinolates and their volatile metabolites (Marquard *et al.* 1968) with a well-developed saturation kinetic, i.e. the sulfur demand for shoot yield is lower than that for the glucosinolates (Figure 15.1). Insufficient sulfur supply can reduce crop and seed yield. The glucosinolate profile is very species- and cultivar-specific (Kirkegaard and Sarwar 1998) and determines the flavour of many cruciferous crops like Brussels sprouts, cabbage, cauliflower, kale, kohlrabi and radish. In seeds of oilseed rape (*Brassica napus*) not only a low but also a high sulfur application can result in a poor seed meal quality. A high content of glucosinolates in seeds may have anti-nutritional and toxic effects on animals, especially by affecting the iodine metabolism (Schone *et al.* 1997).

To tackle this foodchain problem a pragmatic selection has resulted in single low (low in erucic acid, a fatty acid) and double low (low in erucic acid, low in glucosinolate) varieties of oilseed rape. Although the glucosinolate content in the seeds of the double low varieties is diminished by 80 - 90% (Josefsson and Appelqvist 1968), the sulfur demand of these varieties, however, is not diminished (Schnug and Haneklaus 1994; Zhao *et al.* 1993a). A physiological analysis of this discrepancy has shown that a metabolic block in the biosynthesis of glucosinolates is responsible for the low glucosinolate content of the seeds. The unchanged sulfur uptake by the roots and translocation to the shoot result in a large accumulation of inorganic sulfate not only in the pod walls of fruits (Zhao *et al.* 1993b), but also in leaves (Blake-Kalff *et al.* 1998). The allocation and metabolisation of sulfur in leaves of cruciferous crops follows the general pattern of higher plants, i.e. an increased SO_4^{2-}-S content in leaves with maturity (Ernst 1997). The low turnover rate of vacuolar sulfur (Bell *et al.* 1995) builds up a huge above-ground sulfur pool during the growing season. This pool is either removed with the harvested crop or returns after harvest to the soil for remineralization. Sulfate storage in the vacuoles does not help to increase crop yield. Therefore it is necessary to promote the selection of S efficiency in crops, thus diminishing the sulfate storage and solving many aspects of S deficiency.

In plant nutrition the availability of one nutrient has consequences for other nutrients. Because of the well-known regulatory interaction between nitrate and sulfate uptake and assimilation (for a review see Brunold 1993) it is not surprising that an increase of nitrogen fertilization increases the sulfur uptake by 10 - 15 kg S ha^{-1} by oilseed rape (Schnug and Haneklaus 1994; Zhao *et al.* 1993a). An enhanced N supply affects the quality of the glucosinolate profile by stimulating the hydroxylation step from gluconapin to progoitrin within the alkenyl glucosinates not only in oilseed rape (Zhao *et al.* 1994), but also Brussels sprouts (*Brassica oleracea* var. *gemmifera*) and kohlrabi (*Brassica oleracea* var. *gongylodes*) (Fischer 1992; Heany *et al.* 1983). The N:S ratio is often used to determine critical plant levels of both elements. In grasses the critical N:S ratio in leaves varies from 10 in *Avena sativa* to 17 in *Triticum aestivum* (Rasmussen *et al.* 1977; Saalbach 1970). An N:S ratio above 6 can disrupt the transfer of sulfur to the seeds of oilseed rape, thus deregulating seed quality (Fismes *et al.* 1999). This reaction pattern emphasizes the importance of a well-balanced supply of the various nutrients to oilseed rape and to other plant species.

15.2.2 Release of organic sulfur compounds

Another aspect of a plant's sulfur metabolism is the release of volatile sulfur from the standing crop, often enhanced by herbivorous animals, or during and after harvest. The sulfur-based volatiles are related to the above-mentioned secondary metabolites. The emission of such organic sulfur compounds is the reason that onion fields and other *Allium* vegetation can be smelled at a great distance. All *Allium* species emit a mixture of organic sulfur compounds in a species specific pattern with often a dominance of one or two components. Dipropenyldisulfide and methylpropenyldisulfide contribute to 60 % to the odour of the wild onion *Allium ursinum* (Puxbaum and König 1997). Methylsulfide is mainly emitted from Chinese chive (*Allium tuberosum*) having the highest emission rate of all investigated *Allium* crops with 2 mg S kg^{-1} dry mass per day. Propenyl sulfide is the main component in the odour of leek (*Allium fistulosum*) and onion (*Allium cepa*; Kanda and Tsuruta 1995). The emission rate of *Allium fistulosum* is 25% higher than that of *Allium ursinum*, emitting 1.5 mg S m^{-2} d^{-1} as organic sulfur compounds (Puxbaum and König 1997). Even more sulfur compounds are volatilized at the time of harvest, when onions are dried on the field for nearly a week. Together with the S-compounds in the harvested plant parts, the annual sulfur removal by an *Allium* crop may exceed 15 kg S ha^{-1} yr^{-1}. It is not known if the stimulation of the synthesis of the secondary sulfur compounds in *Allium* species by sulfur fertilisation (Haneklaus *et al.* 1997) will also increase the emission of sulfur volatiles from field-grown *Allium* species.

The sulfur taken up by sugar cane (*Saccharum officinarum*) can also partly be lost by volatilization. Sugar cane belongs to the few grasses that can synthesize 3-dimethylsulfoniopropionate (DMSP) from S-methyl-L-methionine and release it as dimethylsulfide (Paquet *et al.* 1994). Although the synthesis of DMSP in sugar cane is only 10% of that in the salt marsh grass *Spartina alterniflora*, the high biomass of a sugar cane field and the enrichment of DMSP in mature leaves may result in an annual sulfur loss of 25 kg S ha^{-1} yr^{-1}, when assuming a similar DMS emission rate as from *Spartina alterniflora* (Steudler and Peterson 1984).

The release of organic sulfur compounds by a standing crop of *Brassica* species is small, but can increase considerably after injury from herbivores and pathogens (Visser 1986). A new source of sulfur losses is introduced by the growing tendency to use harvest remnants of *Brassica* species as green manure and as biocides, i.e. as biofumigants for suppression of soil-borne pathogens (Kirkegaard and Sarwar 1998). At decomposing root and shoot tissue, the glucosinolates are hydrolysed by the myrosinase enzyme. A range of volatile isothiocyanates are released to the soil and finally to the atmosphere. This sulfur loss from the soil has to be added to the sulfur removal by the crop.

Although many plant species belonging to taxonomic groups other than the above-mentioned ones can emit organic sulfur compounds, the impact on sulfur fertilisation has not yet been investigated.

15.3 SULFUR SUPPLY TO AVOID DEFICIENCY

Sulfur inputs by SO_2-S emission will further decrease together with an increase of the application of S-free urea and triple superphosphate fertilizer (Ceccotti 1996) and a decrease of sulfur-containing pesticides (Tabatabai 1984). Therefore intensive agriculture of sulfur-demanding crops like oilseed rape will result in sulfur deficiency in the long term if sulfur is not replenished.

15.3.1 Sulfur supply by fertilizers and industrial wastes

The oldest form of crop fertilization is the application of compost and organic manure. In both substrates sulfur is bound to organic matter and therefore not immediately available to higher plants. It has to be mineralized via biological processes (McGill and Cole 1981; Tabatabai and Bremner 1970). Such slow-release sulfur compounds are less suitable for satisfying the sulfur requirement of crops with a short growing period, such as oilseed rape (harvest within 4-5 months), as shown in the field experiments that have lasted more than 100 years at Askov in Denmark (Eriksen and Mortensen 1999).

Sulfur availability to crops from gypsum, another 'classic' sulfur fertilizer, is more effective than manure (Eriksen *et al.* 1995). The solubility of gypsum in water, however, is low ($18.3 \cdot 10^{-3}$ mol L^{-1}), supplying equal amounts of calcium and sulfate:

$$CaSO_4 2H_2O \leftrightarrow Ca^{2+} + SO_4^{2-} + 2\,H_2O$$

Gypsum has the advantage that in addition to the slow release of sulfur, it increases the calcium supply of the crop. Therefore, it is favoured to be applied on acid and/or marginal soils in less-industrial countries (Chauhan *et al.* 1995).

The agricultural demand for sulfur fertilizer has resulted in various forms of industrially produced sulfur. Formerly, single superphosphate (12% S) and ammonium sulphate (24% S) together with sulfur in rainfall, have met the sulfur requirement of crops. With the increasing demand for high-analysis materials, sulfur disappeared from fertilizers such as urea, or was strongly diminished in fertilizers such as triple superphosphate (1% S) and ammoniumphosphates (2% S). This results in a stagnant sulfur supply at a doubled nitrogen application (Ceccotti 1996; Ceccotti and Messick 1997). Thus without a compensation of the S removal from high-analysis fertilizers, the balance of the various nutrients in plant metabolism is disturbed. Therefore, it is to be expected that an imbalance of the N:S and P:S application will increase the sulfur deficiency. In the world, single superphosphate and ammonium sulfate are still the most common forms of sulfur supply, followed by compound fertilizers. In Europe, however, a preference for the latter product is growing (Ceccotti and Messick 1997). One of the problems of sulfate fertilizers is the high leachability of sulfate ions (Schachtschabel *et al.* 1992). Therefore the fertilizer industries are promoting elemental sulfur (S^o) as the most concentrated sulfur fertilizer source with high antifungal properties (Jolivet 1993). S^o is a slow-release fertilizer which cannot be taken up by the plants, but has to be oxidized to sulfate by microorganisms. Addition of these microorganisms to elemental sulfur stimulates formation of sulfate and thus the crop yield.

Other sulfur fertilizers are by-products of industrial processes. Phosphogypsum is an acidic by-product of the phosphate fertilizer industry. Owing to its content of fluoride, cadmium and some radionuclides originating for the phosphate rocks, its application can have undesired environmental side-effects (Smith *et al.* 1994), comparable to those already known from the increase of cadmium in superphosphate-fertilized fields. As a consequence of clean air legislation sulfur-rich coal has to be desulfurized, resulting in another industrial form of gypsum, i.e. flue gas desulfurization (FDG) gypsum (Clark *et al.* 1997). FDG gypsum makes a positive contribution to the sulfur supply of crops (Stehouwer *et al.* 1996), but its elevated content of boron (B) and molybdenum (Mo) may cause B and Mo toxicity in crops sensitive to these elements (Kukier and Sumner 1996), thus making it less attractive for agricultural applications.

As a result of clean water acts sewage is widely treated with sludge as a voluminous remnant. The use of sewage sludge was formerly believed to be very suitable as fertilizer owing to its high concentrations of N and P. Its sulfur concentration was taken for granted. But the considerable concentration of heavy metals in sewage sludge, even if free from industrial contamination, has resulted in soil contamination (Hemkes *et al.* 1980; Mcbride *et al.* 1997). Consequently, the strong limitations of sewage sludge application in agriculture by European Union guidelines contributes to the current sulfur deficiency of agricultural soils. For fear of increase and dissemination of endocrine disruptors the use of sewage sludge is further restricted or even cancelled, at least in The Netherlands.

15.3.2 Sulfur supply by atmospheric pollution

Sulfur dioxide emission from volcanoes adds around 13 Mt yr^{-1} to the

Earth's atmosphere. Together with the biogenic atmospheric sulfur emitted as DMS from the biota of oceans and wetlands the atmospheric annual input of sulfur ranges from 10 to 20 kg S ha^{-1} (McGrath and Zhao 1995). Ecosystems near the coast get some more sulfur by sea spray. In the industrial world the natural S input was strongly enhanced by SO_2 emission from industrial processes and traffic exhausts. Therefore it was possible that the sulfur supply to and consumption by agricultural crops was unrecognized for a long time, in contrast to the increasing demand of NPK fertilizers. On many sites the degree of SO_2 emission, however, was not only beneficial to plants, but caused damage in grasslands, forests and agricultural crops (Garber 1967). As a consequence of such an environmental pressure, high SO_2 concentrations have favoured the selection of SO_2-resistant grasses such as *Lolium perenne* in England and *Phleum pratense* in Canada (Clapperton and Reid 1994), and SO_2-resistant crops such as cucumber (*Cucumis sativus*) and soybean (*Glycine max*) in the USA. In addition, the high sulfur availability has stimulated the expansion of many cruciferous species, thus also promoting the increase of agricultural weeds such as *Arabidopsis thaliana*, *Brassica nigra* and *Sinapis alba* (Ernst 1993, 1999).

Because of the adoption of pollution control measures in industrial countries the decrease of SO_2 emissions between 1980 and 1987 (Figure 15.2a) varied from 22% in Poland to 64% in France (Ceccotti 1996). Several countries switched from coal-fired to gas-fired industries at the end of the 1960s so that the decrease of the SO_2 emission was even more than that presented in Figure 15.2b. For instance, the SO_2-emission decreased in UK from 2.1 Mt SO_2-S in 1970 to 1 Mt SO_2-S in 1995 (Zhao *et al.* 1997), and at the Woburn experimental site a decrease from 72 kg S ha^{-1} in 1968/70 to 15 kg S ha^{-1} in 1990/92 has been reported (McGrath *et al.* 1996). In Eastern Europe, however, the SO_2-S deposition is still high, in the vicinity of 100 kg S ha^{-1} yr^{-1} (Erkenberg *et al.* 1996).

The improvement in air quality was beneficial to natural ecosystems, especially for SO_2-sensitive organisms and crops. But from the late 1980s onwards, the decreased S supply resulted in a widespread S deficiency in several highly S-demanding crops, especially oilseed rape and cereals in Denmark, England, Northern Germany and Scotland (McGrath and Zhao 1995; Schnug 1991; Withers and O'Donnell 1994).

SO_2-S is predominantly taken up by leaves via the stomata and facilitated by the high solubility and rapid hydration of SO_2 in the aqueous phase (Pfanz *et al.* 1987). As long as the SO_2 concentration remains below the phytotoxic level, plants can benefit from the atmospheric SO_2-S supply

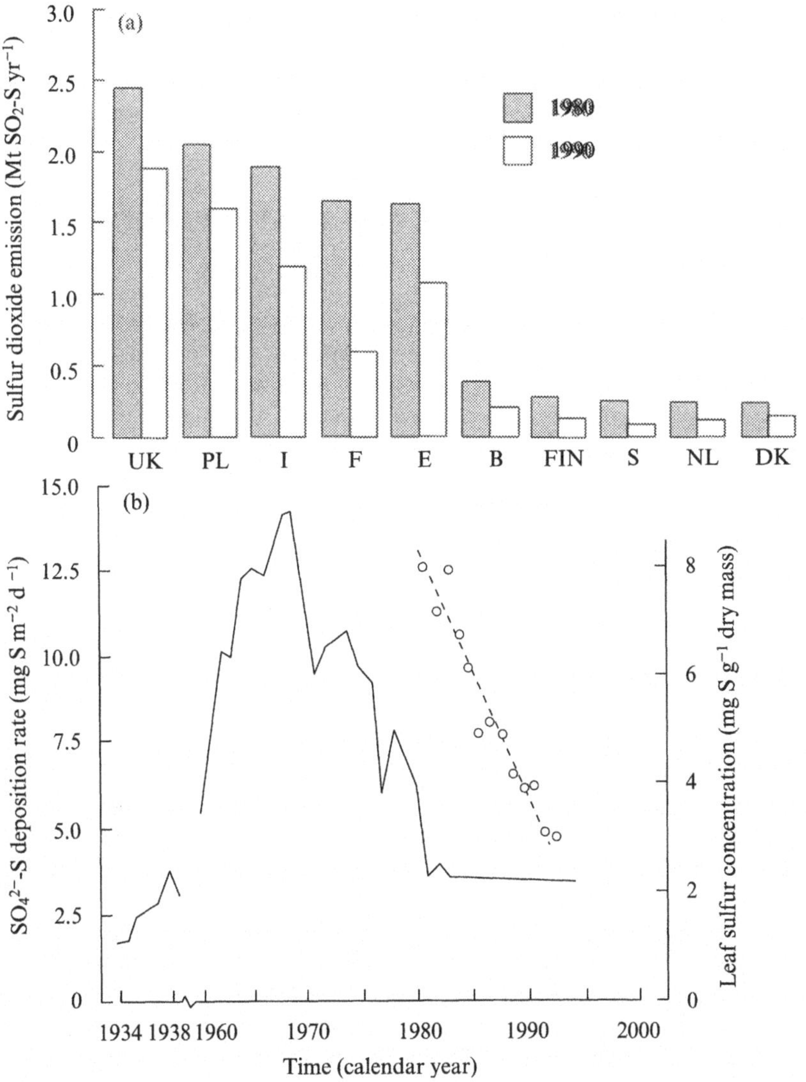

Figure 15.2. (a) The decrease of SO_2 emission in European countries from 1980 to 1990 (Data from UNEP (1993) in Ceccotti and Messick 1997); (b) the change of SO_4^{2-} deposition rate in the western part of The Netherlands from 1934 to 1994 (modified after Ernst (1993), with additional data from Zonneveld (1995) and the decrease of sulfur in leaves of *Brassica napus* in Northern Germany where the decrease in SO_2 emission started nearly ten years later than in The Netherlands (circles, data from Schnug and Haneklaus 1994).

when the sulfur concentration of the soil is insufficient for normal plant development (Cowling and Lockyer 1978). At low soil sulfur supply, non-injurious levels of H_2S can have a similar beneficial effect on the sulfur metabolism of plants as SO_4^{2-} due to the fast incorporation in organic sulfur compounds (De Kok *et al.* 1997).

15.4 PHYTOREMEDIATION OF SULFUR-ENRICHED SOILS

A range of geological and climatological processes can cause an enrichment of rocks and soils with sulfur, either in the reduced or oxidized form. Habitats rich in sulfur and poor in other major nutrients range from those with reduced sulfur (saline marshes) up to those with highly oxidized sulfur (gypsum soils, solfataras). As a consequence, plants have evolved various metabolic routes to cope with excess sulfur and to channel sulfur to such chemical forms in the cell that do not hamper normal plant performance (Ernst 1990). In addition to these natural processes, irrigation of soils in semi-arid areas and water-management in partly flooded soils has strongly increased the area of sulfur-enriched soils. The desulfurization of such soils demands technical and/or biological means. The latter are summarized by the term 'phytoremediation'. Phytoremediation can be done with different options (Table 15.2), ranging from stabilization of the soil by (re)vegetation up to extraction of the element(s) in surplus.

Sulfur-deficiency after diminishing atmospheric sulfur supply to the soil (see above) indicates that plants with high sulfur demand and uptake will be very suitable for the type of phytoremediation that tries to reduce sulfur in the soil by phytoextraction, identical with the term phytomining (Brooks *et al.* 1998). In many sulfur-enriched environments, not sulfur itself, but the counter cation is the problem for plant growth. It demands a resistance of the plants to the cation in surplus. Phytoextraction in an economic perspective demands genotypes of plants with a high uptake preference of these elements in surplus and thus a simultaneous selection for uptake preference and for resistance. From the plant species with high sulfur demands, members of the genus *Allium* (Liliaceae) are not very suitable because they have not evolved the ability to cope with an excess of aluminium, heavy metals and sodium, in contrast to plant species belonging to the family Brassicaceae. The genus *Lepidium* has a high resistance to sodium and sulfur (Albert and Popp 1977; Duvigneaud and Denaeyer-De Smet 1968). *Alyssum*, *Brassica*, *Cardaminopsis*, *Cochlearia* and *Thlaspi* have a high resistance to heavy metals such as cadmium, copper, lead, nickel and

Table 15.2. Types of phytoremediation

Type	Effect
Phytorestoration	A long-term process to repair the damaged vegetation
Phytostabilization	Revegetation of disturbed soils to prevent erosion and contamination of the surroundings and the groundwater
Phytoextraction (=Phytomining)	Removal of the elements in surplus through uptake by plants

zinc (Ebbs *et al.* 1997; Ernst 1974; Robinson *et al.* 1997). Resistance to a surplus of calcium together with a surplus of sulfur is present in many taxa of different families.

15.4.1 Gypsiferous soils

Gypsiferous soils containing high concentrations of gypsum ($CaSO_4.2H_2O$) occur very frequently in arid and semi-arid regions. In these soils the dissolution of gypsum significantly affects the concentration of dissolved salts. As mentioned above, mineral gypsum can help to ensure the sulfur supply to crops. On the other hand, gypsum soils have adverse effects on agriculture and are often not suitable for the growth of crops. Because a surplus of calcium does not demand a very sophisticated resistance mechanism, a surplus of sulfur and calcium can be managed by many plant species. The vegetation of these soils is characterized by plant species with moderate to high sulfur concentrations ranging from 0.20% S in *Ononis natrix* to 7.4% S in *O. tridentata* (Boukhris and Lossaint 1972). The latter legume seems to be very suitable for sulfur extraction, and the symbiosis with nitrogen-fixing *Rhizobium*-bacteria may deliver a reasonable green manure. The high gypsum concentration of the soil and the gypsum precipitation rate from evaporating groundwater, however, will make gypsum extraction an ineffective approach for phytoextraction (Ernst 1998a).

15.4.2 Solonchaks

Solonchaks (white alkali soils) are saline soils in semi-arid and arid climates. They are characterized by enhanced concentrations of sodium, chloride, magnesium, calcium and sulfur which may be present in amounts toxic to plants. The genesis of solonchaks (Schachtschabel *et al.* 1992) may be natural by enrichments of salts in the subsoil (hidden solonchak) or in the topsoil (primary solonchak). Alternatively, it can be caused by anthropogenic activity such as irrigation (cultosolonchak). The duration of inundation determines the redox potential of the soils and the chemical forms of sulfur, ranging from sulfide to sulfate. Reclamation has been achieved by leaching the salt out of the soil and growing moderately salt-resistant plants (Overstreet *et al.* 1955). The leaching process demands, however, a lot of fresh water which is scarce in most (semi-)arid areas. At low water availability, these soils can be revegetated with salt-resistant plants, mostly grasses such as Bermuda grass (*Cynodon dactylon*) and Salt grass (*Distichlis* species) or with indigenous herbs and shrubs such as *Reaumuria*, *Salsola* and *Tamarix* species (phytostabilization).

15.4.3 Thionic fluvisols including acid sulfate soils

Thionic fluvisols are the result of postglacial land uplift in Northern Europe (Palko and Weppling 1994). Acid sulfate soils are formed by drainage and aeration of previously submerged sulfate-enriched soils in many parts of the world (Carvalho and van Raij 1997; Konsten *et al.* 1994). In both soil types, oxidation of pyrites results in the release of a high amount of acidity:

$$2\ FeS_2 + 7\ O_2 + 2\ H_2O \rightarrow 2\ Fe^{2+} + 4\ SO_4^{2-} + 4\ H^+$$

The resulting low pH increases the bioavailability of iron and aluminium, which is the major constraint to plant growth. By application of gypsum and liming, the soil pH value and the calcium supply can be increased, thus diminishing the bioavailability of aluminium and iron, and increasing crop yield (Moore *et al.* 1990; Sumner *et al.* 1986). Liming, however, does not enhance the sulfur extraction, at least by such crops as maize with a moderate sulfur demand (Figure 15.3). In many situations the soil pH will remain so low that the bioavailable aluminium concentration will be too high for plant growth. Therefore, selection of aluminium-resistant crops is necessary. In that case, care has to be taken to keep the transfer of aluminium into the crop below the acceptable daily intake level for animals and man (Palko and Weppling 1994).

15.4.4 Metal-enriched soils

In mineralized soils, high sulfur levels may co-occur with high contents of heavy metals in many combinations (Ernst and Nelissen 1999). For phytostabilization (Ernst 1998b) or phytomining (Robinson *et al.* 1997), this demands application of very metal-resistant plant species such as the Zn- and Cd-hyperaccumulator *Thlaspi caerulescens*, the nickel hyperaccumulator *Alyssum bertolonii* or metal-resistant genotypes of moderately

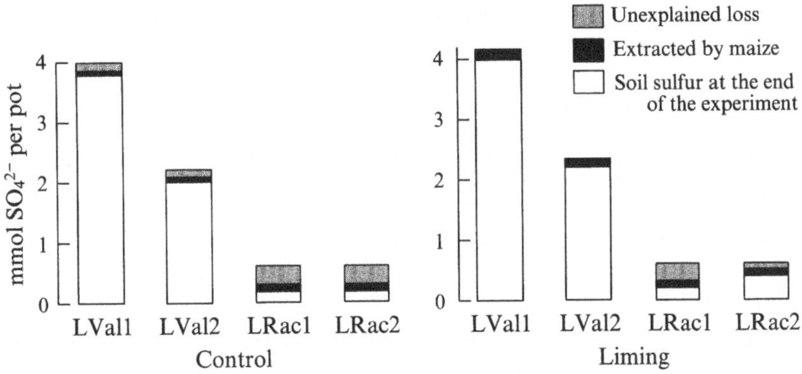

Figure 15.3. Sulfur extraction by maize grown on four acidic sulfate soils in Sao Paulo with and without liming. In the limed latosols (LVal) more sulfur was present at the end than at the start of the experiment. The soil types were classified as alic red-yellow latosol (LVal) and acric dusky-red latosol (LRac). (Data from Carvalho and van Raij 1997).

metal-accumulating species such as *Silene vulgaris*. Agriculture in the neighbourhood of such mineralized soils has to cope with plant injury and reduced yields. Moreover, the runoff of metals from the mineralized soils hampers phytoremediation of such sites (Ernst 1974).

Application of metal-contaminated sludges, industrial sewage effluents and metal-based pesticides to agricultural and horticultural soils (Gadallah 1994; Hemkes *et al.* 1980; Lepp *et al.* 1984; Magalhaes *et al.* 1985) and contamination by industrial emissions (Ernst 1972; Leita *et al.* 1998; Xie and Huang 1998) have resulted in low to moderate metal contamination of agricultural soils, not necessarily associated with high sulfur concentrations. Metal extraction can be achieved by plants with a very moderate metal resistance such as *Brassica* species and maize, with or without diminishing metal availability by liming (Ernst 1995). The major obstacle facing agriculture on such anthropogenically metal-contaminated soils is the metal contamination of the animal and/or human food chain (Xie and Huang 1998). If the metal-contaminated crops are taken out of the food chain, the metals can be concentrated by chemical or biological mineralization of the crop (Chapter 14). The technical means for such a process, however, are not well developed so that storage of contaminated crops or ashes outside the agricutural field is the practice instead of a real industrial recycling.

15.5 REFERENCES

Albert, R. and Popp, M. (1977) Chemical composition of halophytes from the Neusiedler Lake region in Austria. *Oecologia* **27**, 157-170.

Bell, C., Fordham, M., Richardson, P., Cram, J. and Tomos, D. (1995) Cellular and subcellular compartmentation of sulphate in leaves in relation to low sulphur mobility. *Z. Pflanzenern. Bodenk.* **158**, 63-65.

Blagrove, R.J., Gillespie, J.M. and Randall, P.J. (1976) Effect of sulphur supply on the seed globulin composition of Lupinus angustifolius. *Aust. J. Plant Physiol.* **3**, 173-184.

Blake-Kalff, M.M.A., Harrison, K.R., Hawkesford, M.J., Zhao, F.J. and McGrath, S.P. (1998) Distribution of sulfur within oilseed rape leaves in response to sulfur deficiency during vegetative growth. *Plant Physiol.* **118**, 1337-1344.

Bohlmann, H. (1993) Significance of sulfur-rich proteins in seeds and leaves. In *Sulfur Nutrition and Assimilation in Higher Plants. Regulatory, Agricultural and Environmental Aspects* (eds. De Kok, L.J., Stulen, I., Rennenberg, H., Brunold, C. and Rauser, W.E.), pp. 211-219, SPB Academic Publishing, The Hague.

Boukhris, M. and Lossaint, P. (1972) Spécificité biogéochimique des plantes gypsophiles de Tunisie. *Oecol. Plant.* **7**, 45-68.

Brooks, R.R., Chambers, M.F., Nicks, L.J. and Robinson, B.H. (1998) Phytomining. *Trends Plant Sci.* **3**, 359-362.

Brunold, C. (1993) Regulatory interactions between sulfate and nitrate assimilation. In *Sulfur Nutrition and Assimilation in Higher Plants. Regulatory, Agricultural and Environmental Aspects* (eds. De Kok. L.J., Stulen, I., Rennenberg, H., Brunold, C. and Rauser, W.E.), pp. 61-75, SPB Academic Publishing, The Hague.

Carvalho, M.C.S. and Van Raij, B. (1997) Calcium sulphate, phosphogypsum and calcium carbonate in the amelioration of acid subsoils for root growth. *Plant Soil* **192**, 37-48.

Ceccotti, S.P. (1996) Plant nutrient sulphur - a review of nutrient balance, environmental impact and fertilizers. In *Fertilizers and Environment* (ed. Rodriguez-Barrueco, C.), pp. 185-193, Kluwer Academic Publishers, Dordrecht.

Ceccotti, S.P. and Messick, D.L. (1997) A global review of crop requirements, supply, and environmental impact on nutrient sulphur balance. In *Sulfur Metabolism in Higher Plants. Molecular, Ecophysiological and Nutritional Aspects* (eds. Cram, W.J., De Kok, L.J., Stulen, I., Brunold, C. and Rennenberg, H.), pp. 155-163, Backhuys Publishers, Leiden.

Chauhan, D.R., Paroda, S. and Singh, D.P. (1995) Effect of biofertilizer, gypsum and nitrogen on growth and yield of raya (*Brassica juncea*). *Ind. J. Agron.* **40**, 639-642.

Clapperton, M.J., Reid, D.M. (1994) Effects of sulphur dioxide (SO_2) on growth and flowering of SO_2-tolerant and non-tolerant genotypes of *Phleum pratense*. *Environ. Pollut.* **86**, 251-254.

Clark, R.B., Zeto, S.K., Ritchey, K.D. and Baligar, V.C. (1997) Maize growth and mineral acquisition on acid soil amended with flue gas desulfurization by-products and magnesium. *Commun. Soil Sci. Plant Anal.* **28**, 1441-1459.

Cole, R.A. (1997) The relative importance of glucosinolates and amino acids to the development of two aphid pests *Brevicoryne brassicae* and *Myzus persicae* on wild and cultivated *Brassica* species. *Entomol. Exp. Appl.* **85**, 121-133.

Cowling, D.W. and Lockyer, D.R. (1978) The effect of SO_2 on *Lolium perenne* L. grown at different levels of sulphur and nitrogen nutrition. *J. Exp. Bot.* **29**, 257-265.

Cram, W.J. (1990) Uptake and transport of sulfate. In *Sulfur Nutrition and Sulfur Assimilation in Higher Plants. Fundamental, Environmental and Agricultural Aspects* (eds. Rennenberg, H., Brunold, Ch., De Kok, L.J. and Stulen, I.), pp. 3-11, SPB Academic Publishing, The Hague.

Daimon, H. and Mii, M. (1995) Plant regeneration and thiophene production in hairy root cultures of *Rudbeckia hirta* L. used as an antagonistic plant to nematodes. *Jap. J. Crop. Sci.* **64**, 650-655.

De Kok, L.J., Stuiver, C.E.E., Rubinigg, M., Westerman, S. and Grill, D. (1997) Impact of atmospheric sulfur deposition on sulfur metabolism in plants: H_2S as sulfur source for sulfur deprived *Brassica oleracea* L. *Bot. Acta* **110**, 411-419.

Duvigneaud, P. and Denaeyer-De Smet, S. (1968) Essai de classification chimique (éléments minéraux) des plantes gypsicoles du Bassin de l'Ebre. *Bull. Soc. r. Bot. Belg.* **101**, 279-291.

Ebbs, S.D., Lasat, M.M., Brady, D.J., Cornish, J., Gordon, R., Kochian, L.V. (1997) Phytoextraction of cadmium and zinc from a contaminated soil. *J. Environ. Qual.* **26**, 1424-1430.

Eriksen, J., Mortensen, J., Kjellerup, V.K. and Kristjansen, O. (1995) Forms and plant availability of sulfur in cattle and pig slurry. *Z. Pflanzenern. Bodenk.* **158**, 113-116.

Eriksen, J. and Mortensen, J.V. (1999) Soil suphur status following long-term annual application of animal manure and mineral fertilizers. *Biol. Fert. Soils* **28**, 416-421.

Erkenberg, A., Prietzel, J. and Rehfuess, K.E. (1996) Sulfur status of selected European forest soils as dependent on the atmospheric S deposition. *Z. Pflanzenern. Bodenk.* **159**, 101-109.

Ernst, W. (1972) Zink- und Cadmium-Immissionen auf Böden und Pflanzen in der Umgebung einer Zinkhütte. *Ber. Deutsch. Bot. Ges.* **85**, 295-300.

Ernst, W.H.O. (1974) *Schwermetallvegetation der Erde.* G. Fischer Verlag, Stuttgart.

Ernst, W.H.O. (1990) Ecological aspects of sulfur metabolism. In *Sulfur Nutrition and Sulfur Assimilation in Higher Plants. Fundamental, Environmental and Agricultural Aspects* (eds. Rennenberg, H., Brunold, C. De Kok, L.J. and Stulen, I.), pp. 131-144, SPB Academic Publishing, The Hague.

Ernst, W.H.O. (1993) Ecological aspects of sulfur in higher plants: the impact of SO_2 and the evolution of the biosynthesis of organic sulfur compounds on populations and ecosystems. In *Sulfur Nutrition and Assimilation in Higher Plants. Regulatory, Agricutural and Environmental Aspects* (eds. De Kok, L.J., Stulen, I., Rennenberg, H., Brunold, C. and Rauser, W.E.), pp. 295-313, SPB Academic Publishing, The Hague.

Ernst, W.H.O. (1995) Decontamination or consolidation of metal-contaminated soils by biological means. In *Heavy Metals. Problems and Solutions* (eds. Salomons, W., Förstner, U. and Mader, P.), pp. 141-149, Springer Verlag, Berlin.

Ernst, W.H.O. (1997) Life-history syndromes and the ecology of plants from high sulfur-habitats. In *Sulphur Metabolism in Higher Plants. Molecular, Ecophysiological and Nutritional Aspects* (eds. Cram, W.J., Kok, L.J., Stulen, I., Brunold, C. and Rennenberg, H.), pp. 131-146, Backhuys Publishers, Leiden.

Ernst, W.H.O. (1998a) Sulfur metabolism in higher plants: potential for phytoremediation. *Biodegradation* **9**, 311-318.

Ernst, W.H.O. (1998b) The origin and ecology of contaminated, stabilized and non-pristine soils. In *Metal-contaminated Soils: In situ Inactivation and Phytorestoration* (eds. Vangronsveld, J. and Cunningham, S.D.), pp. 17-29, Landes Bioscience, Austin.

Ernst, W.H.O. (1999) Expansion of *Brassica nigra* populations is not due to sulfur demand, but sulfur resistance. *Fragm. Agron.* (in press).

Ernst, W.H.O. and Nelissen, H.J.M. (1999) Life-cycle phases of a zinc- and cadmium-resistant ecotype of *Silene vulgaris* in testing phytotoxicity of mine soils. *Environ. Pollut.* (in press).

Fischer, J. (1992) The influence of different nitrogen and potassium fertilization on the chemical flavour composition of kohlrabi (*Brassica oleracea* var. *gongylodes* L.). *J. Sci. Food Agric.* **60**, 465-470.

Fismes, J., Vong, P.C. and Guckert, A. (1999) Use of labeled sulphur-35 for tracing sulphur transfers in developing pods of field-grown oilseed rape. *Commun. Soil Sci. Plant Anal.* **30**, 221-234.

Gadallah, M.A.A. (1994) Effects of industrial and sewage waste waters on the concentration of soluble carbon, nitrogen, and some mineral elements in sunflower plants. *J. Plant Nutr.* **17**, 1369-1384.

Garber, K. (1967) *Luftverunreinigung und ihre Wirkungen.* Gebr. Bornträger, Berlin.

Gayler, K.R., Halloran, G.M., Eagles, H.A. and Howard, K.A. (1997) Cereal storage proteins: characteristics and responses to nutrition and environment. In *Sulphur Metabolism in Higher Plants. Molecular, Ecophysiological and Nutritional Aspects* (eds. Cram, W.J., De Kok, L.J., Stulen, I., Brunold, C. and Rennenberg, H.), pp. 95-107, Backhuys Publishers, Leiden.

Giovanelli, J. (1990) Regulatory aspects of cysteine and methionine biosynthesis. In *Sulfur Nutrition and Sulfur Asimilation in Higher Plants. Fundamental, Environmental and Agricultural Aspects* (eds. Rennenberg, H., Brunold, Ch., De Kok, L.J. and Stulen, I.), pp. 33-48, SPB Academic Publishing, The Hague.

Godzik, S. and Krupa, S.V. (1982) Effects of sulfur dioxide on the growth and yield of agricultural and horticultural crops. In *Effects of Gaseous Air Pollution in Agriculture and Horticulture* (eds. Unsworth, M.H. and Ormrod, D.P.), pp. 247-265, Butterworth Scientific, London.

Grill, E., Löffler, S., Winnacker, E.L. and Zenk, M.H. (1989) Phytochelatins, the heavy-metal-binding peptides of plants, are synthesized from glutathione by a specific y-glutamylcysteine dipeptidyl transpeptidase (phytochelatin synthase). *Proc. Natl. Acad. Sci. USA* **86**, 6338-6342.

Haneklaus, S., Evans, E. and Schnug, E. (1992) Baking quality and sulphur content of wheat. I. Relation between sulphur- and pro-tein content and loaf volume. *Sulphur Agric.* **16**, 31-34.

Haneklaus,S., Hoppe, L., Bahadir, M. and Schnug, E. (1997) Sulphur nutrition and alliin concentrations in *Allium* species. In *Sulphur Metabolism in Higher Plants. Molecular, Ecophysiological and Nutritional Aspects* (eds. Cram, W.J., De Kok, L.J., Stulen, I., Brunold, C. and Rennenberg, H.), pp. 331-334, Backhuys

Publishers, Leiden.
Hanson, A.D. and Gage, D.A. (1991) Identification and determination by fast atom bombardment mass spectrometry of the compatible solute choline-O-sulfate in *Limonium* species and other halophytes. *Aust. J. Plant Physiol.* **18**, 317-327.
Hanson, A.D., Rivoal, J., Paquet, L. and Gage, D.A. (1994) Biosynthesis of 3-dimethylsulfoniopropionate in *Wollastonia biflora* (L.) DC. Plant Physiol. **105**, 103-110.
Heaney, R.K., Spinks, E.A. and Fenwick, G.R. (1983) The glucosinolate content of Brussels sprouts: Factors affecting their relative abundance. *Z. Pflanzenzücht.* **91**, 219-226.
Hemkes, O.J., Kemp, A. and Van Broekhoven, L.W. (1980) Accumulation of heavy metals in the soil due to annual dressings with sewage sludge. *Neth. J. Agric. Sci.* **28**, 228-237.
Jacobs, J.J.M.R., Engelberts, A., Croes, A.F. and Wullems, G.J., (1994) Thiophene synthesis and distribution in young developing plants of *Tagetes patula* and *Tagetes erecta*. *J. Exp. Bot.* **45**, 1459-1466.
Jolivet, P. (1993) Elemental sulfur in agriculture. In *Sulfur Nutrition and Assimilation in Higher Plants. Regulatory, Agricultural and Environmental Aspects* (eds. De Kok, L.J., Stulen, I., Rennenberg, H., Brunold, C. and Rauser, W.E.), pp. 193-206, SPB Academic Publishing, The Hague.
Jones, M.B. (1964) Effect of sulfur applied and date of harvest on yield, sulfate sulfur concentration, and total sulfur uptake by five annual grassland species. *Agron. J.* **25**, 235-237.
Josefsson, E. and Appelqvist, L.A. (1968) Glucosinolates in seed of rape and turnip rape as affected by variety and environment. *J. Sci. Food Agric.* **19**, 564-570.
Kakes, P. (1989) An analysis of the costs and benefits of the cyanogenic system in *Trifolium repens* L. *Theor. Appl. Genetics* **77**, 111-118.
Kanda, K. and Tsuruta, H. (1995) Emissions of sulfur gases from various types of terrestrial higher plants. *Soil Sci. Plant Nutr.* **41**, 321-328.
Kirkegaard, J.A. and Sarwar, M. (1998) Biofumigation potential of brassicas. *Plant Soil* **201**, 71-89.
Konsten, C.J.M., Van Breemen, N., Suping, S., Aribawa, I.B. and Groenenberg, J.E. (1994) Effects of flooding on pH of rice-producing, acid sulfate soils in Indonesia. *Soil Sci. Soc. Amer.* J. **58**, 871-883.
Kortt, A.A., Cladwell, J.B., Lilley, G.G. and Higgins, T.J.V. (1991) Amino acid and cDNA sequences of a methionine-rich 2S protein from sunflower seed (*Helianthus annuus* L.). *Eur. J. Biochem.* **195**, 329-334.
Kukier, U. and Sumner, M.E. (1996) Boron availability to plants from coal combustion by-products. *Water Air Soil Pollut.* **87**, 92-110.
Lancaster, J.E. and Collin, H.A. (1981) Presence of alliinase in isolated vacuoles and of alkyl cysteine sulphoxides in the cytoplasm of bulbs of onion *(Allium cepa)*. *Plant Sci. Lett.* **22**, 169-176.
Lancaster, J.E., McCallion, B.J. and Shaw, M.L. (1986) The dynamics of the flavour precursors, the S-alk(en)yl-L-cysteine sulphoxides, during leaf blade and scale development in the onion (*Allium cepa*). *Physiol. Plant.* **66**, 293-297.
Leita, L., Mondini, C., De Nobili, M., Simoni, A. and Sequi, P. (1998) Heavy metal content in xylem sap (*Vitis vinifera*) from mining and smelting areas. *Environ. Monit. Assessm.* **50**, 189-200.

Lepp, N.W., Dickinson, N.M. and Ormand, K.L. (1984) Distribution of fungicide-derived copper in soils, litter and vegetation of different aged stands of coffee (*Coffea arabica* L.) in Kenya. *Plant Soil* **77**, 263-270.

Macfarlane Smith, W.H., Wynne Griffiths, D. and Boag, B. (1991) Overwiter variation in glucosinolate content of green tissue of rape (*Brassica napus*) in response to grazing by wild rabbit (*Orcyctolagus cuniculus*). *J. Sci. Food Agric.* **56**, 511-521.

Magalhaes, M.J., Sequeira, E.M. and Lucas, M.D. (1985) Copper and zinc in vineyards of Central Portugal. *Water Air Soil Pollut.* **26**, 1-17.

Marquard, R., Kuhn, H. and Linser, H. (1968) The effect of the sulphur nutrition on the synthesis of mustard oils. *Z. Pflanzenern. Bodenk.* **121**, 221-230.

Mazelis, M. (1993) Catabolism of sulfur-containing amino acids. In *Sulfur Nutrition and Assimilation in Higher Plants. Regulatory, Agricultural and Environmental Aspects* (eds. De Kok. L.J., Stulen, I., Rennenberg, H., Brunold, C. and Rauser, W.E.), pp. 95-108, SPB Academic Publishing, The Hague.

Mcbride, M.B., Richards, B.K., Steenhuis, T., Russo, J.J. and Sauve, S. (1997) Mobility and solubility of toxic metals and nutrients in soil fiftien years after sludge application. *Soil Sci.* **162**, 487-500.

McGill, W.B. and Cole, C.V. (1981) Comparative aspects of cycling of organic C, N, S and P through soil organic matter. *Geoderma* **62**, 267-286.

McGrath, S.P. and Zhao, F.J. (1995) A risk assessment of sulphur deficiency in cereals using soil and atmospheric deposition data. *Soil Use Manage.* **11**, 110-114.

McGrath, S.P., Zhao, F.J. and Withers, P.J.A. (1996) *Development of Sulphur Deficiency in Crops and its Treatment*, The Fertilizer Society, Peterborough.

Mithen, R.F., Lewis, B.G., Heaney, R.K. and Fenwick, G.R. (1987) Resistance of leaves of *Brassica* species to *Leptosphaeria maculans. Trans. Brit. Mycol. Soc.* **88**, 525-531.

Moore, P.A., Attanandana, T. and Patrick, W.H. (1990) Factors affecting rice growth on acid sulfate soils. *Soil Sci. Soc. Amer. J.* **54**, 1651-1656.

Müntz, K., Christov, V., Jung, R., Saalbach, G., Saalbach, I., Waddell, D., Pickardt, T. and Schieder, O. (1997) Genetic engineering of high methionine proteins in grain legumes. In *Sulphur Metabolism in Higher Plants. Molecular, Ecophysiological and Nutritional Aspects* (eds. Cram, W.J., De Kok, L.J., Stulen, I., Brunold, C. and Rennenberg, H.), pp. 71-86, Backhuys Publishers, Leiden.

Naito, S., Hirai, M.Y., Inaba-Higano, K., Nambara, E., Fujiwara, T., Hayashi, H., Komeda, Y. and Chino, M. (1995) Expression of soybean seed storage protein genes in transgeneic plants and their response to sulfur nutritional conditions. *J. Plant Physiol.* **145**, 614-619.

Overstreet, R., Martin, J.C., Schulz, R.K. and McClutcheon, O.D. (1955) Reclamation of an alkali soil of the Hacienda series. *Hilgardia* **27**, 319-332.

Palko, J. and Weppling, K. (1994) Lime requirement experiments in acid sulphate soils. *Acta Agric. Scand., Sect. B: Soil Plant Sci.* **44**, 149-156.

Paquet, L., Rathinasabapathi, B., Saini, H., Zamir, L., Gage, D.A., Huang, Z.H. and Hanson, A.D. (1994) Accumulation of the compatible solute 3-dimethylsulfonioproponate in sugarcane and its relatives, but not in other gramineous crops. *Aust. J. Plant Physiol.* **221**, 37-48.

Pfanz, H., Martinoia, E., Lange, O.L. and Heber, U. (1987) Flux of SO_2 into leaf cells and the cellular acidification by SO_2. *Plant Physiol.* **85**, 928-933.

Piluk, J., Hartel, P.G. and Haines, B.L. (1998) Production of carbon disulfide (CS_2) from L-djenkolic acid in roots of *Mimosa pudica* L. *Plant Soil* **200**, 27-32.

Protoschill-Krebs, G. and Kesselmeier, J. (1992) Enzymatic pathways for the consumption of carbonyl sulphide (COS) by higher plants. *Bot. Acta* **105**, 206-212.

Puxbaum, H. and König, G. (1997) Observation of dipropenyldisulfide and other organic sulfur compounds in the atmosphere of a beech forest with *Allium ursinum* ground cover. *Atm. Environ.* **31**, 291-294.

Randle, W.M., Block, E., Littlejohn, M.H., Putman, D. and Bussard, M.L. (1994) Onion (*Allium cepa* L.) thiosulfinates respond to increasing sulfur fertility. *J. Agric. Food Chem.* **42**, 2085-2088.

Rasmussen, P.E., Ramic, R.E., Ekin, L.G. and Rohde, C.R. (1977) Tissue analyses guidelines for diagnosing sulfur deficiency in white wheat. *Plant Soil* **46**, 153-163.

Rendig, V.V. (1956) Sulfur and nitrogen composition of fertilized and unfertilized alfalfa grown on a sulfur-deficient soil. *Soil Sci. Soc. Amer. Proc.* **20**, 237-240.

Robinson, B.H., Chiarucci, A., Brooks, R.R., Petit, D., Kirkham, J.H., Gregg, P.E.H., Dedominicis, V. (1997) The nickel hyperaccumulator plant *Alyssum bertolonii* as a potential agent for phytoremediation and phytyming of nickel. *J. Geochem. Expl.* **59**, 75-86.

Saalbach, E. (1970) Über die Bestimmung des Schwefelversorgungs-grades von Hafer. *Pflanzenern. Bodenk.* **127**, 92-100.

Saalbach, I., Pickardt, T., Waddell, D.R., Hillmer, S., Schieder, O. and Müntz, K. (1995) The sulphur-rich Brazil nut 2S albumin is specifically formed in transgenic seeds of the grain legume *Vicia narbonensis*. *Euphytica* **85**, 181-192.

Schachtschabel, P., Blume, H.P., Brümmer, G., Hartge, K.H. and Schwertmann, U. (1992) *Lehrbuch der Bodenkunde*, F. Enke Verlag, Stuttgart.

Schone, F., Groppel, B., Hennig, A., Jahreis, G. and Lange, R. (1997) Rapeseed meals, methimazole, thicyanate and iodine affect growth and thyroid - Investigations into glucosinolate tolerance in the pig. *J. Sci. Food Agric.* **74**, 69-80.

Schnug, E. (1990) Glucosinolates - fundamental, environmental and agricultural aspects. In *Sulfur Nutrition and Sulfur Assimilation in Higher Plants. Fundamental, Environmental and Agricultural Aspects* (eds. Rennenberg, H., Brunold, C., De Kok, L.J. and Stulen, I.), pp. 97-106, SPB Academic Publishers, The Hague.

Schnug, E. (1991) Sulphur nutritional status of European crops and consequences for agriculture. *Sulphur Agric.* **15**, 7-12.

Schnug, E. and Haneklaus, S. (1994) Sulphur deficiency in *Brassica napus* - Biochemistry - Symptomatology - Morphogenesis. *Wiss. Mitt. Bundesforschungsanstalt Landw. Sonderheft* **144**, 1-31. Braunschweig.

Schreiner, R.P. and Koide, R.T. (1993) Mustards, mustard oils and mycorrhizas. *New Phytol.* **123**, 107-113.

Shewry, P.R., Price, J., Tatham, A.S., Smith, D.B, Cantrell, I.C., Davies, J.L. and Harris, N. (1993) Spatial patterns of hordein distribution in relation to barley malting quality. Cereal Quality III. *Aspects Appl. Biol.* **36**, 43-49.

Simonsen, J.L. (1947) *The Terpenes.* Vol 1. Cambridge University Press, Cambridge.

Smith, H.J.C., Loock, A.H. and Van der Merwe, A.J. (1994) Gypsum as a soil and water ameliorant. *S. Afr. J. Plant Soil* **11**, 139-141.

Stanford, G. and Jordan, H.V. (1966) Sulfur requirements of sugar, fiber, and oil crops. *Soil Sci.* **101**, 258-266.

Stehouwer, R.C., Sutton, P. and Dick, W.A. (1996) Transport and plant uptake of soil-applied dry flue gas desulfurization by-products. *Soil Sci.* **161**, 562-574.

Steudler, P.A. and Peterson, B.J. (1984) Contribution of gaseous sulphur from salt marshes to the global sulphur cycle. *Nature* **311**, 455-457.

Sumner, M.E., Shahandeh,H., Bouton, J. and Hammel, J. (1986) Amelioration of an acid soil profile through deep liming and surface application of gypsum. *Soil Sci. Soc. Amer. J.* **50**, 1254-1258.

Tabatabai, M.A. (1984) Importance of sulphur in crop production. *Biogeochemistry* **1**, 45-62.

Tabatabai, M.A. and Bremner, J.M. (1970) Arylsulphatase activity of soils. *Soil Sci. Soc. Amer. Proc.* **34**, 225-229.

Visser, J.H. (1986) Host odor perception in phytophagous insects. *Annu. Rev. Entomol.* **31**, 121-144.

Withers, P.J.A. and O'Donnell, F.M. (1994) The response of double-low winter oilseed rape to fertiliser sulphur. *J. Sci. Food Agric.* **66**, 93-101.

Withers, P.J.A., Tytherleigh, A.R.J. and O'Donell, F.M. (1995) Effect of sulphur fertilisers on the grain yield and sulphur content of cereals. *J. Agric. Sci.* **125**, 317-324.

Xie, Z.M. and Huang, C.Y. (1998) Control of arsenic toxicity in rice plants grown on an arsenic-polluted paddy soil. Commun. *Soil Sci. Plant Anal.* **29**, 2471-2477.

Zhao, F.J., Evans, E.J., Bilsborrow, P.E. and Syers, J.K. (1993a) Sulphur uptake and distribution in double and single low varieties of oilseed rape (*Brassica napus* L.). *Plant Soil* **150**, 69-76.

Zhao, F., Bilsborrow, P.E., Evans, E.J. and Syers, J.K. (1993b) Sulphur turnover in the developing pods of single and double low varieties of oilseed rape (*Brassica napus* L.). *J. Sci. Food Agric.* **62**, 111-119.

Zhao, F., Evans, E.J., Bilsborrow, P.E. and Syers, J.K. (1994) Influence of nitrogen and sulphur on the glucosinolate profile of rapeseed (*Brassica napus* L.). J. Sci. *Food Agric.* **64**, 295-304.

Zhao, F.J., Wither, P.J.A., Evans, E.J., Monaghan, J., Salmon, S.E., Shewry, P.R. and McGrath, S.P. (1997) Sulphur nutrition: An important factor for the quality of wheat and rapeseed. *Soil Sci. Plant Nutr.* **43**, 1137-1142.

Zonneveld, E.A. (1995) Air pollution, emission 1993 and 1994. *Kwartaalber. Milieustat.* **12** (4), 3-9.

16

Biodegradation of sulfonated aromatic compounds

Nico C.G. Tan and Jim A. Field

16.1 INTRODUCTION

Microorganisms are known to play an important role in effecting bio- and geochemical cycles by mineralizing biopolymers and xenobiotic compounds (Lie *et al.* 1998). Only 30 years ago it was recognized that aerobic bacteria were able to mineralize sulfonated aromatic compounds. As a result of this biodegradation, the sulfur moiety of the sulfonate group is able to enter the sulfur cycle (Cain and Farr 1968; Focht and Williams 1970; Ripin *et al.* 1971). Today, our knowledge of the biodegradation of sulfonated aromatic compounds is still rather limited. This chapter reviews the present state of the art of microbial and process technological aspects of the degradation of sulfonated aromatic compounds.

The characteristic functional group of the sulfonated aromatic molecule is the sulfonate group (or sulfonic acid group; $-SO_3H$). Sulfonated aromatic

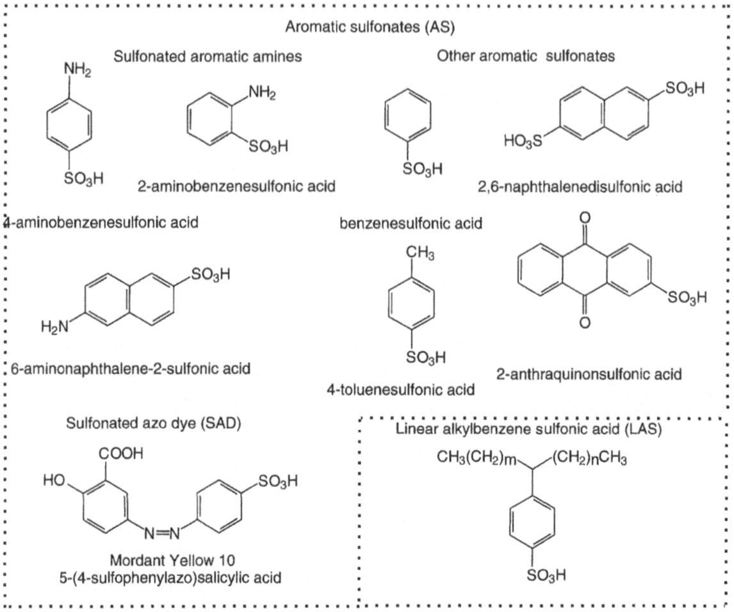

Figure 16.1. Examples of the chemical structure of sulfonated aromatic compounds.

compounds represent 10% of the dissolved organic matter present in the river Rhine (Malle 1978). These compounds are discharged via both industrial (via chemical production and paper pulping) and domestic activities. With the exception of one compound, all sulfonated aromatic compounds are xenobiotic. Phenazinesulfonic acid (Aeruginosin B, 2-amino-6-carboxy-10-methyl-8-sulfo-phenaziniumbetain) is the only reported example of a naturally occurring aromatic sulfonated compound. This pigment is produced by the bacterial strain *Pseudomonas aeruginosa* (Herbert and Holliman 1964).

Sulfonated aromatic compounds can be divided into two main groups (Figure 16.1). The first group comprises the linear alkylbenzene sulfonates (LASs). LAS are used as detergents and mainly emitted via domestic processes. Their fate in the environment has been extensively studied and reviewed (Berna *et al.* 1991; Federle and Ventullo 1990; Jimenez *et al.* 1991; Mampel *et al.* 1998; Thoumelin 1991). The second group of sulfonated aromatic compounds are the aromatic sulfonates (ASs). The group of AS contains, first, the sulfonated azo dyes (SADs) which are used for colouring textiles, paper, food and leather. Second, the group of the ASs contains the

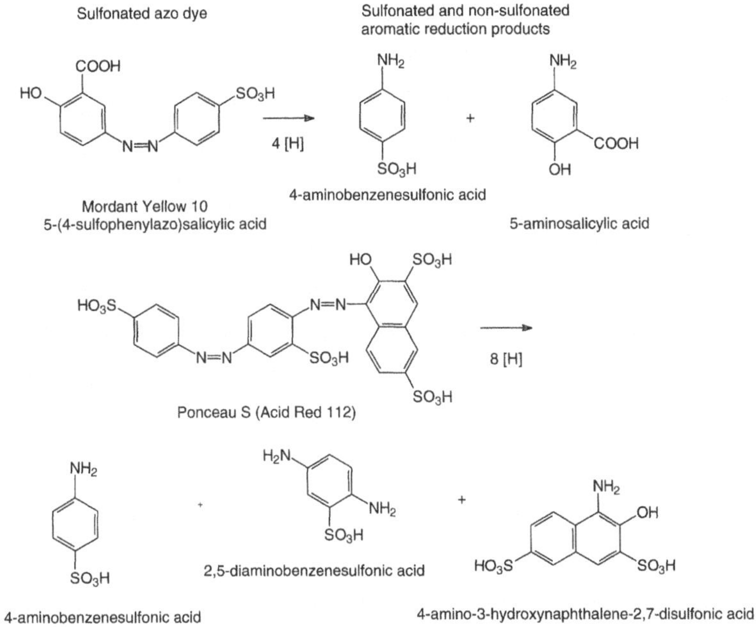

Figure 16.2. Sulfonated azo dyes and their sulfonated and non-sulfonated biodegradation products after reduction and cleavage of the azo linkage.

precursors and biodegradation products of the SADs and mainly contains the sulfonated aromatic amines (Figure 16.2). Third, some other sulfonated aromatic compounds are also in the group of the ASs. These other sulfonated aromatic compounds are mainly used as building blocks for SADs and LASs. Also lignosulfonates, which are produced via sulfite pulping in the paper industry, are a major source of other sulfonated aromatic compounds (Hernandez Perez *et al.* 1998). Only a few cover the biodegradation of ASs (Hooper 1991; Kertesz *et al.* 1994; Cook *et al.* 1998).

The sulfonate group renders compounds containing them to be highly soluble in the aqueous phase. Therefore, ASs are mainly found in the aquatic environment and do not accumulate in sediments. ASs are mainly found in wastewater from chemical industries producing or processing sulfonated compounds, e.g. leather, textile, printing, paper, and pharmaceutical industries. The AS represent structural elements used as building blocks for azo dyes, drugs, detergents, optical brighteners, and artificial sweeteners (Hansen *et al.* 1992; Locher *et al.* 1989). The ASs are also used directly: 4-aminobenzene sulfonic acid (4-ABS) is used as a preservative (Hooper 1991), 3-ABS as a mild oxidant (Locher *et al.* 1989) and 2-benzosulfonic acid (2-BOS) as a wetting agent in toothpaste (Hansen *et al.* 1992).

This chapter will focus on the bacterial biodegradation of ASs and especially the sulfonated aromatic amines will be studied in detail. Sulfonated aromatic amines are mainly released into the environment via either the production or the biodegradation of sulfonated azo dyes (Figure 16.2).

The biodegradation potential of the sulfonated aromatic amines compounds will be reviewed under both anaerobic and aerobic conditions. Furthermore, their biodegradation pathways, with desulfonation as the most important step, will be summarised. At the end of this chapter some applications for the removal of sulfonated aromatic amines and AS from wastewater will be presented.

16.2 BIODEGRADATION

First, the desulfonation as an important step in the biodegradation of the sulfonated aromatic compounds will be focused upon. Second, biodegradation of the sulfonated aromatic amines will be reviewed under both anaerobic and aerobic conditions.

16.2.1 Desulfonation of aromatic sulfonates

The mineralization of sulfonated aromatic compounds results in the release of the sulfur moiety. Under anaerobic conditions, sulfate-reducing bacteria will use this sulfur moiety for the production of sulfide (HS^-). However, until now there is only limited knowledge about the mechanism of anaerobic aromatic desulfonation. Some ASs are used as a sulfur source under anaerobic conditions by fermentative bacteria (Chien *et al.* 1995; Denger *et al.* 1996; Denger and Cook 1997). The pathway and the enzymology involved have not yet been elucidated. The sulfonate group could potentially be used as alternative electron acceptors under anaerobic conditions. However, no sulfonated aromatic compound has yet been shown to serve as an electron acceptor (Cook *et al.* 1998; Lie *et al.* 1996, 1998).

The desulfonation process has mainly been studied under aerobic conditions. Therefore, the mechanism of desulfonation is better understood under aerobic conditions. Aerobically, sulfite (HSO_3^-) will be formed during desulfonation and sulfite can be further oxidised to sulfate (SO_4^{2-}). Three different aerobic patterns of desulfonation are reported in the literature

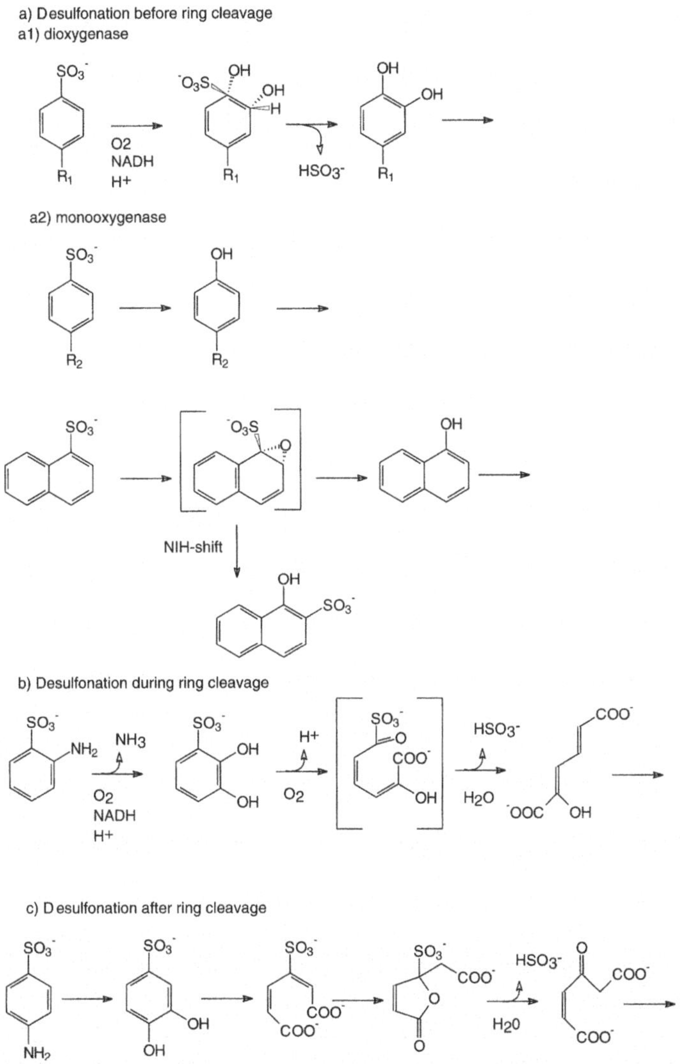

Figure 16.3. Different patterns of aerobic desulfonation of sulfonated aromatic compounds (Feigel and Knackmuss 1988; Kertesz *et al.* 1994; Kneifel *et al.* 1997; Locher *et al.* 1991; Thurnheer *et al.* 1990; Zurrer *et al.* 1987).

(Figure 16.3). Desulfonation can occur (a) before (b) during or (c) after ring cleavage.

The first pattern of desulfonation can be further differentiated in a mono- and a dioxygenase transformation. The dioxygenase is a multi-component mechanism in which an NADH-linked reaction takes place, followed by spontaneous desulfonation (Locher *et al.* 1991). Mono-oxygenase involves the incorporation of one oxygen molecule followed by desulfonation (Zurrer *et al.* 1987). Formation of an epoxide is also possible via a monooxygenase (Kneifel *et al.* 1997). The monooxygenase reactions have only been observed for micro-organisms using the AS as sole sulfur source.

16.2.2 Anaerobic biodegradation

There is scarcely evidence for biodegradation and desulfonation of ASs under anaerobic conditions. Desulfonation of azo dyes under anaerobic conditions is not often found. Two reports showed evidence for usage of one of the sulfonate groups of the sulfonated azo dye Acid Red 1 (disulfonated azo dye) by *Clostridium* strains, which was used as a sulfur source under fermentative and sulfur-limited conditions. However, none of the biodegradation products were identified and therefore the pathway is unclear (Denger *et al.* 1996; Denger and Cook 1997). No evidence for mineralization of the sulfonated aromatic amines has been reported. Two reports were found in which biodegradation was tested under anaerobic conditions and none of the sulfonated aromatic amines tested could serve as carbon and energy sources under the applied methanogenic and sulfate-reducing conditions (Brown and Hamburger 1987; Kuhn and Suflita 1989).

16.2.3 Aerobic biodegradation

The sulfonated aromatic amines can be divided into two groups on the basis of their molecular structure (Figure 16.1): aminobenzenesulfonic acids (ABSs) and aminonaphthalenesulfonic acids (ANSs). The sulfonated aromatic amines are for the most part xenobiotic and have long been considered non-biodegradable (Bretscher 1981). Biodegradation experiments using natural mixed cultures from sludges and sediments, however, have demonstrated that occasionally these compounds are biodegradable. Aerobic biodegradation screening experiments with activated sludge showed that some sulfonated aromatic amines were susceptible to biodegradation (Brown and Hamburger 1987). However, complete mineralization of these compounds was not clearly shown (Brown and Hamburger 1987). Although these compounds are difficult to degrade, some specialised aerobic bacterial strains have been isolated which are able

Table 16.1. Bacterial strains able to degrade ABS (+ biodegradation; - no biodegradation; s compound used only as sulfur source; if not tested open; *Al.* = *Alcaligenes; A.* = *Arthrobacter; P.* = *Pseudomonas; P.a.* = *Pseudomonas acidovorans; H.p.* = *Hydrogenophaga palleronii*)

Bacteria	*Al.* sp.				*P.a.*	*P.* sp.		*A.* sp.	*P.* sp.	*H.p.*
Strain	O-1	O-2	S-1	S-3	M-1	S-313	S-832	DZ-6	DS1304	S1
Reference	a, b	a	a, b	a	b	c	c	c	d	c
2-ABS	+	+	-	-	-	+s	+s	+s		
3-ABS	-	-	-	-	+	+s	+s	+s		
4-ABS	-	-	+	+	-	+s	+s	+s	+	+

References: a (Thurnheer *et al.* 1986); b (Locher *et al.* 1989); c (Zurrer *et al.* 1987); d (Dubeikovskii *et al.* 1992); e (Feigel and Knackmuss 1988, 1993).

Table 16.2. Bacterial strains able to degrade ANS (+ biodegradation; - no biodegradation; s compound used only as sulfur source; if not tested open; *S.* = *Sphigomonas*; *M.* = *Moraxella*; *A.* = *Arthrobacter*; *P.* = *Pseudomonas*; Cc11. = coculture of 11 bacterial strains)

Bacteria	*S.* sp.	Cc	*P.* sp	*P.* sp.	*M.* sp.	*A.* sp.	*P.* sp.		
Strain	BN6	11.	TA-1	TA-2	ASL4	DZ-6	S-313	S-832	Z-63
Reference	a	b	c	d	e	f	f	f	f
2-A-1-NS			+	+					
5-A-1-NS						+s	+s	-	+s
6-A-1-NS	-								
8-A-1-NS						+s	+s	+s	+s
5-A-2-NS	+		-	-					
6-A-2-NS	+	+		-	+				
7-A-2-NS	+								
8-A-2-NS	+								
6-A-4-H-2-NS	+					+s	+s	+s	-
7-A-4-H-2-NS	-					+s	+s	+s	+s
3-A-1,5-NDS						+s	+s	+s	-
2-A-4,8-NDS	-								
4-A-5-H-2,7-NDS						+s	+s	+s	-

References: a (Nortemann *et al.* 1986, 1994); b (Rozgaj and Glancer 1992); c (Ohe and Watanabe 1986); d (Ohe *et al.* 1990); e (Wittich *et al.* 1988); f (Zurrer *et al.* 1987).

to biodegrade these compounds. Table 16.1 and Table 16.2 summarise reported biodegradability experiments of ABSs and ANSs.

16.2.3.1 Aminobenzenesulfonic acids

Of the ABS compounds, only the simple substituted ABS isomers (without any additional substituents) were aerobically biodegraded. No bacteria have been reported that were able to degrade substituted ABS compounds. Four different bacterial strains were isolated, which used the simple ABS as sole energy and carbon sources. These bacteria were isolated after enrichment of inoculum from sewage plants treating sulfonate-containing industrial wastes (Junker *et al.* 1994; Locher *et al.* 1989; Thurnheer *et al.* 1986). One of these isolated bacteria, *Alcaligenes* sp. O-1, which degraded 2-ABS was extensively studied. The degradation pathway of 2-ABS was determined (Figure 16.3b) and enzymes were isolated and characterised. The other three strains isolated were strain M-1, S-1 and S-3. Strain M-1 could degrade 3-ABS and strains S-1 and S-3 were able to degrade 4-ABS. Strain S-1 was tested in our laboratory, but no degradation of 4-ABS was obtained. This may indicate that the degradation capacity was located on a plasmid and was probably lost.

A co-culture composed of two bacterial strains (*Hydrogenophaga palleronii* S1† and *Agrobacterium radiobacter* S2) was also shown to be able to completely mineralise 4-ABS (Feigel and Knackmuss 1988, 1993). 4-ABS was used as sole source of energy, carbon, nitrogen and sulfur. Strain S1 degrades 4-ABS via a dioxygenase to 5-sulfocatechol. In turn, 5-sulfocatechol is utilised by strain S2, which produces essential vitamins for strain S1 (Dangmann *et al.* 1996). The degradation pathway of 4-ABS by the co-culture is depicted in Figure 16.3c.

Afterwards, strain S1 was used for continuous adaptation to obtain strain S5 (from natural mutations) which was able to aerobically degrade a sulfonated azo dye (4-carboxy-4'-azobenzene) as a pure culture (Blumel *et al.* 1998).

Pseudomonas DS1304 is another bacterial strain, which was described as capable of degrading 4-ABS as a pure culture. This strain attacks 4-ABS via an initial dioxygenase which desulfonates the compound (Dubeikovskii *et al.* 1992). Furthermore, three strains have been described which could use the sulfonate group of all three ABS isomers as a sulfur source for growth under sulfur-limiting growth conditions with glucose as the carbon source. These bacteria utilise a monooxygenase for the desulfonation (Zurrer *et al.* 1987).

The bacterial strains B-1, P-2, P-3, *Comamonas testosteroni* T-2, *C. testosteroni* PSB-4 (Thurnheer *et al.* 1986), N-1 (Locher *et al.* 1989), L-1 (Khlebnikov *et al.* 1997), *P. maltophila* BSA56 (Lee and Clark 1993), *Alcaligenes* sp. GA-1 (Takeo *et al.* 1997) and *Agrobacterium radiobacter* S2

† This strain S1 is different to the strain S-1 described by Thurnheer *et al.* (1986).

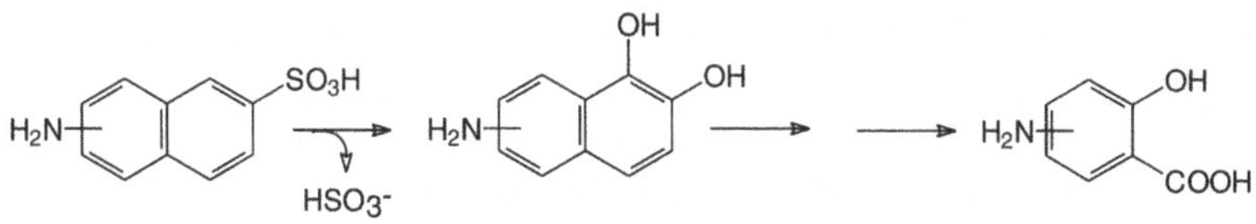

Figure 16.4. Degradation of A-2-NS by *Sphigomonas* sp. BN6.

(Feigel and Knackmuss 1988, 1993) were tested on one or more ABS compounds but showed no biodegradation. However, these strains were able to degrade other sulfonated aromatic compounds, mainly substituted benzene sulfonic acids.

16.2.3.2 Aminonaphthalenesulfonic acids

Several ANSs have been reported to be degraded by specialized aerobic bacteria. The isolated bacterial strains and the compounds which were degraded are summarized in Table 16.2. These results show that 6-A-2-NS is one of the most studied ANS compounds. Two co-cultures and one bacterial strain ASL-4 capable of completely mineralizing 6-A-2-NS are reported (Nortemann *et al.* 1986; Rozgaj and Glancer 1992; Wittich *et al.* 1988).

The strain responsible for the initial degradation of 6-A-2-NS was isolated out of one of the cocultures by Nortemann et al. (1986) and was characterised as a *Sphigomonas* sp. BN6. This strain BN6 could also transform 5-, 6-, 7- or 8-A-2-NS. These compounds were degraded via an initial dioxygenase which desulfonates the compound after ring cleavage, yielding the corresponding aminosalicylic acid (ASA) which accumulates in the growth medium (Figure 16.4). The ASA compounds were further degraded aerobically by another member of the co-culture (Nortemann *et al.* 1986, 1994). The biodegradation of 5-A-2-NS by strain BN6 led to the formation of a dead-end product 5-hydroxy-quinoline-2-carboxylic acid (Nortemann *et al.* 1993). Furthermore, strain BN6 was used to degrade different sulfonated azo dyes in sequential or simultaneous aerobic/anaerobic culture conditions together with an aerobic 5-ASA degrading culture (Haug *et al.* 1991; Kudlich *et al.* 1996).

Another co-culture, which consisted of 11 bacterial strains, was able to degrade 6-A-2-NS and formed a stable co-culture and was enriched from a treatment plant treating industrial wastewater. None of the single strains were able to degrade the 6-A-2-NS individually and each strain played a part in the degradation process. Therefore, this co-culture showed a high degree of co-operation (Rozgaj and Glancer 1992).

The strain ASL-4 originally isolated on 2,6- and 1,6-naphthalenedisulfonic acid (NDS) also showed activity on 6-A-2-NS. The biodegradation is suggested to proceed via an initial dioxygenase, which deaminates the compound, and after ring cleavage, 5-sulfosalicylic acid is formed which is further degraded via an oxygenolitic elimination of the sulfonate group resulting in gentisate (Wittich *et al.* 1988). Two *Pseudomonas* strains were isolated by a Japanese group from soil samples. These strains were able to degrade 2-A-1-NS and utilise this compound as the sole source of carbon and nitrogen. Both bacteria degraded the compound via an initial 1,2-dioxygenase, which lead to the desulfonation and deamination of the molecule (Ohe *et al.* 1990; Ohe and Watanabe 1986).

The same strains, which utilise the sulfonate group from aminobenzenesulfonic acids under sulfate limitation, were able to utilise the sulfonate group of ANS compounds as a sulfur source under sulfate-limited conditions (Zurrer *et al.* 1987). Desulfonation was shown to be caused by a monooxygenase.

16.3 BIOTECHNOLOGICAL REMOVAL OF AROMATIC SULFONATES FROM WASTEWATER

Only a few applications with respect to the removal of ASs are reported in the literature, and these are now reviewed. A summary of the biotechnological approaches is given in Table 16.3.

Thurnheer et al. (1988) used a chemostat with a coculture of five different bacteria to degrade seven sulfonated compounds. The bacterial strains that were used are *Comamonas testosteroni* T-2, *C. testosteroni* PSB-4, *Alcaligenes* sp. O-1, strain M-1, and strain S-1. Suspension of mixtures of these strains in the CSTR could degrade 2-, 3-, 4-ABS, 4-BOS, 4-toluenesulfonic acid (TS), benzenesulfonic acid (BS), and 4-hydroxybenzenesulfonic acid (HBS). A maximum degradation rate of 138 mg C h^{-1} l^{-1} was observed for all compounds added. At the end of the experiment (903 days), 4-chlorobenzenesulfonic acid was added which was also degraded in the reactor (Thurnheer *et al.* 1988).

Treatment of 3-nitrobenzenesulfonic acid (NBS) and 3-ABS in a trickling filter were examined by Kolbener *et al.* (1994). As the inoculum source, six activated sludges were tested. Four activated sludges treated domestic wastewater and two activated sludges treated industrial and textile wastewater. Interestingly, only the last two inoculum sources were able to degrade 3-NBS and 3-ABS, respectively (Kolbener *et al.* 1994).

The removal capacity of two *C. testosteroni* strains to degrade 4-TS was

Table 16.3. Summary of literature in which a biotechnological approach was used for the removal of aromatic sulfonates from wastewater (*Al.* = *Alcaligenes*; *C.t.* = *Comamonas testosteroni*; *P.t.* = *Pseudomonas testosteroni*; *S.* = *Sphigomonas*)

Compounds treated	Reactor type (volume (l))	Inoculum used	Organic loading rate (g COD l^{-1} d^{-1})	Efficiency (%)	Ref.
2-,3-,4-ABS 4-BOS 4-TS BS 4-HBS	chemostat (2.0)	*C.t.* T-2 *C.t.* PBS-4 *Al.* O-1 strain M-1 strain S-1	0.3 - 10.1	100	a
6-AN-2-S	airlift-loop reactor (2.5)	*S.* BN6 BN9 BN11	8.7	99	b
2-NS	airlift-loop reactor (2.5)	*P.t.* A3	13.1	99	b
4-TS	fixed bed biofilm reactor (0.066)	*C.t.* T-2	6.2 - 12.4	70	c
4-TS	fixed bed biofilm reactor (0.066)	strain L-1	6.2 - 12.4	59	c
3-NBS 3-ABS	trickling filter (0.7)	6 activated sludges	not given	removal if industrial sludge used	d
1-, 2-NS 1,5-NDS 1,6-NDS 2,6-NDS	two stage airlift-loop reactors low salt	*P.t.* A3 strain RK3 mixed culture	2.2 - 69.3 first stage 1.0 - 8.1 second stage	84, except no removal of 1,5-NDS	e
1-, 2-NS 1,5-NDS 1,6-NDS 2,6-NDS	two stage airlift-loop reactors high salt (2.5)	*P.t.* A3 strain RK3	2.2 - 69.3 first stage 1.0 - 8.1 second stage	71, except no removal of 1,5-NDS	e
2-NS	airlift (2.0)	*P.t.* A3	11.9	100	f

References: a (Thurnheer *et al.* 1988); b (Diekmann *et al.* 1990); c (Khlebnikov *et al.* 1997); d (Kolbener *et al.* 1994); e (Krull and Hempel 1994); f (Pack and Hempel 1997).

compared in an aerobic down-flow fixed-bed bioreactor (Khlebnikov *et al.* 1997). Strain T-2 had a higher degradation efficiency up to 70 % and reached faster steady-state compared with strain L-1 (59 % degradation efficiency) at a dilution rate of 0.15 h^{-1}. However, the degradation rates in batch were similar i.e. 66 mg C h^{-1} l^{-1} for strain T-2 and 68 mg C h^{-1} l^{-1} for stain L-1. The biomass was mainly attached to the upper part of the reactor.

C. testosteroni T-2 was also used for bioaugmentation of an activated sludge process to remove 4-TS added as a sole substrate. Addition of different percentages of the strain T-2 to different types of activated sludge tested both survival and activity. The removal of 4-TS also occurred in the batch without any addition of strain T-2. However, higher percentages of strain T-2 led to a faster degradation of 4-TS. Therefore, bacterial strain T-2 was able to survive and enhance the removal of 4-TS in a natural complex microcosm (Bokhamy *et al.* 1997). These results are not surprising since 4-TS was dosed as the only carbon source, which created a competitive advantage for strain T-2 relative to the endogenous activated sludge population.

Immobilization of *P. testosteroni* A3 on sand particles was first tested (Wagner and Hempel 1988) before these were used to test sub-optimal environmental conditions (Diekmann *et al.* 1990). A temperature-range between 12 °C and 35 °C and a pH-range between pH 5.7 and pH 8.8 did not influence the complete 2-NS removal. If the system was operated without oxygen for periods between 0.5 and 68 h or with pH shock-loads of 3 and 10, the removal capacity completely recovered after a temporary inhibition. The same results under sub-optimal environmental conditions were observed for 6-AN-2-S and the strain BN6, BN9, and BN11 (Diekmann *et al.* 1990). This mixed culture degrading 6-AN-2-S showed a linear correlation between production of carbon dioxide and growth rate for immobilized and suspended cultures (Diekmann and Hempel 1989).

The influence of heavy metals on the performance of an airlift-loop reactor fed with 2-NS was evaluated by Pack and Hempel (1997). The reactor was inoculated with an immobilized *P. testosteroni* A3 on broken sand particles. Shock loads of nickel (100-400 mg l^{-1}) and zinc (1000 mg l^{-1}) only temporarily affected the 2-NS removal efficiency. Continuous feeding of 20 mg l^{-1} cadmium and nickel completely inhibited the 2-NS removal. The removal recovered completely within 24 h after omission of the heavy metals dosage (Pack and Hempel 1997).

Two airlift-loop bioreactors operated sequentially were tested to treat 1-, 2-naphthalenesulfonic acid (NS), 1,5-, 1,6- and 2,6-NDS, which are all precursors of sulfonated azo dyes (Krull and Hempel 1994). *Pseudomonas testosteroni* A3 and a mixed culture degrading 1-, 2-NS were used as the inoculum source for the first stage. Strain RK3 was used as the inoculum source for the second stage. The bacteria were immobilized on broken sand particles. Since high salt concentrations are used during NS and NDS manufacturing, low (medium nutrients only) and high (6.6 g l^{-1} Na_2SO_4) salt-regimes were evaluated. Feeding the two-stage system wastewater containing the five sulfonated compounds (total concentration of 3127 mg l^{-1}) showed a removal efficiency of 71% with high salt concentrations compared with 84% at low salt conditions. The first stage was used to degrade the readily degradable NS compounds, whereas the second stage was used to degrade the NDS compounds. Since no bacterial strain was added capable to degrade 1,5-NDS, this compound was not degraded. These experiments showed that these NS and NDS compounds were degraded in an immobilized bioaugmented bioreactor (Krull and Hempel 1994).

16.4 REFERENCES

Berna, J.L., Moreno, A. and Ferrer, J. (1991) The behavior of LAS in the environment. *J. Chem. Technol. Biotechnol.* **50**(3), 387-398.

Blumel, S., Contzen, M., Lutz, M., Stolz, A. and Knackmuss, H.J. (1998) Isolation of a bacterial strain with the ability to utilize the sulfonated azo compound 4-carboxy-4′-sulfoazobenzene as the sole source of carbon and energy. *Appl. Environ. Microbiol.* **64**(6), 2315-2317.

Bokhamy, M., Deront, M., Adler, N. and Peringer, P. (1997) Survival and activity of Comamonas testosteroni in mixed population. *Wat. Res.* **31**(11), 2802-2810.

Bretscher, R.B. (1981) Waste disposal in chemical industry. In *Microbial degradation of xenobiotics and recalcitrant compounds*, Academic Press Inc., London, pp 65-74.

Brown, D. and Hamburger, B. (1987) The degradation of dyestuffs: Part III - Investigations of their ultimate degradability. *Chemosphere* **16**(7), 1539-1553.

Cain, R.B. and Farr, D.R. (1968) Metabolism of arylsulfonates by micro-organisms. *Biochem. J.* **106**, 859-877.

Chien, C.C., Leadbetter, E.R. and Godchaux, W., III (1995) Sulfonate-sulfur can be assimilated for fermentative growth. *FEMS Microbiol. Lett.* **129**(2-3), 189-193.

Cook, A.M., Laue, H. and Junker, F. (1998) Microbial desulfonation. *FEMS Microbiol. Rev.* **22**(5), 399-419.

Dangmann, E., Stolz, A., Kuhm, A.E., Hammer, A., Feigel, B., Noisommitrizzi, N., Rizzi, M., Reuss, M. and Knackmuss, H.J. (1996) Degradation of 4-aminobenzenesulfonate by a two-species bacterial coculture - Physiological interactions between *Hydrogenophaga palleronii* S1 and *Agrobacterium radiobacter* S2. *Biodegradation* **7**(3), 223-229.

Denger, K., Kertesz, M.A., Vock, E.H., Schoen, R., Maegli, A. and Cook, A.M. (1996) Anaerobic desulfonation of 4-tolylsulfonate and 2-(4-sulfophenyl) butyrate by a *Clostridium* sp. *Appl. Environ. Microbiol.* **62**(5), 1526-1530.

Denger, K. and Cook, A.M. (1997) Assimilation of sulfur from alkyl- and arylsulfonates by *Clostridium* spp. *Arch. Microbiol.* **167**(2-3), 177-181.

Diekmann, R. and Hempel, D.C. (1989) Production rates of carbon dioxide by suspended and immobilized bacteria degrading 6-amino-2-naphthalenesulfonic acid. *Appl. Microbiol. Biotechnol.* **32**(1), 113-117.

Diekmann, R., Naujoks, M., Gerdes Kuehn, M. and Hempel, D.C. (1990) Effects of suboptimal environmental conditions on immobilized bacteria growing in continuous culture. *Bioprocess Eng.* **5**(1), 13-17.

Dubeikovskii, A.N., Gafarov, A.B. and Naumov, A.V. (1992) Plasmid determined degradation of sulfoaromatic compounds by *Pseudomonas* sp. BS1304. *Genetika* **28**, 34-39.

Federle, T.W. and Ventullo, R.M. (1990) Mineralization of surfactants by the microbiota of submerged plant detritus. *Appl. Environ. Microbiol.* **56**(2), 333-339.

Feigel, B.J. and Knackmuss, H.J. (1988) Bacterial catabolism of sulfanilic acid via catechol-4-sulfonic acid. *FEMS Microbiol. Lett.* **55**, 113-118.

Feigel, B.J. and Knackmuss, H.J. (1993) Syntrophic interactions during degradation of 4-aminobenzenesulfonic acid by a two species bacterial culture. *Arch. Microbiol.* **159**(2), 124-130.

Focht, D.D. and Williams, F.D. (1970) The degradation of p-toluenesulfonate by a *Pseudomonas. Can. J. Microbiol.* **16**, 309-316.

Hansen, C., Fortnagel, P. and Wittich, R.M. (1992) Initial reactions in the mineralization of 2-sulfobenzoate by *Pseudomonas* sp. RW611. *FEMS Microbiol. Lett.* **92**(1), 35-40.

Haug, W., Schmidt, A., Nortemann, B., Hempel, D.C., Stolz, A. and Knackmuss, H.J. (1991) Mineralization of the sulfonated azo dye Mordant Yellow 3 by a 6-aminonaphthalene-2-sulfonate-degrading bacterial consortium. *Appl. Environ. Microbiol.* **57**(11), 3144-3149.

Herbert, R.B. and Holliman, F.G. (1964). Aeroginosin B - a naturally occuring phenazinesulfonic acid(eds). *Proc. Chem. Soc.*, pp. 19

Hernandez Perez, G., Goma, G. and Rols, J.L. (1998) Enhanced degradation of lignosulfonated compounds by *Streptomyces viridosporus. Water Sci. Technol.* **38**(4-5), 289-297.

Hooper, S.W. (1991) Biodegradation of sulfonated aromatics. In *Biological degradation and bioremediation of toxic chemicals*, Chapman & Hall, London, pp 169-182.

Jimenez, L., Breen, A., Thomas, N., Federle, T.W. and Sayler, G.S. (1991) Mineralization of LAS by a four member aerobic bacterial consortium. *Appl. Environ. Microbiol.* **57**(5), 1566-1569.

Junker, F., Field, J.A., Bangerter, F., Ramsteiner, K., Kohler, H.P., Joannou, C.L., Mason, J.R., Leisinger, T. and Cook, A.M. (1994) Oxygenation and spontaneous deamination of 2-aminobenzenesulfonic acid in *Alcaligenes* sp. strain O-1 with subsequent meta ring cleavage and spontaneous desulfonation to 2-hydroxymuconic acid. *Biochem. J.* **300**(2), 429-436.

Kertesz, M.A., Cook, A.M. and Leisinger, T. (1994) Microbial metabolism of sulfur- and phosphorous-containing xenobiotics. *FEMS Microbiol. Rev.* **15**(2-3), 195-215.

Khlebnikov, A., Zhoukov, V. and Peringer, P. (1997) Comparison of p-toluenesulfonic acid degradation by two *Comamonas* strains. *Biotechnol. Lett.* **19**(4), 389-393.

Kneifel, H., Elmendorff, K., Hegewald, E. and Soeder, C.J. (1997) Biotransformation of 1-naphthalenesulfonic acid by the green alga *Scenedesmus obliquus. Arch. Microbiol.* **167**(1), 32-37.

Kolbener, P., Baumann, U., Cook, A.M. and Leisinger, T. (1994) 3-Nitrobenzenesulfonic acid and 3-aminobenzesulfonic acid in a laboratory trickling filter: Biodegradability with different activated sludges. *Wat. Res.*

28(9), 1855-1860.

Krull, R. and Hempel, D.C. (1994) Biodegradation of naphthalenesulfonic acid-containing sewages in a two-stage treatment plant. *Bioprocess Eng.* **10**(5-6), 229-234.

Kudlich, M., Bishop, P.L., Knackmuss, H.J. and Stolz, A. (1996) Simultaneous anaerobic and aerobic degradation of the sulfonated azo dye Mordant Yellow 3 by immobilized cells from a naphthalenesulfonate-degrading mixed culture. *Appl. Microbiol. Biotechnol.* **46**(5-6), 597-603.

Kuhn, E.P. and Suflita, J.M. (1989) Anaerobic biodegradation of nitrogen-substituted and sulfonated benzene aquifer contaminants. *Hazardous Waste and Hazardous Materials* **6**(2), 121-133.

Lee, N.A. and Clark, D.P. (1993) A natural isolate of *Pseudomonas maltophila* which degrades aromatic sulfonic acids. *FEMS Microbiol. Lett.* **107**(2-3), 151-155.

Lie, T.J., Pitta, T., Leadbetter, E.R., GodchauxIII, W. and Leadbetter, J.R. (1996) Sulfonates: novel electron acceptor in anaerobic respiration. *Arch. Microbiol.* **166**, 204-210.

Lie, T.J., Leadbetter, J.R. and Leadbetter, E.R. (1998) Metabolism of sulfonic acids and other organosulfur compounds by sulfate-reducing bacteria. *Geomicrobiology journal* **15**(2), 135-149.

Locher, H.H., Thurnheer, T., Leisinger, T. and Cook, A.M. (1989) 3-Nitrobenzesulfonate, 3-aminobenzenesulfonate, and 4-aminobenzenesulfonate as sole carbon sources for bacteria. *Appl. Environ. Microbiol.* **55**(2), 492-494.

Locher, H.H., Leisinger, T. and Cook, A.M. (1991) 4-Sulfobenzoate 3,4-dioxygenase: Purification and properties of a desulfonative two-component enzyme system from *Comamonas testosteroni* T-2. *Biochem. J.* **274**(3), 833-842.

Malle, K.G. (1978) Wie schmutzig ist der Rhein? Chemie Unserer Zeit **12**, 111-122.

Mampel, J., Hitzler, T., Ritter, A. and Cook, A.M. (1998) Desulfonation of biotransformation products from commercial linear alkylbenzenesulfonates. *Environ. Toxicol. Chem.* **17**(10), 1960-1963.

Nortemann, B., Baumgarten, J., Rast, H.G. and Knackmuss, H.J. (1986) Bacterial communities degrading amino- and hydroxynaphtalene-2-sulfonates. *Appl. Environ. Microbiol.* **52**(5), 1195-1202.

Nortemann, B., Glasser, A., Machinek, R., Remberg, G. and Knackmuss, H.J. (1993) 5-Hydroxyquinoline-2-carboxylic acid, a dead-end metabolite from the bacterial oxidation of 5-aminonaphthalene-2-sulfonic acid. *Appl. Environ. Microbiol.* **59**(6), 1898-1903.

Nortemann, B., Kuhm, A.E., Knackmuss, H.J. and Stolz, A. (1994) Conversion of substituted naphthalenesulfonates by *Pseudomonas* sp. BN6. *Arch. Microbiol.* **161**(4), 320-327.

Ohe, T. and Watanabe, Y. (1986) Degradation of 2-naphthylamine- 1-sulfonic acid by *Pseudomonas* strain TA-1. *Agric. Biol. Chem.* **50**(6), 1419-1426.

Ohe, T., Ohmoto, T., Kobayashi, Y., Sato, A. and Watanabe, Y. (1990) Metabolism of naphthalenesulfonic acids by *Pseudomonas* sp. TA-2. *Agric. Biol. Chem.* **54**(3), 669-676.

Pack, H. and Hempel, D.C. (1997) Effect of mixtures of heavy metals on the biological treatment of sewages containing naphthalenesulfonic acid. *Acta Hydrochimica et Hydrobiologica* **25**(6), 306-310.

Ripin, M.J., Noon, K.F. and Cook, T.M. (1971) Bacterial metalbolism of arylsulfonates. *Appl. Microbiol.* **21**(3), 495-499.

Rozgaj, R. and Glancer, S.M. (1992) Total degradation of 6-aminonaphthalene-2-sulfonic acid by a mixed culture consisting of different bacterial genera. *FEMS Microbiol. Ecol.* **86**(3), 229-236.

Takeo, M., Nagayama, T., Takatani, K., Maeda, Y. and Nakaoka, M. (1997) Mineralization and desulfonation of 3-nitrobenzenesulfonic acid by *Alcaligenes* sp. GA-1. *Journal of Fermentation and Bioengineering* **83**(5), 505-509.

Thoumelin, G. (1991) Presence and behavior of linear alkylbenzenesulfonates (LAS) and alkylphenol ethoxylates (APE) in the aquatic environment: Review. *Environ. Technol.* **12**(11), 1037-1046.

Thurnheer, T., Koehler, T., Cook, A.M. and Leisinger, T. (1986) Orthanilic acid and analogs as carbon-sources for bacteria: growth physiology and enzymic desulfonation. *J. Gen. Microbiol.* **132**, 1215-20.

Thurnheer, T., Cook, A.M. and Leisinger, T. (1988) Co-culture of defined bacteria to degrade seven sulfonated aromatic compounds: efficiency, rates and phenotypic variations. *Appl. Microbiol. Biotechnol.* **29**(6), 605-609.

Thurnheer, T., Zurrer, D., Hoglinger, O., Leisinger, T. and Cook, A.M. (1990) Initial steps in the degradation of benzenesulfonic acid, 4-toluenesulfonic acids, and orthanilic acid in *Alcaligenes* sp. strain O-1. *Biodegradation* **1**(1), 55-64.

Wagner, K. and Hempel, D.C. (1988) Biodegradation by immobilized bacteria in an airlift-loop reactor - influence of biofilm diffusion limitation. *Biotechnol. Bioeng.* **31**(6), 559-566.

Wittich, R.M., Rast, H.G. and Knackmuss, H.J. (1988) Degradation of naphthalene-2,6- and naphthalene-1,6-disulfonic acid by a *Moraxella* sp. *Appl. Environ. Microbiol.* **54**(7), 1842-1847.

Zurrer, D., Cook, A.M. and Leisinger, T. (1987) Microbial desulfonation of substituted naphthalenesulfonic acids and benzenesulfonic acids. *Appl. Environ. Microbiol.* **53**(7), 1459-1463.

17

Metal effects on sulfur cycle bacteria and metal removal by sulfate reducing bacteria

Oliver J. Hao

17.1 INTRODUCTION

The corrosion of sewers and the control of H_2S odour problems have been described in previous chapters. Essentially, the control measures involve the prevention and/or inhibition of sulfide oxidizing bacteria (SOB) and sulfate reducing bacteria (SRB). Conversely, there are many circumstances that require the enhancement of the activities of (i) SRB for anaerobic degradation of certain hazardous organic compounds as well as the removal of sulfate, and of (ii) SOB for sulfide conversion to sulfur or sulfate as well as for metal leaching from sludges, ores, and sediments. For example, SBR may be responsible for the degradation of certain hazardous organic substances in sediments, where long-chain *n*-alkanes (C_{15} - C_{34}) and

Table 17.1. Reaction stoichiometry and thermodynamics of oxyanion and sulfate reduction with butyrate as the electron donor (Tebo and Obraztsova 1998)

Butyrate reaction	$\Delta G^{o\prime}$ (kJ mol^{-1})
$CH_3CH_2CH_2COO^- + 2/3Cr_2O_7^{2-} + 4/3H_2O + 1/3H^+ \rightarrow 2CH_3COO^- + 4/3Cr(OH)_3$	-333
$CH_3CH_2CH_2COO^- + 2MnO_2 + 3H^+ \rightarrow 2CH_3COO^- + 2Mn^{2+} + 2H_2O$	-291
$CH_3CH_2CH_2COO^- + 2UO_2^{2-} + 2H_2O \rightarrow 2CH_3COO^- + 2UO_2 + 5H^+$	-130
$CH_3CH_2CH_2COO^- + 1/4SO_4^{2-} \rightarrow 2CH_3COO^- + 1/2HS^- + 1/2H^+$	-28

chlorophenol could be mineralized to CO_2 (Caldwell *et al.* 1998; Häggblom 1998). The enhancement of sulfate-reducing activity in soils and sediments should also minimize mobilization and toxicity of heavy metals. As will be discussed below, the engineering implication of using these sulfur bacteria for the treatment of metal-contaminated wastes is significant. Thus, quantification of factors affecting the activities on bacteria of sulfur cycle is essential.

In addition to environmental factors (pH, temperature and dissolved oxygen) influencing activities of aerobic SOB and anaerobic SRB, there are many chemical inhibitors of these sulfur bacteria. Many metalloid oxyanions exert specific effects on SRB. For example, the selenate oxyanion (SeO_4^{2-}) is a competitive inhibitor of sulfate reduction (Newport and Nedwell 1988) and the molybdate ion (MoO_4^{2-}) acts on depleting the cellular ATP pool (Taylor and Oremland 1979). On the other hand, some strains of SRB are capable of reducing MoO_4^{2-} to $MoO_{2(s)}$, U(VI) to $UO_{2(s)}$, SeO_4^{2-}/SeO_3^{2-} to elemental selenium (Se^o) or selenide (Se^{2-}), As(V) to As(III), Pd(II) to elemental Pd, Mn(IV) to Mn(II), and Cr(VI) to Cr(III). Tables 1 and 2 provide, respectively, the free energy and redox potential (E_h) values of sulfate and oxyanion reduction half reactions. Microbial-mediated reduction renders these soluble, toxic and oxidized forms into insoluble phases. Hence, the SRB may be partly responsible for the immobilization of these toxic oxyanions in natural ecological systems. Indeed, there is great potential for using SRB strains/consortia to reduce these hazardous metalloids *in situ* or in engineered systems (e.g. Lovley 1993, 1995).

As with most bacteria, SOB or SRB are also sensitive to many microbiocides and toxic organic substances. Tanimoto *et al.* (1989) evaluated some 88 compounds and reported that 10 individual compounds at 10 mg l^{-1} each, including two detergents (dodecylbenzene sulfonic acid and polyethylene glycerol stearylamine), six antibiotics, one dye (crystal violet) and trinitrophenol, were effective inhibitors to a *Desulfotomaculum*

Table 17.2. Equilibrium Eh values (pH 7 and 25 °C) for oxyanion and sulfate reduction half reactions (Tucker *et al.* 1997)

Redox pair	Reaction stoichiometry	Eh (V)
Cr(VI)-Cr(III)	$2CrO_4^{2-} + 10H^+ + 6e^- \rightarrow Cr_2O_{3(s)} + 5H_2O$	0.60
U(VI)-U(IV)	$UO_2(CO_3)_2^{2-} + 2H^+ + 2e^- \rightarrow UO_{2(s)} + 2HCO_3^-$	0.11
Mo(VI)-Mo(IV)	$HMoO_4^- + 3H^+ + 2e^- \rightarrow MoO_{2(s)} + 2H_2O$	−0.12
S(VI)-S(II)	$SO_4^{2-} + 9H^+ + 8e^- \rightarrow HS^- + 4H_2O$	−0.17
Acetate-lactate	$CH_3COO^- + 5H^+ + HCO_3^- + 4e^- \rightarrow CH_3OCH(OH)CO_2^- + 2H_2O$	−0.43

Table 17.3. Solubility product (K_{sp}) of metal sulfide and metal hydroxide precipitates (Zumdahl 1989)

	K_{sp}	
Metal	Metal sulfide	Metal hydroxide
Cu	8.5×10^{-45}	1.6×10^{-19}
Zn	1.2×10^{-23}	8.5×10^{-17}
Pb	3.4×10^{-28}	1.2×10^{-15}
Cd	3.5×10^{-29}	5.9×10^{-15}
Fe	3.7×10^{-19}	1.8×10^{-15}
Ni	1.6×10^{-16}	1.6×10^{-16}
Cr	–	6.7×10^{-31}

strain. Finally, the product toxicity encountered in many biological/ chemical systems is best exemplified by sulfide toxicity towards SRB, which has been described in a previous chapter.

Heavy metals present in different environments above certain threshold levels can be toxic to SRB, thus restricting SRB growth and sulfide generation. However, once sulfide is formed, it readily reacts with heavy metals to form metal sulfide (MeS) precipitates, resulting in a lower sulfide concentration and reduced metal toxicity. These MeS precipitates exhibit extremely low solubilities (Table 17.3), and are relatively stable in environments. The black colour of the sediments as well as primary and digested sludge where sulfate reduction occurs is due to the accumulation of FeS. The practice of adding iron salts in sewers and reactors is common to alleviate sulfide toxicity, reduce sulfide odour problems, and control H_2S gas present in the digester biogas. The role of the MeS sink to reduce metal toxicity is best illustrated in the sediment environments in which the availability of sulfide results in no apparent toxicity to aquatic life, despite the higher metal contents in these sediments (e.g., Pesch *et al.*, 1995). Furthermore, the SRB-mediated enzymatic reactions for the reduction of soluble, oxidized oxyanions to insoluble species are responsible for the accumulation of insoluble metalloids in many environmental settings, e.g. anaerobic sediments (Lovley 1993).

The ability of sulfide to readily react with metals presents an excellent opportunity in using sulfate-reducers to generate sulfide *in situ* for removing metals. Because of the formation of Fe sulfides, the *in situ* bioremediation process has been demonstrated in several natural environments including the Kidd Creek mine tailings impoundment (Fortin and Beveridge 1997). For man-made engineered systems, four such process schemes are shown in Figure 17.1. Wastewater containing organics/metals/sulfate or inorganic wastewater with metal/sulfate supplemented with electron donor is fed into a sulfate-reducing reactor (Figure 17.1a), where sulfide generated *in situ* precipitates out metals. In the scheme (2), metal-rich wastewater is added to a subsequent precipitation reactor, followed by a sulfate reduction reactor (Figure 17.1b). Metals can also be pre-precipitated by recycling either sulfide-laden wastewater (Figure 17.1c), or H_2S-laden gas stream to an inlet metal precipitator (Figure 17.1d). The low cost of the scheme (1), however, may necessitate the influent pretreatment to reduce metal levels and to increase pH to minimize their adverse effects on SRB. Also, higher ratios of sulfate to organics must be provided for SRB to compete more successfully with acid formers and methane-producing bacteria. For the scheme (2), metal-laden wastewater with little organic content is preferred, and the stoichiometric relationship between sulfide and metal is of importance. Scheme (3) requires additional pumping facilities for recycling flow, and the return flow rate depends on the concentrations of sulfide and metals. The last scheme is similar to scheme (3), except that the gas is recycled for metal precipitation, normally in a countercurrent system. The final effluent of these processes may need further treatment for sulfide removal, especially for the last scheme. These treatment schemes appear promising, at least in theory, for the simultaneous removal of sulfate, organics, and metals. Unfortunately, many factors have prevented the widespread application of these techniques, including the sulfide toxicity to SRB, potential odour problems, high metal content due to low sulfide production, lack of adequate organic substrate, and ineffective separation of metal sulfide precipitates from the treated wastewater.

The SOB, such as *Thiobacillus* sp., have been utilized in biological gaseous odour-removal process, aqueous sulfide oxidation, and metal leaching from sludge, sediments, mines, and contaminated soils. Information concerning metal effects on SOB under acidic conditions, however, is rather limited. The fact that metals under such low pH conditions are present in free forms indicates that acidophilic SOB are less subject to metal inhibition. For example, the Cu concentration at 30 mg l^{-1} resulting from bioleaching of sludge at pH 1.5 has no effect on SOB

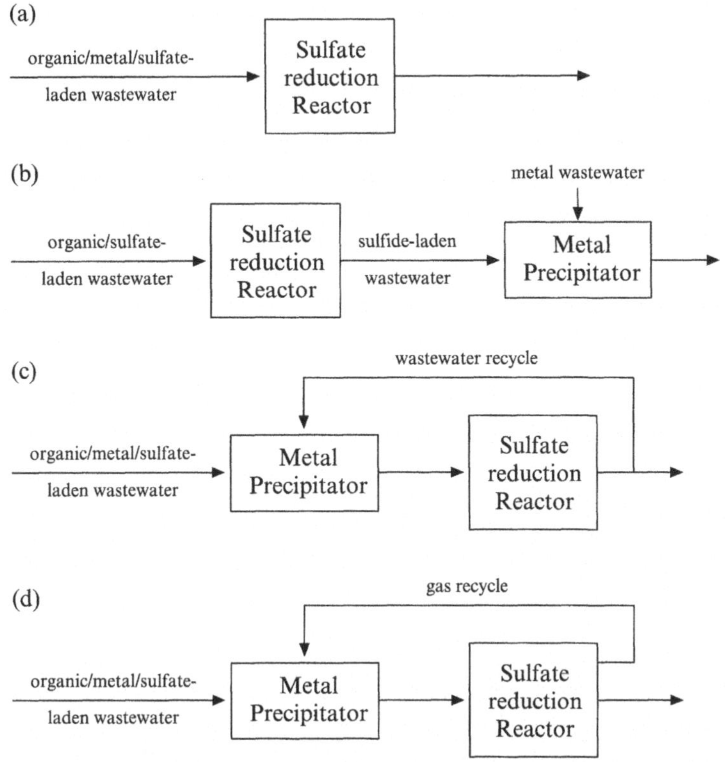

Figure 17.1. Different schemes for metal removal from wastewater in SRB systems.

(Ravishankar *et al.* 1994). Zn up to 1800 mg l^{-1} exhibits no effect on indigenous SOB activities in sediments (Seidel *et al.* 1998); and 10 g l^{-1} of Cn, Ni, Cu, Co or Al have no effect on iron oxidation by *Thiobacillus ferrooxidans* (Tuovinen *et al.* 1971). There is an excellent review of the biology of metal leaching (Rawlings 1998) as well as recent developments in the integrated processes of using both SOB (bioleaching) and SRB (bioprecipitation) for bioremediation of metal-contaminated solids (e.g. White *et al.* 1997, 1998). This chapter only covers metal effects on SRB and metal removal by SRB.

17.2 METAL INHIBITION AND INTERACTION

The reduction of metal concentrations in SRB-containing systems is common in natural and man-made anaerobic environments. The role of sulfide in preventing metal toxicity is well established owing to insoluble MeS precipitates. As mentioned earlier, the external addition of iron salts may be used to precipitate sulfide for odour control (Sercombe 1995), reduce metal/sulfide toxicity (Edwards *et al.* 1997), reduce H_2S gas in anaerobic digesters (Dezham *et al.* 1988), and control dissolved sulfide in sewers (Padival *et al.* 1995). In fact, Jin *et al.* (1998) even add sulfide into a methanogenic system to reduce the effect of copper toxicity.

The high concentrations of different heavy metals will, of course, inhibit SRB. Thus, before using SRB for metal removal, the concentrations of toxic metals need to be known. The degree of metal toxicity depends on many factors, especially in the complicated SRB/metal-interaction system. These factors include operating conditions (e.g., cell concentration, pH, temperature, batch or continuous, suspended or fixed film, and plug flow or complete mixing system) and waste characteristics (e.g., type of carbon sources, type and concentrations of initially present and bio-produced complexing compounds, concentrations of sulfate present and sulfide generated, type and concentration of metals, and the amount of iron present). For example, some strains of SRB can produce exopolymers (Beech and Cheung 1995) that complex metals; the resultant metal complexes are, therefore, less toxic to SRB. Also, the microbial organisms can adapt to adverse environments including metal-contaminated systems. Thus, metals may be effective SRB inhibitors in batch systems, but not in continuous flow systems (Clancy *et al.* 1992).

Toxicity studies are often reported as a function of the initial concentrations of heavy metals. Because of complicated biological processes, some metals may be precipitated out, uptaken by biomass, and/or physically/chemically entrapped within/around/onto the biomass. Thus, the residual soluble metal concentrations should be determined to account for the observed toxicity phenomenon. Furthermore, metals present in the free form, and not the complexed forms, exhibit toxic effects. Different oxidation states of metals, e.g., Av(V) versus As(III) and Se(VI) versus Se(IV) exert different effects. Thus, metal speciation is required to clearly evaluate the toxic effect. Compounding the problems of the metal effects is the so-called synergistic phenomenon; metal mixtures may exert an additional effect compared with the sum of the individual effects. Some strains of SRB are highly tolerant to heavy metals at higher concentrations, e.g., aluminium at 50 mM and lead at 10 mM (Hard *et al.* 1997). In the presence of 3.6 mM Fe(II), a pure culture of *Desulfotomaculum* sp. was resistant to Ni concentration at 9.5 mM (Fortin *et al.* 1994). In fact, some strains of SRB are not only tolerant but also able to reduce high concentrations of metalloid oxyanions, e.g., *Desulfovibrio desulfuricans* species could reduce selenate concentrations up to 10 mM (Tomei *et al.* 1995) and U(VI) at 24 mM (Lovley and Phillips 1992b). A consortium of SRB was able to remove Cr(VI) up to 40 mM (Fude *et al.* 1994), and the Cr(VI) reduction was not affected in the presence of 1.5 mM Zn or 0.4 mM U(VI). Also, the following metals at 0.1 mM concentrations did not affect the Cr(VI) reduction by *Desulfovibrio vulgaris*: Ni, Co, Cu(I), Cu(II), Zn, V, Mo(VI), and Se(IV) (Lovley and Phillips 1994). Except for copper (either the cupric or cuprous state), the above-mentioned compounds at 0.1 mM did not significantly affect U(VI) reduction by *D. desulfuricans* (Lovley and Phillips 1992b). As for Mo(VI) reduction by *D. desulfuricans*, no inhibition was observed with the following compounds at 0.1 mM: Cd, Cu, Ni, Mn, and Zn (Tucker *et al.* 1997). A threshold level of 0.5-0.8 mM molybdate, above which sulfate reduction was inhibited, was reported (Liu and Fang 1997), and a much higher level of 3 mM could cease sulfate reduction (Tanaka and Lee 1997). Again, conflicting literature results indicating much lower inhibitory concentrations, e.g., uranyl nitrate solution at 1 mM for *D. desulfuricans* (Tucker *et al.* 1997), are due to a variety of reasons described before. In general, the inhibitor order of the four oxyanions is of CrO_4^{2-} > MoO_4^{2-} = WO_4^{2-} > SeO_4^{2-} (Taylor and Oremland 1979).

Several inhibitory concentrations of heavy metal towards SRB are summarized in Table 17.4. Again, the wide range of these concentrations is expected due to different experimental conditions, and any reference to this table should be made with care. Nevertheless, two studies (Morton *et al.* 1991; Hao *et al.* 1994) show similar inhibitory concentrations as well as the order of metal inhibition (Cu > Cd > Ni > Zn > Cr > Pb) in test tubes. Unlike test tube experiments, the metal inhibitory concentrations would be much higher for systems with higher biomass concentrations. A sufficient sulfide (effluent 45 mg S/l) was produced to completely remove 100 mg l^{-1} Cu, Mn, Ni, and Zn present in the feed without affecting COD removal efficiencies in both methanogenic and sulfate-reducing reactors (Nedwell and Reynolds 1996). In fact, in a mixture of sewage sludge, leaf mulch, wood chips and sawdust, metal concentrations as high as 600 mg Fe l^{-1}, 480 mg Ni l^{-1} and 135 mg Cd l^{-1} were effectively decreased by over 99% through SRB activities (Waybrant *et al.* 1998). It is expected that metal inhibitory concentrations for SRB are higher than those for acid formers and methane-producing bacteria. For comparison, the relative toxicity of the metals to methanogenic volatile fatty acid degradation is of Cu > Cr > Cd = Zn > Ni >> Pb (Lin and Chen 1999), and for anaerobic sludge digestion is of Ni > Cu > Cd > Cr > Pb (Mueller and Steiner 1992).

Table 17.4. Metal concentrations (mg l^{-1}) reported to be inhibitory to SRB

Metal	References								
	Morton *et al.* (1991)[1]	Bharathi *et al.* (1990)[2]	Ellwood *et al.* (1992)[3]	Hao *et al.* (1994)[1]	Song *et al.* (1998)[4]	Fortin *et al.* (1994)	Karnachuk (1995)[5]	Ueki *et al.* (1991)[6]	Poulson *et al.* (1991)[7]
Cd	6	40		> 4					
Cr(III)	23		10(1)	60					
Cu	6		118(0)	4	100			64	
Pb	25	80	30(0)	> 80					
Ni	13		32(2)	10		> 100		58	10
Zn	25		33(0.5)		40			65	13
CN					20				
Cr(VI)							35-70		

[1] Complete inhibition of SRB in test tubes.
[2] 50% inhibition of growth of an isolated SRB strain.
[3] Metal added into an active culture of *Desulfovibrio* strain; values shown inside the parenthesis are soluble concentrations measured after 5 days.
[4] 50% inhibition of SRB growth in serum bottles.
[5] Minimum inhibition of two strains of SRB; concentrations dependent upon the strain and type of the electron donor.
[6] Complete inhibition in the culture to treat AMD supplemented with cattle waste.
[7] Inhibition concentrations to pure culture of *D. desulfuricans*.

There is a significant presence of SRB in sediments receiving acid mine drainage (AMD) as reported by several investigators (Herlihy and Mills 1985; Blodau *et al.* 1998). The correlation between metal toxicity and sulfide content in sediments has been extensively studied (e.g., Pesch *et al.* 1995) in terms of the ratio of simultaneously extracted metal (SEM) to the acid volatile sulfide (AVS). In general, sediment samples with molar SEM/AVS ratios of heavy metals (e.g., Cd, Zn, Pb, and Cu) greater than one were consistently toxic to the test organisms, whereas samples with ratios less than one were not toxic. The detoxification was due to binding between sulfides and metals, thereby forming insoluble and biologically unavailable MeS. The hypothesis about metal sulfide formation on a mole to mole basis may not be true in the aqueous phase. Van den Berg *et al.* (1998) cautioned that SEM/AVS ratios in mixed homogenized sediment samples might not be suitable for the assessment of metal toxicity of sediments. Formation of organic metal complexes increase on the stoichiometrically predicted ratio of metal to precipitate for a given concentration of sulfide. Furthermore, oxidation of AVS will certainly affect bioavailability of heavy metals (Peterson *et al.* 1996).

17.3 HEAVY METAL REMOVAL

The feasibility of removing sulfate present in sulfate-rich wastewaters has been well demonstrated in a variety of anaerobic bioreactor types (upflow or downward fixed bed, upflow sludge blanket, completely mixed, or gas-lift reactors), including immobilized SRB. Essentially, SRB outcompete acid formers and methane-producing bacteria if environmental factors favour the former, such as higher sulfate/COD ratios and lower hydraulic detention times. To reduce H_2S toxicity, the influent pH could be adjusted to a neutral level, and/or dosed with iron salts. Once sulfide is generated *in situ*, it readily reacts with metals to form MeS precipitates. Despite the potential of using sulfide generated *in situ* for metal removal from wastewaters, applications are mostly related to the treatment of AMD, which has been covered in a previous chapter. In addition to briefly describing metal removal for the AMD pollution abatement in this section, the limited information available for other industrial wastewater treatment, contaminated groundwater, and leachate from landfill and contaminated soils is summarized in Table 17.5.

The potential use of SRB for AMD pollution abatement has long been recognized (Tuttle *et al.* 1969; Wakao *et al.* 1979). Since 1980s, laboratory, pilot scale, and full-scale evaluations have demonstrated the capability of

Table 17.5. Summary of metal removal of metal-contaminated wastewaters subject to SRB activities

Wastewater	Reactor Type	Operation	Influent concentrations (mg l^{-1}, except pH)					Reference
			COD	pH	Metals	SO_4^{2-}-S	Metal removal efficiency (%)	
Landfill leachate	Anaerobic filter; surface area of media = 206 m^2/m^3	Upflow with effluent recycle; HRT = 34 d	11,630	6.5	430 (Fe) 16 (Zn) 5.6 (Cu)	103	97 (Fe) 94 (Zn) 74 (Cu)	DeWalle *et al.* (1979)
Gold mine effluent supplemented with molasses	Anaerobic filter with dolomite pebbles (diameter = 2-3 mm)	Upflow; HRT = 10 h	3,400	6.0	6.7 (Ni) 3.2 (Co) 2.9 (Mn)	830	93 (Ni) 72 (Co) 52 (Mn)	Maree *et al.* (1987)
Ni plating waste supplemented with sewage	Anaerobic filter with stones (diameter = 1.5 mm)	Upflow; HRT = 24 h	230	7.0	53 (Ni)	60	99 (Ni)	Gundry *et al.* (1989)
Hydrometallurgical waste supplemented with molasses	Packed bed with crushed dolomite (2-3 mm)	Upflow with effluent recycle; T = 30 °C; HRT = 4.8 d	24,000	4.1	63 (Fe) 3 (Zn) 630 (Mn)	2,400	97 (Fe) 87 (Zn) 32 (Mn)	Somlev and Tishkov (1994)
Synthetic waste (lactate)	Anaerobic filter with plastic pall rings	Upflow; T = 20 °C; HRT = 12 h			45 (Cr) 18 (Pb) 400 (Cu) 500 (Cd)		80 (Cr) > 90 (Pb) > 90 (Cu) > 99 (Cd)	Wijaya *et al.* (1993)
Plating effluent supplemented with food processing waste	Anaerobic filter	Upflow; T = 20 °C; HRT = 24 h	variable		9 (Ni) 30 (Zn) 10 (Cu)		98 (Ni) 95 (Zn) 99 (Cu)	Bewtra *et al.* (1995)

Groundwater supplemented with ethanol	Sludge blanket (9 m^3)	Upflow; T = 21 °C; HRT = 11 h	33,300	4.9	1070 (Zn) 18 (Cd) 6.8 (Cu)	1,000	> 99 (Zn) > 99 (Cd) > 99 (Cu)	Barnes *et al.* (1991)
Soil leachate supplemented with ethanol and nutrients	Suspended system with biomass recycle	HRT = 10 h, SRT = 500 h	3,200	6.5	7.1 (Cd) 6.3 (Co) 5.3 (Cr) 4.2 (Cu) 83 (Mn) 6.3 (Ni) 1.7 (Zn)	1,600	98 (Cd) 92 (Co) 97 (Cr) 94 (Cu) 90 (Mn) 87 (Ni) 97 (Zn)	White *et al.* (1998)
Synthetic waste (glucose, acetate and Cu)	Suspended system with biomass recycle	Fed-batch operation; effluent recycle to a pre-metal stripper; T = 35 °C; HRT = 7 d	10,400		250 (Cu)	1,000	> 99 (Cu)	Haas and Polprasert (1993)

HRT: hydraulic residence time. SRT: solids residence time.

using SRB to remove sulfate and metals from AMD. Hammack and Edenborn (1992) used laboratory columns containing mushroom compost for Ni removal. The simulated mine waste (670 mg SO_4^{2-}-S l^{-1}, 50-1000 mg Ni l^{-1}, 3500 mg lactate l^{-1}, and pH 4.5) was pumped through the column at the flow rate of 15-25 ml h^{-1}. The maximum Ni removal rate was 540 mg Ni kg^{-1} d^{-1} for columns receiving 1000 mg Ni l^{-1}. Pilot-scale reactors, consisting of three 200 litre reactors in series with spent mushroom compost (50-60% organic matter and 10-15% pulverized limestone), were used to treat metal-contaminated water from a coal mine dump, and a single 4500-liter reactor for treatment of the smelting residues (Dvorak *et al.* 1992). Typically, the removal efficiencies of Al, Cd, Fe, Mn, No, and Zn were more than 95%, and system pH increased to about 6.4. Although most metals precipitate as metal sulfides, the metals of Al, Mn, and to some extent, Zn, were retained as hydroxides or carbonates. Christensen *et al.* (1996), based on the reduction of dissolved concentrations of Cu, Zn, Fe and Al in bench-scale of cylinder reactors, reported that it might be feasible to use SRB for treatment of AMD *in situ* in water-filled mines and open pits. The amount of organics added should be balanced with the metal content, and the type of organics used certainly exhibits a significant impact on sulfate reduction rate (White and Gadd 1996).

The low pHs associated with AMD may not hinder the sulfate reduction process. First, some SRB strains can withstand acidic conditions, e.g., Lyew *et al.* (1994) reported a reduction of 90% dissolved metals in AMD with pH at 4.8 at a hydraulic detention time of 14 days in a 40 litre reactor containing sediments. Recent studies further demonstrate that two isolated strains of SRB could even grow at pH < 4.5 (Hard *et al.* 1997) and SRB are capable of sulfate reduction and production of alkalinity at pH values as low as 3.3 (Elliott *et al.* 1998). Secondly, the sulfate reduction process generates bicarbonate alkalinity, and the increased pH may not significantly affect SRB activity in the otherwise acidic conditions. Finally, the addition of limestone used in many studies will further increase alkalinity to neutralize acidity present in the AMD.

Because iron salts are normally present in large quantities in AMD, FeS precipitation normally occurs. The interaction between other metals and FeS typically follows the following exchange reaction: $Me^{2+} + FeS(s) \rightarrow MeS(s) + Fe^{2+}$. The kinetic exchange between different metals and FeS has been investigated (Davis *et al.* 1994) and the preference order of the divalent metals is: Cu > Zn > Pb > Cd. The released Fe is then re-adsorbed by other metal sulfide precipitates. Peiffer *et al.* (1992) further provide an evidence of a direct oxidation of sulfide by iron hydroxides. In addition to MeS precipitates, biosorption of heavy metals by SRB needs to be evaluated (El Bayoumy *et al.* 1997), especially with the dead biomass.

Because sulfate reduction may be significant in artificial or natural wetlands, there has been some keen interest in using wetlands for the treatment of the AMD, owing to low-cost and low-maintenance techniques. A survey conducted in 1989 indicated that a total of 142 wetlands in the eastern USA were constructed specifically for AMD treatment (Wieder 1989). The effectiveness of these wetland treatments of AMD has been variable and unpredictable (Wieder 1989), partly due to complicated ecology of wetland systems. A mass balance of sulfur in a constructed wetland indicates that the concentrations of the wetland effluent range from 3 - 30 mg S l^{-1}, and the average emission rate from the surface of the substrate is about 50 mg l^{-1} m^{-2} d^{-1} in the summer (Machemer *et al.* 1993). The loss of sulfide into the atmosphere is insignificant in reducing the amount of sulfide available for metal precipitation. As would be expected, AVS present on wetland substrates increases with time. Since rates of metal removal may not correlate with sulfide generation (Webb *et al.* 1998), other biological mechanisms of metal removal in addition to sulfate reduction are also involved. Furthermore, Fe(III)-reducing bacteria may generate additional alkalinity for the observed pH increases in wetland systems (Vile and Wieder 1993).

Hammack *et al.* (1994) used a similar system shown in Figure 17.1d to treat mine waste, with a limestone neutralization reactor between the metal precipitator and sulfate reduction reactor. The countercurrent metal precipitator removed metal and the limestone reactor increased the pH to 6. Sodium lactate was injected directly to the sulfate reduction reactor. The process removed more than 99% of the initial concentrations of Fe (620 mg l^{-1}), Cu (178 mg l^{-1}), Zn (530 mg l^{-1}), and Al (278 mg l^{-1}), in addition to 91% Mn (191 mg l^{-1}).

Porous and permeable reactive walls, installed in the path of contaminated waste, may provide a promising alternative for remediating metal-contaminated wastes. Based on batch tests using simulated mine wastes, different reactive wall mixtures were constructed with a permeability ranging from 10^{-4} to 10^{-2} cm s^{-1} (Waybrant *et al.* 1998). These systems were able to remove significant amounts of dissolved metals, e.g., concentrations of 480 mg l^{-1} Ni and 135 mg l^{-1} Cd decreased to below 0.05 mg l^{-1} within 10 days. A full-scale retaining wall (3.6 m deep, 15 m long, and 4 m wide) was installed into an aquifer at a mine site near Sudbury, Ontario (Benner *et al.* 1997) for prevention of AMD migration. The reactive wall was composed of 40% municipal compost, 50% leaf compost and 20% wood

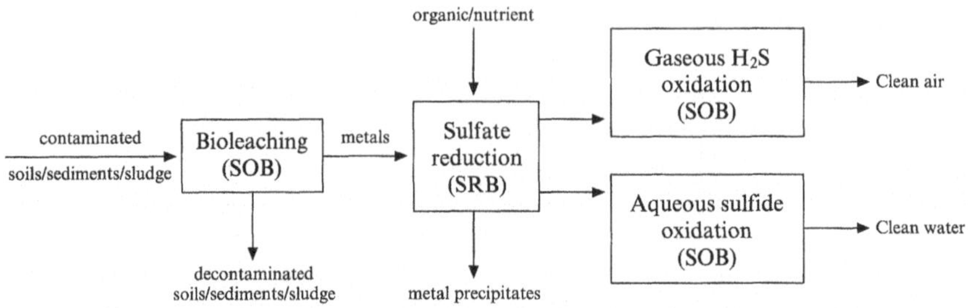

Figure 17.2. Application of SOB and SRB for metal decontamination.

chips. This particular operation was successful: Fe concentrations decreased from 250-1300 to 10-40 mg l^{-1}, pH increased from 5.8 to 7.0, and alkalinity increased from 50 to 600 - 2000 mg $CaCO_3$ l^{-1}. The reactive wall has the potential to remain effective for 15 years. Several patents were granted for *in situ* systems that remove dissolved metals (Blowes and Ptacek 1994; Geraghty and Millers 1996). The inventions essentially comprise the formation of *in situ* anaerobic reactive zones in the path of migrating groundwater to precipitate dissolved metals as MeS.

In short, the SRB and SOB certainly cause some environmental problems, including odour, AMD, sewer corrosion, and corrosion in different industrial settings, such as heat exchangers. However, the activities of these bacteria may also alleviate some environmental problems in nature and in man-made engineered systems, specifically for the minimization of metal mobility and toxicity because of insoluble metal sulfide precipitates. As biotechnology advances with the aid of biogenetic engineering, together with improvement of the reactor design, a widespread utilization of SRB in removing heavy metals is envisioned for the future. Unlike metal immobilization by SRB, bioleaching or solubilization of metals from ores, contaminated sediments, and sludge will also render SOB ecologically important species. The utilization of both SOB and SRB for ultimate metal decontamination is conceptually illustrated in Figure 17.2.

17.4 METALLOID OXYANION REDUCTION

Many strains of SRB are capable of removing metalloid oxyanions by their dissimilatory metabolism. Under anaerobic conditions, these bacteria reduce the soluble oxidized oxyanions (e.g., MoO_4^{2-}) to insoluble phases (e.g., MoS_2 or MoO_2). Growth of these microbes may occur with these oxyanions as the sole electron acceptors (Lovley 1993; Newman *et al.* 1997; Tebo and Obraztsova 1998), although some SRB strains cannot grow with oxyanions as sole electron acceptors (Lovley and Phillips 1994). In the latter case, they also require sulfate as the electron acceptor. The so-called dissimilatory metal reduction occurs where electrons are transferred from organic compounds to the oxidized forms of the oxyanions (Lovley 1993). The enzymatically mediated reactions are responsible for oxyanion reduction. For example, both hydrogenase and cytochrome *c* were responsible for Cr(VI) reduction (Lovley and Phillips 1994) and c_3 cytochrome is the U(VI) reductase (Lovley *et al.* 1993). Some studies, however, indicate the important role of sulfide (Mohagheghi *et al.* 1985; Fude *et al.* 1994), and, to a lesser extent, SRB adsoprtive capability (Mohagheghi *et al.* 1985) in removing these oxidized oxyanions.

Reduction of U(VI) to U(IV) by indigenous bacteria in contaminated groundwater in the presence of sulfate has been documented (Abdelouas *et al.* 1998). The SRB (*Shewanella putrefaciens,* also known as Fe(III) reducers) were able to reduce U(VI) up to 235 mg U l^{-1} to less than 0.2 mg l^{-1}. The electron diffraction analysis indicated poorly crystallized solid of $(U.Ca)O_2$. The washed cells of *D. desulfuricans* and *Desulfovibrio vulgaris* were able to reduce Mo(VI) to Mo(IV) with either lactate or H_2 as the electron donor (Tucker *et al.* 1997). In the presence of sulfide, the reduction occurred more rapidly and efficiently resulting in a black precipitate of $MoS_{2(s)}$. The reduction of CrO_4^{2-}, MoO_4^{2-}, SeO_4^{2-}, and $UO_2(CO_3)_2^{2-}$ by intact cells of *D. desulfuricans* immobilized in polyacrylamide gel was demonstrated in batch reactors (Tucker *et al.* 1998a). Figure 17.3 illustrates the reduction of several oxyanions with both free and immobilized cells of *D. desulfuricans.* Metal removal efficiencies of 86 - 96% were achieved for initial concentrations of 1 mM Mo(VI), Se(VI), and U(VI) and 0.5 mM Cr(VI); the highest metal removal occurred at 37 °C and pH 7. The order of the pseudo first-order rates of metalloid reduction was: U(VI) > Cr(VI) > Se(VI) > Mo(VI). Insoluble metal precipitates (MoS_2, Se^o, UO_2) were attached on the surface and interior of the polyacrylamide gel. Tucker *et al.* (1998b) further reported successful removals of more than 99% of 5 mg $UO_2(CH_3COO)_2$ as U l^{-1} and 10 mg MoO_4^{2-}–Mo l^{-1} of simulated contaminated groundwater in column reactors with the immobilized SRB. Formate or lactate was used as the electron donor and the hydraulic detention time ranged from 24 to 36 hrs. Sulfate concentrations up to 670 mg SO_4^{2-}-S l^{-1} did not inhibit the reduction of U(VI) or Mo(VI), but nitrate at 50 mg l^{-1} began to inhibit U(VI) reduction (50% inhibition at 300 mg l^{-1} nitrate). Tomei *et al.* (1995) used *D. desulfuricans* strain for Se(VI) or Se(IV) reduction. The high initial concentrations of 10 mM selenate and 0.1 mM selenite did not inhibit the microbial growth in the presence of formate

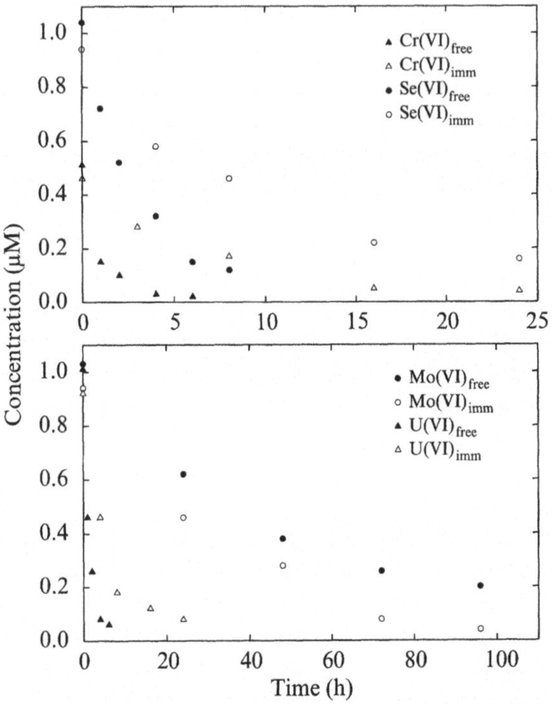

Figure 17.3. Oxyanion metal removal by free or immobilized cells of *Desulfovibrio desulfuricans* with lactate as the electron donor (From: Tucker *et al.* 1998a).

(electron donor) and either sulfate or fumarate as the electron acceptor. The elemental selenium, confirmed by electron microscopy with energy-dispersive X-ray, was accumulated inside the cells. For the system containing chromium, the Cr(VI) reduction rates may be enhanced in the presence of Pb, Cu, Cd, and Ni, possibly due to the formation of metal chromate complexes (e.g., $MeCrO_4$), which react faster with sulfide than free chromate (Pettine *et al.* 1998).

Kinetic coefficients for *D. desulfuricans* in the presence of sulfate (electron acceptor) and pyruvate (electron donor) for U(VI) removal were determined from chemostat studies at dilution rates varying from 0.03 to 0.11 d^{-1} at 28 °C at the initial U(VI) concentration of 1 mM (Tucker *et al.* 1996). The maximum pyruvate utilization rate was 4.7 d^{-1}, the half saturation coefficient 127 mg l^{-1}, the yield coefficient 0.021 g mol^{-1} of pyruvate, and the endogenous decay constant 0.072 d^{-1}. The pseudo first-order rate constants of U(VI), Cr(VI) and Mo(VI) removal in suspended

D. desulfuricans cells with lactate as the electron donor in the bicarbonate buffer were approximately 0.8 hr^{-1} (Lovley and Phillips 1992a), 0.15 hr^{-1} (Fude *et al.* 1994), and 0.02 hr^{-1} (Tucker *et al.* 1997), respectively. Also, the yield of U(VI) reduction by *D. desulfuricans* is strongly pH dependent (Panak *et al.* 1998).

Recently, co-culture of *D. desulfuricans* and an anaerobic photoautotroph (*Chromayiu, vinosum*) was used to convert SeO_4^{2-} into intercellular Se globules (Nelson *et al* 1996). Since *D. desulfuricans* reduce sulfate to sulfide and selenate to selenide (H_2Se_{aq}), the newly formed sulfide is then oxidized by *C. vinosum* to form intercellular S^o globules. The internal S^o is eventually replaced by selenide to produce Se^o (Nelson *et al.* 1996):

$$S^o\ (s) + HSe^-\ (aq) \rightarrow Se^o\ (s) + HS^-\ (aq) \qquad (17.1)$$

This particular process has considerable practical applications for the treatment of Se contaminated wastewaters. As laboratory experiments are transformed into pilot and full-scale works, microbial transformation of hazardous oxyanions such as As(V)/As(III), Se(VI)/Se(IV), Cr(VI), Mo(VI), U(VI), etc. by microbial-mediated reactions would place SRB as one of the most important microbial species in bioremediation.

17.5 REFERENCES

Abdelouas, A., Lu, Y., Lutze, W. and Nuttall, H. E. (1998) Reduction of U(VI) to U(IV) by indigenous bacteria in contaminated ground water. *J. Contam. Hydrol.* **35**(1/3), 217-233.

Barnes, L. J., Janssen, F. J., Sherren, J., Versteegh, J. H., Koch, R. O. and Scheeren, P. J. H. (1991) A new process for the microbial removal of sulfate and heavy metals from contaminated waters extracted by a geohydrological control system. *Trans. Inst. Chem. Eng.* **69**(A3), 184-186.

Beech, I. B. and Cheung, C. W. S. (1995) Interactions of exopolymers produced by sulfate-reducing bacteria with metal-ions. *Int. Biodeterior. Biodegrad.* **35**(1/3), 59-72.

Benner, S. G., Blowes, D. W. and Ptacek, C. J. (1997) A full-scale porous reactive wall for prevention of acid mine drainage. *Ground Water Monitoring and Remediation* **17**(4), 99-107.

Bewtra, J. K., Biswas, N., Henderson, W. D. and Nicell, J. A. (1995) Recent advances in treatment of selected hazardous wastes. *Wat. Pollut. Res. J. Can.* **30**(1), 115-125.

Bharathi, P. A. L., Sathe, V. and Chandramohan, D. (1990) Effect of lead, mercury, and cadmium on a sulfate-reducing bacterium. *Environ. Pollut.* **67**(4), 361-374.

Blodau, C., Hoffmann, S., Peine, A. and Peiffer, S. (1998) Iron and sulfate reduction in the sediments of acid mine lake 116 (Brandenburg, Germany): rates and geochemical evaluation. *Wat. Air Soil Pollut.* **108**(3/4), 249-270.

Blowes, D. W. and Ptacek, C. J. (1994) Systems for treating contaminated groundwater. US Patent 5514279.

Caldwell, M. E., Garrett, R. M., Prince, R. C. and Suflita, J. M. (1998) Anaerobic biodegradation of long-chain n-alkanes under sulfate-reducing conditions. *Environ. Sci. Technol.* **32**(14), 2191-2195.

Christensen, B., Laake, M. and Lien, T. (1996) Treatment of acid mine water by sulfate-reducing bacteria: results from a bench scale experiment. *Wat. Res.* **30**(7), 1617-1624.

Clancy, P. B., Venkataraman, N. and Lynd, L. R. (1992) Biochemical inhibition of sulfate reduction in batch and continuous anaerobic digesters. *Wat.Sci. Technol.* **25**(7), 51-60.

Davis, A. P., Hao, O. J. and Chen, J. M. (1994) Kinetics of heavy metal reactions with ferrous sulfide. *Chemosphere* **28**(6), 1147-1164.

DeWalle, F. B., Chian, E. S. K. and Brush, J. (1979) Heavy metal removal with completely mixed anaerobic filter. *J. Wat. Pollut. Control Fed.* **51**(1), 22-36.

Dezham, P., Rosenblum, F. and Jenkins, D. (1988) Digester gas H_2S control using iron salts. *J. Wat. Pollut. Control Fed.* **60**(4), 514-517.

Dvorak, D. H., Hedin, R. S., Edenborn, H. M. and McIntire, P. E. (1992) Treatment of metal-contaminated waster using bacterial sulfate reduction: results from pilot scale reactors. *Biotechnol Bioeng.* **40**(5), 609-616.

Edwards, M., Courtney, B., Heppler, P. S. and Hernandez, M. (1997) Beneficial discharge of iron coagulation sludge to sewers. *J. Environ. Eng.* **123**(10), 1027-1032.

El Bayoumy, M. A., Bewtra, J. K., Ali, H. I. And Biswas, N. (1997) Biosorption of lead by biomass of sulfate reducing bacteria. *Can. J. Civ. Eng.* **24**(5), 840-843.

Elliott, P., Ragusa, S. and Catcheside, D. (1998) Growth of sulfate-reducing bacteria under acidic conditions in an upflow anaerobic bioreactor as a treatment system for acid mine drainage. *Wat. Res.* **32**(12), 3724-3730.

Ellwood, D. C., Hill, M. J. and Watson, J. H. P. (1992) Pollution control using microorganisms and magnetic separation. In *Microbial Control of Pollution* (Eds. J.C. Fry, G.M. Gadd, R.A. Herbert, C.W. Jones, and I.A. Watson-Craik), pp. 89-112, Cambridge University Press, England.

Fortin, D. and Beveridge, T. J. (1997) Microbial sulfate reduction with sulfides mine tailings: formation of diagenetic Fe sulfides. *Geomicrobiol.* **14**(1), 1-21.

Fortin, D., Southam, G. and Beveridge, T. J. (1994) Nickel sulfide, iron-nickel sulfide and iron sulfide precipitation by a newly isolated *Desulfotomaculum* species and its relation to nickel resistance. *FEMS Microbiol. Ecol.* **14**(2), 121-132.

Fude, L., Harris, B., Urrutia, M. M. and Beveridge, T. J. (1994) Reduction of Cr(VI) by a consortium of sulfate reducing bacteria (SRB III). *Appl. Environ. Microbiol.* **60**(5), 1525-1531.

Geraghty and Miller, Inc. (1996) In situ anaerobic reactive zone for in situ metals precipitation and to achieve denitrification. US Patent 5554290.

Gundry, M. J., Henry, J. G. and Prasad, D. (1989) Treatment electroplating wastewater using an anaerobic filter. *Proc. 44th Purdue Waste Conf.*, pp. 279-285, Lewis Publishers, Chelsea, MI.

Hao, O. J., Huang, L., Chen, J. M. and Buglass, R. L. (1994) Effects of metal addition on sulfate reduction activity in wastewaters. *Toxicol. Environ. Chem.*

46(4), 197-212.

Haas, C. N. and Polprasert, C. (1993) Biological sulfide prestripping for metal and COD removal. *Wat. Environ. Res.* **65**(5), 645-649.

Häggblom, M. M. (1998) Reductive dechlorination of halogenated phenols by a sulfate-reducing consortium. *FEMS Microbiol. Ecol.* **26**(1), 35-41.

Hammack, R. W. and Edenborn, H. M. (1992) The removal of nickel from mine waters using bacterial sulfate reduction. *Appl. Microbiol. Biotechnol.* **37**(5), 674-678.

Hammack, R. W., Edenborn, H. M. and Dvorak, D. H. (1994) Treatment of water from an open-pit copper mine using biogenic sulfide and limestone: a feasibility study. *Wat. Res.* **28**(11), 2321-2329.

Hard, B. C., Friedrich, S. and Babel, W. (1997) Bioremediation of acid mine water using facultatively methylotrophic metal-tolerant sulfate-reducing bacteria. *Microbiol. Res.* **152**(1), 65-73.

Herlihy, A. T. and Mills, A. L. (1985) Sulfate reduction in freshwater sediments receiving acid mine drainage. *Appl. Environ. Microbiol.* **49**(1), 179-186.

Jin, P., Bhattacharya, K., Williams, C. J. and Zhang, H. (1998) Effects of sulfide addition on copper inhibition in methanogenic systems. *Wat. Res.* **32**(4), 977-988.

Karnachuk, O. V. (1995) Influence of hexavalent chromium on hydrogen sulfide formation by sulfate-reducing bacteria. *Microbiol.* **64**(3), 262-265.

Lin, C-Y, and Chen, C-C (1999) Effect of heavy metals on the methanogenic UASB granule. *Wat. Res.* **33**(2), 409-416.

Liu, Y. and Fang, H. H. P. (1997) Effects of molybdate on sulfate reduction and benzoate degradation. *J. Environ. Eng.* **123**(9), 949-953.

Lovley, D. R. (1993) Anaerobes into heavy metals: Dissimilatory metal reduction in anoxic environments. *Trends Ecol. Evo.* **8**(6), 213-217.

Lovley, D. R. (1995) Bioremediation of organic and metal contaminants with dissimilatory metal reduction. *J. Ind. Microbiol.* **14**(2), 85-93.

Lovley, D. R. and Phillips, E. J. P. (1992a) Reduction of uranium by *Desulfovibrio desulfuricans. Appl. Environ. Microbiol.* **58**(3), 850-856.

Lovley, D. R. and Phillips, E. J. P. (1992b) Bioremediation of uranium contamination with enzymic uranium reduction. *Environ. Sci. Technol.* **26**(11), 2268-2234.

Lovley, D. R. and Phillips, E. J. P. (1994) Reduction of chromate by *Desulfovibrio vulgaris. Appl. Environ. Microbiol.* **60**(2), 726-728.

Lovley, D. R. Widman, P. K., Woodward, J. C. and Phillips, E. J. P. (1993) Reduction of uranium by cytochrome c_3 of *Desulfovibrio vulgaris. Appl. Environ. Microbiol.* **59**(11), 3572-3576.

Lyew, D., Knowles, R. and Sheppard, J. (1994) Biological treatment of acid mine drainage under continuous flow conditions in a reactor. *Trans. Inst. Chem Eng.* Part B, **72**(1), 42-47.

Machemer, S. D., Reynolds, J. S., Laudon, L. S. and Wildeman, T. R. (1993) Balance of S in a constructed wetland built to treat acid mine drainage, Idaho Springs, Colorado, U.S.A. *Appl. Geochem.* **8**(6), 587-603.

Maree, J. P., Gerber, A. and Hill, E. (1987) An integrated process for biological treatment of sulfate-containing industrial effluents. *J. Wat. Pollut. Control Fed.* **59**(12), 1069-1074

Mohagheghi, A., Updegraff, D. M. and Goldhaber, M. B. (1985) The role of sulfate-reducing bacteria in the deposition of sedimentary uranium ores. *Geomicrobiol. J.* **4**(2), 153-173.

Morton, R. L., Yanko, W. A., Graham, D. W. and Arnold, R. G. (1991) Relationships between metal concentration and crown corrosion in L.A. county sewers. *Res. J. Wat. Pollut. Control Fed.* **63**(5), 789-798.

Mueller, R. F. and Steiner, A. (1992) Inhibition of anaerobic digestion caused by heavy metals. *Wat. Sci. Technol.* **26**(3/4), 835-846.

Nedwell, D. B. and Reynolds, P. J. (1996) Treatment of landfill leachate by methanogenic and sulfate-reducing digestion. *Wat. Res.* **30**(1), 21-28.

Nelson, D. C., Casey, W. H., Sison, J. D., Mack, E. E., Ahmad, A. and Pollack, J. S. (1996) Selenium uptake by sulfur-accumulating bacteria. *Geochimica et Cosmochimica Acta* **60**(18), 3531-3539.

Newman, D. K., Kennedy, E. K., Coates, J. D., Ahmann, D., Ellis, D. J., Lovley, D. R. and Morel, F. M. M. (1997) Dissimilatory arsenate and sulfate reduction in *Desulfotomaculum auripigmentum* sp. nov. *Arch. Microbiol.* **168**(5), 380-388.

Newport, P. J. and Nedwell, D. B. (1988) The mechanism of inhibition of *Desulfovibrio* and *Desulfotomaculum* species by selenate and molybdate. *J. Appl. Bacteriol.* **65**(5), 419-423.

Padival, N. A., Kimbell, W. A. and Redner, J. A. (1995) Use of iron salts to control dissolved sulfides in truck sewers. *J. Environ. Eng.* **121**(11), 824-829.

Panak, P., Hard, B. C., Pietzsch, K., Kutschke, S., Roske, K., Selenska-Pobell, S., Bernhard, G. and Nitsche, H. (1998) Bacterium from uranium mining waste pile: interactions with U(VI). *J. Alloy. Compd.* **271**, 262-266.

Peiffer, S., Afonso, M. D. S., Wehril, B. and Gachter, R. (1992) Kinetics and mechanisms of the reaction of H_2S with Lepidocrocite. *Environ. Sci. Technol.* **26**(12), 2408-2413.

Pesch, C. E., Hansen, D. J., Boothman, W. S., Berry, W. J. and Mahony, J. D. (1995) The role of acid-volatile sulfide and interstitial water metal concentrations in determining bioavailability of cadmium and nickel from contaminated sediments to the marine polychaete *Neanthes arenaceodentata*. *Environ. Toxicol. Chem.* **14**(1), 129-141.

Peterson, G. S., Ankley, G. T. and Leonard, E. N. (1996) Effect of bioturbation on metal-sulfide oxidation in surficial freshwater sediments. *Environ. Toxicol. Chem.* **15**(12), 2147-2155.

Pettine, M., Barra. L., Campanella, L. and Millero, F. (1998) Effect of metals on the reduction of chromium (VI) with hydrogen sulfide. *Wat. Res.* **32**(9), 2807-2813.

Poulson, S. R., Colberg, P. J. S. and Drever, J. I. (1997) Toxicity of heavy metals (Ni, Zn) to *Desulfovibrio desulfuricans. Geomicrobiol.* **14**(1), 41-49.

Ravishankar, B. R., Blais, J. F., Benmoussa, H. and Tyagi, R. D. (1994) Bioleaching of metals from sewage sludge: elemental sulfur recovery. *J. Environ. Eng.* **120**(2), 462-470.

Rawlings, D. E. (1998) Industrial practice and the biology of leaching of metals from ores. The 1997 Pan Labs Leture. *J. Ind. Microbiol. Biotechnol.* **20**(5), 268-274.

Seidel, H., Ondruschka, J., Morgenstern, P. and Stottmeister, U. (1998) Bleaching of heavy metals from contaminated aquatic sediments using indigenous sulfur-oxidizing bacteria: a feasibility study. *Wat. Sci. Technol.* **37**(6/7), 387-394.

Sercombe, D. C. W. (1995) The control of septicity and odours in sewerage systems and at sewage treatment works operated by Anglian water services limited. *Wat. Sci. Technol.* **31**(7), 283-292.

Somlev, V. and Tishkov, S. (1994) Anaerobic corrosion and bacterial sulfate reduction: applications for the purification of industrial wastewater. *Geomicrobiol. J.* **12**(1), 53-60.

Song, Y-C, Piak, B-C, Shin, H-S, and La, S-J. (1998) Influence of electron donor and toxic materials on the activity of sulfate reducing bacteria for the treatment of electroplating wastewater. *Wat. Sci. Technol.* **38**(4/5), 187-194.

Tanaka, S. and Lee, Y-H (1997) Control of sulfate reduction by molybdate in anaerobic digestion. *Wat. Sci. Technol.* **36**(12), 143-150.

Tanimoto, Y., Tasaki, M., Okamura, K., Yamaguchi, M. and Minami, K. (1989) Screening growth inhibitors of sulfate-reducing bacteria and their effects on methane fermentation. *J. Ferment. Bioeng.* **68**(5), 353-359.

Taylor, B. F. and Oremland, R. S. (1979) Depletion of adenosine triphosphate in *Desulfovibrio* by oxyanions of Group VI elements. *Current Microbiol.* **3**(2), 101-103.

Tebo, B. M. and Obraztsova, A. Y. (1998) Sulfate-reducing bacterium grows with Cr(VI), U(VI), Mn(IV), and Fe(III) as electron acceptors. *FEMS Microbiol. Lett.* **162**(1), 193-198.

Tomei, F. A., Barton, L. L., Lemanski, C. L., Zocco, T. G., Fink, N. H. and Sillerud, L. O. (1995) Transformation of selenate and selenite to elemental selenium by *Desulfovibrio desulfuricans*. *J. Ind. Microbiol.* **14**(3/4), 329-336.

Tucker, M. D., Barton, L. L. and Thomson, B. M. (1996) Kinetic coefficients for simultaneous reduction of sulfate and uranium by *Desulfovibrio desulfuricans*. *Appl. Microbiol. Biotechnol.* **46**(1), 74-77.

Tucker, M. D., Barton, L. L. and Thomson, B. M. (1997) Reduction and immobilization of molybdenum by *Desulfovibrio desulfuricans*. *J. Environ. Qual.* **26**(4), 1146-1152.

Tucker, M. D., Barton, L. L. and Thomson, B. M. (1998a) Reduction of Cr, Mo, Se and U by *Desulfovibrio desulfuricans* immobilized in polyacrylamide gels. *J. Ind. Microbiol.* **20**(1), 13-19.

Tucker, M. D., Barton, L. L. and Thomson, B. M. (1998b) Removal of U and Mo from water by immobilized *Desulfovibrio desulfuricans* in column reactors. *Biotechnol. Bioeng.* **60**(1), 88-96.

Tuovinen, O. H., Niemelä S. I. and Gyllenberg, H. G. (1971) Tolerance of *Thiobacillus ferrooxidans* to some metals. *Antonie van Leeuwenhoek* **37**(4), 489-496.

Tuttle, J. H., Dugan, P. R. and Randles, C. I. (1969) Microbial sulfate reduction and its potential utility as an acid mine water pollution abatement procedure. *Appl. Microbiol.* **17**(2), 297-302.

Ueki, K., Ueki, A., Itoh, K., Tanaka, T. and Satoh, A. (1991) Removal of sulfate and heavy metals from acid mine water by anaerobic treatment with cattle waste: effect of heavy metals on sulfate reduction. *J. Environ. Sci. Health* **A26**(8), 1471-1489.

Van den Berg, G. A., Loch, J. P. G., Van der Heijdt, L. M. and Zwolsman, J. J. G. (1998) Vertical distribution of acid-volatile sulfide and simultaneously extracted metals in a recent sedimentation area of the River Meuse in the Netherlands.

Environ. Toxicol. Chem. **17**(4), 758-763.
Vile, M. A. and Wieder, R. K. (1993) Alkalinity generation by Fe(III) reduction versus sulfate reduction in wetlands constructed for acid mine drainage treatment. *Wat. Air Soil Pollut.* **69**(3/4), 425-441.
Wakao, N., Takahashi, T., Sakurai, Y. and Shiota, H. (1979) A treatment of acid mine waster using sulfate-reducing bacteria. *J. Ferment. Technol.* **57**(5), 445-452.
Waybrant, K. R., Blowes, D. W. and Ptacek, C. J. (1998) Selection of reactive mixtures for use in permeable reactive walls for treatment of mine drainage. *Environ. Sci. Technol.* **32**(13), 1972-1979.
Webb, J. S., McGinness, S. and Lappin-Scott, H. M. (1998) Metal removal by sulfate-reducing bacteria from natural and constructed wetlands. *J. Appl. Microbiol.* **84**(2), 240-248.
White, C. and Gadd, G. M. (1996) A comparison of carbon/energy and complex nitrogen sources for bacterial sulfate-reduction: potential applications to bioprecipitation of toxic metals as sulfides. *J. Ind. Microbiol.* **17**(2), 116-123.
White, C. and Gadd, G. M. (1997) An internal sedimentation bioreactor for laboratory-scale removal of toxic metals from soil leachates using biogenic sulphide precipitation. *J. Ind. Microbiol. Biotechnol.* **18**(6), 414-421.
White, C., Sayer, J. A. and Gadd, G. M. (1997) Microbial solubilization and immobilization of toxic metals: key biogeochemical processes for treatment of contamination. *FEMS Microbiol. Rev.* **20**(3/4), 503-516.
White, C., Sharman, A. K. and Gadd, G. M. (1998) An integrated microbial process for the bioremediation of soil contaminated with toxic metals. *Nat. Biotechnol.* **16**(6), 572-575.
Wieder, R. K. (1989) A survey of constructed wetlands for acid mine drainage treatment in the Eastern US. *Wetlands* **9**(2), 299-315.
Wijaya, S., Henderson, W. D., Bewtra, J. K. and Biswas, N. (1993) Optimization of dissolved metals removal using sulfate reducing bacteria. *Porc. 48th Purdue Ind. Wastewater Conf.*, pp. 469-481, Lewis Publishers, Chelsea, MI.
Zumdahl, S. S. (1989) *Chemistry*, D.C. Heath, Lexington, MA.

18

Interactions of the sulfur and nitrogen cycles: microbiology and process technology

Philippe M. Chazal and Piet N.L. Lens

18.1 INTRODUCTION

Microorganisms play a key role in the biogeochemical cycling of numerous elements. This is particularly the case for the elements of living biomass such as carbon, hydrogen, oxygen, nitrogen, sulfur and phosphorus. When studying these cycles, the interactions between the different elements are of utmost importance. This chapter describes the interactions between the sulfur and nitrogen cycles, with particular attention paid to processes allowing nitrogen removal from polluted environments, using sulfur cycle conversions.

18.1.1 The nitrogen cycle

The nitrogen cycle largely depends on microbial activities (Figure 18.1). Gaseous N_2 from the air is reduced in soil or water to ammonium by bacterial nitrogen fixation. These bacteria possess the *Nif* genes that code for a specialized enzyme, nitrogenase, which catalyses this N_2 reduction under well defined conditions. In soil, it occurs in free-living microorganisms such as *Azotobacter* sp. and *Klebsiella* sp. or in symbiotic bacteria such as *Rhizobium* sp. that grow in association with plants. Nitrogen fixation also occurs in aquatic habitats by Cyanobacteria. Ammonium thus formed is incorporated into organic matter (e.g. proteins and nucleic acids) by plants, animals and humans. Organic matter released by these eukaryotes in the environment is degraded by other bacteria such as members of the Myxobacteriales. During the decay ammonium is released (ammonification).

Ammonium is then oxidized by nitrifying bacteria. In a first step *Nitrosomonas* sp. and related genera oxidize ammonia to nitrous acid (HNO_2). In a second step *Nitrobacter* sp. and related genera oxidize nitrites to nitrates. Nitrates are the most common form of nitrogenous compounds used by plants. Nitrates present in excess to plants' requirements can be reduced to gaseous forms of nitrogen (mainly N_2, but also NO, N_2O, etc) by several denitrifying bacteria such as *Pseudomonas* sp., *Bacillus* sp., *Thiobacillus* sp. and *Spirillum* sp.

The presented nitrogen cycle (Figure 18.1) is a synoptic overview and has been extensively and periodically reviewed during the past 20 years. Recent data suggest that this cycle is much more complex (Verstraete and Philips 1998) because of newly isolated bacterial species, numerous microbial interactions in microbial communities and interferences between (well known as well as newly discovered) metabolic pathways.

For instance, methanotrophs (see Amaral *et al.* 1995) and even thermophilic bacteria (Pel *et al.* 1997) could also be implied in nitrification or denitrification reactions. Nitrifiers can communicate cell to cell because of the release of pheromone like molecules sech as homoserine derivatives (Batchelor *et al.* 1997). This would allow nitrifying bacteria to form biofilms that could help them to survive better when the environmental conditions are unfavourable, for instance in the case of substrate starvation (Batchelor *et al.* 1997). Nitrites can be reduced to dinitrogen gas in the presence of ammonia via hydroxylamine and hydrazine in the so-called Annamox process (Van der Graaf *et al.* 1996). The microorganism implied in this bioconversion has not yet been fully characterized. Binswanger *et al.* (1997) reported that ammonium can be converted to a mixture of nitrite and N_2

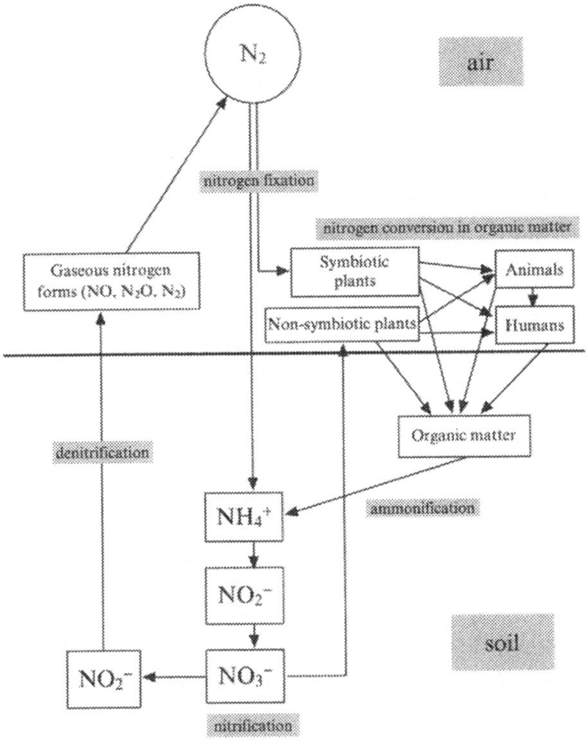

Figure 18.1. Schematic representation of the nitrogen cycle.

via hydroxylamine. Part of the nitrites are reduced by NADH + H^+ released during ammonia oxidation. A complete review of these new aspects of the nitrogen cycle has been presented by Verstraete and Philips (1998).

18.1.2 Environmental problems of nitrate

Nitrates and nitrites, which are the main nitrogen forms on which we focus in this chapter, are implied in the pathogenicity of mammals and humans. Nitrates can be reduced to nitrites in the gastrointestinal tract. Nitrites can thus combine with amino-acids to release nitrosamines, which are presumed to be carcinogenic (Esquian and Simon 1990). Nitrites also combine with haemoglobin to form methaemoglobin, a protein unable to carry oxygen in blood like haemoglobin (De Saint-Blanquat and Fritsch 1994). This can lead to serious forms of anaemia, which can even be lethal. Therefore, in the European Community the maximum admissible levels of nitrates and nitrites in drinking water are nowadays, respectively, 50 and 0.1 mg l^{-1} (respectively 11.3 mg NO_3^--N l^{-1} and 0.03 mg NO_2^--N l^{-1}) whereas the nitrate contamination level of groundwater or surface waters can be up to 80–100 mg l^{-1} (18–22.6 mg NO_3^--N l^{-1}). For nitrates, the European Guide Level is 25 mg l^{-1} (5.6 mg NO_3^--N l^{-1}).

Northern European countries such as Belgium, The Netherlands, Germany and Denmark are severely affected by nitrate contamination of surface and groundwater. This is also the case for numerous French areas where animal culture and breeding are carried out at intensive levels, such as Bretagne, Nord-Pas-de-Calais, Beauce and Brie. The magnitude of this problem gradually increases. Three main nitrate pollution sources that continuously become more important have been identified: urban wastewaters, industrial wastewaters, and agricultural and animal breeding wastes.

18.1.3 Nitrate pollution abatement techniques

Preventive procedures have been developed to limit nitrate contamination in the environment. Nevertheless, this prevention policy is restricted to nitrates that are released nowadays. One must take into account that the nitrates polluting surface or groundwaters now were released in soils 15-20 years ago (Mariotti 1992). This is why the development of remedial procedures to treat pollution is also necessary.

Both physico-chemical and biological procedures have been developed for nitrate removal. Physico-chemical procedures (denitratation) are mainly based on ion exchange resins (Sibony 1979) osmosis and electrodialysis (Richard and Leprince 1982; Philipot and Sibony 1986). These procedures are very effective for nitrate removal and produce drinking water that meets legal standards. However, they also present disadvantages. They induce high production costs and all methods accumulate nitrates: e.g. saline eluates of ion exchange resins. These highly concentrated nitrate solutions also need proper treatment.

Biological procedures are nowadays also used for nitrate removal even at industrial scale. They are mostly based on the use of denitrifying heterotrophic bacteria. Denitrifiers are strict aerobes which are able to use alternative electron acceptors such as nitrate or nitrite in the absence of oxygen. In anaerobiosis, but in the presence of nitrates, oxidative metabolism can be achieved (Figure 18.2). The electrons coming from the oxidation of the organic substrate are transferred to nitrate which is thus reduced to N_2 via different intermediary forms such as nitrites, NO and

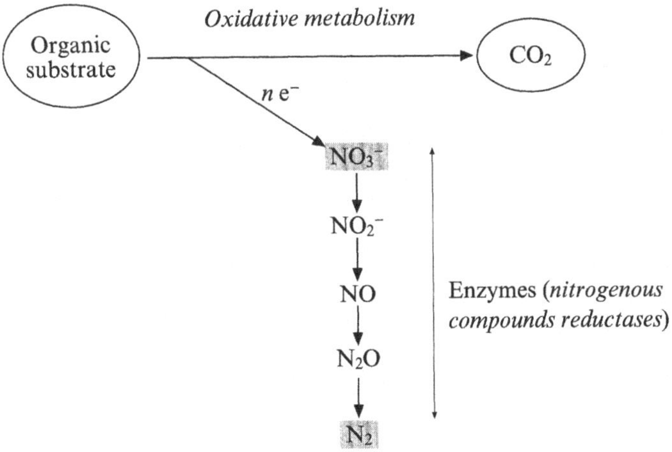

Figure 18.2. Scheme of nitrate reduction in anaerobiosis by heterotrophs coupled to the oxidation of organic substrate.

N_2O. Two organic substrates (carbon source and electron donor) are frequently used at industrial scale: ethanol and acetic acid. The degradation of these compounds releases carbon dioxide.

For drinking water production, several reactor configurations have been developed to completely denitrify, e.g. Nitrazur (Richard and Partos 1986) Biodenit (Phillipot and Patte 1982) Denipor (Roennefahrt 1986) and Nitraflux (Guerber *et al.* 1998). The crude water is recirculated in a reactor containing solid material (Biolite, Biodagene, sand, pouzzolane...). Ethanol or acetic acid are added at the appropriate concentration to provide enough carbon for bacterial growth. During start-up, the denitrifying bacteria colonize the filter. When enough biomass has colonized the support material (10^{11}–10^{12} g biomass per g of material) the reactor is fed with the nitrate-rich water in a continuous way. In the first part of the filter, dissolved oxygen is consumed without nitrate removal. When oxygen is depleted, nitrate reduction takes place in the anaerobic part of the filter.

Interesting data concerning the influence of different physico-chemical parameters on technological aspects of heterotrophic denitrification can be found in the book of Haslay and Leclerc (1993). Optimized denitrification plants can achieve groundwater denitrification at volumetric denitrification rates of 1–1.5 kg NO_3-N /m^3 d^{-1} (Ravarini *et al.* 1988; Richard and Partos 1986; Guerber *et al.* 1998).

18.2 MICROBIAL ASPECTS OF OXIDATIVE PROCESSES

The sulfur and nitrogen cycles can interact via the sulfur-oxidizing bacteria. These interactions are encountered during denitrification of wastewaters containing sulfurous compounds, drinking water production or during natural denitrification in aquifers or in sediments located in well-defined geographical areas. The sulfur-oxidizing bacteria belong to a very heterogeneous bacterial group of which the members share the ability to use reduced sulfur compounds as the energy source (Robertson and Kuenen 1992). This group has been named in 1890 by Winogradsky the "Colourless sulfur bacteria", because these bacteria contain no pigments. They are able to oxidize reduced sulfur forms (equations 18.1 to 18.4):

$$S^{2-} \rightarrow S^{\circ} + 2e^- \quad (18.1)$$
$$S^{\circ} + 4H_2O \rightarrow SO_4^{2-} + 8H^+ + 6e^- \quad (18.2)$$
$$S_2O_3^{2-} + 5H_2O\ 2\rightarrow SO_4^{2-} + 10H^+ + 8e^- \quad (18.3)$$
$$S_4O_6^{2-} + 10H_2O \rightarrow 4SO_4^{2-} + 20H^+ + 14e^- \quad (18.4)$$

These oxidative reactions are coupled to the reduction of O_2 or NO_3^- (equations 18.5 and 18.6)

$$O_2 + 4e^- + 4H^+ \rightarrow 2H_2O \quad (18.5)$$
$$2NO_3^- + 12H^+ + 10e^- \rightarrow N_2 + 6H_2O \quad (18.6)$$

Equation 18.6 integrates the reduction of nitrates to nitrites and the reduction of nitrites to N_2 via NO and N_2O.

When both oxygen and nitrates are present, the reduction of oxygen occurs preferentially because of the more favorable Gibb's free energy (ΔG^0). In anaerobic conditions, nitrate reduction can occur in the presence of denitrifying bacteria. The colourless sulfur bacteria have a versatility of growth characteristics but are mainly aerobic. However, *Thiobacillus denitrificans* is able to reduce nitrate under anaerobic conditions. This has resulted in its taxonomic subdivisions (Kuenen *et al.* 1992).

18.2.1 *Thiobacillus denitrificans*, a denitrifying "colourless sulfur bacterium"

Taylor and Hoare (1971) developed a selective growth medium for *Thiobacillus denitrificans* which is still in use today even if under a slightly

Table 18.1. Morphological and growth characteristics of *Thiobacillus denitrificans* (After: Kuenen 1989; Kelly and Harrison 1989)

Characteristic	Description
Shape	Gram negative rod 0.5 × 1–3 μm
Motility	Motile with polar flagellum
Aspect of colonies on agar slants containing $S_2O_3^{2-}$ and NO_3^-	Opalescent
Reduced sulfur sources that support growth	S^{2-}, S°, $S_2O_3^{2-}$, $S_4O_6^{2-}$
Denitrification in the presence of	S^{2-}, S°, $S_2O_3^{2-}$
Mean growth rate	0.06–0.11 h^{-1}

modified composition (Koenig and Liu 1996; Trouvé *et al.* 1998a). The most accurate way to determine genera and species is the sequencing of 16S RNA, but traditional dichotomic keys remain available, although they are less precise (Kuenen 1989; Kelly and Harrison 1989). The principle of this determination key is described in Box 18.1. Its main morphological and growth characteristics are given in Table 18.1. Tables 18.2 and 18.3 summarize, respectively, the main enzymatic activities and growth rates of *Thiobacillus denitrificans.*

Box 18.1. Determination keys for Thiobacillus denitrificans

A1 The microorganism is a prokaryote that uses H_2S, S^{2-}, polysulfides, S°, $S_2O_3^{2-}$, (poly)thionates and SO_3^{2-} as energy source. The main oxidation by-products are S° and SO_4^{2-} even if polysulfides, $S_2O_3^{2-}$, SO_3^{2-} or polythionates can be produced.

A2 This microorganism does not accumulate intracellularly S°, even when grown in the presence of O_2 and S^{2-}.

A3 The optimum growth temperature is below 55°C. The microorganism growth is strictly chemolithotrophic.

A4 Rod type cellular morphology, motile or non motile.

→ **Genus *Thiobacillus*** (Kuenen 1989)

B1 Neutrophilic microorganism (optimum growth pH between 6 and 8, but potentially initiated at pH 3)

B2 Strictly autotrophic and chemolithotrophic microorganism. Grows on media with $S_2O_3^{2-}$ as the sole energy source. Unable to use glucose as sole energy source. Fitted out with ubiquinone Q-8

B3 Facultative anaerobe, produces N_2 from nitrate, nitrite and N_2O reduction in the absence of O_2. GC% content in DNA varies from 63 to 68%.

→ ***Thiobacillus denitrificans*** (Kelly and Harrison 1989)

Table 18.2. Enzymatic activities of *Thiobacillus denitrificans* (for the complete thiosulfate metabolism see Figure 18.3)

Type of metabolism	Enzymes (purified or simply evidenced)	References
Carbon assimilation (acetate can also be assimilated but its tribute to the total cellular amount never exceeds 10%)	Benson-Calvin enzymes Uncomplete tricarboxylic acids cycle	Baalsrud and Baalsrud (1954) Taylor and Hoare (1971) Kelly (1989)
Nitrate reduction	Nitrate reductase, nitrite reductase, nitrous oxide reductase	Adams *et al.* (1971) Le Gall *et al.* (1979) Hole *et al.* (1996)
Sulfur oxidation	APS reductase, sulfite reductase	Kelly (1982) Taylor (1989) Schedel and Trüper (1980) Sawhney and Nicolas (1977)

Table 18.3. Growth yields of *Thiobacillus denitrificans*

	Growth yield	References
Carbon assimilation	0.38 fixed CO_2 mole / mole $S_2O_3^{2-}$	Hoor (1976, 1981)
Nitrate reduction	7.95 g dry mass / mole NO_3^- 8.55 g dry mass / mole NO_3^-	Justin and Kelly (1978a; b) Claus and Kutzner (1985)
Thiosulfate oxidation	11.37 g dry mass / mole NO_3^- 9.52 g dry mass / mole NO_3^-	Justin and Kelly (1978a; b) Claus and Kutzner (1985)

Several physico-chemical parameters influence denitrification by *Thiobacillus denitrificans* using thiosulfate as the electron donor (Claus and Kutzner 1985).

These authors determined optimum pH values between 7.5 and 8. The optimum temperature was located between 25 and 35 °C, and no denitrification was noticed below 7–8 °C. Denitrification inhibition by nitrite started at approximately 100 mg l^{-1} and 100% inhibition occurred at 2 g l^{-1}. Sulfates and chlorides also inhibited denitrification, respectively, at 4 and 22 g l^{-1}. The electron donor and acceptor also inhibited denitrification at high concentration: in the case of both nitrate and thiosulfate, inhibition started at 12 g l^{-1}.

Numerous work remains to be done to obtain a better knowledge of the physiology of *Thiobacillus denitrificans*. Some contradictory results have been published in the literature. For instance, according to Baalsrud and Baalsrud (1954) *Thiobacillus denitrificans* is unable to use nitrates as a cellular nitrogen source. In contrast, the growth of some strains of this bacterium has been demonstrated in the absence of ammonium, but in the presence of nitrate as the sole nitrogen source (Baldensperger and Garcia 1975; Claus and Kutzner 1985). This suggests that, at least in the case of

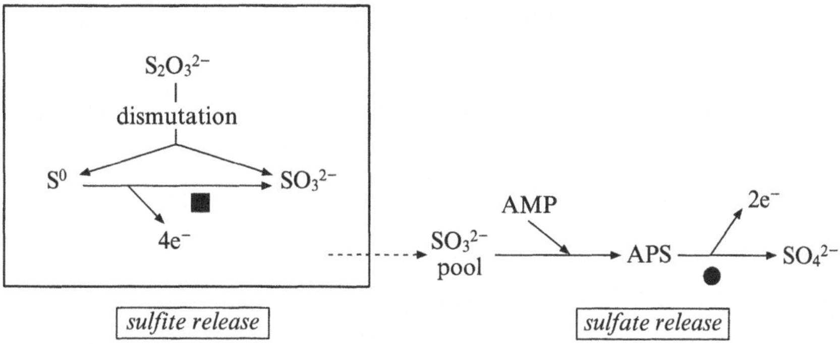

Figure 18.3. Schematic representation of the thiosulfate oxidation pathway by *Thiobacillus denitrificans*. ■, Sulfite reductase; ●, APS reductase.

such strains, nitrate metabolism should not only imply their denitrification to N_2 (the nitrates being the electrons acceptors) but also nitrate reduction to ammonia. The NH_4^+ thus produced is incorporated in cellular molecules, for instance amino acids, proteins, nucleotides and nucleic acids.

Trouvé *et al.* (1998a) recently isolated, using the procedure of Taylor (1971) strains of *Thiobacillus denitrificans* from the environment that denitrified in non-standard physico-chemical conditions compared with the results published by Claus and Kutzner (1985). In the case of strain 96.C2 isolated by Trouvé *et al.* (1998a) a 100% denitrification yield was obtained between pH 5.5 and 8, and this yield was still 83% at pH 5. The pH range thus appears to be much wider than the one determined by Claus and Kutzner (1985). For temperature effects, the nitrate removal rate (determined as the amount of nitrate removed per litre and per hour) was optimum at 10 °C and was only half at 5 °C. These results significantly differ from those published by Claus and Kutzner (1985) who found an optimum temperature of 25–35 °C. The question arises if this bacterium is a new strain of *Thiobacillus denitrificans* or another bacterium. This warrants further research using molecular biological methods such as 16S ribosomal RNA sequencing (Kuenen *et al.* 1992).

18.2.2 Other sulfur-oxidizing and nitrate-utilizing microorganisms

Although less cited in the technological literature other microorganisms present in different ecological habitats, such as sediments or microbial mats, are able to oxidize reduced sulfur forms along with nitrate reduction.

Below, some of these bacterial species are listed although this list is not exclusive.

Two other *Thiobacillus* species have been reported to exhibit nitrate reduction coupled to sulfide oxidation: *Thiobacillus versutus* and *Thiobacillus thyasiris* (Kuenen *et al.* 1992). If the GC% content (percentage of GC base pairs in DNA) of *Thiobacillus thyasiris* is 52, this percentage is between 65–68 for the species *T. versutus* which is close to the 63–68% determined for *Thiobacillus denitrificans.*

Thiomicrospira denitrificans (GC% : 36) shares the same properties (Kuenen *et al.* 1992). This bacterium has a spiral shaped morphology that avoids confusion with the rod-shaped *Thiobacillus* sp.

Among the other sulfur bacteria, the gliding filamentous *Beggiatoa* sp. are often found to grow at the surface of sediments (freshwater or sulfide-rich marine water). When growing in such environmental conditions, they develop conspicuous white mats in which sulfide oxidation is coupled to nitrate reduction to N_2 (Sweerts *et al.* 1990).

Thiosphaera panthotropha is also able to denitrify using reduced sulfur compounds as the electron donor (Kuenen *et al.* 1992). This mixotrophic bacterium possesses a complex metabolism. Denitrification is also obtained with hydrogen or a wide range of organic compounds (except those containing one carbon) as electron donors. Evidenced as a heterotrophic nitrifier, *Thiosphaera panthotropha* can also denitrify aerobically (at an oxygen concentration of 3.8 mg l^{-1}) and heterotrophically.

Thiomargarita namibiensis, a 0.75 mm long bacterium was isolated in Africa and in South America (Schultz *et al.* 1999) This "giant" bacterium is of interest because of its large size compared with that of "usual" bacteria (about 1 μm) but also because of its denitrifying capacity in the presence of reduced sulfur compounds. This bacterium is found in an environment containing high hydrogen sulfide levels which are toxic for most animal life. Another peculiarity of this bacterium is that it accumulates nitrates in vacuoles that are consumed during starvation periods. Microbiologists are awaiting new developments in the physiology of this microorganism for potential applications.

18.2.3 Denitrification in natural environments and at oxygen interfaces

The literature provides few data about the quantitative contribution of *Thiobacillus denitrificans* and sulfur-oxidizing bacteria to natural denitrification phenomena. Firstly, its presence has not always been correlated with natural denitrification. Because the nitrate concentration in the environment is generally low, *Thiobacillus denitrificans* is only present in low numbers. Moreover, it is strongly adsorbed to the rocky matrix where its nutrients are located. Secondly, it appears difficult to correlate the presence of this type of microorganism in a given habitat to its quantitative contribution to the total denitrification.

Rödelsperger (1989) reported that denitrification occurs in a German aquifer (in a forest area near Hannover, Germany) containing high pyrite levels: the nitrate level is less than 0.4 mg N l^{-1} and during the past ten years the sulfate concentration has increased by approximately 100 mg l^{-1}, probably involving autotrophic bacteria. Le Bideau *et al.* (1994) reported a similar phenomenon in a French aquifer of the same type near Poitiers, and the presence of *Thiobacillus denitrificans* was noticed.

Golterman (1991) noticed denitrifying activity in sediments from the Camargue area (France) and this activity increased after the addition of FeS. Garcia-Gil and Golterman (1993) demonstrated that the denitrification kinetics in these sediments after FeS addition were of the Michaelis–Menten type. Furthermore, they observed that denitrification was accompanied by sulfate release according to the stoichiometry of autotrophic denitrification.

Sweerts *et al.* (1990) reported data implying mixotrophic *Beggiatoa* sp. in natural denitrification at the surface of freshwater sediments. Using microsensors, Sweerts *et al.* (1990) demonstrated that neither oxygen nor nitrate penetrated the sediment below 600 μm of the surface of the *Beggiatoa* sp. mat. They also demonstrated that anaerobic sulfide oxidation was coupled to nitrate reduction in the mat. This type of interaction occurs at oxygen interfaces present in microbial mats or, generally speaking, biofilms. Unravelling of the exact microbial composition of the denitrifying population present in both natural and man-made ecosystems warrants further research using specific analytical techniques, e.g. microelectrodes for nitrate and ammonium in combination with ribosomal RNA probes (Schramm *et al.* 1996).

18.3 TECHNOLOGICAL ASPECTS OF SULFIDE OXIDATION

18.3.1 Process conditions

When planning to use *Thiobacillus denitrificans* or other sulfur-oxidizing bacteria in a technological process, it is a prerequisite to pay special attention to some parameters that affect the growth of these bacteria: (i) the sulfide concentration (when it is the sulfur source); (ii) the oxygen concentration; (iii) temperature and (iv) sterile conditions.

Among the reduced sulfur forms, sulfide exhibits an inhibitory effect on the growth of *Thiobacillus denitrificans*. So, the process should be operated under sulfide concentrations below the inhibitory levels. The latter has to be determined for each bacterial strain used in a specific application. Sublette *et al.* (1998) reported that an increase in tolerance can be obtained by repeated exposure to increasing sulfide concentrations (under the Na_2S form). The *Thiobacillus denitrificans* wild type strain was inhibited by sulfide concentrations between 0.1–0.2 mM, whereas the resistant F strain could resist up to 2.5 mM (over ten times more) sulfide.

Oxygen and nitrates are both electron acceptors of the oxidation of reduced sulfur compounds (see equations 18.5 and 18.6). In the presence of oxygen, sulfur oxidation is preferentially coupled with oxygen reduction. Sublette *et al.* (1998) reported that denitrification using sulfide as electron donor cannot occur until O_2 levels are below 1 mg l^{-1}.

The temperature under which *Thiobacillus denitrificans* usually denitrifies and oxidizes sulfur is in the mesophilic range (20–30 °C). Recent works, however, showed that it is possible to select *Thiobacillus denitrificans* strains able to denitrify and oxidize reduced sulfur compounds at temperatures around 5 °C (Trouvé *et al.* 1998a).

Sterile conditions can be obtained in the laboratory but the maintenance of such conditions at the industrial level is illusory. Even if the influent organic matter level is poor, the heterotrophic counts can be up to 10^8 colony forming units (CFU) ml^{-1} (Sublette *et al.* 1998). The autotrophic growth of *Thiobacillus denitrificans* may indeed release organic carbon as well as cell lysis material. With high organic matter concentrations (wastewater) one cannot predict the influence of heterotrophs on the contribution of *Thiobacillus denitrificans* to denitrification. This should be considered separately for each type of influent. A way to proceed should be to compare the theoretical and experimental ratio of the [reacted oxidized products]/[reacted reduced substrates] for sulfur and the [remaining oxidized forms]/[initial oxidized forms] for nitrate. Nevertheless, further research is needed to control and minimize heterotrophic growth in a *Thiobacillus denitrificans* bioreactor.

18.3.2 Process design

18.3.2.1 Wastewater treatment

The coupling of the sulfur and nitrogen cycles has been used for the treatment of different wastewaters, either with sulfide or nitrate removal as the objective. It appears difficult to establish comparisons between all the experimental processes, because of differences concerning the objective, and the versatility of the operating conditions. In general (with the exception of a few cases) a reduced sulfur source has to be added when nitrate removal is the main purpose. This can be a solid (filter beds containing elemental sulfur) or soluble (e.g. sulfide or thiosulfate) sulfur source. In contrast nitrates have to be added when sulfide removal is the objective (Sublette *et al.* 1998). In the latter case, one must consider that there is no net sulfur removal because sulfide is converted into the highly water soluble sulfur. Below, some technological approaches for nitrate removal are presented. Table 18.4 summarizes the characteristics of some of the main processes developed in this field.

Filter beds containing elemental sulfur

In 1976, Sikora and Keeney used *Thiobacillus denitrificans* for the treatment of nitrified septic tank effluent. They used a fixed-bed pilot reactor containing S° and dolomite (a natural mixture of calcium and magnesium carbonates, also called maërl) chips. The inoculation was done using a collection strain of *Thiobacillus denitrificans* (ATCC 23642). At 23 °C and after a seeding period of 3 days, they reached and maintained a denitrification yield of about 100% at a volumetric loading rate of 290 g NO_3^--N/m^3 per day. The problem with this type of technology was the low solubility of elemental sulfur and the release of high sulfate concentrations (90 μg SO_4^{2-}-S l^{-1}).

Batchelor and Lawrence (1978) used an elemental sulfur filter, inoculated with *Thiobacillus denitrificans* for wastewater treatment. They reached a 99.5% denitrification yield in steady-state, continuous flow reactors. They modelled the system using Equation 18.7:

$$1.10\ S° + NO_3^- + 0.4\ CO_2 + 0.08\ HCO_3^- + 0.08\ NH_4^+ \rightleftharpoons 1.10SO_4^{2-} + 0.5\ N_2 + 0.08\ C_5H_7O_2N + 1.28\ H^+ \quad (18.7)$$

According to the stoichiometry of Equation 18.7, removal of 1 mg l^{-1} nitrate produces 2.5 mg l^{-1} SO_4^{2-} and 0.09 mg l^{-1} of H^+, which corresponds to an alkalinity consumption of 4.5 mg $CaCO_3$ l^{-1}. Equation 18.7 also shows that denitrification of 1 kg NO_3^- consumes 2.5 kg elemental sulfur. If the

Table 18.4. Characteristics of some processes using *Thiobacillus denitrificans* for the autotrophic denitrification of wastewater

Nature of the influent	Reactor type and sulfur source	Efficiency	Reference
Septic tank effluent diluted with tap water (1:36) and enriched with 40 mg NO_3^--N l^{-1}	Elemental sulfur mixed with dolomite (ratio 1:1, wt/wt). *Thiobacillus denitrificans* ATCC 23642 Continuous flow at 23 °C	Complete denitrification with a 3.3 h residence time at 23 °C. Release of 90 mg SO_4^{2-}-S l^{-1}. The authors assume that complete denitrification should occur at 5 °C in 1.5–2 days residence time	Sikora and Keeney (1976)
Synthetic wastewater enriched with 25–50 mg NO_3^--N l^{-1}	Continuous flow sulfide/limestone, sulfur and sulfur/limestone packed bed reactors. Enriched cultures of *Thiobacillus denitrificans* (origin unknown) in a synthetic thiosulfate medium. Temperature: 20 °C	Complete denitrification can be achieved in the presence of both sulfide and sulfur. When using sulfur the particles size strongly influence the minimum hydraulic retention time. Limestone provides alkalinity, the consumption of which being inherent to autotrophic denitrification	Driscoll and Bisogni (1978)
Synthetic wastewater enriched with 30 mg NO_3^--N	Continuous slurry reactor system using elemental sulfur (under continuous stirring) between 12 and 30 °C. Enriched cultures of *Thiobacillus denitrificans* (origin unknown) in a synthetic thiosulfate medium	Complete nitrate removal	Batchelor and Lawrence (1986)
Synthetic wastewater	Fluidized bed reactor using thiosulfate or elemental sulfur at 20 °C. Inoculation by an effluent sewage and start-up using a synthetic thiosulfate medium	Thiosulfate denitrification was a stable biological process but denitrification with elemental sulfur was unstable	Matsui and Yamamoto (1986)
Synthetic wastewater	Continuous flow reactor containing elemental sulfur (under continuous stirring) at 25 °C. Inoculation with activated sludge acclimated to denitrifying S° oxidation	High percentages of denitrification (over 95%) with a rate of 0.19–0.24 mg NO_3^--N/mg TOC.day	Hashimoto *et al.* (1987)
Municipal waste-water type influent	Ananox process: an elaborate multichambers design. *Thiobacillus denitrificans* oxidizes in an anoxic chamber sulfides released in an anaerobic chamber. The sulfide rich effluent is mixed with nitrates coming from an oxic nitrifying chamber	Complete denitrification. Denitrification rates varied between 3.6 and 2.84 mg NO_3^--N/g VSS/hour according to the anoxic chamber height.	Xiushan *et al.* (1993)
Ion exchange saline eluates	Elemental sulfur in fixed bed reactor at 20°C. Inoculation by enrichment of environmental *Thiobacillus denitrificans* in thiosulfate synthetic medium	Potential application for saline influents: 90% denitrification observed for NaCl concentration up to 30 g l^{-1}, for a volumetric loading rate of 1 kg NO_3^-/m^3.day	Sant'Anna *et al.* (1996)

average value of a nitrified household wastewater is 25 mg l^{-1} (Kuai and Verstraete 1999) up to 62.5 mg of sulfur are consumed to denitrify 1 m^3 wastewater. Thus, the initially added sulfur should be sufficient for continuous operation during lasting months. Moreover, fresh sulfur can be added to the reactor regularly, e.g. once a month, if required.

Batchelor and Lawrence (1978) further emphasized an additional advantage of autotrophic denitrification: it would be more stable under transient loading conditions. They also showed that the maximum nitrate removal rate is a linear function of the ratio between the sulfur (S) and biomass (X) concentration over the S/X range: 45–194 (expressed as mg S / mg organic N). The stoichiometry for autotrophic denitrification was relatively constant over the temperature range of 12–30°C and for the S/X range previously cited. In general, biomass production is low in an autotrophic denitrification reactor, compared with a heterotrophic reactor. However, the production of $CaSO_4$ might cause problems during practical applications, because of extra (chemical) sludge production and clogging of pipes.

Kuai and Verstraete (1999) compared autotrophic (using an elemental sulfur lime bed) to heterotrophic denitrification for the treatment of wastewaters from small scale treatment systems (fewer than 10,000 inhabitant equivalents). The sulfur lime bed removed 75% of the total oxidisable nitrogen at a loading rate of 0.2 total oxidisable nitrogen/l.day and a hydraulic retention time of 3 h. The filter bed also retained some COD (about 10%) and reduced faecal indicator organisms with about 1 log unit. The major drawback of the sulfur lime filter was its poor phosphorus removal.

Sant'Anna *et al.* (1996) studied how to valorize exchange ion resin eluates using a fixed-bed reactor containing elemental sulfur (Figure 18.4). NaCl is abundant in this type of eluate (25 g l^{-1} as mean value). They examined the influence of this salt on the denitrification yield. The modelling procedure they developed allowed the determination of an inhibition constant (K_i) of 41 g NaCl l^{-1}. This opens perspectives for future developments of this technology, as the NaCl concentration in these eluates is usually far below this K_i. For a 30 g l^{-1} NaCl concentration in eluates, a 90% denitrification yield was obtained at 20 °C at a nitrate volumetric loading rate of 1 kg $NO_3^-/S°$ m^3 per day.

Addition or use of soluble sulfur sources

Bisogni and Driscoll (1977) studied and demonstrated the efficiency of $S_2O_3^{2-}$ and S^{2-} as soluble sulfur energy sources for denitrification of synthetic wastewaters. They modelled the systems by equations 18.7 (thiosulfates) and 18.8 (sulfides) where $C_5H_7O_2N$ is the formula of cellular biomass.

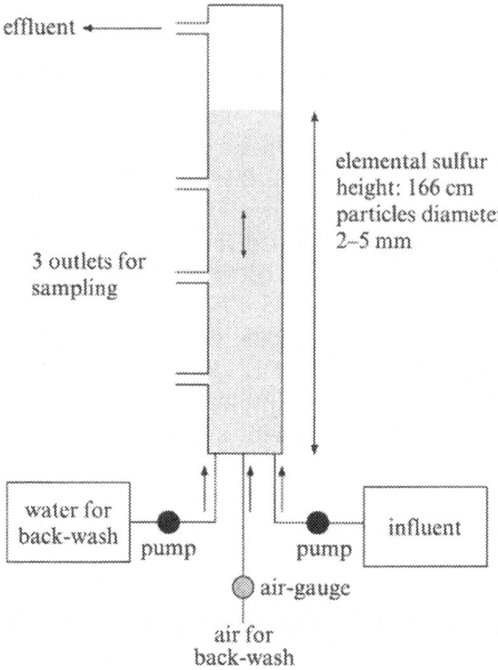

Figure 18.4. Pilot reactor scheme for autotrophic denitrification of ion exchange resin eluates, using an elemental sulfur filter bed (After: Sant'Anna *et al.* 1996).

$$0.844\ S_2O_3^{2-} + NO_3^- + 0.347\ CO_2 + 0.0865\ HCO_3^- + 0.0865\ NH_4^+ + 0.434\ H_2O \rightleftharpoons 1.689\ SO_4^{2-} + 0.5\ N_2 + 0.0865\ C_5H_7O_2N + 0.697\ H^+ \quad (18.8)$$

$$0.421\ S^{2-} + 0.421\ HS^- + NO_3^- + 0.346\ CO_2 + 0.0865\ HCO_3^- + 0.0865\ NH_4^+ \rightleftharpoons 0.842\ SO_4^{2-} + 0.5\ N_2 + 0.0865\ C_5H_7O_2N + 0.697\ H^+ + 0.413\ H_2O \quad (18.9)$$

The same authors confirmed the efficiency of S^{2-} addition compared with S° in an upflow reactor for wastewater treatment (Driscoll and Bisogni 1978). They suggest the use of limestone as an alkalinity source. The alkalinity consumption rate was 3.74 mg $CaCO_3$ per milligram of NO_3^--N reduced. Furthermore, they underlined the importance of the support surface. They suggested that the greater the surface of the support material compared with its volume, the more important the biofilm surface was. Thus, at a high surface ratio, the same denitrification yield can be reached for lower residence times, and thus at higher volumetric loading rates.

When using natural inocula from various origins, wastewaters were also successfully treated by autotrophic denitrification. The next related works give some information about the environmental inocula and the enrichment methods for denitrification in the presence of thiosulfate (Matsui and Yamamoto 1986) or elemental sulfur (Hashimoto *et al.* 1987). *Thiobacillus denitrificans* has also been employed in an industrial plant for the removal of organic matter and nitrogen from municipal waste water, the Ananox® process (Xiushan *et al.* 1993). Denitrification is integrated in a complex treatment system involving different processes such as methanogenesis, sulfate reduction, sulfide oxidation, nitrification and denitrification.

More recently the denitrification of mining wastewater enriched with nitrates was studied by Bouffard *et al.* (1998). Those effluents can be contaminated by high levels of nitrates (up to 1 g l^{-1}) owing to the use of nitrate-based explosives and nitric acid leaching for metal recovery. The anaerobic closed-cycle system they are currently developing is based on the co-culture of *Thiobacillus denitrificans* and sulfate-reducing bacteria. The latter bacteria generate "*in situ*" sulfide, which is then used by *Thiobacillus denitrificans* for denitrification.

18.3.2.2 Drinking water production

Table 18.5 summarizes the characteristics of some of the main autotrophic denitrification processes for drinking water production using *Thiobacillus denitrificans.*

Filter beds containing elemental sulfur

Blécon *et al.* (1983) studied a fixed-bed reactor with upflow feeding. In the reactor, dolomite was used as bacterial growth support and also as a carbon and alkalinity source. In alkaline waters the dolomite content should not exceed 10%. Higher dolomite concentrations lowered the denitrification yield. Several works followed this preliminary study. A mathematical model of the procedure was proposed in continuous flow and taking into account the variations of volumetric or hydraulic loading rates (Le Cloirec *et al.* 1985). A 100% denitrification yield was reached for a volumetric nitrate loading rate of 0.18 kg NO_3^--N/S° m^3 per day. The biomass production amounted 0.23 g dry mass per gram nitrate removed in a reactor containing half elemental sulfur and half maërl. The nitrate removal kinetics were determined as first-order reactions. These authors also determined the experimental values of the kinetic constants k_1 and k_2 of the denitrification phenomenon, according to Equation 18.10:

Table 18.5. Characteristics of some of the main autotrophic denitrification processes using *Thiobacillus denitrificans* for drinking water production

Influent	Reactor type and sulfur source	Efficiency	Reference
Alkaline water ($NaHCO_3$: 420 mg l^{-1}) The water contained 18 mg NO_3^-- N l^{-1}	Elemental sulfur in fixed bed reactor at 20 °C Inoculation by enrichment of environmental *Thiobacillus denitrificans* in thiosulfate synthetic medium	Potential application for saline influents: 90% denitrification observed for NaCl concentration up to 30 g l^{-1}, for a volumetric loading rate of 1 kg NO_3^-/m^3.day	Blécon *et al.* (1983)
Tap water enriched with nitrate Nitrate loading rate of 0.18 kg NO_3^--N/S°-m^3.day	Fixed bed reactor with elemental sulfur and dolomite (1 : 1 ratio)	100% yield was obtained. The authors determined that the reactions were first order kinetics	Le Cloirec *et al.* (1985)
Groundwater enriched with nitrate (15.8 mg NO_3^--N l^{-1})	Slow sulfur/limestone filtration (1 : 1 ratio) Filter operating at a rate of 25 m h^{-1}	The nitrate content was reduced from 15.8 to less than 5.6 mg NO_3^--N l^{-1} (the European Guide Level). No backwashing required for a period of time of over 9 months because of slow filtration	Schippers *et al.* (1987)
Groundwater Nitrate volumetric loading rate 9 g NO_3^--N/m^3.h	Sulfur/limestone filtration process Temperature of the groundwater between 10–12 °C	Recommended sulfur : limestone ratio is 1 : 2 (in order to simplify maintenance) Long running times: 300–400 days (for nitrate concentrations below the European Guide Level)	Van der Hoek *et al.* (1992)
Crude water (chemically well defined) containing about 11 mg NO_3^--N l^{-1}	The first reactor is filled with iron wool and the second contains biolite (fixed bed type)	The process allows the production of drinking water with nitrate levels close to the European Guide Level. Few maintenance is required	Montiel and Welté (1994)
Tap water enriched with nitrate Nitrate volumetric loading rate of 1.5 kg NO_3^--N/m^3.day	Fixed bed reactor with a mixture of pouzzolane and Neutralg (3 : 1 ratio). Thiosulfate as the sole energy source. Pilot scale experiments at low temperature 10 °C	100% denitrification yield	Trouvé et al. (1999)

$$NO_3^- \xrightarrow{k_1} NO_2^- \xrightarrow{k_2} N_2. \tag{18.10}$$

They obtained values for k_1 and k_2 amounting to 3.15×10^{-3} / mg h^{-1} and 2.75×10^{-1} / mg h^{-1}, respectively.

During the same period of time a research programme was developed in The Netherlands to adapt this process to small-size drinking water production units. The purpose of this project was to elaborate a cheap procedure that did not require excessive maintenance. Schippers *et al.* (1987) developed a sulfur/limestone process (ratio 1:1) functioning at a capacity of 35 m^3/h. The nitrate concentration range in the influent and in the effluent was, respectively, 14.7–15.8 and 0–5.6 mg NO_3^--N l^{-1}. A closely related process was also elaborated by Kruithof *et al.* (1988). Van der Hoek *et al.* (1992) also developed a successful autotrophic sulfur/limestone denitrification process. Different sulfur:limestone ratios were studied and the maximal nitrate loading rate was 9 g NO_3^--N/m^3.h (to avoid high nitrite effluent concentrations). Although the pilot performance appeared to be independent on the sulfur:limestone ratio, they recommended a 1:2 ratio in order to simplify the maintenance of the filter. The reactor worked for about 400 days. All along this operation time, no back-washing was required. According to the authors, a low back-washing frequency is inherent to the slow filtration procedure. It can be an advantage for small drinking water production units which require little maintenance.

Hijnen *et al.* (1988) used a reactor similar to the one developed by Schippers *et al.* (1987) for drinking water production. They also studied methods for post treatment of the denitrified effluent that contained high bacterial and easily assimilable organic carbon (AOC) levels inherent to the process nature. For this purpose the denitrified effluent was fully aerated and infiltrated in soil (infiltration well technology) where it stayed for about 12 weeks before further extraction by pumping in an aquifer located near the infiltration area. The bacteriological water quality of the extracted water was in good agreement with Dutch legislation of 1983, and the AOC was similar to that measured in the influent before denitrification.

Lampe and Zhang (1996) reported the possibility of artificial *in situ* denitrification for superficial water at a laboratory scale by adding a mixture of S°:maërl (ratio 3:1) to natural sediments (the sediment wet weight : sulfur-limestone weight was 2:1).

Addition of soluble sulfur sources

Montiel and Welté (1994) developed a laboratory scale biological denitrification process, based on the existence of synergetic relations between chemolithotrophic bacteria able to coexist in fixed beds. These

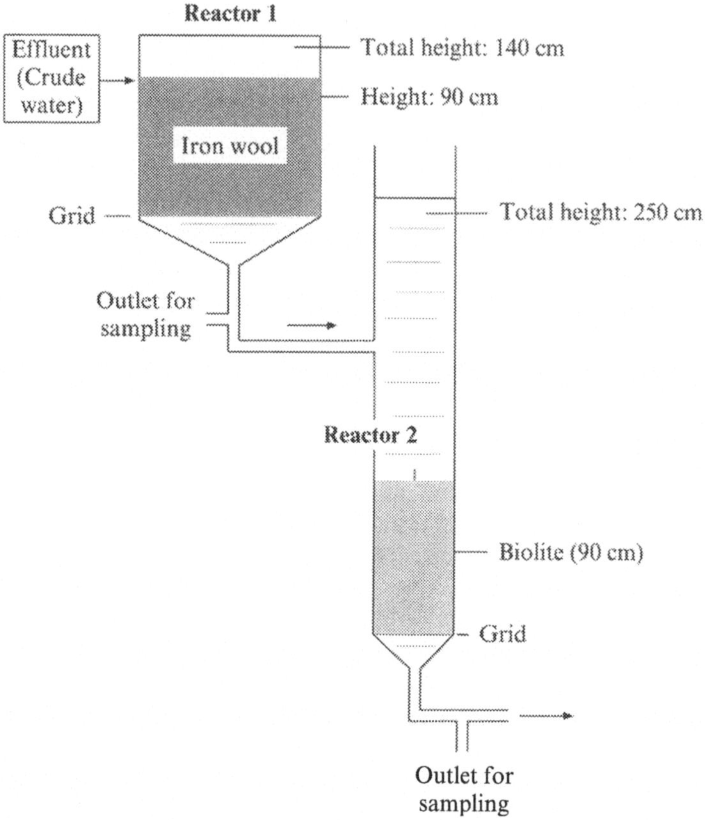

Figure 18.5. Scheme of the pilot reactor for autotrophic denitrification (used by Montiel and Welté 1994).

microorganisms originated from the natural microbiota of the influent. Iron, sulfur and nitrogen cycling was successively combined in this process. The pilot consisted of two main modules: the first module was heavily loaded with iron wool and the second was filled with biolite (Figure 18.5). In the first module metal iron of the iron wool was oxidized to ferrous iron by contact with incoming water (which also contained nitrates). Thus, the E_h value and the dissolved oxygen concentration decreased. This phenomenon was immediately followed by the reduction of incoming sulfates in sulfides by sulfate-reducing bacteria. This led to the formation of ferrous sulfide, which was then reoxidized in sulfates, accompanied by the reduction of NO_3^- into N_2 by denitrifying thiobacilli. In the second module, iron bacteria oxidized ferrous iron into insoluble ferric iron (FeOOH) that remained in the bed. The influent nitrate concentration was decreased from 43 mg/l to 22 mg/l in the effluent. The sulfate concentration (11 mg l^{-1}) was not affected. The residence time (under optimization) was 12 h. According to the authors this process is suitable for the treatment of water for small denitrification plants owing to its low costs.

Entrapment of both *Thiobacillus denitrificans* and nitrifying bacteria in alginate beads has also been investigated as a process allowing full nitrification/denitrification (Lewandowski *et al.* 1987). In addition, $CaCO_3$ was incorporated, thus increasing the bead stability (after Ca^{2+} release) and providing the carbon source for *Thiobacillus denitrificans*. This process was developed to remove any soluble form of nitrogen and obtained an efficiency of 4.8 mg N removed per liter and per hour. Beads were stable and active for 3 months in a continuous flow system.

Lee and Sublette (1990) evidenced that *Thiobacillus denitrificans* can also remove gaseous nitrogen compounds, such as NO in the presence of thiosulfate and carbon dioxide. They demonstrated that NO can act as an electron acceptor for this bacterium. Up to 96% of NO removal was observed with an average NO / $S_2O_3^{2-}$ ratio of 4 : 1.

More recently, Trouvé *et al.* (1998a, b) made some advances in this research field. Some *Thiobacillus denitrificans* strains achieved good denitrification at low temperature (5 °C) and at acid (pH 5) or basic (pH 8) pH in batch conditions (Trouvé *et al.* 1998a). Attempts made in the presence of ferrous sulfide from microbial origin (released by sulfate reducing bacteria) did not allow as efficient denitrification as that obtained in the presence of ferrous sulfide from chemical origin. This last experiment was done at pilot scale under continuous flow (Trouvé *et al.* 1998b). In the presence of ferrous sulfide from chemical origin, a 100% denitrification yield was obtained at a residence time of 1 hour and allowed the removal of a volumetric nitrate loading rate of 0.32 kg $N\text{-}NO_3^-$ per cubic meter and per day. Furthermore, using the same type of strains Trouvé *et al.* (1999) attempted denitrification of crude water to which only thiosulfates and nitrates were added at low temperature (10 °C). This temperature value seems to be a limit for classical heterotrophic denitrification processes. In the presence of *Thiobacillus denitrificans* 96.C2, a pilot scale fixed bed reactor (containing a mixture of pouzzolane and Neutralg) was operated in continuous flow (Figure 18.6). A 100% denitrification yield was observed at a 30 minutes residence time and a volumetric nitrate loading rate of 1.5 kg $N\text{-}NO_3^-$ per m^3 and per day. The experimental [sulfate]/[thiosulfate], [thiosulfate]/[nitrate] and [sulfate]/[nitrate] ratios were close to the theoretical ones (Driscoll and Bisogni 1978) and were, respectively 2, 0.845

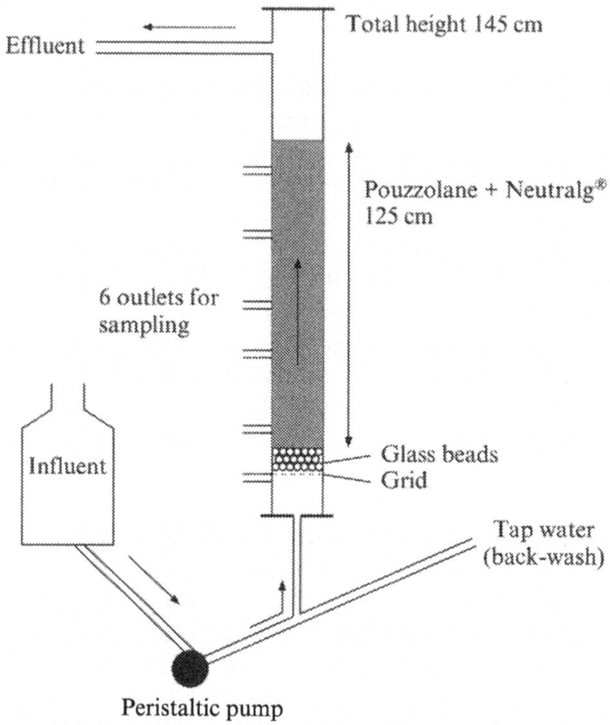

Figure 18.6. Scheme of the fixed bed pilot reactor for autotrophic denitrification used by Trouvé *et al.* (1999).

and 1.689. Figure 18.7 shows the evolution of the concentrations of nitrogenous and sulfurous compounds as a function of the column height in this reactor.

18.3.3 Optimization of the fixed bed reactor design

Some problems can affect autotrophic denitrification by *Thiobacillus denitrificans* in fixed-bed reactors, some of them may have the same origin:

- dysfunctioning releasing excess nitrate and nitrite concentrations in the effluent;
- biomass accumulation and need to optimize the back wash frequency;
- gaseous oversaturation due to the release of N_2 or other gaseous forms of nitrogen (NO_x);

- hydraulic short-circuits;
- production of unwanted sulfurous compounds.

Nitrate release in the effluent can be due to intensive bacterial proliferation (Kruitoff *et al.* 1988): biomass in excess limits mass transfer of nitrate inside the biofilm (Batchelor and Lawrence 1978b). The other consequence of this phenomenon is that excessive N_2 release leads to gaseous oversaturation and bubble formation. This is the origin of hydraulic short-circuits in the filter beds that significantly modify the residence time, and thus affect the denitrification yield.

Nitrite release has been explained differently according to different authors. Van der Hoek *et al.* (1992) noticed that the presence of nitrite was also accompanied by excess nitrate release. According to Haider *et al.* (1988) the presence of nitrites in the effluent is a function of the fixed bed height for a given volumetric loading rate. For Driscoll and Bisogni (1978) nitrites appeared when the sulfur source was limiting in the growth medium, or in other words, if the ratio between electron donor and acceptor was unbalanced, as noticed by Blécon *et al.* (1983). The decrease of residence time and the rapid increase of the loading rate have also been implied

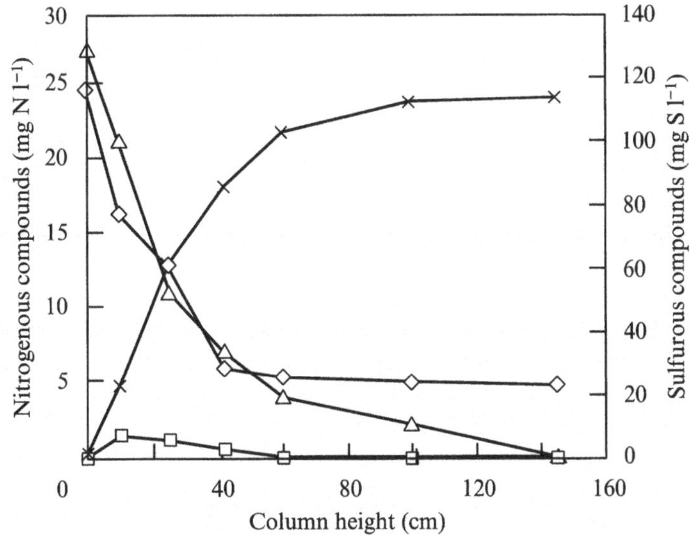

Figure 18.7. Autotrophic denitrification by *Thiobacillus denitrificans* at 10 °C. Evolution of the concentrations of NO_3^--N (Δ), NO_2^--N (□), $S_2O_3^{2-}$-S (◇) and SO_4^{2-}-S (×) as a function of the reactor height, at a nitrate loading rate of 1.5 kg NO_3^--N/ m^3.d and a HRT of 30 min (After: Trouvé *et al.* 1999).

(Sikora and Keeney 1976; Le Cloirec and Martin 1988) because k_1 (kinetic constant for nitrate reduction to nitrite) is superior to k_2 (kinetic constant for nitrite reduction to N_2). According to Lazarova *et al.* (1994) nitrite release is independent of (i) operational parameters (O_2, pH, temperature, etc) (ii) the hydraulic/volumetric loading rates, and (iii) the biofilm thickness. Lazarova *et al.* (1994) postulated that nitrite release is governed by (i) the type of the bacterium used in the filter and (ii) the density and the homogeneity of the biofilm developed by this bacterium. It also depends on the ratio between proteins and polysaccharides within the biofilm (Lazarova *et al.* 1994).

Sulfides may also be released by autotrophic denitrification processes utilizing *Thiobacillus denitrificans* using S° or $S_2O_3^{2-}$. The complete removal of nitrates in a limited part of the biofilter creates a locally anaerobic area in which high sulfate concentrations can accumulate. In these anaerobic microniches, sulfate or sulfur reducing bacteria such as *Desulforomonas acetoxidans* can grow using organic matter from cellular lysis (Van der Hoek *et al.* 1992). Sulfide can inhibit denitrification by *Thiobacillus denitrificans* (Xiushan *et al.* 1993) (see 18.3.1). They may also induce clogging or increase effluent turbidity when further oxidized to S° (Hijnen *et al.* 1988).

The preliminary gas removal (using a vacuum system) from the influent just before the injection in the reactor has been proposed by Schippers *et al.* (1987) and Soares *et al.* (1988). O_2 removal prevents aerobic bacterial growth. Moreover, aerobic oxidation of S° which releases more sulfates is discarded, thus lowering the effluent sulfate concentration. Furthermore, the risk of gaseous oversaturation inside the reactor is greatly decreased, thus minimizing bubble and gas bag accumulation and related short-circuits.

Control of biomass growth can be done by different means. According to Blécon *et al.* (1983) the plants were given a back wash every 10 days. This procedure lowered the biomass concentration, thus providing back the initial hydraulic characteristics which guaranted good operating conditions. Koch and Siegrist (1997) noticed that mainly total suspended solids (TSSs) retention and accumulation of biomass and N_2 contributed to the development of a headloss. N_2 solubility is a function of temperature and hydrostatic pressure. Its solubility decreases when the temperature is lowered and when the headloss increases. The gas bags have to be removed, and the required removal frequency increases all along the operational running time. This frequency can be determined by headloss measurements. Washing procedures mainly remove opportunistic biomass and TSS deposits. The biofilm, mainly comprising denitrifying microorganisms, remains in the reactor and is thus made more performant by the washing (Boller *et al.* 1997).

Numerous authors prevented clogging and short-circuits by using alternative reactor concepts. The latter ones have been developed to treat larger volumetric loading rates, while avoiding nutrient diffusion problems independently of the biofilm thickness. The following reactor designs can be cited: biofilm airlift suspension reactors (Picioreanu *et al.* 1997; Van Benthum *et al.* 1997) membrane reactors (Watanabe *et al.* 1997) upflow anaerobic sludge blanket (Tarre *et al.* 1995; Rouse *et al.* 1999) reactors with both microaerophilic and anaerobic areas (Richter and Krüner 1994).

When using fixed beds, the nature of the support material inside the reactor influences the process. According to Boehler and Haldenwang (1992) this material has to be porous and wrinkled. The more porous the material is, the greater the biofilm surface becomes, which can develop outside and inside the support. Furthermore, the same bacterial concentration dispatched on a greater surface induces thinner, more performing biofilms (Lazarova *et al.* 1994). The material roughness has to be calibrated to allow good bacterial adhesion. It also influences the residual fixed bacterial concentration after a washing procedure (Hanus and Bernard 1988). Nevertheless, a lot of work remains to be done before fixed beds processes can be applied at the industrial scale, although of great interest from a technological point of view.

18.4 MICROBIAL AND TECHNOLOGICAL ASPECTS OF REDUCTIVE REACTIONS

The versatility of bioconversions in the sulfur cycle also allows the integration of nitrogen and sulfur removal from wastewater. Nitrogen elimination has been combined with sulfur reclamation from concentrated photographical wastewater (Rooden *et al.* 1995). Thiosulfate is converted into sulfide in an upflow anaerobic sludge bed (UASB) reactor, and S° is produced in a next reactor. In the aerobic post treatment, nitrogen and residual COD are removed by combined nitrification/denitrification.

In this respect, the sulfide present in the effluent of the anaerobic stage can be toxic to nitrifiers (Hooper and Terry 1973) and thus can upset nitrification in the aerobic post treatment. On the other hand, sulfide can contribute to nitrate removal, when its reoxidation to S° or sulfate by *Thiobacillus denitrificans* can be scaled up. Garuti *et al.* (1992) reported that sulfide oxidation in the denitrification reactor reduced the need for external carbon source.

In anaerobic process technology, direct integration of denitrification and methanogenesis in a single reactor has been suggested as an alternative method to the classical concept of anaerobic organic carbon removal followed by nitrification/denitrification for nitrogen removal (Akunna *et al.* 1992; Hendriksen and Ahring 1996). In addition, the combination of denitrification and methanogenesis has also been considered in drinking water production. Both immobilized mixed cultures and bioreactors have been investigated to remove excess carbon source, mostly methanol, from denitrified effluents by methanogenesis (Lin and Chen 1995).

In high-rate sulfidogenic reactors (Hulshoff Pol *et al.* 1998) which are optimised for sulfate reduction, acetate removal is the rate limiting step. A way to optimise the acetate removal of these reactor systems is to increase the activity of facultative (an)aerobic bacteria as acetate scavengers. This can be done by dosing minor amounts of alternative electron acceptors, e.g. oxygen or nitrate, into the reactor system (Lens *et al.* 1998). Nitrate addition is an attractive option, as nitrate is much more soluble in water (several grams per litre) than oxygen and acetate is an excellent carbon source for denitrifying bacteria (Ganaye *et al.* 1996).

In addition to denitrification (nitrate reduction to nitrogen gas) also nitrate reduction to ammonium (ammonification) can occur in anaerobic environments. Nitrite is an intermediate of both pathways. Ammonification can be expected in sulfidogenic reactors, as (i) pure cultures of sulfate reducing bacteria convert nitrate to ammonium (Widdel 1988) (ii) anaerobic inocula contain large populations of fermentatives and obligate anaerobes, i.e. ammonium formers, compared with true denitrifiers (Kasper *et al.* 1981) and (iii) certain carbon sources, e.g. glycerol and glucose favour ammonification (Akunna *et al.* 1993). With acetate as the carbon source, nitrate is totally converted into nitrogen gas via denitrification (Akunna *et al.* 1993; Guynot *et al.* 1998; Lens *et al.* 1999). This is very beneficial, as nitrate reduction to ammonium is of no interest from a wastewater treatment point of view.

In both methanogenic (Percheron *et al.* 1998) and sulfidogenic (Lens *et al.* 1999) granular sludge, sulfide disappears upon the addition of nitrate. Percheron *et al.* (1998) showed that denitrifiers can use sulfide as the electron donor. In sulfidogenic granular sludge, however, also SRB can be responsible for the combined sulfide and nitrate removal. Indeed, SRB are able to completely reoxidize sulfide to sulfate in the presence of either oxygen or nitrate as electron acceptors. The latter phenomenon has been reported for pure cultures of *Desulfovibrio desulfuricans* and *Desulfobulbus propionicus* (Dannenberg *et al.* 1992) as well as for anaerobic fresh water sediments (Brunet and Garcia-Gil 1996). However, the complete reoxidation of sulfide to sulfate is coupled to the ammonification of nitrate or nitrite, which contrasts the observed constant ammonium concentration (Lens *et al.* 1999). The constant ammonium concentration also indicated the absence of the anammox process, i.e. the concomitant removal of nitrate/nitrite and ammonium (Van de Graaf *et al.* 1995) in the studied sulfidogenic granular sludge. However, a possible turnover of nitrogen compounds in the system which cannot be detected by the nitrogen determinations applied cannot be excluded. Further research using for example ^{15}N-labelled substrates (Van de Graaf *et al.* 1995) is needed for the elucidation of the exact prevailing nitrogen conversions in methanogenic and sulfidogenic granular sludge.

18.5 REFERENCES

Adams M.E., Warnes G.M. and Nicolas D.J.D. (1971) A sulfite dependent nitrate reductase from *Thiobacillus denitrificans*. *Biochim. Biophys. Acta* **235**, 398-406

Akunna J.C., Bizeau C. and Moletta R. (1992) Denitrification in anaerobic digesters: possibilities and influence of wastewater COD/NOx-N ratio. *Environ. Technol.* **13**, 825-836.

Akunna J.C., Bizeau C. and Moletta R. (1993) Nitrate and nitrite reductions with anaerobic sludge using various carbon sources: glucose, glycerol, acetic acid, lactic acid and methanol. *Wat. Res.* **27**, 1303-1312.

Amaral J.A., Archambaut C., Richards S.R and Knowless R. (1995) Denitrification associated with groups I and II methanotrophs in a gradient enrichment system. *FEMS Microbiology and Ecology* **18**, 289-298.

Baalsrud K. and Baalsrud K.S. (1954) Studies on *Thiobacillus denitrificans*. *Archiv. für Mikrobiologie* **20**, 34 - 62.

Baldensperger J. and Garcia J.L. (1975) Reduction of oxidized inorganic nitrogen compounds by a new strain of *Thiobacillus denitrificans*. *Archives of Microbiology* **103**, 31-36

Batchelor S.E., Cooper M., Chhabra S.R, Glover L.A., Stewart G.S.A.B., Williams P. and Prosser J.I. (1997) Cell density regulated recovery of starved biofilm populations of ammonia oxidizing bacteria. *Applied and Environmental Microbiology* **63**, 2281-2286

Batchelor B. and Lawrence A.W. (1978) Autotrophic denitrification using elemental sulfur. *Journal WPCF* **50**(8), 1986-2001.

Binswanger S., Siegrist H. and Lais P. (1997) Simultane Nitrifikation/ Denitrifikation von stark ammonium-belasteten Abwässern ohne organische Kohlenstoffquellen. *Korrespondenz Abwasser* **44**, 1573-1580.

Bisogni J.J. and Driscoll C.T. (1977) Denitrification using thiosulfate and sulfide. *Journal of Environmental Engineering Division, American Society of Civil Engineers* **103**, 593-604.

Blecon G., Gillet M., Martin G. and Philipot J.M. (1983) Autotrophic denitrification process by *Thiobacillus denitrificans* on sulfur-Maërl. *Revue*

Française des Sciences de l'Eau **2**, 267-279.

Boehler E. and Haldenwang L. (1992) The NEBIO tube reactor process for nitrate removal from drinking water. *Water Supply* **10**(3), 71-77.

Boller M., Tschui M. and Gujer W. (1997) Effects of transient nutrient concentrations in tertiary biofilm reactors. *Wat. Sci. Technol.* **36**(1), 101-109.

Bouffard S.C., Hackl R.P. and Sobolewski A. (1998) Novel nitrate / sulphate closed-cycle biological treatment of mining effluents. *European Conference on new advances in biological nitrogen anf phosphorus removal for municipal or industrial wastewaters Narbonne, France.* October 12 - 14, 279 -282.

Brunet R.C. and Garcia-Gil L.J. (1996) Sulfide-induced dissimilatory nitrate reduction to ammonia in anaerobic freshwater sediments. *FEMS Microbiol. Ecol.* **21**, 131-138.

Claus G., Kutzner H.J. (1985) Physiology and kinetics of autotrophic denitrification by *Thiobacillus denitrificans, Appl. Microbiol. Biotechnol.* **22**, 283-288

Dannenberg S., Kroder M., Dilling W. and Cypionka H. (1992) Oxidation of H_2, organic compounds and inorganic sulfur compounds coupled to reduction of O_2 or nitrate by sulfate-reducing bacteria. *Arch. Microbiol.* **158**, 93-99.

De Saint Blanquat G. and Fritsch P. (1994) Nitrates - Nitrites - Nitrosamines. Aspects chimiques environnementaux et approches toxicologiques. In *Les Nitrates*. Coll. Santé et Société **1**, 20-52.

Driscoll C.T. and Bisogni J.J. (1978) The use of sulfur and sulfide in packed bed reactors for autotrophic denitrification. *J. WPCF* **50**, 569-577.

Esquian A. and Simon J.C. (1990) Danger des nitrates, A la pointe de l'élevage - Spécial fourrages et nitrates, 6-8.

Ganaye V., Fass S., Urbain V., Manem J. and Block J.C. (1996) Biodegradation of volatile fatty acids by three species of nitrate-reducing bacteria. *Environ. Technol.* **17**, 1145-1149.

Garcia-Gil L.J. and Golterman H.L. (1993) Kinetics of FeS-mediated denitrification in sediments from the Camargue. *FEMS Microbiology Ecology* **13**, 85-92.

Garuti G., Dohanyos M. and Tilche A. (1992) Anaerobic-aerobic combined process for the treatment of sewage with nutrient removal: the ananox process. *Wat. Sci. Tech.* **25**, 383-394.

Golterman H.L. (1991) Influence of FeS on denitrification in shallow waters. *Verh. Int. Verein. Limnol.* **24**, 3025-3028.

Guerber F., Gerbier C. and Chazal P.M. (1998) Le Nitraflux, un nouveau procédé pour l'élimination des ions nitrate des eaux destinées à la potabilisation. *Techniques, Sciences et Méthodes, L'eau* **4** (Avril), 107-114.

Guynot M.E., Toribio A., Quevedo M. and Muxi L. (1998) Microflora of dissimilative nitrate reduction in a denitrifying reactor. *Appl. Microbiol. Biotechnol.* **50**, 396-400.

Haider N., Morvan J., Le Cloirec P. and Martin G. (1988) Denitrification in a reactor packed with iron pyrite: laboratory studies. *Environmental Technology Letters* **9**(5), 411-420.

Hanus F. and Bernard C. (1988) Dénitrification des eaux potables dans un réacteur biologique à élimination continue de la biomasse en excès : un procédé inédit. *T.S.M.* 243-246.

Hashimoto S., Furukawa K. and Shioyama M. (1987) Autotrophic denitrification using elemental sulfur. *Journal of Fermentation Technology* **65**(6), 683-692.

Haslay C. and Leclerc H. (1993) Microbiologie des eaux d'alimentation. *Technique et Documentation*. Editions Lavoisier. 495 p.

Hendriksen H.V. and Ahring B.K. (1996) Integrated removal of nitrate and carbon in an upflow anaerobic sludge blanket (UASB) reactor: operating performance. *Wat. Res.* **30**, 1451-1458.

Hijnen W.A.M., Koning D., Kruithof J.C. Van der Kooij D. (1988) The effect of bacteriological nitrate removal on the concentration of bacterial biomass and easily assimilable organic carbon compounds in ground water. *Wat. Supply* **6**, 265-273.

Hole U.H., Vollack K.U., Zumft W.G., Eisenman E., Siddiqui R.O., Friedrich B., Kroneck P.M.H. (1996) Characterization of the membranous denitrification enzimes nitrite reductase (cytochrome cd1) and copper containing nitrous oxide reductase from *Thiobacillus denitrificans*. *Archives of Microbiology* **165**(1), 55-61.

Hooper A.B. and Terry K.R. (1973) Specific inhibitors of ammonia oxidation in *Nitrosomonas*. *J. Bacteriol.* **115**, 480-485.

Hoor A.T. (1976) Energetic aspect of the metabolism of reduced compounds in *Thiobacillus denitrificans*. *Antonie van Leewenhoek* **42**, 483-492

Hoor A.T. (1981) Cell yields and bioenergetics of Thiomicrospira *denitrificans* compared with *Thiobacillus denitrificans*, *Antonie van Leewenhoek* **47**, 231-243

Hulshoff Pol L., Lens P., Stams A.J.M., Lettinga G. (1998) Anaerobic treatment of sulfate-rich wastewaters. *Biodegradation* **9**, 213-224.

Justin P. and Kelly D.P. (1978a) Growth kinetics of *Thiobacillus denitrificans* in anaerobic and aerobic chemostat culture. *J. Gen. Microbiol.* **107**, 123-130.

Justin P. and Kelly D.P. (1978b) Metabolic changes in *Thiobacillus denitrificans* accompanying the transition from aerobic to anaerobic growth in continuous chemostat culture. *The Journal of General Microbiology* **107**, 131-137.

Kasper H.F., Tiedje J.M. and Firestone R.B. (1981) Denitrification and dissimilatory nitrate reduction to ammonium in digested sludge. *Can. J. Microbiol.* **27**, 878-885.

Kelly D.P. (1982) Biochemistry of the chemolithotrophic oxidation of inorganic sulphur. *Phil. Trans. R. Soc. Lond.* **B298**, 499-528

Kelly D.P. (1989) Physiology and biochemistry of unicellular sulfur bacteria. In *Autotrophic bacteria*. Springer-Verlag. Science Tech publisher. 193-217.

Kelly D.P. and Harrison A.P. (1989) The genus *Thiobacillus*. In Staley JT, Bryant MP, Pfennig N and Holt JD (Eds) *Bergey's Manual of systematic Bacteriology* **3**, 1842-1858

Koch G. and Siegrist H. (1997) Denitrification with methanol in tertiary filtration at wastewater treatment plant Zürich-Werdhölzli. *Wat. Sci. Technol.* **36**(1), 165-172.

Koenig A. and Liu L.H. (1996) Autotrophic denitrification of landfill leachate using elemental sulphur. *Wat. Sci. Technol.* **34**(5-6), 469-476.

Kruithof J.C., Van Bennekom C.A., Dierx H.A.L., Hijnen W.A.M., Van Paassen J.A.M. and Schippers J.C. (1988) Nitrate removal from groundwater by sulphur / limestone filtration. *Wat. Supply* **6**, 207-217.

Kuai L. and Verstraete W. (1999) Autotrophic denitrification with elemental sulfur in small-scale wastewater treatment facilities. *Environmental Technology* **20**, 201-209.

Kuenen J.G. (1989) Colorless sulfur bacteria. In Staley JT, Bryant MP, Pfennig N and Holt JD (Eds) *Bergey's Manual of systematic Bacteriology* **3**, 1834-1837.

Kuenen J.G., Robertson L.A., and Tuovinen O.H. (1992) The genera *Thiobacillus*, *Thiomicrospira* and *Thiosphaera*. In Balows A, Trüper HG, Dworkin M, Harder W & Schlefer KH (Eds) *The Prokaryotes*, Vol III, 2638-2657. Springer-Verlag, New York.

Lampe D.G. and Zhang T.C. (1996) Evaluation of sulfur-based autotrophic denitrification. In HSRC / WERC joint conference on the environment. Albuquerque, New Mexico, May 96.

Lazarova V., Capdeville B. and Nikolov L. (1994) Influence of seeding conditions on nitrite accumulation in a denitrifying fluidized bed reactor. *Wat. Res.* **28**(5), 1189-1197.

Le Bideau L., Lachaise M. and Beaulieu G. (1994) Phénomène de dénitrification naturelle sur le site d'exploitation de Beuxes-Comprigny. 11ème congrès JIE, Poitiers 1994.

Le Cloirec P., Martin G., Benbarka B and Leroux A.Y. (1985) Un modèle mathématique de la dénitrification sur des filtres soufre-carbonate de calcium. *Chem. Eng. J.* **31**, B9-B18.

Le Cloirec P. and Martin G. (1988) Simulation of the biological denitrification in a sulfur-calcium carbonate reactor. *Environ. Technol. Lett.* **9**, 207-218.

Lee K.H. and Sublette K.L. (1990) Reduction of nitric oxide to elemental nitrogen by *Thiobacillus denitrificans*. *Appl. Biochem. Biotechnol.* **24/25**, 441-445.

Le Gall J., Payne W.J., Morgan T.V. and Der Vartanian D. (1979) On the purification of nitrite reductase from *Thiobacillus denitrificans* and its reaction with nitrite under reducing conditions. *Biochem. and Biophys. Res. Comm.* **87**, 2, 355-362.

Lens P.N.L., Visser A., Janssen A.J.H., Hulshoff Pol L.W. and Lettinga G. (1998) Biotechnological treatment of sulfate rich wastewaters. *Crit. Rev. Environ. Sci. Technol.* **28**, 41-88.

Lens P., Sipma J., Hulshof Pol L. and Lettinga G. (1999) Effect of staging and nitrate addition on sulfidogenic acetate removal. *Wat. Res.* **34**, 31-42.

Lewandowski Z., Bakke R. and Characklis W.G. (1987) Nitrification and autotrophic denitrification in calcium alginate beads. *Wat. Sci. Tech.* **19**(1/2), 175-182.

Lin Y.-F. and Chen K.-C. (1995) Denitrification and methanogenesis in a co-immobilized mixed culture system. *Wat. Res.* **29**, 35-43.

Mariotti A. (1992) La dénitrification *in situ* dans les eaux souterraines, processus naturels ou provoqués. *Les dossiers de la cellule environnement* **4**, 41-62.

Matsui S. and Yamamoto R. (1986) A new method of sulphur denitrification for sewage treatment by a fluidized bed reactor. *Wat. Sci. Tech.* **18**, 355-362.

Montiel A. and Welté B. (1994) Nouveau procédé de dénitrification des eaux par le fer métallique. 11ème congrès JIE, Poitiers 1994.

Pel R., Olhenduis R., Brand W., VOLS A., Gottschal J.C. and Zwart K.B. (1997) Stable isotope analysis of a combiend nitrification-denitrification sustained by thermophilic methanotrophs under low-oxygen conditions. *Appl. Environ. Microbiol.* **49**, 1131-1141.

Percheron G., Michaud S., Bernet N., and Moletta R. (1998) Nitrate and nitrite reduction of a sulfide-rich environment. *J. Chem. Technol. Biotechnol.* **72**,

213-220.

Philipot J.M. and Patte A. (1982) La dénitrification des eaux par voie biologique, *Techniques, Sciences et Méthodes, L'eau* **77**(4), 165-172.

Philipot J.M. and Sibony J. (1986) L'élimination des nitrates dans les eaux potables. 16e Cong. Inter. & Expo. Distr. D'Eau, Rome, 11-15.

Picioreanu C., Van Loosdrecht M.C.M. and Heijnen J.J. (1997) Modelling the effect of oxygen concentration on nitrite accumulation in a biofilm airlift suspension reactor. *Wat. Sci. Technol.* **36**(1), 147-156.

Ravarini P., Coutelle J. and Damez F. (1988) L'usine de Dennemont. Une unité de dénitrification-nitrification à grande échelle. *Techniques, Sciences et Méthodes, L'eau* **83**(4), 235-240.

Richard Y. and Leprince A. (1982) Pollution par les nitrates : traitements disponibles. *Trib. Cebedeau* **458**(35), 21-23.

Richard Y. and Partos J. (1986) Elimination biologique des nitrates en vue de la production d'eau potable. *Techniques, Sciences et Méthodes, L'eau* **81**(3), 141-146

Richter K.-U. and Krüner G. (1994) Elimination of nitrogen in two flooded and statically packed bed biofilter with aerobic and anaerobic microsites. *Wat. Res.* **28**(3), 709-716.

Robertson L.A. and Kuenen J.G. (1992) The colourless sulfur bacteria. In Balows A., Trüper H.G., Dworkin M., Harder W. & Schlefer K.H. (Eds) *The prokaryotes*, vol III, 385-413 Springer-Verlag New-York.

Rödelsperger M. (1989) Natural denitrification processes in the aquifer. In *Contaminant Transport in Groundwater*. Edited by Kobus H.E. and Kinzelbach W. 159-161. Balkema, Rotterdam.

Roennefhart K.W. (1986) Nitrate elimination with heterotrophic aquatic microorganisms in fixed bed reactors with buoyant carriers. *Aqua* **5**, 283-285.

Rooden A., Bosma W.W., Buisman C.N.L. and De Munck P.R. (1995) The treatment of photographic waste streams. *Med. Fac. Landbouww. Univ. Gent.* **60**, 2729-2734.

Rouse J.D., Sumida K., Kida K. and Furukawa K. (1999) Maintainability of denitrifying granular sludge in soft to marginally hard waters in an upflow sludge-blanket reactor. *Environmental Technology* **20**, 219-225.

Sant'Anna F.P.S., Martin G. and Taha S. (1996) Effect of sodium chloride on the denitrification of ion exchange resin eluates by a sulphur-oxidizing chemoautotrophic bacterium. *Revue des Sciences de l'Eau* **3**, 333-350.

Sawhney V. and Nicolas D.J.D. (1977) Sulphite-and NADH-dependent nitrate reductase from *Thiobacillus denitrificans. The Journal of General Microbiology* **100**, 49-58.

Schedel M. and Trüper H.G. (1980) Anaerobic oxidation of thiosulfate and elemental sulfur in *Thiobacillus denitrificans*. *Arch. Microbiol.* **124**, 205-210.

Schippers J.C., Kruithof J.C., Mulder F.G. and Van Lieshout J.W. (1987) Removal of nitrate by slow sulphur / limestone filtration. *Aqua* **5**, 274-280.

Schramm A., Larsen L.H., Revsbech N.P., Ramsing N.B., Amann R. and Schleifer K.-H. (1996) Structure and function of a nitrifying biofilm determined by *in situ* hybridisation and the use of microelectrodes. *Appl. Environ. Microbiol.* **62**, 4641-4647.

Schultz H.N., Brinkhoff T., Ferdelman T.G., Hernandez Marine M., Teske A.,

Jorgenson B.B. (1999) Dense populations of a giant sulfur bacterium in Namibian shelf sediments. *Science* **284**, 493-495.

Sebillote J. (1992) La pollution nitratée des eaux: l'évolution de l'action sous l'impulsion européenne. *C.R. Acad. Agric. Fr.* **78**(7), 105-110.

Sibony J. (1979) traitement physico-chimique des nitrates et de l'amoniaque pour la production d'eau potable. *TSM l'Eau* **6**, 355-359.

Sikora L.J. and Keeney D.R. (1976) Evaluation of a sulfur-*Thiobacillus denitrificans* nitrate removal system. *Jour. Environ. Qual.* **5**(3), 298-302.

Soares M.I.M., Belkin S. and Abeliovich A. (1988) Biological groundwater denitrification : laboratory studies. *Wat. Sci. Tech.* **20**(3), 189-195.

Sublette K.L., Kolhatkar R. and Raterman K. (1998) Technological aspects of the microbial treatment of sulfide-rich wastewaters: A case study. *Biodegradation* **9**, 259-271.

Sweerts J.P.R.A., De Beer D., Nielsen L.P., Verdouw H., Van Den Heuvel J.C., Cohen Y. and Cappenberg T.E. (1990) Denitrification by sulphur oxidizing *Beggiatoa* spp. mats on freshwater sediments. *Nature* **344**,762-763.

Tarre S., Armon R., Shelef G. and Green M. (1995) Effects of water characteristics on granular sludge formation in a USB reactor for denitrification of drinking water. *Wat. Sci. Tech.* **30**(9), 141-147.

Taylor B.F. and Hoare D.S. (1971) *Thiobacillus denitrificans* as an obligate chemolithotroph - II. Cell suspension and enzymic studies. *Arch. Microbiol.* **80**, 262-276.

Taylor B.F. (1989) Thermotolerance of adenylsulfate reductase from *Thiobacillus denitrificans. FEMS Mic. Lett.* **59**, 351-354.

Trouvé C., Chazal P.M., Guéroux B. and Sauvaître N. (1998a) Denitrication by new strains of *Thiobacillus denitrificans* under non standard physico-chemical conditions. Effect of temperature, pH and sulphur source. *Environmental Technology* **19**, 601 - 610.

Trouvé C., Chazal P.M. and Guéroux B. (1998b) Influence of the nature of ferrous sulphide upon denitrification by *Thiobacillus denitrificans* at a pilot scale. *Journal Européen d'Hydrologie* **29**(1), 43-62.

Trouvé C. and Chazal P.M. (1999) Autotrophic denitrification by *Thiobacillus denitrificans* with thiosulphate as sole energy source. Pilot scale experiments at low temperature. *Environmental Technology*, in press.

Van Benthum W.A.J., Van Loosdrecht M.C.M. and Heijnen J.J. (1997) Process design for nitrogen removal using nitrifying biofilm and denitrifying suspended growth in a biofilm airlift suspension reactor. *Wat. Sci. Technol.* **36**(1), 119-128.

Van de Graaf A.A., Mulder A., de Bruijn P., Jetten M.S.M., Robertson L.A. and Kuenen J.G. (1995) Anaerobic oxidation of ammonium is a biologically mediated process. *Appl. Environ. Microbiol.* **61**, 1246-1251.

Van de Graaf A.A., de Bruijn P., Robertson L.A., Jetten M.S.M. and Kuenen J.G. (1996) Autotrophic growth of anaerobic, ammonium-oxidising micro-organisms in a fluidized bed reactor. *Microbiology* **142**, 2187-2196.

Van Der Hoek J.-P., Hijnen W.A.M., Van Benekomm C.A. and Mijnarends B.J., (1992) Optimization of the sulfur-limestone filtration process for nitrate removal from groundwater. *J. Water SRT-Aqua* **41**(4), 209-218.

Verstraete W. and Philips S. (1998) Nitrification-denitrification processes and technologies in new contexts. *Environmental pollution* **102**(S1), 717-726.

Watanabe Y., Kimura K., Okabe S., Ozawa G. and Ohkuma N. (1997) A novel biofilm-membrane reactor for ammonia oxidation at low concentrations. *Wat. Sci. Technol.* **36**(1), 51-60.

Widdel F. (1988) Microbiology and ecology of sulfate- and sulfur reducing bacteria. In: *Biology of anaerobic micro-organisms* (Edited by Zehnder A.J.B.) pp. 469-586. John Wiley & Sons, New York.

Xiushan Y., Garuti G. and Tilche A. (1993) Denitrification with *Thiobacillus denitrificans* in the ANANOX® process. *Biotechnology Letters* **15**(5), 531-536.

19

Sulfur-storing bacteria and bulking of activated sludge

Dick H. Eikelboom

19.1 INTRODUCTION

The activated sludge process is the most common biological wastewater treatment method in the world (Figure 19.1). After presettling, which is not always applied, the wastewater is mixed up with biomass in the aeration tank. The biomass is present as so-called activated sludge flocs (aggregated bacterial cells, etc.). The flocs are separated from the final effluent by gravity settling in the secondary clarifier. Consequently, good settling properties of the flocs (≥ 1 m/h) are very important for stable plant operation, namely achieving a good quality final effluent. If nutrient removal (nitrogen and phosphorus) is aimed for as well, anoxic and anaerobic zones are incorporated in the process (Wanner 1997).

The biomass is usually called "sludge", suggesting a rather simple material. However, activated sludge is a very complex community

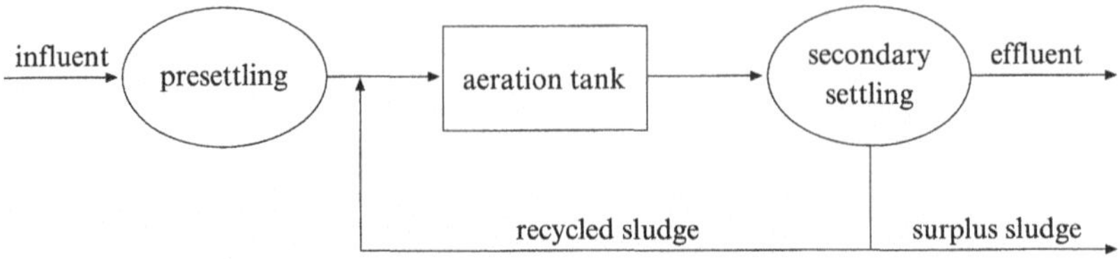

Figure 19.1. Schematic outline of the activated sludge process (nutrient removal not incorporated).

composed of living microorganisms (mainly bacteria), protozoa/metazoa, dead cells and cell residues, entrapped (in)organic particles and precipitated compounds.

Any treatment plant is an artificial ecosystem. The composition of the biomass in a specific plant mainly depends upon:

- quality of the influent (domestic/industrial, stale/fresh, raw/presettled)
- sludge load applied, i.e. the amount of substrate available for the microorganisms
- plant configuration, e.g. completely mixed versus plug flow
- aerobic/anoxic/anaerobic zones (nutrient removal)
- temperature and pH

In almost all activated sludge plants the amount of available nutrients is limited, resulting in a strong competition between the microorganisms for this scarce substrate. The winner(s) dominate(s) in the sludge population. It must be emphasised that the population is dynamic in nature. It not only changes continuously, owing to for example fluctuations of the temperature or the oxygen concentration, but the population can also be affected deliberately by altering the operational conditions.

Filamentous organisms belong to the normal sludge population. They may even contribute to a better sludge quality. However, if their number increases the settling properties of the biomass deteriorate. An activated sludge is 'bulking' if the sludge volume index, a measure for the settling characteristics of activated sludge, exceeds a value of 150 ml/g. Besides, some filamentous species (often *Microthrix parvicella* or *Actinomyces* species) may cause foaming of activated sludge as well (scum formation). Figure 19.2 shows sludge flocs with good settling properties; an example of bulking sludge is presented in Figure 19.3.

Owing to poor settling properties, bulking may result in a severe loss of solids with the final effluent. Moreover, bulking causes a deterioration of the dewatering characteristics of surplus sludge.

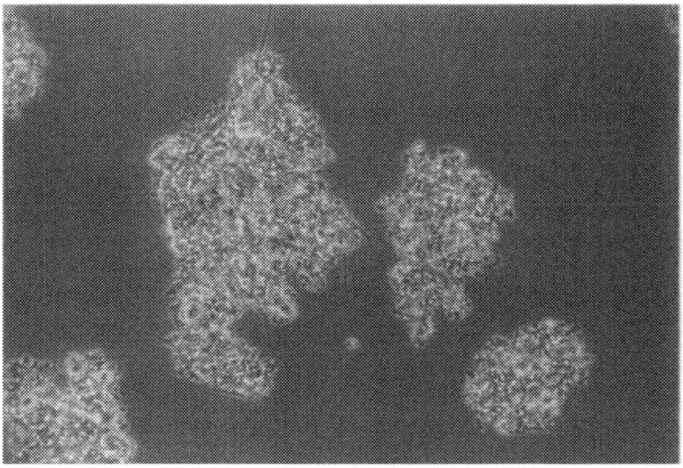

Figure 19.2. Flocs with good settling properties (150 ×).

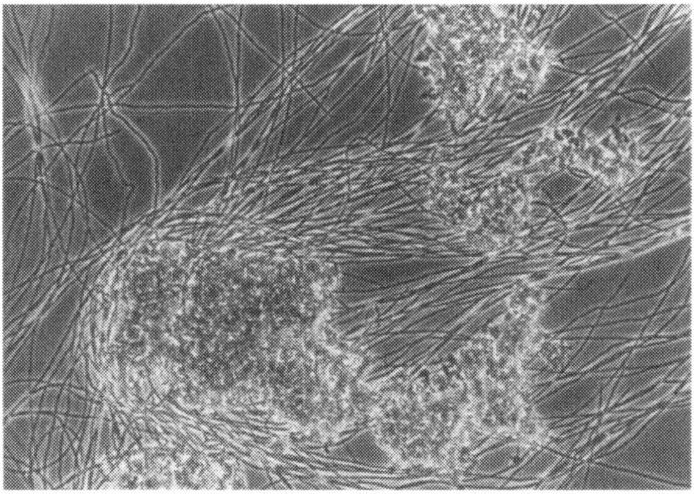

Figure 19.3. Bulking activated sludge (150 ×).

Bulking is a serious problem in 40 - 50% of all activated sludge plants (e.g. Blackbeard *et al.* 1986; Eikelboom 1987; Andreasen and Sigvardsen 1989). In plants treating an industrial wastewater this percentage is even higher.

19.2 FILAMENTOUS MICROORGANISMS IN ACTIVATED SLUDGE

Over 30 filamentous organisms, mainly bacteria, have been observed in activated sludge (Eikelboom 1975; La Rivière and Eikelboom 1986). Since most of these filamentous microorganisms were/are totally unknown, many species are still referred to by a number. A manual about this subject, including *in situ* identification keys, was compiled by Eikelboom and Van Buijsen in 1979 and was recently completely updated (Eikelboom 1999). These 30 types do not all occur with the same frequency. Only those species that can be frequently observed in bulking sludge are inserted in Table 1, which shows the effect of the sludge load applied on the population of filamentous bacteria in domestic treatment plants in The Netherlands. Although there are (small) differences in ranking of the various species, more or less similar results have been obtained in other countries (Wagner 1982; Strom and Jenkins 1984; Blackbeard *et al.* 1986; Andreasen and

Table 19.1. Effect of the sludge load applied on the filamentous population in domestic plants with bulking sludge

Filamentous species	Sludge load (kg BOD/kg MLSS.day)		
	< 0,1 [1)]	0,1 - 0,2 [2)]	> 0,2 [2)]
M. parvicella [3)]	+++	+++	+
Type 0041	++	+	
Type 0092	++	+	
Type 0803/0914	+	+	
Type 0581	+		
Type 1851	+	+	
Actinomyces [3)]	+		
N. limicola [3)]	+	++	+
Type 021 N [3)]	++	++	+++
H. hydrossis [3)]	+	++	+++
Type 1701 [3)]			+++
S. natans [3)]			+++
Thiothrix sp. [3)]		+	++
Type 1863 [3)]			++
Type 0411			++

[1)] Mainly plants with (a) raw influent, (b) complete nitrification and (c) simultaneous denitrification (not optimized).
[2)] Presettled influent and sometimes incomplete nitrification.
[3)] Isolated in pure culture and data about the physiology available.
+++, abundant in more than 40% of all plants; ++, abundant in 15-40% of all plants; +, abundant in less than 15% of all plants.

Sigvardsen 1989). The introduction of nutrient removal conditions causes a shift in the filamentous population as it favours *M. parvicella* above other filamentous species (Eikelboom *et al.* 1998).

19.2.1 Sulfur storing filamentous bacteria in activated sludge

From four filamentous species, it is known that they can store elemental sulfur granules inside their cells: *Thiothrix spp.*, *Beggiatoa spp.* (not included in Table 19.1), Type 021 N and Type 0914 (Figures 19.4 - 19.8). The fact that *Beggiatoa* is not included in Table 19.1 indicates that this bacterium is rather uncommon in activated sludge plants. Except for Type 0914, which has not been isolated in pure culture so far, it is confirmed in pure culture studies that these microorganisms derive additional energy from the oxidation of reduced sulfur compounds (H_2S, etc.). Elemental sulfur is stored as an intermediate product in the cells. The bright granules are clearly observable by phase contrast microscopy and therefore a great help for the *in situ* identification of these species in activated sludge. The other characteristics that are used for the *in situ* identification are listed in Table 19.2. The identification procedure includes a sulfur deposit test (Eikelboom 1999), as S-granules are not always present in *Thiothrix* (and Type 021 N) filaments. These two species are strongly related to each other (Williams and Unz 1985, 1989). However, Type 021 N stores less sulfur granules than *Thiothrix*. There is strong evidence that Type 0914 and Type 0803 (Figure 19.9) are different phenotypes of the same species (Eikelboom *et al.* 1998). Type 0803, however, never stores sulfur with the S-deposit test. *In situ*

Table 19.2. Characteristics used for the *in situ* identification of sulfur storing filamentous bacteria in activated sludge (Eikelboom, 1999)

Characteristics	*Thiothrix*	*Beggiatoa*	Type 021 N	Type 0914
branching of the filaments	rarely	no	no	no
motile filaments	no	yes	no	no
shape filaments	straight	bent	bent	bent
attached growth	no	no	no	no
sheath	sometimes	no	no	no
septa clearly visible [1)]	yes	no	yes	no
cell diameter (μm)	0.5 - 2.5	1.0 - 3.0	1.0 - 2.0	0.6 - 0.8
shape of the cells [1)]	rectangular	-	variable	-
S granules *in vivo*	often	always	no	always
S granules after S-test	always	-	some	-
Gram staining	negative	negative	negative	positive
Neisser staining	negative	negative	negative	negative

1) Septa and shape of the cells are only visible in filaments without stored S-granules.

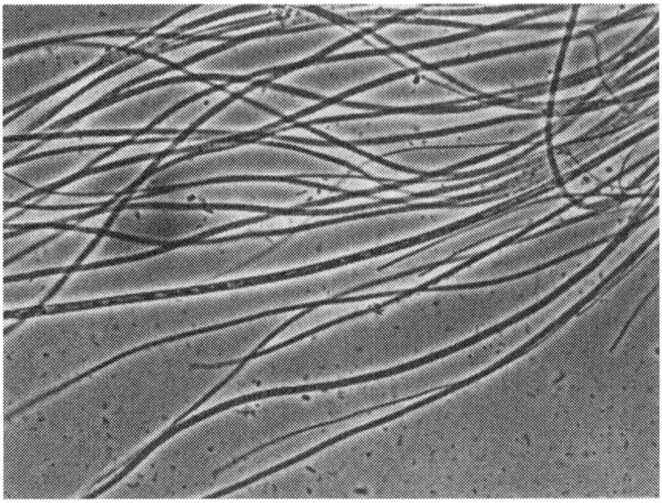

Figure 19.4. *Thiothrix* filaments without S granules (625 ×).

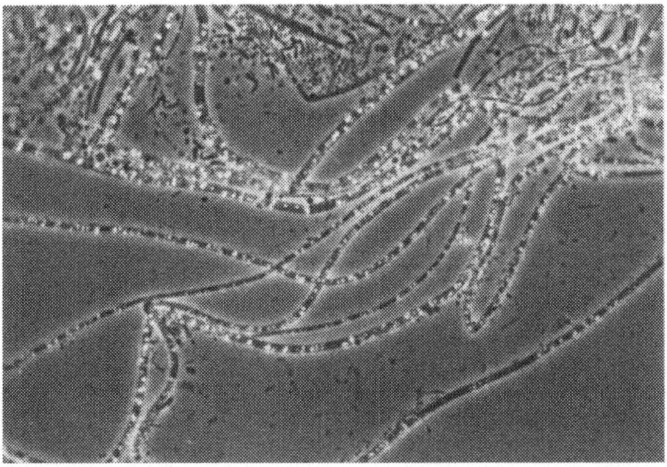

Figure 19.5. *Thiothrix* filaments filled with S granules (625 ×).

identification by application of molecular probes is now possible for Type 021 N and *Thiothrix spp.* (Wagner and Aman 1997).

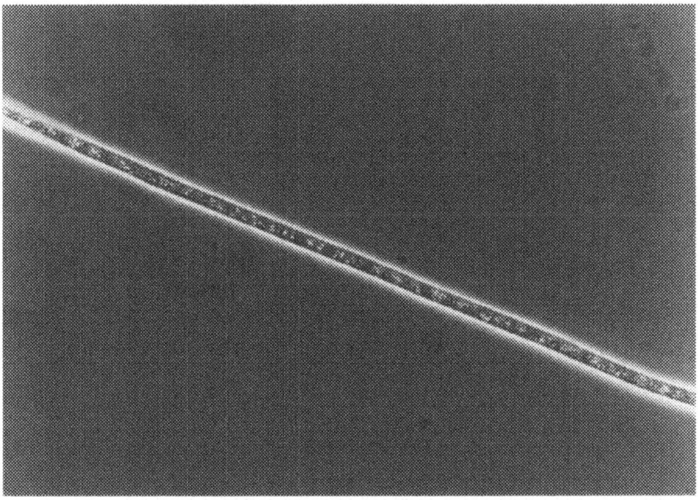

Figure 19.6. *Beggiatoa* (625 ×).

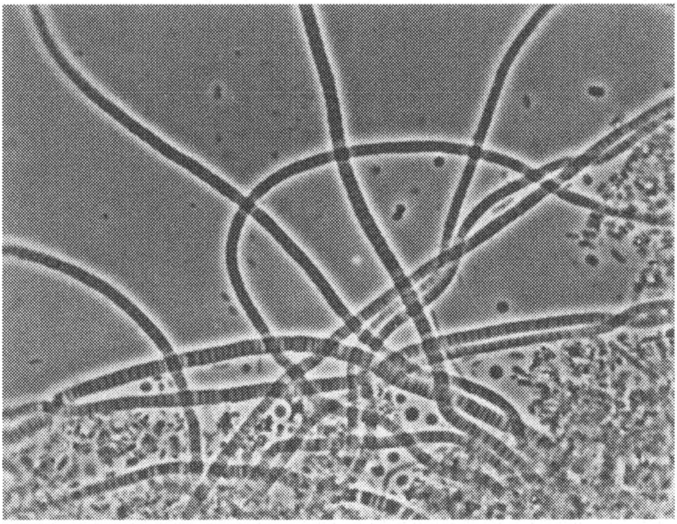

Figure 19.7. Type 021 N filaments (1250 ×).

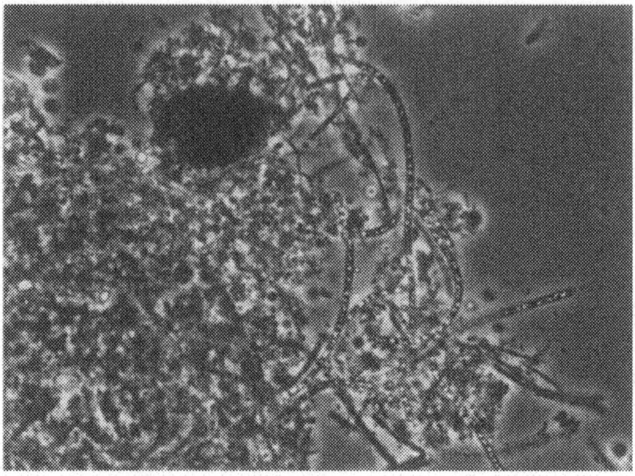

Figure 19.8. Type 0914 filaments (1625 ×).

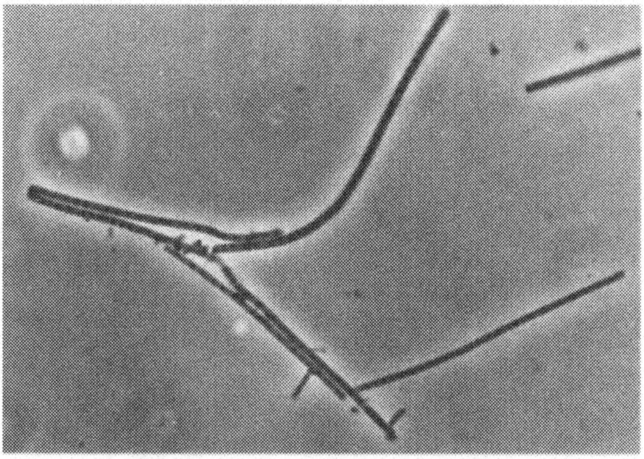

Figure 19.9. Type 0803 filaments (625 ×).

19.3 SELECTION/GROWTH OF FILAMENTOUS BACTERIA IN ACTIVATED SLUDGE

Table 19.1 indicates which filamentous species have been isolated so far. Information about their nutritional requirements is presented in Van Veen (1973), Van Veen *et al.* (1973), Van Veen *et al.* (1982), Slijkhuis (1983),

Lemmer (1985), Williams and Unz (1985, 1989), Diekman and Saby (1988), Nowak and Brown (1990), Seviour *et al.* (1997) and Andreasen and Nielsen (1998). For most species (including *Thiothrix*, Type 021 N and *Beggiatoa*), low molecular mass compounds like acetic acid are the main carbon source. So, they use the soluble organic fraction present in the influent of the plant for their growth. There is evidence that the same is true for Type 0914/0803 (Eikelboom *et al.* 1998). Compounds entering the plant in the particulate fraction are metabolised by just a few filamentous species: *M. parvicella* and probably the Types 0041, 0092 and 1851 as well.

It must be emphasised that for most filamentous sulfur bacteria, reduced sulfur compounds are (just) an additional energy source. They can use low molecular organic compounds as well. It might be, however, that the competition with the floc-forming bacteria for S^{2-} is less strong than for the organic compounds provided with the wastewater.

Quantitatively, namely compared with the amount of available carbon sources, reduced sulfur compounds are of minor importance in most domestic treatment plants. Owing to the aerobic conditions, the S^{2-} concentration is usually less than 1 mg/l in gravity sewers. In pressure mains the concentration may increase to 5-10 mg S^{2-}-l (Nielsen *et al.* 1992). Much higher concentrations (40-60 mg S^{2-}/l) were only determined in pressure mains with extreme long hydraulic retention times of 2-4 days (Clemens and Snaterse 1999). In some industrial effluents, however, particularly if the wastewater is anaerobically pretreated, high sulfide concentrations might be present.

Except for *Nostocoida limicola*, all isolated species are strictly aerobic, i.e. they do not grow at all if molecular oxygen is completely absent. The information about the nutritional requirements of the species involved, is not sufficient to explain the excessive filamentous growth in many plants, however.

Table 19.3 shows kinetic growth constants for some filamentous and floc-forming microorganisms. The former are characterised in general by relatively low values of the maximum specific growth rate. However, at substrate limiting conditions this selection factor hardly plays a role. Their strong competitive advantage consists of the low values of the saturation constants for substrate (K_s) and/or oxygen (K_{DO}).

These data strongly support the hypothesis that growth-limiting conditions in particular favour the filamentous organisms in the mixed sludge population. This hypothesis is in good agreement with the results of bulking control experiments done in the past (e.g. Chudoba *et al.*, 1973; Heide and Pasveer 1974; Rensink 1974) and observations in full scale plants

(Chambers and Tomlinson 1982). Nowadays, it is generally accepted that bulking is especially caused by specific nutritional deficiencies, as N, P, O_2 or even C (in plants with completely mixed aeration tanks). The effects of simultaneous limitations (e.g. N and O_2), are cumulative which explains the often severe bulking problems in industrial plants.

If the process conditions favour filamentous microorganisms, the influent composition, the configuration of the plants and the load applied mainly determine which species will proliferate (see, for example, Table 1). Sulfur oxidising filamentous species are of course stimulated if the wastewater treated contains relatively high concentrations of sulfides. This fact can be used the other way round as well: the presence of sulfur oxidising filamentous species in activated sludge indicates that sulfides are present in the wastewater entering the aeration tank.

Because their relative low μ_{max} values, filamentous organisms cannot compete successfully anymore with the floc-forming bacteria at a very high sludge load (greater than approx. 1.0 kg BOD/kg MLSS.day). Therefore, filamentous bulking hardly ever occurs in the high-loaded first stage of a two-stage treatment plant (e.g. the A/B process).

19.4 CONTROL OF FILAMENTOUS BULKING BY FAVOURING THE FLOC-FORMING POPULATION

In recent decades various biological remedial/preventive bulking sludge control strategies have been developed. They are all based on the same principle: create conditions that allow floc-forming bacteria to take up a great percentage of the organic substrate provided. This can be accomplished in two different ways (Wanner 1997):

kinetic selection: favouring floc-forming bacteria by eliminating growth-limiting conditions;

metabolic selection: prevent substrate uptake by filamentous organisms through maintaining anoxic or anaerobic conditions until the bulk of the substrate is taken up by the floc-forming population.

As filamentous sulfur bacteria not only metabolise organic compounds but also sulfides, controlling their growth needs an additional step aimed at removal of the sulfides from the wastewater entering the plants. This can be accomplished by pre-aeration of the influent or by the addition of compounds which oxidise (e.g. H_2O_2) or precipitate (e.g. Fe^{2+}) sulfides.

Table 19.3. Kinetic growth constants for filamentous and floc-forming microorganisms

Parameter	*S. natans*	1701	Type 021 N		*H. hydrossis*	*M. parvicella*	Floc-forming organisms
			N_2 [1]	N_7 [1]			
μ_{max}	6.5	2.6	3.8	1.4	1.2 - 2.2	1.4	1 - 9.2
K_s	10	2	1	1	5	very low	1.8 - 5.0
K_{DO}	0.033	0.014				0.016	0.073 - 0.15
Y	0.53	0.44	0.71	0.56	0.59 - 0.42	1.41	0.55
K_e	0.065	0.06	0.55	0.16			0.15
Substrate	Glucose	Glucose	Glucose	Glucose	Glucose	Oleic acid	
References		Richard *et al.* (1985a, b)			Krul (1977); Van Veen *et al.* (1982)	Slijkhuis (1983)	Richard *et al.* (1985a); Krul (1977); Van Veen *et al.* (1982)

[1] Two isolated strains.

u_{max}, maximum specific growth rate (d^{-1}); *Y*, yield coefficient (mg cells.mg substrate^{-1}); *K*, saturation constant (mg.l^{-1}); k_e, decay or endogeneous coefficient (d^{-1}).

19.4.1 Kinetic selection

Limitations of N, P or O_2 can be eliminated rather easily. Solving bulking problems caused by a carbon limitation, namely a low carbon concentration in the aeration tank, is more complicated, particularly in existing plants.

It has been ascertained that the competition between floc-forming and filamentous bacteria is strongly affected by the substrate concentration during mixing of influent and (return) sludge; see, for example, Houtmeyers *et al.* (1980), Chambers and Tomlinson (1982), Van den Eynde *et al.* (1983), Jenkins *et al.* (1984), Van Niekerk (1985) and Chudoba (1985). Figure 19.10 shows several modifications of the activated sludge process and the resulting substrate concentration in the aeration tank. In completely mixed systems (A), the influent is diluted with the total content of the aeration tank. The substrate concentration is low as well and equals that in the final effluent. Such circumstances are beneficial to microorganisms that still grow relatively well at carbon-limited conditions. Therefore, bulking frequently occurs in completely mixed systems. In the modifications B, C and D a short period with a high substrate concentration is followed by a long period without fresh substrate supply. If certain other conditions are fulfilled, this mode of operation is selective for floc-forming bacteria. In modification C a separate selector is incorporated. In B the first part of the aeration tank acts as a selective zone, whereas a fill-and-draw operation (D) is selective by itself.

Figure 19.11 illustrates what is actually happening in the high-loaded selective zone. A certain percentage of the substrate available adsorbs almost instantaneously to the floc. These nutrients are also not longer available for the filamentous bacteria. This is a physico-chemical process, independent of the oxygen concentration in the selective zone. The fraction removed instantaneously mainly consists of non-soluble/particulate substrate. This removal process therefore hardly contributes to controlling filamentous species using the soluble influent fraction for their growth.

Any further substrate removal in the selective zone is the result of active uptake of nutrients by microorganisms. It has been found that aerobic conditions result in a selection of floc-forming bacteria with a large storage capacity for soluble substrate. The substrate removal rate (β in Figure 19.11) thus increases if a selector/selective zone is incorporated in the process. From results obtained in full-scale domestic plants, it was concluded that a sludge load of 2-5 kg BOD/kg MLSS per day should be maintained in the selector for effective bulking control (Eikelboom 1987). As the total sewage flow reaches domestic plants in about 10 hours, the selective conditions are better expressed by a load per hour of wastewater

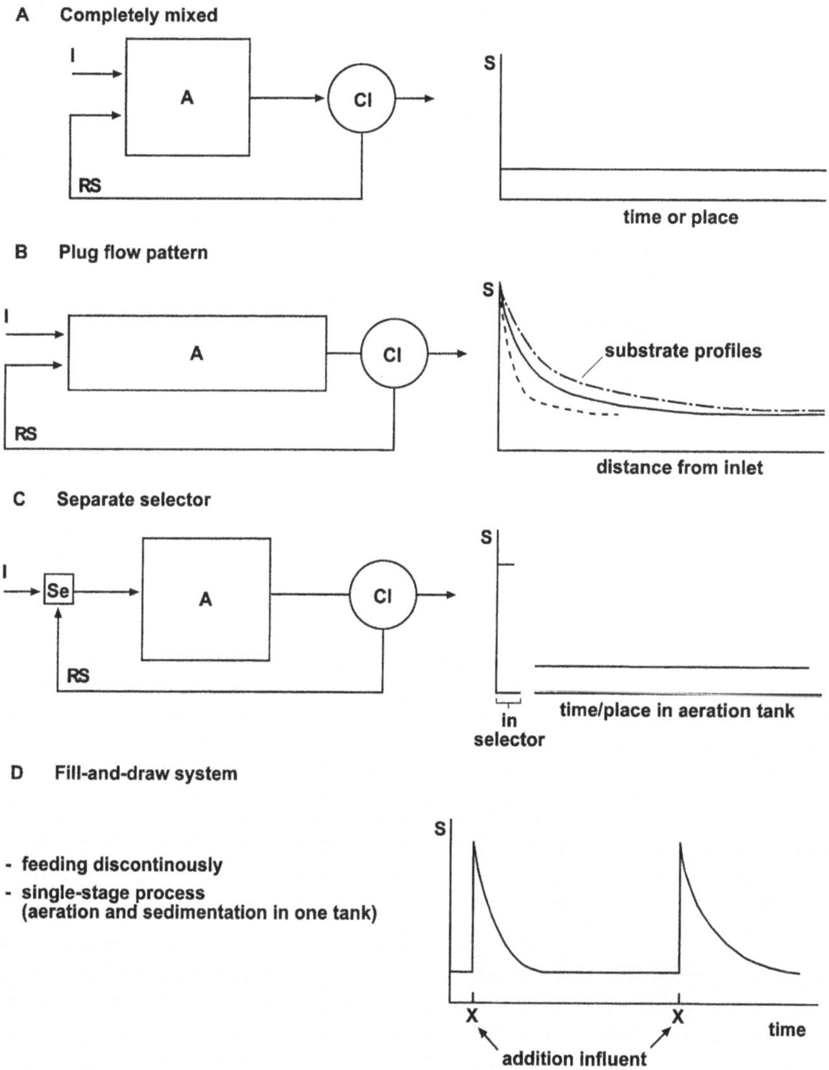

Figure 19.10. Substrate concentration (S) in four modifications of the activated sludge process. I = influent; RS = return sludge; Se = selector; A = aeration tank; Cl = final clarifier.

supply: 0.2-0.5 kg BOD/kg MLSS per hour. Because of the high load, particulate material is removed simultaneously. The term "biosorption" refers to both removal processes. Table 19.4 shows the biosorption by two

Table 19.4. Biosorption of substrate by two different sludges; sludge A originates from a treatment plant with an aerobic selector

Sludge	Influent	COD removed in 10 min. (%)
A	A	85
A	B	92
B	B	36
B	A	24

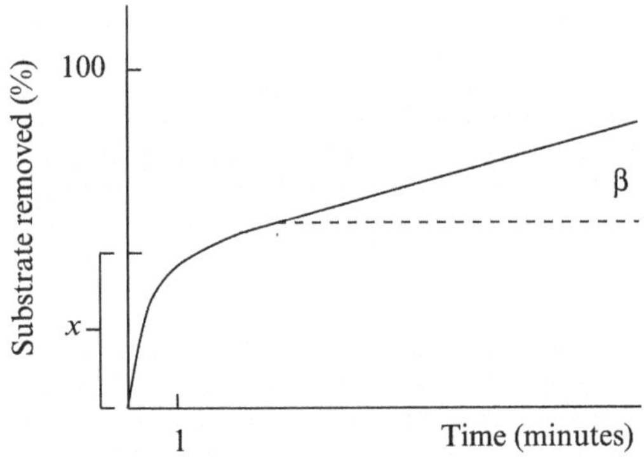

Figure 19.11. Substrate removal in a high loaded selective zone.
X = physico-chemical; β = biological processes.

sludges: one of them originates from a treatment plant with an aerobic selector.

The substrate taken up in the selector is metabolised further in the aeration tank. It is important that the solids retention time in the aeration tank is long enough to allow complete metabolisation of the substrate store (Chudoba *et al.* 1982). Otherwise the biosorption capacity is not completely regenerated when the biomass re-enters the selector. A reduction of the removal percentage in the selective zone will result again in filamentous growth in the aeration tank. At normal operational conditions the solids retention time in the aeration tank is long enough if the load of the treatment plant remains below about 0.5 kg BOD/kg MLSS per day. However, this critical level is lower if the oxygen concentration in the aeration tank is below 2.0 mg O_2^{-1}.

Because of the high substrate removal rate, a short hydraulic retention time (10-30 min.) in the selector is sufficient. A plug flow pattern in the selective zone maximizes the substrate concentration at the inlet. Therefore, this configuration in generally applied nowadays. For more detailed information about this subject, the reader is referred to Prendl (1997).

It is quite clear now that aerobic selectors are effective tools in controlling an excessive growth of many filamentous species, in particular organisms that require soluble compounds for growth. However, some questions still remain concerning the effect on the growth of organisms that use particulate substrate for their growth such as *M. parvicella*.

19.4.2 Metabolic selection

The removal of excess N and P by biological methods requires the incorporation of anoxic and/or anaerobic zones in a treatment plant. During the development of this technology it was noticed that anoxic/anaerobic zones sometimes resulted in an improved settleability of the sludge as well. These observations suggested that filamentous organisms cannot grow at all if molecular oxygen is completely lacking. Studies with pure cultures largely confirmed this hypothesis. The only exception so far is *N. limicola* (Nowak and Brown 1990).

Just like in aerobic selectors, anaerobic/anoxic conditions will only select for floc-forming bacteria if a large percentage of the substrate available is removed in this zone (Wanner *et al.* 1987a, b). Removal of carbon compounds by denitrification and/or biological phosphate removal are the

Table 19.5. Dominating filamentous species in European nutrient removal plants during spring 1994

	The Netherlands				Denmark			Germany	Greece
	O.D.	Car.	Schr.	V.C. 1)	Bio-P	Denitro	RC		
Species					plants			V.C. 2)	V.C. 3)
M. parvicella	58	66	44	67	74	28	43	75	55
Type 0041	6		11		32	38	20		
Type 0092	3				5	9			10
Type 1851			11			3			
Type 0803					5	3		12	
Type 0914					5				
N. limicola					16	3	10		
Actinomyces									20
Type 021N		3	11			6			
Thiothrix sp.						6			
S. natans	3		11						
n plants	31	32	9	21	19	32	30	8	11

1) incl. 4 bio-P plants, 2) incl. 3 bio-P plants, 3) incl. 2 bio-P plants.

O.D., oxidation ditch; Car., carroussel; Schr., Schreiber; V.C., various configurations; R.C., recirculation plants.

Figures: percentage of the total number of plants where a specific species was present (Eikelboom *et al.* 1998).

only alternatives if O_2 is lacking. This implies that bulking sludge control and biological nutrient removal processes can be linked to each other. Table 19.5 confirms that filamentous species, using soluble carbon compounds for their growth, are of minor importance in nutrient removal plants.

19.5 REFERENCES

Andreasen, K., and L. Sigvardsen (1989) Letslam. Report Firm Krüger, Denmark.

Andreasen, K., and P.H. Nielsen (1998) *In situ* characterization of substrate uptake by *Microthrix parvicella* using micro-autoradiography. *Wat. Sci. Tech.* **37**, 19-26.

Blackbeard, J.R., G.A. Ekama and G. van R. Marais (1986) A survey of filamentous buking and foaming in activated sludge plants in South Africa. *Water Pollut. Contr.* **85**, 90-100.

Chambers, B. and E.J. Tomlinson (1982) Bulking of activated sludge. Preventative and remedial methods. Ellis Horwoord Publ., Chichester.

Chudoba, J. (1985) Control of activated sludge filamentous bulking. VI Formulation of basic principles. *Wat. Res.* **19**, 1017-1022.

Chudoba, J., M. Dohanyos and P. Grau (1982) Control of activated sludge filamentous bulking. IV Effects of sludge regeneration. *Water Sci. Techn.* **14**, 73-93.

Chudoba, J., P. Grau and V. Ottavà (1973) Control of activated sludge filamentous bulking. II Selection of microorganisms by means of a selector. *Wat. Res.* **7**, 1389-1406.

Clemens, F.H.L.R. and C. Snaterse (1999) Personal communication.

Diekmann, H. und S. Saby (1988) Mischkulturversuche mit dem Fadenbildner *Haliscomenobacter hydrossis* und dem Flockenbildenden *Arthrobacter globiformis.* II. Einfluss des Phosphorangebots und der Salzkonzentration. *GWF (Wasser-Abwasser)* **129**, 422-425.

Eikelboom, D.H. (1975) Filamentous organisms in observed in activated sludge. *Wat. Res.* **2**, 365-388.

Eikelboom, D.H. (1987) Licht-slibhandboek (Bulking sludge control manual) *MT-TNO, Report R 87/358.*

Eikelboom, D.H. and H.J.J. van Buijsen (1979/1983) Microscopic sludge investigation manual. *MT-TNO, Report A94a*, 2nd Ed.

Eikelboom, D.H., A. Andreadakis and K. Andreasen (1998) Survey of filamentous populations in nutrient removal plants in four European countries. *Wat. Sci. Tech.* **37**, 281-289.

Eikelboom, D.H. (1999) Process control of activated sludge plants by microscopic sludge investigation. *TNO Report R 99/057* (English edition in press).

Eynde, E. van den, J. Geerts, B. Maes and H. Verachert (1983) Influence of the feeding pattern on the glucose metabolism of *Arthrobacter sp* and *Sphaerotilus natans*; Growth in chemostat culture, simulating activated sludge bulking. *European J. Appl. Microbiol.* **17**, 35-43.

Heide, B.A. and A. Pasveer (1974) Oxidation ditch: Prevention and control of filamentous sludge. *H_2O* **7**, 373-377.

Houtmeyers, J., E. van den Eynde, R. Poffè and H. Verachtert (1980) Relations between substrate feeding pattern and development of filamentous bacteria in activated sludge processors. Part I. Influence of process parameters. *Eur. J. Appl. Microb. Biotechnol.* **9**, 63-77.

Jenkins, D., M.G. Richards and J.B. Neething (1984) Causes and control of activated sludge bulking. *Water Pollut. Contr.* **83**, 455-472.

Kroiss, H. (1985) Blähschlammprobleme bei der Zuckerfabrik Leopoldsdorf. *Wiener Mitteilungen* **56**, 81-107.

Krul, J.M. (1977) Experiments with *Haliscomenobacter hydrossis* in continuous culture with and without *Zoogloea ramigera. Wat. Res.* **11**, 197-204.

Lemmer, H. (1985) Mikrobiologische Untersuchungen zur Bildung von Schwimmschlamm auf Kläranlagen. *Thesis TU München.*

Nielsen, P.H., K. Raunkjaer, N.H. Norsker, N.A. Jensen and Th. Hritved-Jacobsen (1992) Transformation of wastewater in sewer systems - A review. *Wat. Sci. Tech.* **25**, 17-31.

Niekerk, A. van (1985) Competitive growth of flocculant and filamentous microorganisms in activated sludge systems. *Thesis Univ. Of California*, Berkeley.

Nowak, G. and G.D. Brown (1990) Characteristics of *Nostocoida limicola* and its activity in activated sludge suspension. *Research J. Wat. Pollut. Contr. Fed.* **62**, 137-142.

Prendl, L. (1997) Ein Beitrag zur Verständnis und Anwendung aeroben Selektoren für die Blähschalammvermeidung. *Wiener Mitteilungen* Vol. **139**.

Rensink, J.H. (1974) New approach to preventing bulking sludge. *J. Wat. Pollut. Contr. Fed.* **46**, 1889-1894.

Richard, M., O. Hao and D. Jenkins (1985a) Growth kinetics of *Sphaerothilus species* and their significance in activated sludge bulking. *J. Wat. Pollut. Contr. Fed.* **15**, 68-81.

Richard, M.G., G.P. Shimizu and D. Jenkins (1985b) The growth physiology of the filamentous organisms type 021N and its significance to activated sludge bulking. *J. Wat. Pollut. Contr. Fed.* **57**, 1152-1162.

Rivière, J.W.M. la and D.H. Eikelboom (1986) Role of specific microbial groups in wastewater treament; filamentous organisms as an example. *Proc. of the fourth Int. Symp. on microbial Ecology*, Ljubljana, 59-65.

Seviour, E.M., L.L. Blackall, C. Christensson, P. Hugenholtz, M.A. Cunningham, D. Bradford, H.M. Stratton and R.J. Seviour (1997) The filamentous morphotype Eikelboom Type 1863 is not a single genetic entity. *Journ. Of Appl. Microb.* **82**, 411-421.

Slijkhuis. H. (1983) The physiology of the filamentous bacterium *Microthrix parvicella. Thesis, Landbouw Hogeschool Wageningen.*

Strom, P.F. and D. Jenkins (1984) Identification and significance of filamentous microorganisms in activated sludge. *J. Wat. Pollut. Control. Fed.* **56**, 449-459.

Veen, W.L. van (1973) Bacteriology of activated sludge, in particular the filamentous bacteria. *Ant. Van Leeuwenhoek* **39**, 1989-205.

Veen, W.L. van, D. van der Kooy, E.C.W.A. Geuze and A.W. van der Vlies (1973) Investigation on the sheathed bacterium *Haliscomenobacter hydrossis* gen. n., sp.n, isolated from activated sludge. *Ant. Van Leeuwenhoek* **39**, 207-216.

Veen, W.L. van, J.M. Krul and C.J.E.A. Bulder (1982) Some growth parameters of *Haliscomenobacter hydrossis* (syn. *Streptothrix hyalina*) A bacterium occuring in bulking activated sludge. *Wat. Res.* **16**, 531-534.

Wagner, F. (1982) Ursachen, Verhindering und Bekämpfung der Bläschlammbildung in Belebungsanlagen. *Stuttgarter Berichte zur Siedlungswasserwirtschaft* **76**, 1-220.

Wagner, M. and R. Amann (1997) Molecular techniques for determining microbial community structures in activated sludge. In: T.E. Cloethe and N.Y.O. Muyima (Eds.) Microbial commmunity analysis. IAWQ, Scientific and Technical Report No. 5, 61-72.

Wanner, J., J. Chudoba, K. Kucman and L. Proske (1987a) Control of activated sludge filamentous bulking - VII. Effect of anoxic conditions. *Wat. Res.* **21**, 1447-1451.

Wanner, J., K. Kucman. V. Ottovà and P. Grau (1987b) Effect of anaerobic conditions on activated sludge filamentous bulking in laboratory systems. *Wat. Res.* **21**, 1541-1546.

Wanner, J. (1997) Microbial population dynamics in biological wastewater treatment plants. In: Cloete, T.E. and N.Y.O. Muyima (Eds): Microbial community analysis. *IAWQ Scientific and Technical report No. 5*, 35-59.

Williams, T.M. and R.F. Unz (1985) Filamentous sulphur bacteria of activated sludge: Characterization of *Thiothrix*, *Beggiatoa* and Eikelboom Type 021N strains. *Appl. Environ. Microbiol.* **49**, 887-898.

Williams, T.M. and R.F. Unz (1989) The nutrition of *Thiothrix*, Type 021N, *Beggitoa* and *Leucothrix* strains. *Wat. Res.* **23**, 15-22.

20

Sulfur problems in anaerobic digestion

Vincent O'Flaherty and Emer Colleran

20.1 INTRODUCTION

High concentrations of oxidised sulfur compounds, such as sulfate and sulfite are found in many wastewater streams for which anaerobic digestion is considered an economically attractive treatment method. Examples include wastewaters from the pulp and paper, fermentation and the edible oil production industries (Colleran *et al.* 1995). During anaerobic treatment of these wastewaters, sulfate-reducing bacteria (SRB) will competitively interact with the other anaerobic bacteria involved in methanogenesis, resulting in the formation of H_2S rather than methane. This can present several problems for the anaerobic digestion process. The main manifestations of these problems include:

1. The inhibitory effect of H_2S on many bacterial trophic groups involved in anaerobic digestion, thus reducing reactor performance (Widdel 1988; Koster *et al.* 1986; Hilton and Oleskiewicz 1988).

2. A reduction of the methane yield, and thus less energy recovery.
3. Malodour, corrosion of piping, pumps, etc., necessity for scrubbing of the biogas and posttreatment of effluents to meet discharge standards (Rinzema 1988).

The presence of sulfate can also have some beneficial effects during anaerobic treatment of wastewaters. Most methanogens lack assimilatory sulfate reductases (Daniels *et al.* 1986) and their sulfur requirements are satisfied by a combination of inorganic sulfide and organic sulfur compounds, such as cysteine, glutathione, etc. Consequently, the production of sulfide by dissimilatory bacterial species during anaerobic digestion may enhance methanogenesis by satisfying the sulfur growth requirements of methanogens. The production of sulfide during anaerobic digestion can also be beneficial as the formation of sulfide precipitates can alleviate the toxicity of many heavy metal ions, such as cobalt and lead (Lawrence and McCarty 1965; Tursman and Cork 1989).

20.2 INHIBITION PHENOMENA DURING ANAEROBIC DIGESTION OF WASTESTREAMS WITH HIGH LEVELS OF OXIDISED SULFUR COMPOUNDS

The inhibition phenomena encountered as a result of the presence of sulfur compounds in wastewaters can be divided into three categories: inhibition by sulfide, by sulfite and by cations. Of these, inhibition by sulfide is the most important and will therefore be the principal focus of this chapter.

20.2.1 Sulfide toxicity during anaerobic digestion

The activity of SRB results in the production of sulfide which is the most stable form of sulfur under anaerobic conditions. Sulfide is highly reactive, corrosive and toxic to microorganisms, plants and man. Hydrogen sulfide dissociates in water according to the following equations (Garrels and Christ 1965):

$$H_2S \Leftrightarrow H^+ + HS^- \qquad (k_1 = 1.0 \times 10^{-7})$$

$$HS^- \Leftrightarrow H^+ + S^{2-} \qquad (k_2 = 1.0 \times 10^{-14})$$

The toxicity of sulfide is regarded as being pH dependent because only the neutral undissociated hydrogen sulfide (H_2S) molecule can pass through the cell membrane (Speece 1983). The pK_a value of the dissociation equilibrium of H_2S is about 7.04 at 18 °C (Weast 1981). Above pH 8.0-9.0, virtually all dissolved sulfide is present in its ionised form. At the neutral pH values typical of methanogenic systems, 20-50% of the dissolved sulfide is present as H_2S, depending on the temperature. The exact mechanism of H_2S toxicity is unclear. One possible mechanism is native protein denaturation through formation of sulfide and disulfide cross links between polypeptide chains. H_2S may also interfere with coenzymes A and M through the formation of sulfide linkages. This may result in interference with the acetyl coenzyme A pathway for CO_2 fixation which is common to both SRB and methanogens (Widdel 1988). H_2S may also affect the internal cell pH.

It is assumed that the inhibitory form of sulfide is the undissociated form: H_2S. Thus, one would expect a direct correlation between the H_2S concentration and the extent of inhibition. However, this relationship is not always straightforward and other parameters, i.e. the total sulfide concentration, can correlate better with the observed inhibition. It has indeed been reported by several authors that inhibition of SRB and methane-producing bacteria (MPB) is related to the total sulfide concentration in the pH range 7.0-9.0 (Koster *et al.* 1986; Visser 1995; O'Flaherty *et al.* 1998a). On the other hand, Oleskiewicz *et al.* (1989) and McCartney and Oleskiewicz (1993) concluded that inhibition of methanogens in suspended sludge was related solely to the undissociated H_2S concentration. Hilton and Oleskiewicz (1988) found that inhibition of SRB correlated with the total sulfide concentration, whereas inhibition of the MPB correlated with the free H_2S concentration. O'Flaherty *et al.* (1998a) found that sulfide inhibition of a range of methanogens, syntrophs and SRB was related to the undissociated H_2S concentration between pH 6.8 and 7.2, whereas above pH 7.2 the inhibition was related to the total sulfide concentration (Figure 20.1). These observations suggest that H_2S may become more toxic at higher pH levels, possibly because of the development of strong pH gradients across the cell membrane, which may affect the diffusion properties of the H_2S molecule.

Alternatively, both total sulfide and H_2S may exert an inhibitory effect. The organisms, therefore, may have two inhibition thresholds, one for of H_2S and one for total sulfide. The levels of undissociated H_2S required for 50% inhibition of the different bacterial groups (the so-called IC_{50} value) were found to be much lower than the total sulfide IC_{50} value by O'Flaherty *et al.* (1998a), indicating that H_2S was clearly the most toxic form of sulfide. It was also found that propionate degrading SRB had a much lower threshold for total sulfide than the other bacteria studied. By contrast, the

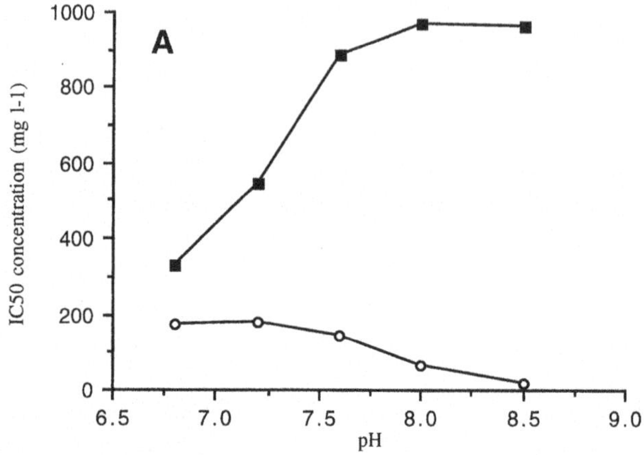

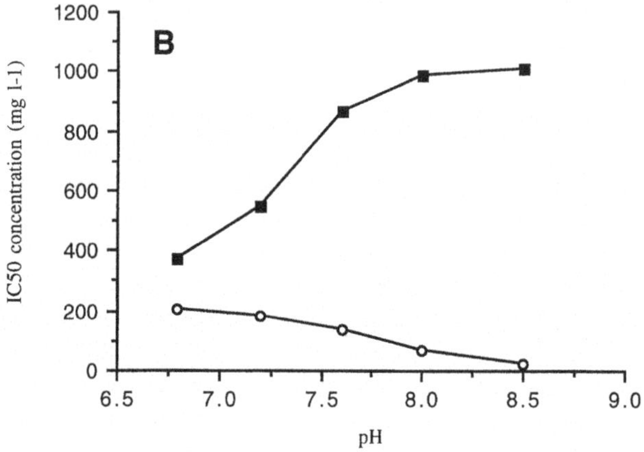

Figure 20.1. IC_{50} values for H_2S and total sulfide inhibition of pure cultures of MPB and SRB. (A) *M. soehngenii* and (B) *D. acetoxidans*. H_2S (-O-), total sulfide (-■-). (After: O'Flaherty *et al.* 1998a).

level of undissociated H_2S required to cause 50% inhibition of growth was similar to that of other SRB and MPB (O'Flaherty *et al.* 1998a).

20.2.2 Sulfite inhibition

Few data are available on sulfite inhibition of anaerobic bacteria. It has been shown in batch experiments that sulfite induces a lag phase in methane production of variable length. Yang *et al.* (1979) observed lag phases of between 60 hours and 12 days after dosing 25 and 75 mg l^{-1} SO_3^{2-}, respectively, to an acetoclastic methanogenic enrichment culture. The effects of sulfite inhibition appear to be far less severe with adapted sludges, however, probably owing to the presence of SRB which will reduce sulfite to sulfide. Eis *et al.* (1983) reported no lag phase after addition of 100 mg l^{-1} SO_3^{2-} to an adapted sludge. Maaskant and Hobma (1981) found a 50% inhibition of the methanogenic activity at about 150-200 mg l^{-1} SO_3^{2-}, an effect that declined upon repeated dosing of SO_3^{2-} to the sludge.

It is generally considered that sulfite inhibition will be insignificant in continuous cultures and during reactor operation as in these cases SRB populations that will eliminate the sulfite will develop (Rinzema and Lettinga 1988; Visser 1995). It has been shown that wastewaters with sulfite concentrations of up to 800 mg l^{-1} SO_3^{2-} can be treated satisfactorily in anaerobic reactors (Eis *et al.* 1983; Ferguson *et al.* 1982; Särner 1986; Särner 1989).

20.2.3 Sodium inhibition

The high cation (both Na^+ and Ca^{2+}) concentrations present in sulfate-rich wastewaters can inhibit anaerobic bacteria. The effect of sodium on methanogenic digestion has been studied extensively. As for sulfide inhibition thresholds, the literature shows many inconsistencies as indicated by reported values ranging from 6 to 40 g l^{-1} for the 50% inhibition of methanogenic bacteria by sodium (Kugelman and Mc Carty 1964; Van den Berg *et al.* 1976; Lettinga and Vinken 1981; de Baere *et al.* 1984). These differences can be attributed to the history of the sludge, antagonistic and synergistic effects and the test method used.

The presence of other cations, such as potassium, causes antagonistic or synergistic effects, resulting in a significant change in the sodium sensitivity of anaerobic bacteria (McCarty and McKinney 1961; Kugelman and Chin 1971; Rinzema *et al.* 1986; Mendez *et al.* 1995). The sulfidogenic activity of granular sludge adapted to sodium levels of 1.5-2 and 5.5-6 g l^{-1}, respectively, was inhibited at sodium concentrations exceeding 11 g l^{-1}. A 50% inhibition of the activity was observed at about 15 g l^{-1} of sodium for both sludges (Visser 1995). Such high sodium concentrations are essential for the growth of many marine SRB, but are inhibitory to freshwater SRB (Widdel 1988). In practice, the sodium content of a sulfate-rich wastewater is very unlikely to cause process failure.

20.2.4 Calcium inhibition

Although the calcium ion does not exert a severe direct toxic effect, $CaCO_3$ and/or $Ca_3(PO_4)_2$ precipitates can indirectly upset the reactor performance by scaling. These precipitates are entrapped in the reactor biomass, where they gradually accumulate and ultimately result in a complete loss of the activity of the sludge granules owing to a calcium layer which can completely block substrate transport (Langerak *et al.* 1998). Serious scaling of biomass by calcium precipitates may occur at Ca^{2+} concentrations as low as 400 mg l^{-1}. Clogging problems can also arise from precipitates in the piping system. Moreover, calcium phosphate precipitation can cause phosphate deficiency and thus limit microbial activity (Callander and Barford 1983; Lettinga 1995; Langerak *et al.* 1998).

20.3 TECHNIQUES FOR QUANTIFICATION OF SULFIDE TOXICITY ON MICROBIAL POPULATIONS INVOLVED IN ANAEROBIC DIGESTION

There are two principal approaches for analysis of the toxic effects of sulfide on the microbial populations involved in anaerobic digestion:

1. Sludge activity/toxicity tests.
2. Determination of kinetic growth properties of microbial populations.

The main aim of both of these approaches is to determine the effect of various concentrations of a toxicant, i.e. sulfide, on the trophic groups involved in methanogenesis or sulfate reduction. This information will be valuable in predicting the levels of in-reactor sulfide which will be acceptable for successful anaerobic digestion and also in determining the outcome of competition between MPB and SRB.

20.3.1 Sludge activity/toxicity tests

20.3.1.1 Specific methanogenic activity/toxicity tests

Numerous toxicity testing strategies have been developed to monitor the effects of inhibitory compounds on anaerobic bacteria, and in particular on methanogens. These tests are based on the effect of inhibitors on sludge/biomass activity compared with a control. Anaerobic toxicity tests can be classified on the basis of their operating conditions. They can be conducted under batch (Owen *et al.* 1979; Field and Lettinga 1989; Colleran *et al.* 1992; O'Flaherty *et al.* 1998a) or continuous conditions (Speece and Parkin 1983; Govind *et al.* 1991; Britz *et al.* 1992). Tests can be done directly with reactor sludge, with enrichments from reactor sludge (Wang *et al.* 1991) or with pure cultures of anaerobic bacteria (O'Flaherty *et al.* 1998a). The parameters that indicate inhibition are:

1. a drop in methane yield;
2. volatile fatty acid (VFA) accumulation;
3. a decrease in the chemical oxygen demand (COD) removal efficiency.

Biogas or methane production can be monitored by water displacement (Field and Lettinga 1989), by the displacement of a plunger of glass syringes (Owen *et al.* 1979) or by electronic pressure transducers (Colleran and Pistilli 1994). The methane percentage of the biogas produced and the VFA levels are usually monitored by gas chromatography (GC).

Specific sludge activities are typically analysed using batch experiments. Sludge samples are placed in glass serum bottles in the presence of a basal medium which may or may not contain nutrients and a buffering agent (Visser 1995; O'Flaherty *et al.* 1998b, 1999). Substrates of interest are added in appropriate concentrations to control vials whereas inhibitor (sulfide) is included at various concentrations in parallel tests for an assessment of activity as a function of the sulfide concentration. The pH of the vials can be corrected by addition of acid or base. All vials are prepared under anaerobic conditions by flushing with an N_2/CO_2 mixture. The methanogenic activity can be determined by measuring the methane concentration in the headspace of the serum vials over time (Visser 1995). Alternatively, the gas pressure increase (or decrease in the case of H_2/CO_2) in the headspace can be measured with a pressure transducer device and the concentration of methane can then be determined by GC (Colleran and Pistilli 1994; O'Flaherty *et al.* 1998b). In both cases, the activity is calculated from the slope of the progress line (either relative methane concentration or millilitres of methane). The volatile suspended solids (VSS) present in the serum vial are determined on completion of the test and the specific methanogenic activity is expressed per unit VSS. The effect of added sulfide concentrations can be expressed as a percentage inhibition of control activity and thus a 50% inhibition concentration value (IC_{50}) can be calculated (Figure 20.2). The same procedure can be done in the presence of bromoethane sulphonic acid (BES) and sodium molybdate (Mb), specific inhibitors for MPB and SRB, respectively (Oremland and Capone 1988). As well as the methane production, the degradation of substrates of interest can also be measured directly, normally by GC, and the effect of added sulfide on the substrate degradation rates can be determined (Figure 20.3).

In general, batch toxicity tests of the type described above generate acute

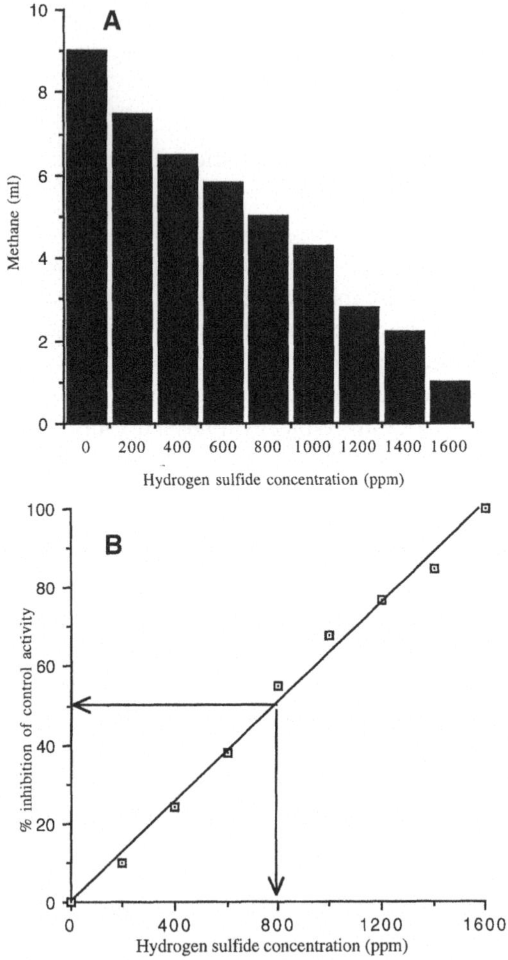

Figure 20.2. Effect of increasing H_2S concentrations on (A) cumulative methane production from acetate after 5 h at pH 8.0 and (B) percentage inhibition by H_2S of control methanogenic activity with acetate at pH 8.0.

toxicity data about the effects of a single dose of an inhibitor, such as sulfide, a short time after the administration of the dose. These data can then be used to determine the dose response relationship and toxicity thresholds. These values are useful when making comparisons between the toxicity of compounds or between the sensitivities of different trophic groups of organisms. There are also several toxicity testing protocols that

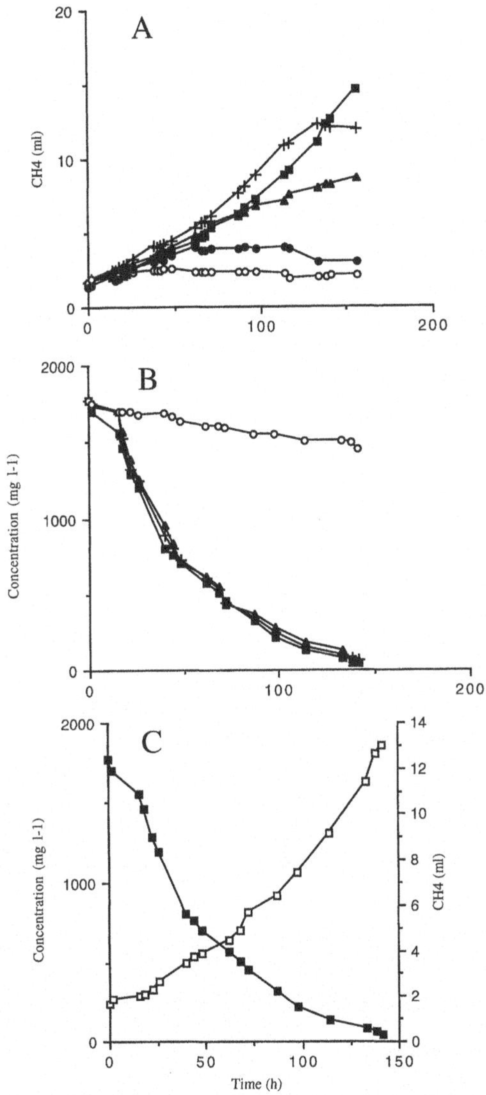

Figure 20.3. (A) Methane production by full-scale sulfate-adapted sludge samples from acetate (-■-) and from acetate in the presence of sulfate (+); sulfate and molybdate (-▲-); sulfate and BES (-○-); blank (-●-). (B) Acetate degradation in the presence and absence of sulfate and specific inhibitors (symbols as above). (C) Correlation between methane production (-□-) and acetate utilisation (-■-) by full-scale sulfate adapted sludge in the absence of sulfate or inhibitors. (After: O'Flaherty *et al.* 1998b).

incorporate both batch and continuous flow strategies to determine IC_{50} values for toxic compounds. These protocols have also been used to monitor the ability of the biological system to recover from, or to acclimatise to, a toxic dose (Soto *et al.* 1993; Young and Tabac 1993).

20.3.1.2 Specific sulfidogenic activity/toxicity tests

In this case, the basal medium used in the activity tests is supplemented with sulfate, at an appropriate concentration, so that sulfate will never become limiting (O'Flaherty *et al.* 1998b). Methanogenic activity can be eliminated with BES. The substrate degradation can be measured directly, e.g. by GC analysis (Visser 1995; O'Flaherty *et al.* 1998b). The sulfidogenic activity can then be calculated from the slope of substrate degradation and the VSS content (Visser 1995; O'Flaherty *et al.* 1998b). Toxicity thresholds are determined by the dose response relationship, as described above.

20.3.2 Determination of kinetic growth properties of microbial populations

The growth and kinetic properties of microorganisms can be determined by simple procedures in batch culture (Visser 1995; O'Flaherty *et al.* 1998a). Typically, the properties of most interest are the specific methanogenic or sulfidogenic growth rates and the affinities for substrate/sulfate. These parameters are of importance in studies to determine the likely outcome of competition between SRB and other anaerobic bacteria, and also to determine the effect of sulfide on the growth of microbial populations. Normally in these experiments the parameters are determined using substrate (or sulfate) depletion curves (Visser 1995; O'Flaherty *et al.* 1998a).

20.4 SULFIDE TOXICITY IN ANAEROBIC DIGESTION

20.4.1 Toxicity towards the bacterial trophic groups involved in methanogenesis

The literature on sulfide toxicity is highly complex and often contradictory. Information on the prevailing pH is often not included, which creates difficulties in determining the role of undissociated H_2S. As regards inhibition of the different trophic groups involved in methane production (Table 20.1), Koster *et al.* (1986) reported inhibition of acetoclastic methanogenesis at 250 mg l^{-1} H_2S. Rinzema and Lettinga (1988) found that in granular sludge adapted to sulfate, propionate conversion was severely

Table 20.1. IC_{50} (mg l^{-1}) values for H_2S inhibition of methanogenesis in anaerobic sludges (After: Visser 1995)

Biomass	Substrate	°C	pH	H_2S (mg l^{-1})	Ref.
Suspended sludge	distillery wastewater	37	7.0-7.2	130	1
Suspended sludge	lactate	35	7.0	100	2
			8.0	100	
Suspended sludge	acetate	-	-	50	3
Suspended sludge	acetate	35	6.5-7.4	125	4
			7.7-7.9	100	
Granular sludge	acetate	30	6.2-6.4	246	5
			7.0-7.2	252	
			7.8-8.0	90	
Granular sludge	acetate	30	7.2-7.4	184	6
			8.1-8.3	38	

1. Karhadkar *et al.* (1987); 2. McCartney and Oleskiewicz (1993); 3. Kroiss and Plahl-Wabnegg (1983); 4. Oleskiewicz *et al.* (1989); 5. Koster *et al.* (1986); 6. Visser (1995).

inhibited at H_2S concentrations exceeding 100 mg l^{-1}, with an IC_{50} of 140 mg l^{-1} H_2S at pH 7.0-7.4. Oleskiewicz *et al.* (1989) found that the most sensitive trophic group was the propionate oxidisers, with inhibition increasing for electron donors as follows: lactate, butyrate, acetate, propionate. McCartney and Oleskiewicz (1991, 1993) found that, during the degradation of lactate at a COD:sulfate ratio of 3.7:1, propionate accumulation occurred at 110 mg l^{-1} H_2S. However, at lower COD:sulfate ratios, no propionate accumulation was seen at H_2S concentrations of up to 325 mg l^{-1} (McCartney and Oleskiewicz 1991, 1993). This was apparently due to a direct oxidation of lactate to acetate by SRB, followed by acetoclastic methanogenesis at lower ratios. By contrast, an indirect oxidation of lactate to propionate and acetate by non-SRB occurred at a ratio of 3.7:1. The propionate formed in this way was then oxidised to acetate by SRB, and finally acetate was converted to methane by acetoclastic methanogens. At the higher COD:sulfate ratio, the step involving the SRB was the rate-limiting step with respect to sulfide inhibition, whereas at lower ratios acetoclastic methanogenesis was the rate-limiting step. This suggests that propionate-oxidising SRB are more sensitive to H_2S inhibition than lactate-oxidising SRB or acetoclastic methanogens.

Visser (1995) found an IC_{50} value for acetoclastic methanogens of 184 mg l^{-1} H_2S at pH 7.2-7.4 and 38 mg l^{-1} H_2S at pH 8.1-8.3 and a 50% inhibition of the growth rate at 248 mg l^{-1} H_2S and 20 mg l^{-1} H_2S at pH 7.0

Table 20.2. IC_{50} (mg l^{-1}) values in the form of total sulfide concentration for methanogenic bacteria in a range of anaerobic sludges over the pH range 6.8-8.5 (undissociated H_2S concentration in parentheses). Values presented are the mean of triplicates (After: O'Flaherty *et al.* 1998a)

Sludge type	Substrate	IC_{50} (mg l^{-1}) at varying pH				
		6.8	7.2	7.6	8.0	8.5
Lab-scale	acetate	327	543	889	970	963
non-sulfate adapted		(176.5)	(178)	(142.5)	(68)	(19)
	H_2/CO_2	483	638	1050	1075	1110
		(261)	(207)	(168)	(75)	(22)
Lab-scale	acetate	350	630	851	977	1000
sulfate adapted		(189)	(205)	(136)	(68)	(20)
	H_2/CO_2	470	712	1056	1064	1089
		(254)	(232)	(169)	(74.5)	(22)
Full-scale	acetate	568	650	880	935	877
sulfate-adapted*		(258)	(211)	(141)	(65.5)	(17.5)
	H_2/CO_2	467	734	1103	1245	1167
		(252)	(239)	(177)	(87)	(23.5)

*Upflow fixed-film reactor treating citric acid production wastewater (O'Flaherty *et al.* 1998b).

and 8.0, respectively. It was observed that MPB in anaerobic sludges were less sensitive to inhibition by sulfide than MPB in pure culture. No significant difference was observed between sulfate- and non-sulfate-adapted sludges. Consequently, it is likely that the higher levels of sulfide are required for inhibition in the anaerobic sludges reflect granular or floc diffusion gradients for sulfide. The levels reported in the literature for inhibition of MPB are quite varied, with IC_{50} values of 50-125 mg l^{-1} H_2S at pH 7-8 for suspended sludge and 250 and 90 mg l^{-1} H_2S at pH values of 6.4-7.2 and 7.8-8.0, respectively, for sludge granules (Kroiss and Plahl-Wabnegg 1983; Koster *et al.* 1986; Oleskiewicz *et al.* 1989; McCartney and Oleskiewicz 1993).

O'Flaherty *et al.* (1998a) found that syntrophic bacteria were less susceptible to sulfide inhibition than MPB and that their toxicity thresholds were comparable to those of the SRB. However, some evidence has been obtained indicating that inhibition of the syntrophic organisms was irreversible, unlike SRB (O'Flaherty *et al.* 1999). This would leave syntrophic organisms at a competitive disadvantage with SRB in anaerobic sludge. The available data for sulfide inhibition of methanogenic and syntrophic bacteria in anaerobic sludges are summarised in Tables 20.2-20.3, respectively. Data obtained from pure culture studies are presented in Table 20.4.

Table 20.3. IC_{50} (mg l^{-1}) values in the form of total sulfide concentration for syntrophic bacteria in a range of anaerobic sludges over the pH range 6.8-8.5 (undissociated H_2S concentration in parentheses). Values presented are the mean of triplicates (After: O'Flaherty *et al.* 1998a)

		IC_{50} (mg l^{-1}) at varying pH				
Sludge type	Substrate	6.8	7.2	7.6	8.0	8.5
Lab-scale non-sulfate adapted	propionate	567 (306)	861 (280)	1206 (193)	1550 (108.5)	1569 (31.5)
	butyrate	527 (286)	847 (275)	973 (156)	920 (64.5)	1030 (21)
	ethanol	389 (210)	787 (256)	1522 (243.5)	1678 (117.5)	1500 (30)
Lab-scale sulfate adapted	propionate	467 (252)	828 (269)	1250 (200)	1610 (113)	1623 (32.5)
	butyrate	574 (310)	880 (286)	915 (146.5)	943 (66)	1065 (21)
	ethanol	451 (243.5)	900 (292.5)	1486 (238)	1707 (119.5)	1721 (36.5)
Full-scale sulfate-adapted*	propionate	-	-	-	-	-
	butyrate	344 (186)	636 (207)	965 (154.5)	1000 (70)	1050 (21)
	ethanol	366 (197.5)	654 (212.5)	1299 (208)	1340 (94)	1380 (28)

*Upflow fixed-film reactor treating citric acid production wastewater (O'Flaherty *et al.* 1998b).

20.4.2 Effect of sulfide toxicity on SRB

Few data are available on the sensitivity of SRB to sulfide toxicity, and the published data are quite contradictory (Table 20.5). Isa *et al.* (1986) concluded that SRB growing in a fixed film reactor were not affected by high levels of sulfide. By contrast, Widdel (1988) reported inhibition of pure cultures of *Desulfotomaculum acetoxidans* at H_2S concentrations of 85 mg l^{-1}. Hilton and Oleskiewicz (1988) found that, during the degradation of lactate, inhibition of SRB was directly related to the total sulfide concentration, whereas inhibition of methanogenesis was related to the H_2S concentration. Reis *et al.* (1992) reported complete inhibition of lactate conversion by SRB at H_2S concentrations of 547 mg l^{-1} at pH 6.2-6.6. Okabe *et al.* (1992) found an 50% inhibition of the growth of *Desulfovibrio desulfuricans* on lactate at a H_2S level of 250 mg l^{-1} at pH 7.0.

Stucki *et al.* (1993) reported process failure in a sulfidogenic fixed-bed reactor treating a mixture of acetate and sulfate at H_2S concentrations of 50 mg l^{-1}. Visser (1995) reported an IC_{50} value for acetate-utilising SRB in

Table 20.4. IC_{50} (mg l^{-1}) values in the form of total sulfide concentration for a range of pure cultures of SRB and MPB over the pH range 6.8-8.5 (undissociated H_2S concentration in parentheses). Values presented are the mean of triplicates (After: O'Flaherty *et al.* 1998a)

		IC_{50} (mg l^{-1}) at varying pH				
Bacteria	Substrate	6.8	7.2	7.6	8.0	8.5
D. magnum	acetate	443	671	660	659	708
		(239)	(218)	(105.6)	(46)	(14)
D.acetoxidans	acetate	487	775	1360	1500	1450
		(263)	(252)	(218)	(105)	(29)
D. vulgaris	H_2/CO_2	554	840	1343	1499	1400
		(299)	(273)	(215)	(105)	(28)
D. sapovorans	butyrate	513	796	1133	1170	1215
		(277)	(259)	(181)	(82)	(24.3)
D. postgatei	acetate	583	926	1248	1119	1290
		(315)	(301)	(200)	(78)	(26)
D. multivorans	ethanol	498	851	1383	1488	1570
		(269)	(277)	(221)	(105)	(31.5)
D. propionicus	propionate	223	355	500	525	570
		(120.5)	(115)	(80)	(37)	(11.5)
M. barkeri	H_2/CO_2	226	363	681	1114	1000
		(122)	(118)	(109)	(78)	(20)
M. hungatei	H_2/CO_2	278	514	937.5	1042	980
		(150)	(167)	(150)	(73)	(40)
M. mazei	acetate	130	175	287.5	621	690
		(70)	(57)	(46)	(43.47)	(13.8)
M. soehngenii	acetate	295	317	744	857	880
		(123)	(103)	(119)	(60)	(17.5)

Table 20.5. IC_{50} (mg l^{-1}) values for H_2S inhibition of sulfate reduction (After: Visser 1995)

Biomass	Substrate	°C	pH	H_2S (mg l^{-1})	Ref.
Desulfovibrio sp.	lactate	37	6.2-6.6	450	1
Suspended sludge	lactate	35	7.2-7.6	80	2
Desulfovibrio desulfuricans	lactate	35	7.0	250	3
Suspended sludge	lactate	35	7.0	>300	4
			8.0	185	
Granular sludge	acetate	30	7.2-7.4	171	5
	acetate	30	8.1-8.3	57	

1. Reis *et al.* (1992); 2. Okabe *et al.* (1992); 3. McCartney and Oleskiewicz (1991); 4. McCartney and Oleskiewicz (1993); 5. Visser (1995).

granular sludge of 171 mg l^{-1} H_2S (615 mg l^{-1} total sulfide) at pH 7.2-7.4 and 57 mg l^{-1} H_2S (1125 mg l^{-1} total sulfide) at pH 8.1-8.3. Visser (1995) found that, for acetate-utilising SRB and MPB, the sensitivity of both groups of bacteria was similar between pH 7.0-7.5. However, at higher pH levels, the SRB were found to be considerably less sensitive to sulfide inhibition than the MPB.

The results of a comprehensive study by O'Flaherty *et al.* (1998a) show that there was a considerable variation among different groups of SRB with respect to sulfide inhibition. The propionate-utilising SRB were considerably more sensitive to inhibition than the other SRB under test (Table 20.6).

Table 20.6. IC_{50} (mg l^{-1}) values in the form of total sulfide concentration for SRB bacteria in a range of anaerobic sludges over the pH range 6.8-8.5 (undissociated H_2S concentration in parentheses). Values presented are the mean of triplicates (After: O'Flaherty *et al.* 1998a)

		IC_{50} (mg l^{-1}) at varying pH				
Sludge type	Substrate	6.8	7.2	7.6	8.0	8.5
Lab-scale	acetate	-	-	-	-	-
non-sulfate adapted	H_2/CO_2	446	723	1397	1670	1698
		(241)	(235)	(223.5)	(117)	(34)
	butyrate	489	769	911	935	960
		(264.5)	(250)	(146)	(65.5)	(19)
	ethanol	544	849	978	1030	1124
		(294)	(276)	(156.5)	(72)	(22.5)
Lab-scale	acetate	-	-	-	-	-
sulfate adapted	H_2/CO_2	474	729	977	1343	1340
		(256)	(237)	(156)	(94)	(27)
	propionate	-	-	-	-	-
	butyrate	467	802	941	965	988
		(252)	(261)	(151)	(67.5)	(20)
	ethanol	500	788	990	1019	1004
		(270)	(256)	(158.5)	(71)	(21)
Full-scale	acetate	374	550	867	990	1011
sulfate-adapted*		(202)	(179)	(139)	(69)	(20)
	H_2/CO_2	505	760	1127	1243	1246
		(273)	(247)	(180)	(87)	(25)
	propionate	328	410	595	572	559
		(177)	(134)	(95.2)	(40)	(11)
	butyrate	593	900	1875	2005	2059
		(320)	(292.5)	(300)	(140)	(41)
	ethanol	561	880	878	1130	1164
		(303)	(286)	(140.5)	(79)	(23)

*Upflow fixed-film reactor treating citric acid production wastewater (O'Flaherty *et al.* 1998b).

This may explain the commonly reported finding that propionate degradation is the rate-limiting step during anaerobic treatment of sulfate-containing organic wastewaters.

20.4.3 Effect of sludge aggregation

A further difficulty in the interpretation of literature data regarding sulfide toxicity is the fact that major differences are observed between data from experiments with suspended sludge systems (e.g. continuously stirred tank reactors) and those carried out on granular sludge systems. For suspended sludge systems, values for the inhibition of methanogenesis correlate well with the H_2S concentration in a pH range of 6.5-8.0 (Kroiss and Plahl-Wabnegg 1983; Koster *et al.* 1986; Karhadkar *et al.* 1987; Oleskiewicz *et al.* 1989; McCartney and Oleskiewicz 1993) with IC_{50} values of 100-130 mg l^{-1} H_2S being reported. In the case of granular sludge, inhibition of methanogenesis appears to be mediated in a more complex manner. Koster *et al.* (1986) reported that, at pH levels above 7.8, the inhibition caused by H_2S was more severe than at lower pH levels (6-7). In the lower pH range, the observed inhibition correlated with the H_2S concentration whereas in the higher range the inhibition appeared to correlate with the total sulfide concentration. The IC_{50} values reported by Koster *et al.* (1986) are 250 mg l^{-1} H_2S at pH 6.2-7.2 and 825 mg l^{-1} total sulfide (90 mg l^{-1} H_2S) at pH 7.8-8.0. The IC_{50} values for granular and suspended sludges are very similar at higher pH values, but granular sludge is less inhibited at lower pH values than suspended sludge. Visser *et al.* (1993) reported a similar trend for thermophilic sludges. Other studies by Maillacheruvu *et al.* (1993) clearly show that sulfide toxicity is mediated at lower concentrations in suspended growth systems than in the attached biofilm of fixed bed systems. Factors, such as substrate transport within biofilms/flocs/granules, the site of sulfate reduction and its proximity to the site of methanogenesis, the diffusion of H_2S and dissolved sulfide and pH gradients, etc., clearly play an important role in the ultimate degree of inhibition.

20.5 EFFECT OF SULFIDE ON PROCESS OPERATION

20.5.1 Modelling the effect of sulfide toxicity in anaerobic digestion

The primary difficulty in applying anaerobic treatment technologies to sulfate-containing wastewaters arises from the production of H_2S by SRB and from its toxicity towards the various trophic groups of bacteria involved in the process (Oude-Elferink *et al.* 1994; Colleran *et al.* 1995). Inhibition decreases the efficiency of reactor performance and can even lead to complete process failure (Koster *et al.* 1986; Hilton and Oleskiewicz 1988). In general, wastewaters with a COD:sulfate ratio higher than 10 do not pose problems for methanogenic treatment (Rinzema and Lettinga 1988). So far, no models have been developed that allow prediction of the conditions that result in process failure of digesters treating wastewaters with a COD:sulfate ratio lower than 10. The outcome of competitive interactions between SRB and other anaerobic bacteria, such as syntrophs and methanogens, in digester sludges will determine the amount of sulfide in the digester, and ultimately the extent of process failure. The nature of this competition is discussed in Chapter 22 and reviewed by several authors (Oude-Elferink *et al.* 1994; Colleran *et al.* 1995; O'Flaherty *et al.* 1998a).

An important consideration in determining the likely outcome of anaerobic treatment of a sulfate-containing wastewater the origin and microbial composition of the seed sludge: i.e. whether it has been acclimatised to sulfate or not. Many studies have shown that SRB should outcompete other anaerobes for substrate based on kinetic properties (Widdel 1988; Oude-Elferink *et al.* 1994; Colleran *et al.* 1995; O'Flaherty *et al.* 1998a). This is supported by studies at both lab-scale (Visser *et al.* 1993; McCartney and Oleskiewicz 1991, 1993) and full-scale (O'Flaherty *et al.* 1998b) using sulfate-adapted sludges with large active populations of SRB. On the other hand O'Flaherty *et al.* (1999) and O'Flaherty and Colleran (1999) showed that the addition of 4 g l^{-1} sulfate to the influent of a mesophilic anaerobic hybrid reactor, seeded with unadapted sludge treating a mixture of volatile fatty acids, resulted in severe process disturbance, with propionate and acetate degradation, in particular, being inhibited. The authors demonstrated that the likely cause of the process disturbance was sulfide inhibition of acetate degradation and subsequent inhibition of propionate degradation by high levels of acetate. In this case, the absence of propionate or acetate-degrading SRB in the reactor sludge was very significant for the overall effect of sulfide toxicity.

The study of O'Flaherty and Colleran (1999) also reported an oscillating pattern of sulfate and sulfide concentrations in the effluent of the anaerobic hybrid reactor, indicative of a reversible inhibition of SRB by sulfide in the reactor sludges (Figure 20.4). This oscillating pattern of sulfide inhibition is similar to that reported by Vavilin *et al.* (1994) where inhibition was also correlated with pH oscillation. This pattern was absent in other studies and probably reflects that only hydrogen-utilising SRB had established to any

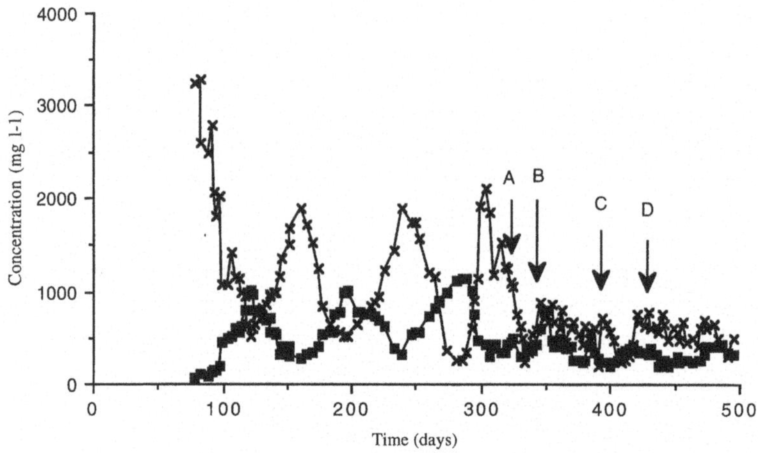

Figure 20.4. Sulfate (-×-) and total sulfide (-■-) concentrations in a sulfate-fed reactor effluent over a 495 day trial period. Arrows: (A) introduction of 70 l d^{-1} nitrogen scrubber; (B) introduction of 150 l d^{-1} nitrogen scrubber; (C) re-inoculation with seed sludge; (D) inoculation with full-scale sludge adapted to sulfate rich wastewater. (After: O'Flaherty and Colleran 1999).

great extent in the sludge studied by O'Flaherty and Colleran (1999), in contrast to the situation observed in many sulfidogenic sludges (Omil *et al.* 1997; O'Flaherty *et al.* 1998b), where SRB populations less sensitive to sulfide inhibition may be present. These SRB compete for sulfate, thus masking the oscillating pattern of sulfide inhibition. This is supported by the fact that the oscillating pattern of SRB inhibition and the process disturbance was eliminated in the anaerobic hybrid reactor studied by O'Flaherty and Colleran (1999) by a nitrogen gas scrubber to reduce sulfide concentrations and bioaugmentation with a sulfate adapted sludge, respectively (Figure 20.4).

20.5.2 Alleviating problems of sulfide toxicity during anaerobic digestion

Although sulfide toxicity can result in difficulties for anaerobic treatment, the use of appropriate operational strategies has allowed many examples of successful anaerobic treatment of sulfate-containing wastewaters, including full-scale plants (see Chapters 7-9). Successful strategies include dilution of the wastewater, decreasing the unionised sulfide concentration (e.g. by scrubbing or precipitation) and employing a multistep anaerobic digestion process. An outline of the different process configurations that can be used to alleviate the problems of sulfide toxicity has been presented by Visser (1995). These include:

Anaerobic digestion in two stages: i.e. a pre-acidification step with sulfate reduction followed by a methanogenic stage. The sulfide can be removed in the first stage or between the two stages.

The precipitation of sulfide in the anaerobic digester by the use of metals, for example, iron. The drawback of the use of iron is the accumulation of FeS precipitate in the reactor, resulting in a reduced VSS:TSS ratio of the sludge and increased total sludge production.

Removal of sulfide from the effluent of the reactor combined with recirculation of the effluent. The removal of the sulfide can be by chemical precipitation or by chemical or biological oxidation.

Stripping of the sulfide from the anaerobic reactor with the biogas, using gas scrubbing and recirculation. This process was described by Särner (1989). A gas-washing system was established using a solution of ferric ions. The Fe^{3+} ions react with the H_2S in the biogas to produce elemental sulfur. A chelating agent was added to the liquid to prevent FeS, $Fe(OH)_2$ or $Fe(OH)_3$ precipitation. After separation of the sulfur, the Fe^{2+} was oxidised to Fe^{3+} using oxygen in air. In this way, the iron could be re-used for gas-washing.

Biological oxidation of sulfide to elemental sulfur. This process is based on the biological conversion of sulfide by the colourless sulfur bacteria (Buisman *et al.* 1993). The process also results in the recovery of elemental sulfur and has been applied at full scale (Buisman *et al.* 1993).

20.6 REFERENCES

Britz, T.J., Van der Merwe, M. and Riedel, K.H.-J. (1992) Influence of phenol additions on the efficiency on an anaerobic hybrid digester treating landfill leachate. *Biotechnol. Lett.* **14**(4), 323-328.

Buisman, C.N.J., de Boer, J., Boonstra, J., Paaschens, C.W.M. and Sopjes, A. (1993) New biotechnological method of H_2S removal from biogas. *Paper presented at the TAPPI Environmental Conference*, Boston, USA.

Callender, I.J. and Barford, J.P. (1983) Precipitation, chelation and the availability of metals as nutrients in anaerobic digestion I Methodology, II Applications. *Biotechnol. Bioeng.* **25**, 1947-1972.

Colleran, E., Concannon, F., Golden, T., Geoghan, F., Crumlish, B., Killilea, J.E., Henry, M. and Coates, J. (1992) Use of methanogenic activity tests to characterise anaerobic sludges, screen for anaerobic biodegradability and determine toxicity thresholds against individual anaerobic trophic groups and species. *Wat. Sci. Technol.* **25**, 31-40.

Colleran, E. and Pistilli, A. (1994) Activity test system for determining the toxicity of xenobiotic chemicals to the methanogenic process. *Ann. Microbiol. Enzymol.* **44,** 1-20.

Colleran, E., Finnegan, S. and Lens, P. (1995) Anaerobic treatment of sulfate-containing waste streams. *Antonie van Leeuwenhoek* **67,** 29-46.

Daniels, L., Belay, N. and Rajagopal, B.S. (1986) Assimilatory reduction of sulfate and sulfite by methanogenic bacteria. *Appl. Environ. Microbiol.* **51,** 703-709.

De Baere, L.A., Devocht, M., van Assche, F. and Verstraete, W. (1984) Influence of high NaCl and NH_4Cl salt levels on methanogenic associations. *Wat. Res.* **18,** 543-548.

Eis, B.J., Ferguson, J.F. and Benjamin, M.M. (1983) The fate and the effect of bisulfate in anaerobic treatment. *J. Wat.Poll.Cont.Fed.* **55,** 1355-1365.

Ferguson, J.F., Eis, B.J. and Benjamin, M.M. (1982) Neutralisation in anaerobic treatment of an acidic waste. In *Preprints presented at the IAWPRC specialised seminar "Anaerobic Treatment of Wastewater in Fixed Film Reactors"*, Copenhagen, Denmark, 16-18 June, 339-355.

Field, J.A. and Lettinga, G. (1989) The effect of oxidative coloration on the methanogenic toxicity and anaerobic biodegradability of phenols. *Biological Wastes* **29,** 161-179.

Garrels, R.M. and Christ, C.L. (1965) *Solutions, minerals and equilibria,* Harper and Row, New York.

Govind, R., Flaherty, P.A. and Dobbs, R.A. (1991) Fate and effects of semivolatile organic pollutants during anaerobic digestion of sludge. *Wat. Res.* **25**(5), 547-556.

Hilton, B.L. and Oleskiewicz, J.A. (1988) Sulfide-induced inhibition of anaerobic digestion. *J. Environ. Engin.* **114,** 1377-1391.

Isa, Z., Grusenmeyer, S. and Verstraete, W. (1986) Sulfate reduction relative to methane production in high-rate anaerobic digestion: Technical aspects. *Appl. Environ. Microbiol.* **51,** 572-579.

Karhadkar, P.P., Audic, J.-M., Faup, G.M. and Khanna, P. (1987) Sulfide and sulfate inhibition of methanogenesis. *Wat. Res.* **21,** 1061-1066.

Koster, I.W., Rinzema, A., de Vegt, A.L. and Lettinga, G. (1986) Sulfide inhibition of the methanogenic activity of granular sludge at different pH levels. *Wat. Res.* **12,** 1561-1567.

Kroiss, H. and Plahl-Wabnegg, F. (1983) Sulfide toxicity with anaerobic wastewater treatment. In *Proceedings of the European Symposium on Anaerobic Wastewater Treatment (AWWT),* TNO Corporate Communications Department, The Hague, The Netherlands, 72-78.

Kugelman, I.J. and McCarty, P.L. (1964) Cation toxicity and stimulation in anaerobic wastewater treatment. In *Proceedings of the 19th Industrial Waste Conference,* Purdue University, West Lafayette, Indiana, U.S.A., 667-686.

Kugelman, I.J. and Chin, K.K. (1971) Toxicity, synergism and antagonism in anaerobic wastewater treatment processes. In *Advanced Chemistry Series 105,* American Chemical Society, Washington, U.S.A., 667-686.

Langerak, van E.P.A., Gonzalez-Gil, G., Aeslt, van A., van Lier, J.B., Hamelers, H.V.M. and Lettinga, G. (1998) Effect of high-calcium concentrations on the development of methanogenic sludge in UASB reactors. *Wat. Res.* **32,** 1255-1263.

Lawrence, A.W. and McCarty, P.L. (1965) The role of sulfide in preventing heavy metals toxicity in anaerobic treatment. *J. Wat. Poll. Cont. Fed.* **37,** 392-409.

Lettinga, G. and Vinken, J.N. (1981) Feasibility of the upflow anaerobic sludge blanket (UASB) process for the treatment of low strength wastes. In *Proc. 35th Industrial Waste Conference.,* Purdue University, West Lafayette, Ind., U.S.A., 625-634.

Lettinga, G. (1995) Anaerobic wastewater treatment systems. *Antonie van Leeuwenhoek* **67,** 3-28.

Maaskant, W. and Hobsma, S.W. (1981) The influence of sulfite on anaerobic treatment. *H_2O* **14**(25), 596-598.

Maillacheruvu, K.Y., Parkin, G.F., Peng, C.Y., Kuo, W.-C., Oonge, Z.I. and Lebduschka, V. (1993) Sulfide toxicity in anaerobic systems fed sulfate and various organics. *Wat. Env. Res.* **65,** 100-109.

McCartney, D.M. and Oleskiewicz, J.A. (1991) Sulfide inhibition of anaerobic degradation of lactate and acetate. *Wat. Res.* **25,** 203-209.

McCartney, D.M. and Oleskiewicz, J.A. (1993) Competition between methanogens and sulfate reducers: effect of COD:sulfate ratio and acclimatization. *Wat. Env. Res.* **65,** 655-664.

McCarty, P.L. and McKinney, R.E. (1961) Salt toxicity in anaerobic digestion. *J. Wat. Poll.Cont. Fed.* **33,** 399-415.

Mendez, R., Lema, J.M. and Soto, M. (1995) Treatment of seafood-processing wastewaters in mesophilic and thermophilic anaerobic filters. *Wat. Environ. Res.* **67,** 33-45.

O'Flaherty, V., Mahony, T., O'Kennedy, R. and Colleran, E. (1998a) Effect of pH on growth kinetics and sulfide toxicity thresholds of a range of methanogenic, syntrophic and sulfate-reducing bacteria. *Process Biochem.* **33**(5), 555-569.

O'Flaherty, V., Lens, P., Leahy, B. and Colleran, E. (1998b) Long term competition between sulfate-reducing and methane-producing bacteria during the full-scale anaerobic treatment of citric acid production wastewater. *Wat. Res.* **32**(3), 815-825.

O'Flaherty, V. and Colleran, E. (1999) Effect of sulfate addition on volatile fatty acid and ethanol degradation in an anaerobic hybrid reactor. I. process disturbance and remediation. *Biores. Technol.* **68,** 101-107.

O'Flaherty, V., Colohan, S., Mulkerrins, D. and Colleran, E. (1999) Effect of sulfate addition on volatile fatty acid and ethanol degradation in an anaerobic hybrid reactor. II. microbial interactions and toxic effects. *Biores. Technol.* **68,** 109-120.

Okabe, S., Nielson, P.H. and Characklis, W.G. (1992) Factors affecting microbial sulfate reduction by *Desulfovibrio desulfuricans* in continuous culture: Limiting nutrients and sulfide concentrations. *Biotechnol. Bioeng.* **40,** 725-734.

Oleskiewicz, J.A., Marstaller, T. and McCartney, D.M. (1989) Effects of pH on sulfide toxicity to anaerobic processes. *Environ. Technol. Lett.* **10,** 815-822.

Omil, F., Oude-Elferink, S.J.W.H., Lens, P., Hulshoff-Pol, L.W. and Lettinga, G. (1997) Effect of inoculation with *Desulforhabdus amnigenus* and pH or O_2 shocks on the competition between sulfate reducing and methanogenic bacteria in an acetate fed UASB reactor. *Biores. Technol.* **60,** 113-122.

Oremland, R.S. and Capone, D.G. (1988) Use of "specific" inhibitors in biogeochemistry and microbial ecology. In *Advances in Microbial Ecology, Vol*

8, Plenum Press, New York, 295-383.

Oude-Elferink, S.J.W.H., Visser, A., Hulshoff Pol, L.W. and Stams, A.J.M. (1994) Sulfate reduction in methanogenic bioreactors. *FEMS Microbiol. Rev.* **15,** 119-136.

Owen, W.F., Stuckey, D.C., Healy, J.B., Young, L.Y. and McCarty, P.L. (1979) Bioassay for monitoring biochemical methane potential and anaerobic toxicity. *Wat. Res.* **13,** 485-493.

Reis, M.A.M., Almeida, J.S., Lemos, P.C. and Carrondo, M.J.T. (1992) Effect of hydrogen sulfide on the growth of sulfate reducing bacteria. *Biotechnol. Bioeng.* **40,** 593-600.

Rinzema, A. (1988) Anaerobic treatment of wastewater with high concentrations of lipid or sulfate. Ph.D. Thesis, Wageningen Agricultural University, The Netherlands.

Rinzema, A., Paardekooper, A., de Vegt, A. and Lettinga, G. (1986) Anaerobic treatment of edible oil refinery wastewater in granular UASB reactors. In *Proceedings of the IAWPRC conference "Anaerobic treatment- a grown-up technology",* Amsterdam, The Netherlands, 205-217.

Rinzema, A. and Lettinga, G. (1988) The effect of sulfide on the anaerobic degradation of propionate. *Environ. Technol. Lett.* **9,** 83-88.

Särner, E. (1986) Influence and control of H_2S on full scale plants and pilot plant experiments. In *Proceedings of the IAWPRC conference "Anaerobic Wastewater Treatment - a grown-up technology",* Amsterdam, The Netherlands, 189-204.

Särner, E. (1989) Removal of sulfate and sulfite in an anaerobic trickling (antric) filter. In *Proceedings of the Conference on Recent Developments: Technological Advances in Biofilm Reactors,* Nice, France, 4-6 April, 405-414.

Soto, M., Mendez, R. and Lema, J.M. (1993) Methanogenic and non-methanogenic activity tests - theoretical basis and experimental set-up. *Wat. Sci. Technol.* **27,** 1361-1376.

Speece, R.E. (1983) Anaerobic biotechnology for industrial wastewater treatment. *Env. Sci. Technol.* **17,** 416A-427A.

Speece, R.E. and Parkin, G.F. (1983) The response of methane bacteria to toxicity. In *Proceedings of the Third International Symposium on Anaerobic Digestion,* Evans and Faulkner Inc., Watertown, Massachusetts, U.S.A., 23-35.

Stucki, G., Hanselman, K.W. and Hürzeler, A. (1993) Biological sulfuric acid transformation: reactor design and process optimisation. *Biotechnol. Bioeng.* **41,** 303-315.

Tursman, J.F. and Cork, D.J. (1989) Influence of sulfate and sulfite-reducing bacteria on anaerobic digestion technology. In *Biological Waste Treatment,* Alan R Liss Inc., New York, 273-281.

Van den Berg, L., Patel, G.B., Clarck, D.S. and Lentz, C.P. (1976) Factors affecting rate of methane formation from acetic acid by enriched methanogenic cultures. *Can. J. Microbiol.* **22,** 1312-1319.

Vavilin, V.A., Vasilev, V.B., Rytov, S.V. and Ponomerev, A.V. (1994) Self-oscillating co-existence of methanogens and sulphate reducers under hydrogen sulphide inhibition and the pH regulating effect. *Biores. Technol.* **49,** 105-119.

Visser, A., Nozhevnikova, A.N. and Lettinga, G. (1993) Sulfide inhibition of methanogenic activity at various pH levels at 55°C. *J. Chem. Technol. Biotechnol.* **57,** 9-14.

Visser, A. (1995) The anaerobic treatment of sulfate-containing wastewater. PhD Thesis, Wageningen Agricultural University, The Netherlands.

Wang, Y.-T., Gabbard, D.H. and Pai, P.-C. (1991) Inhibition of acetate methanogenesis by phenols. *J. Environ. Eng.* **117**(4), 487-500.

Weast, R.C. (1981) *Handbook of Chemistry and Physics, 62th ed.,* CRC Press Inc., Boca Raton, U.S.A.

Widdel, F. (1988) Microbiology and ecology of sulfate- and sulfur-reducing bacteria. In *Biology of Anaerobic Microrganisms,* John Wiley and Sons, New York, 469.

Yang, C.H.J., Parkin, G.F. and Speece R.E. (1979) Recovery of anaerobic digestion after exposure to toxicants. Final report to the U.S. Department of Energy, Contract No. EC-77-S-02-4391.

Young, J.C. and Tabac, H.H. (1993) Multilevel protocol for assessing the fate and effect of toxic chemicals in anaerobic treatment processes. *Wat. Environ. Res.* **65**(1), 34-45.

21

Corrosion and sulfur bacteria

Brenda J. Little, Richard I. Ray and Robert K. Pope

21.1 INTRODUCTION

Most microbiologically influenced corrosion (MIC) takes place in the presence of microbial consortia in which many different physiological types of bacteria interact in complex ways within the structure of biofilms (Pope *et al.* 1984; Little *et al.* 1991). Microbiologically mediated oxidation and reduction reactions of sulfur and sulfur compounds within biofilms are important mechanisms contributing to MIC. Sulfur and sulfur compounds, including sulfides (S^{2-}), bisulfides (HS^-), hydrogen sulfide (H_2S), thiosulfates ($S_2O_3^{2-}$), and sulfuric acid (H_2SO_4), may be produced and trapped in biofilms causing corrosion of materials. H_2S and H_2SO_4 may be transported in a gaseous or aqueous phase before reacting with materials.

Sulfur and sulfur compounds can produce pitting, crevice corrosion, dealloying, stress corrosion cracking and stress-oriented hydrogen induced cracking of susceptible metals and alloys. Despite the recognition that sulfides and sulfuric acid can be extremely corrosive there are no correlations between numbers and types of sulfur-related organisms and the probability/rate of corrosion. In the following chapter, reactions of sulfur and sulfur compounds resulting in corrosion will be discussed in the context of other environmental processes that may be equally important to corrosion. For example, the most aggressive corrosion by obligate anaerobic SRB does not occur under anaerobic conditions. In strictly anaerobic environments, SRB produce a uniform sulfide layer with accompanying short-term accelerated corrosion. In the absence of mechanical removal of the sulfide layer by turbulence, erosion or the introduction of oxygen, corrosion is not particularly aggressive.

Determination of specific mechanisms for corrosion due to microbiologically mediated oxidation and reduction of sulfur and sulfur compounds is complicated by (1) the variety of potential metabolic/energy sources and by-products, (2) the coexistence of reduced and oxidized sulfur species, (3) competing reactions with inorganic and organic compounds, and (4) the versatility and adaptability of microorganisms in biofilms. The microbial ecology of sulfur-rich environments is poorly understood because of the association of aerobes and anaerobes and the mutualism or succession of heterotrophs to autotrophs. The physical scale over which the sulfur cycle influences corrosion varies with environment. The complete sulfur cycle of oxidation and reduction reactions can take place in macroenvironments, including sewers and polluted harbours or within the microenvironment of biofilms (Figure 21.1).

21.2 REDUCTION REACTIONS

21.2.1 Sulfide generation and corrosion

21.2.1.1 Elemental sulfur and thiosulfate reduction

Crolet and Magot (1996) described a group of bacteria isolated from an oilfield production facility capable of reducing thiosulfate ($S_2O_3^{2-}$), not sulfate, to sulfide. The non-SRB fermentative strains produced H_2S and organic acids from $S_2O_3^{2-}$ and peptides. Corrosion penetration rates of carbon steel in the presence of these organisms exceeded 1 cm per year. Sulfides produced by microbial reduction of sulfur and thiosulfate react with metals to produce metal sulfides and catalyse the penetration of hydrogen into steel, a process known as H^2S-induced cracking or sulfide stress cracking.

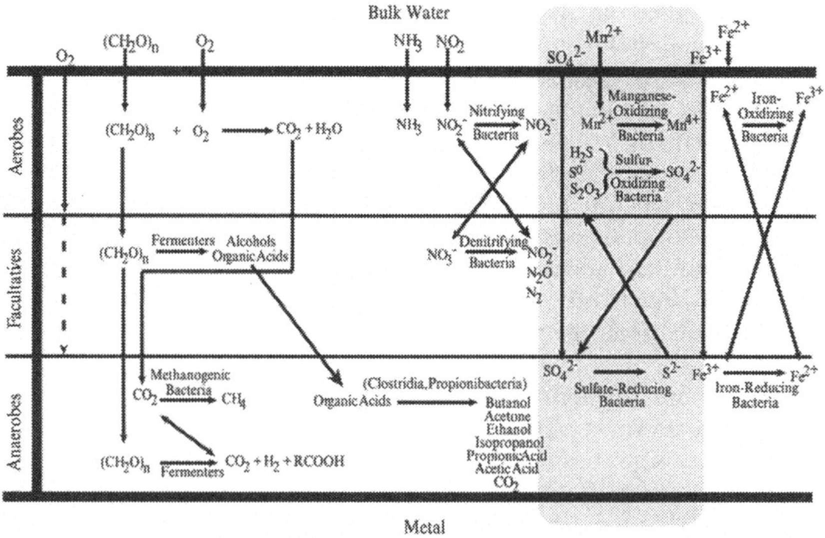

Figure 21.1. Possible reactions within a biofilm. Sulfur cycle is highlighted.

21.2.1.2 Sulfate reduction

SRB are the organisms most closely identified with sulfide production and MIC. They are ubiquitous and easy to culture. Several diagnostic kits are available for detection and quantification of SRB based on hydrogenase, adenosine diphosphate reductase and sulfide reactions with metals (Little and Wagner 1992).

21.2.1.3 Corrosion mechanisms

21.2.1.3.1 Cathodic depolarization

The early work of Von Wolzogen Kuhr and Van der Vlugt (1934) suggested the following electrochemical reactions for the involvement of SRB in MIC of iron:

$$4Fe \Rightarrow 4Fe^{2+} + 8e^- \quad \text{(anodic reaction)} \quad (21.1)$$

$$8H_2O \Rightarrow 8H^+ + 8OH^- \quad \text{(water dissociation)} \quad (21.2)$$

$$8H^+ + 8e^- \Rightarrow 8H \text{ (adsorbed)} \quad \text{(cathodic reaction)} \quad (21.3)$$

$$SO_4^{2-} + 8H \text{ (adsorbed)} \Rightarrow S^{2-} + 4H_2O \quad \text{(consumption by SRB)} \quad (21.4)$$

$$Fe^{2+} + S^{2-} \Rightarrow FeS \quad \text{(corrosion product)} \quad (21.5)$$

$$3Fe^{2+} + 6OH^- \Rightarrow 3Fe(OH)_2 \quad \text{(corrosion product)} \quad (21.6)$$

$$4Fe + SO_4^{2-} + 4H_2O \Rightarrow 3Fe(OH)_2 + FeS + 2\,OH^- \quad (21.7)$$

They described the overall process as cathodic depolarization, based on the theory that SRB remove atomic hydrogen accumulated at the cathode. When the rate of electron generation at the anode is not matched by a corresponding consumption of electrons at the cathode, the cathode is said to be "polarized." Removing hydrogen from the cathode allows the cathodic reaction to proceed, a "depolarization". Removal of cathodic hydrogen in step (3) forces iron to dissolve at the anode in step (1). Although the overall reaction may be described correctly in Equation 21.7, it is doubtful that individual reaction steps (1)–(6) proceed in the manner proposed by Von Wolzogen Kuhr and Van der Vlugt (1934). It is unlikely that a layer of atomic hydrogen exists on the metal surface as postulated in step (3).

Many strains of SRB metabolize gaseous H_2 because of a reversible hydrogenase. Boivin *et al.* (1990) proposed that under anaerobic conditions the reduction of protons at the cathode produced hydrogen gas that could be used as a source of energy for hydrogenase positive ($Hase^+$) organisms. Removal of H_2 would depolarize the cathode and increase metal loss at the anode. The relationship between numbers of $Hase^+$ bacteria and rate or extent of corrosion has never been demonstrated.

21.2.1.3.2 Production of metal sulfides

$Hase^+$ and hydrogenase-negative strains of SRB can stimulate corrosion by producing sulfide minerals. Some metal oxides can be destabilized and act as sources of metal ions to react with sulfide. Experimental work with cultures of SRB has shown that metal ions sorbed to bacterial cells tend to be more chemically reactive toward sulfide than those in solution (Mohagheghi *et al.* 1985). The following is a summary of mineralogical products formed during SRB-influenced corrosion reactions (McNeil and Odom 1994):

- *Ag* - acanthite (Ag_2S).
- *Ag-Cu alloys* - acanthite, argentite (the high temperature polymorph of Ag_2S) or jalpaite (Ag_3CuS_2).
- *Cu* - complex suites of sulfide minerals: the most common product is chalcocite (Cu_2S). Final product in many cases is covellite (CuS_{1+x}).
- *Cu-Ni alloys* - Sulfide corrosion products similar to those of Cu but with significant djurleite ($Cu_{31}S_{16}$). No Ni sulfides observed.
- *Cu-Sn alloys* - Corrosion products similar to those in Cu.
- *Fe (carbon steel)* - Final product is pyrite (FeS_2) with numerous intermediates.
- *Fe (stainless steel alloys)* - Rates are slower than for pure Fe or

carbon steel. No Ni minerals have been detected. Stainless steels with 6% or more Mo appear to be very resistant.

- *Ni* - millerite (NiS).
- *Pb* - galena (PbS).

McNeil and Odom (1994) developed a thermodynamic model to predict metal susceptibility to corrosion by SRB. If the reaction to produce the sulfide from the oxide has a negative Gibbs free energy, the reaction will take place. If the value is positive, the metal is immune to derivation by sulfides and will not be vulnerable to corrosion by SRB. The model is limited to thermodynamic predictions and does not consider metal toxicity to the organisms, tenacity of the resulting sulfide or others factors that influence corrosion rate.

21.2.2 Metal sulfide interactions

In the following sections, SRB sulfide mineral production is reviewed for iron, copper, copper alloys, silver, zinc and lead. The metal interface under the biofilm and corrosion layers is referred to as base metal to differentiate it from layers of minerals and metal ions derived from corrosion reactions. Mineralogical data, thermodynamic stability diagrams (Pourbaix 1966; Wagman *et al.* 1982) and the simplexity principle for precipitation reactions (McNeil *et al.* 1991) will be used to interpret corrosion product mineralogy in fresh and saline water and to demonstrate the action of SRB.

Iron. The corrosion rate of iron in the presence of H_2S is accelerated by the formation of iron sulfide minerals (Wikjord *et al.* 1980) that stimulate the cathodic reaction. Once electrical contact is established, mild steel behaves as an anode and electron transfer occurs through the iron sulfide. In the absence of oxygen, the metabolic activity of SRB causes accumulation of sulfide near metal surfaces. This is particularly evident when metal surfaces are covered with biofilms as indicated by concentration profiles of sulfide and oxygen in a biofilm accumulated on the surface of a mild steel corrosion coupon (Figure 21.2). The concentration of sulfide is highest near the metal surface where iron sulfide forms quickly and covers the steel surface if both ferrous and sulfide ions are available. At low ferrous ion concentrations (0–10 mg L^{-1}), adherent and temporarily protective films of iron sulfides form on the steel surface with a consequent reduction in anodic and cathodic currents. Cathodic and anodic current densities up to 3.0 mA cm^{-2} and 3.5 mA cm^{-2}, respectively, were maintained only in high concentrations of ferrous ion (> 60 mg L^{-1} (Lee and Characklis 1993)).

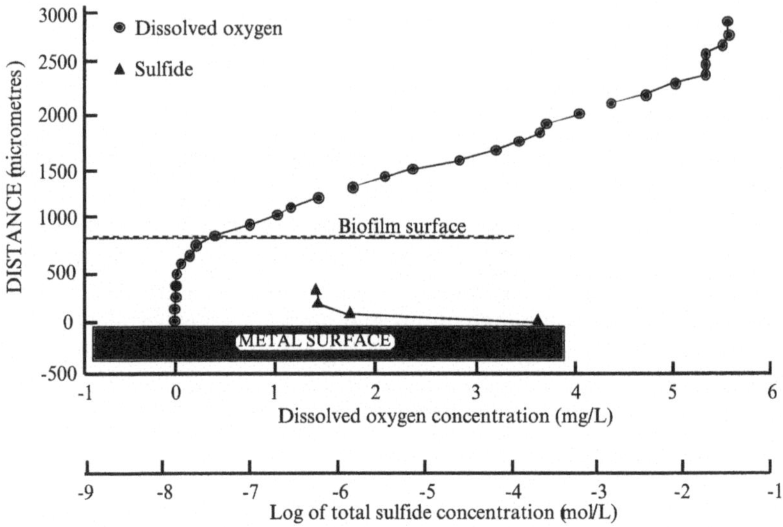

Figure 21.2. Concentration profiles of sulfide and oxygen on a biofilm on mild steel (Lee *et al.* 1993 a,b).

Figure 21.3 is a stability diagram for an iron–water-reduced sulfur system with lines for 10^{-6} M ferrous iron and 10^{-2} M sulfide. For clarity, pyrite (cubic FeS_2) and mackinawite (tetragonal FeS_{1-x}) are the only sulfides indicated. Parallelograms superimposed on the diagram are bounded by the highest and lowest pH values commonly found in natural fresh and saline surface waters. The upper portion of the hatched area applies to waters less than 10 m from the surface. The lower portion (reverse hatched) represents waters at depths greater than 10 m. Conditions in the upper hatched parallelogram represent those readily achieved in stagnant waters. The lower portion indicates conditions not found in near-surface environments. Mackinawite is unstable above 150 °C and cannot be produced by conventional techniques.

Thermodynamic analyses indicate in natural surface waters that only pyrite should be stable. The region of stability of mackinawite is wholly outside the region defined by surface water conditions, excluding waters influenced by peat bogs, coal mines, volcanic activity and some industrial effluents. Furthermore, pyrite forms relatively easily in non-biological corrosion, so the preferential formation of less stable sulfides is difficult to attribute to slow pyrite formation kinetics. The region of stability for pyrite is accessible under severe conditions.

Other potential iron–sulfur compounds not shown on the stability

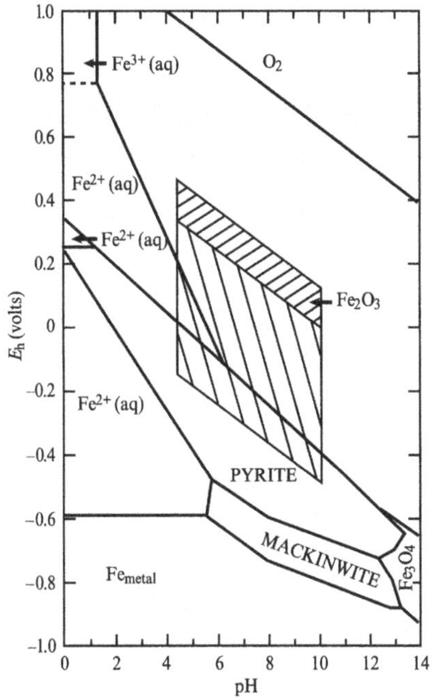

Figure 21.3. Iron stability diagram: water without chloride ions; total sulfide 10^{-2}M.

diagram (Figure 21.3) include the thiospinel greigite (Fe_3S_4), the hexagonal compound smythite (Fe_9S_{11}), and cubic FeS. By applying H_2S pressures in the range of one atmosphere, Berner (1969) produced "tetragonal FeS", which has the same symmetry as mackinawite but contains somewhat less sulfur and has slight but systematic differences in lattice parameter. Presumably, further increases in H_2S pressure could produce material equivalent to natural mackinawite.

During corrosion of iron and steel in the presence of SRB, a thin (approximately 1 μm), adherent layer of "tarnish" is first formed. This was originally termed "kansite", but has since been identified as mackinawite. As this layer thickens, it becomes less adherent. If ferrous ion concentration in the electrolyte is low, mackinawite alters to greigite. This alteration is not observed in non-biological systems. If ferrous ion concentration is high, mackinawite is accompanied by green rust 2, a complex ferrosoferric oxyhydroxide.

In summary, mackinawite is produced from iron and iron oxides by consortia of microorganisms that include SRB. The presence of mackinawite in corrosion products formed in shallow water environments

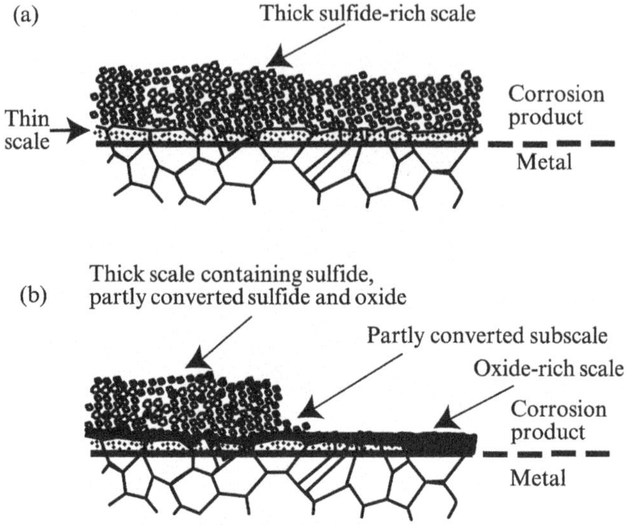

Figure 21.4. Schematic representation of (a) thick sulfide-rich scale on copper alloy and (b) disruption of sulfide film (Syrett 1981).

with the exclusions previously delineated is proof that the corrosion was SRB-induced. Recent work indicates that on continued exposure to SRB mackinawite alters to greigite, smythite and finally to pyrrhotite (FeS) (McNeil and Little 1990). SRB in thin biofilms on pottery surfaces (Duncan and Ganiaris 1987; Heimann 1989) can produce pyrite in iron-rich waters. Pyrite is not a typical iron corrosion product, but SRB can produce pyrite from mackinawite in contact with elemental sulfur (Berner 1969). Abiotic aqueous synthesis of these minerals, with the possible exception of pyrite, requires H_2S pressures higher than those found in shallow waters.

Copper. Cuprite (Cu_2O), the first product of copper corrosion, forms as a direct reaction product of copper with dissolved O_2 or with water molecules (North and Pryor 1970). Cuprite has a high electrical conductivity and permits transport of copper ions through the oxide layer so they can dissolve in water and reprecipitate. If the water chemistry approximates that of seawater, copper ions reprecipitate as botallackite ($Cu_2(OH)_3Cl$); which can alter in minutes or hours to either paratacamite or atacamite (other crystal structures, depending on local water chemistry (Pollard *et al.* 1989)). A porous layer of cuprous sulfide with the general stoichiometry $Cu_{2-X}S$, $O<X<1$ forms in the presence of sulfide ions (Syrett 1980). Copper ions migrate through the sulfide layer, react with more sulfide, and produce a thick, black scale (Figure 21.4).

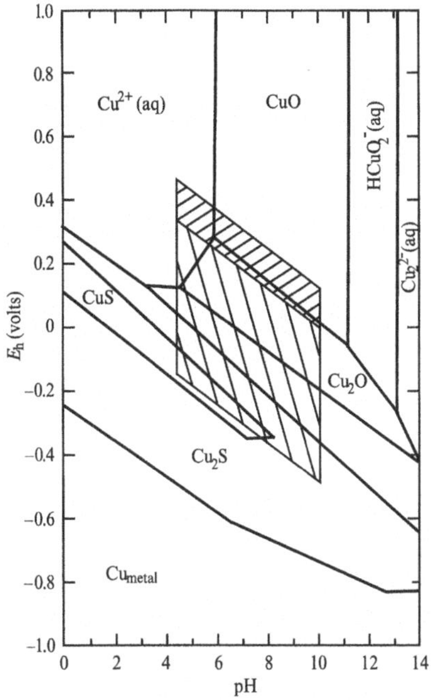

Figure 21.5. Copper stability diagram: water without chloride ions; total sulfide 10^{-2}M.

Figure 21.5 is a stability diagram for copper and its minerals drawn for 10^{-6} M total dissolved copper and 10^{-2} M total sulfide. Parallelograms superimposed on the diagram are similar to those described for Figure 21.3 and are appropriate for the analysis of corrosion mineralogy under non-hydrothermal conditions. McNeil *et al.* (1991) used Figure 21.5 to interpret results from laboratory experiments. They exposed mixed cultures known to contain SRB to copper and copper/nickel alloys in a variety of natural and synthetic waters containing sulfates for 150 days. The pH values of the waters, measured after two weeks, were between 5.5 and 6.8. All copper-containing metals exposed to SRB in isolated cultures and in the natural augmented waters were covered with black sulfur-rich deposits (Figure 21.6A). The thickness and tenacity of the surface deposits varied among the metals and cultures. Corrosion products on commercially pure copper were consistently non-adherent. Corrosion products on copper alloys were more adherent and in some cases difficult to scrape from the surface. In all cases, bacteria were closely associated with sulfur-rich deposits (Figure 21.6B). Most scanning electron microscopy (SEM) micrographs of SRB on copper surfaces indicate a monolayer of cells overlaying a sulfide layer.

(A)

(B)

Figure 21.6. (A) Black sulfide film on copper foil after 4-month exposure to SRB; (B) Micrograph of bacteria with corrosion products on copper-containing foils.

Transmission electron microscopy (TEM) has been used to demonstrate that bacteria are intimately associated with sulfide minerals. TEM further demonstrated that on copper-containing surfaces, bacterial cells were attached to base metal and were located between layers of corrosion products (Blunn 1986).

Biomineralogy of copper sulfides has been studied for over a century (Daubree 1862; de Gouvernain 1875; Baas-Becking and Moore 1961; Mor and Beccaria 1975; Syrett 1977, 1980, 1981; McNeil and Little 1992). The complexity of the resulting observations reflects the complexity of the copper-sulfur system, especially in the presence of alloying elements or iron in the environment (Ribbe 1976; Kostov and Minceva-Stefanova 1981).

Both high- and low-temperature polytypes of chalcocite (Cu_2S), digenite (Cu_9S_5), djurleite ($Cu_{1.93}S$–$Cu_{1.97}S$), anilite (Cu_7S_4), spionkopite ($Cu_{39}S_{28}$), geerite (Cu_8S_5), and covellite (CuS, generally blue-remaining) have been reported. In long-term corrosion where waters contain significant iron, chalcopyrite ($CuFeS_2$) is a common product (McNeil and Mohr 1993). Chalcopyrite films can be formed abiotically in high sulfur concentrations (Cuthbert 1962), but chalcopyrite and most other copper sulfides are not generally found as products of abiotic corrosion.

Detailed kinetics of individual reactions are not fully understood, and the consequences for corrosion depend on many factors, including mineral morphology and variations of redox potential and pH with time (McNeil and Mohr 1993). Discussions of alteration kinetics are contained in several papers (Baas-Becking and Moore 1961; Roseboom 1966; Craig and Scott 1976; Putnis 1977). Microbial consortia that include SRB produce anoxic, sulfide-rich environments in which the conversion of copper to copper sulfides is thermodynamically favoured at a concentration of 10^{-2} M total sulfur. Initial sulfur-poor compounds are converted to sulfur-rich compounds. In experiments lasting 150 days with excess of copper over available sulfur, chalcocite with little or no covellite is formed (McNeil *et al.* 1991). Covellite is produced if excess sulfide is available, either deliberately provided (Baas-Becking and Moore 1961) or naturally available (Daubree 1862, Mor and Beccaria 1975).

The presence of dissolved iron leads to other complications. Not only has chalcopyrite been observed, but also digenite (Baas-Becking and Moore 1961; Mor and Beccaria 1975; North and MacLeod 1986; McNeil *et al.* 1991), djurleite (Macdonald *et al.* 1979; McNeil *et al.* 1991), and the hexagonal high-temperature polytype of chalcocite (McNeil *et al.* 1991). Impurities tend to stabilize high-entropy, high-temperature polytypes (Goldschmidt 1953). Digenite stability is promoted by iron (Craig and Scott 1976). Nickel stabilizes djurleite on copper-nickel and the stabilization of djurleite means that the sulfide layer is more tenacious. McNeil *et al.* (1991) observed that corrosion layers showing strong digenite lines were never observed on pure copper, but frequently on copper–nickel alloys. It has been argued that if the copper sulfide layer were djurelite, the sulfide layer would be protective (North and Macleod 1986). Even if such a sulfide film were technically passivating, the mechanical stability is so poor that sulfide films on copper surfaces are useless for corrosion protection. In the presence of turbulence, loosely adherent sulfide films are removed, exposing fresh copper surfaces to react with sulfide ions. For these reasons turbulence-induced corrosion and sulfide attack of copper alloys cannot easily be decoupled. In the presence of oxygen, the possible corrosion reactions in a copper sulfide system are extremely complex because of the large number of stable copper sulfides (Keevil 1989), their differing electrical conductivities, and catalytic effects. Transformations between sulfides, or conversions of sulfides to oxides, result in changes in volume that weaken the attachment scale and oxide subscale leading to spalling.

Two attempts have been made to develop a diagnostic for SRB corrosion of copper alloys: mineralogical fingerprints and sulfur isotope fractionation. Many sulfides under near-surface conditions can only be produced by microbiological action on specific precursor materials such as metals. If a corrosion process can be shown to have taken place in a pH-Eh range typical of near-surface conditions and no compelling kinetic arguments can be reached, then mineralogical and geochemical data indicate that the presence of these minerals as corrosion products implies SRB activity. McNeil *et al.* (1991) demonstrated that djurleite, spionkopite and the high temperature polymorph of chalcocite appear to be mineralogical fingerprints for the SRB corrosion of copper-nickel alloys. The stable isotopes of sulfur (^{32}S and ^{34}S) naturally present in a sulfate source are selectively metabolized during sulfate reduction by SRB and the resulting sulfide is enriched in ^{32}S (Chambers and Trudinger 1979). The ^{34}S isotope accumulates in the starting sulfate as the ^{32}S is removed and concentrated in the sulfide. Little *et al.* (1993) demonstrated sulfur isotope fractionation by SRB in sulfide corrosion products on a copper alloy.

Silver. Laboratory data on sulfide and silver derivatives can be summarized as follows: (1) corrosion of silver by reduced sulfides, whether H_2S (Sinclair 1982; Volpe and Peterson 1989) or organic sulfides (Sinclair 1982) produces acanthite (monoclinic Ag_2S), (2) the corrosivity of organic sulfides appears to be controlled by transport mechanisms and thus by vapour pressures, and (3) the rate of sulfidation is strongly affected by NH_3 and dissolved iron (Biestek and Drys 1987). Abiotic aqueous corrosion of silver in the presence of reduced sulfur species produces acanthite in bulk (Birss and Wright 1981; Campbell *et al.* 1982). Argentite (body-centered cubic Ag_2S) is observed when objects made of impure silver (e.g., coins) are corroded in sediments over archaeological periods (Gettens 1963; North and MacLeod 1986). If Cl^- is present acanthite or argentite combined with cerargyrite (AgCl) is formed.

These observations support the hypothesis that formation of argentite is limited by sulfide precipitation. Argentite formation occurs when an object made of silver-copper alloy, e.g., jewellery or coinage, is in a water-saturated deposit containing SRB in a biofilm capable of maintaining reducing conditions, and bacteria (perhaps ammonia producers) capable of solubilizing silver and copper atoms. A layer of sand or soil restricts the ability of the metal ions to escape, so that concentrations of copper and silver ions within the biofilm rise to levels which cause precipitation. Jalpaite ($Ag_{1.5}Cu_{0.5}S$) forms in regions where the copper concentration is high.

Other metals. SRB-induced corrosion of zinc produces sphalerite (ZnS) (Baas-Becking and Moore 1961). SRB on lead carbonates produce galena (PbS) (McNeil and Little 1990). Galena has been found more recently as a lead corrosion product in SRB-induced corrosion of lead-tin alloys (McNeil and Mohr 1993).

21.2.3 Practical aspects

MIC failures due to SRB have been reported for mild steel piping and equipment exposed in the marine environment, soil, oil refining industry, fossil fuel, nuclear power plants, and process industries. Two volumes edited by Dexter (1986) and Kobrin (1993) provide case histories.

Sanders and Hamilton (1986) analysed microbial corrosion in North Sea oil exploration and defined two distinct forms of SRB-mediated corrosion:

1. pitting caused by SRB growing in the biofilm on metal surfaces, and
2. sulfide-induced stress corrosion cracking, hydrogen induced cracking or blistering caused by hydrogen permeation in high dissolved sulfide conditions.

Bibb (1986) presented three case histories of pipe failures in South African power plants including failure of mild steel pipework handling raw water, an epoxy-lined seawater cooling pipe and galvanized hot water pipework. The presence of the organism *Desulfovibrio desulfuricans* was confirmed in all failures. Honneysett *et al.* (1985) documented MIC of carbon steel in a cooling system for a casting machine process due to SRB. The onset of the problem coincided with the use of reclaimed sewage water. Tatnall (1981) documented MIC failures and sulfur cycling (Figure 21.1) in a paper mill in which corroded areas were typically covered with aerobic slime-producing bacteria and SRB. In addition to the premature corrosion failures, bacteria produced noxious gases, including H_2S, methane and mercaptans.

The impact of oxygen on obligate anaerobic SRB was examined by Hardy and Bown (1984) using mild steel and weight loss measurements. Successive aeration–deaeration shifts caused variation in the corrosion rate. The highest corrosion rates (129 mg /dm^{-2} /d^{-1} $\approx$ 5 cm per year) were observed during air-sparging (50 ml min^{-1} for 5 h) of steel foils that had been previously incubated anaerobically with SRB. In the absence of air sparging corrosion rates were low (1.45 mg /dm^{-2}/ d^{-1}). Lee *et al.* (1993a,b) determined that corrosion of mild steel could not be initiated by SRB in the absence of ferrous ion. King *et al.* (1986) demonstrated that weight loss of steel in the presence of SRB was proportional to the concentration of ferrous sulfide and depended on the stoichiometry of the particular ferrous sulfide minerals.

The impact of biogenic sulfides on the corrosion of copper alloys has received considerable attention. Little *et al.* (1988, 1989, 1990) published several reports documenting localized corrosion of copper alloys by SRB in estuarine environments. Others (Rowlands 1965; Gudas and Hack 1979) reported the failure of copper alloys in polluted seawater containing waterborne sulfides as a result of pitting and stress corrosion cracking. CDA 706 (90:10, Cu:Ni) sustained accelerated corrosion in seawater containing 0.01 ppm sulfide after 1 day exposure.

MIC has also been documented for copper and copper alloys used in potable water applications. Alanis *et al.* (1986) described a case of localized corrosion in underground brass pipes of low zinc content (8.22%) used for drinking water distribution. Perforation was due to SRB in the soil.

Monel 400, a nickel alloy containing 66.5% nickel, 31.5% copper and 1.25% iron, is prone to pitting in chloride-containing environments where the passive film can be disturbed. Under stagnant conditions chlorides penetrate the passive film at weak points and cause pitting attack. Sulfides can cause either a modification of the oxide layer as described for copper or breakdown of the oxide film of nickel alloys. Pit initiation and propagation depend on depth of exposure, temperature and presence of surface deposits. Little *et al.* (1990) reported selective dealloying of nickel from Monel 400 in the presence of SRB in an estuarine environment.

Pope (1986) reported a case study from nuclear power plants in which Monel heat exchanger tubes were found to have discrete deposits under which severe pitting corrosion was observed. Deposits formed by iron- and manganese-depositing bacteria in association with SRB contained large amounts of iron and copper, significant amounts of manganese and silicon and reduced amounts of nickel.

Several investigators have demonstrated that there is no direct correlation between numbers of sulfate-reducing bacteria and the likelihood that corrosion has or will occur. Jack *et al.* (1994) prepared a review of 30 months of electrochemical, weight-loss data, water chemistry and microbiological data for an oilfield waterflood operation in which produced brine was injected to displace oil from the reservoir. They concluded that SRB numbers could be used as an index of biocide performance in these field systems. No other correlations between corrosion measurements and microbial numbers were found. In a two-year study of MIC in natural gas pipeline facilities Pope *et al.* (1988) found no relation between numbers of SRB and extent of corrosion in carbon steel.

21.3 OXIDATION REACTIONS

Corrosion associated with sulfur oxidation reactions involves autotrophic organisms. Specific oxidation reactions leading to the production of sulfuric acid varies with the starting species of reduced sulfur. Elemental sulfur, thiosulfates, metal sulfides, H_2S, and tetrathionates can be oxidized to sulfuric acid.

21.3.1 Oxidation of hydrogen sulfide

Corrosion in sewers and other concrete structures is often caused by oxidation of sulfides generated by the activities of SRB and may occur in many steps. Concrete is a moderately porous mixture of highly alkaline inorganic precipitates and mineral aggregates. Strong acids react with concrete materials thereby destroying its structural integrity. Anaerobic conditions in sewage support SRB that convert sulfate to H_2S, which volatilizes to the sewer atmosphere and redissolves in condensate on the sewer crown (Islander *et al.*, 1991). A second community of microorganisms, including thiobacilli, at the crown oxidizes the sulfide to corrosive sulfuric acid. *Thiobacillus concretivorus*, 'concrete eater' (Harris 1962), may be *T. thiooxidans* adapted to a specific environment. Islander *et al.* (1991) reported mutualism between autotrophs and heterotrophs and succession of acidophiles on concrete sewer crowns.

Eashwar *et al.* (1993) reported corrosion of mild steel in seawater due to the coexistence of oxygen, hydrogen sulfide, and acids during the putrefaction of the marine alga *Ulva lactuca*. Mittleman and Danko (1995) determined that cycling of sulfur, i.e., sulfate reduction and sulfide oxidation, by microorganisms was responsible for concrete and carbon steel deterioration in a dam in South America. Cleland (1995) demonstrated the co-existence of elemental sulfur and sulfate in the presence of iron sulfide corrosion products in ballast tanks of ships. Tatnall (1981) discussed cases of pitting caused by bacteria in cooling towers associated with commercial air conditioning systems. Pitting was attributed to hydrogen sulfide production by SRB and subsequent reoxidation to sulfuric acid by thiobacilli within biofilms.

21.3.2 Oxidation of metal sulfides

Corrosion of carbon steel coal mining equipment, shipping containers (Brozal *et al.* 1997), and pipelines through coal mining areas is often caused by microbial oxidation of metal sulfides. Formation of oxidation products from metal sulfides depends on the type and composition of the mineral. The monosulfides e.g. sphalerite (ZnS), greenockite (CdS), covellite (CuS), pyrrhotite (FeS), galena (PbS) and millerite, (NiS), react with acid to form H_2S (Parsons and Ingraham 1970), the reactivity increasing with increasing solubility. Upon contact with air, H_2S will oxidize rapidly to form sulfur, thiosulfate, sulfite, and, ultimately but more slowly, sulfate. The disulfides do not form H_2S upon reaction with acids. Disulfides, e.g., pyrite and marcasite (FeS_2), form both elemental sulfur and thiosulfate upon oxidative attack in acid solutions, depending on pH, concentration of oxidants, and temperature. The occurrence of H_2S, sulfur, and thiosulfate in mine waste environments indicates either sulfate reduction or acid dissolution of monosulfides. Minerals such as molybdenite (MoS_2) and chalcopyrite ($CuFeS_2$) are not disulfides because the sulfur atoms are bonded to metal atoms, not to other sulfur atoms.

Coal is a generic name, and the solids to which it is applied can be dissimilar. Variation of coal types is related to the biological starting materials, variations in conditions during coal genesis and geological age. Inorganic and organic sulfur compounds are present in coal. Inorganic sulfur occurs predominantly as ferrous sulfide (FeS_2) in its mineral forms pyrite and marcasite, whereas organic sulfur compounds are part of the molecular configuration of the coal. If FeS_2-containing coals are exposed to moisture and oxygen, spontaneous FeS_2 oxidation starts, producing ferric iron and sulfuric acid (Equation 21.8):

$$4FeS_2 + 15O_2 + 2H_2O \rightarrow 2Fe_2(SO_4)_3 + 2H_2SO_4 \qquad (21.8)$$

The pH of the water phase will drop during the oxidation process. Because rates of spontaneous pyrite oxidation decrease as pH falls, the process will stop at pH 4. There is little difference in the corrosion rate of steel in natural waters having pH values between 4.5 and 9.5 (Boyer and Gall 1988). Corrosion of steel cannot be attributed to spontaneous chemical pyrite oxidation. Acidophilic pyrite-oxidizing bacteria, indigenous in coals, thrive at low pH and continue the oxidation to pH values lower than 2. The ferric iron produced in these reactions acts as an oxidizing agent to solubilize other metal sulfides. The ferrous iron can be reconverted biologically to ferric iron (Bos and Kuenen 1990; Bos *et al.* 1992). Reviews on pyrite oxidation have been published by Lowson (1982), Nordstrom (1982), and Evangelou (1995). The process is extremely selective, occurring at exposed pyrite inclusions and at pyrite crystals and along topographical defects in coal (Bos and Kuenen 1990). The more electroconductive the metal sulfide, the slower the rate of oxidation, whereas the less conductive sulfides exhibit increased reaction rates as long as they are in electronic contact with the more conductive sulfide (Nordstrom and Southam 1997).

Microbial succession in coal spoil has been studied in the laboratory (Harrison 1978). Heterotrophic bacteria dominated freshly exposed samples followed by a bloom of autotrophic sulfur-oxidizing bacteria after 20 weeks and finally a dominance of acidophilic heterotrophs. Many acidophilic organisms capable of attacking mineral sulfides have been isolated from coal sources (Rawlings and Woods 1995). These organisms are ubiquitous in nature and are ideally suited to growth in inorganic mineral environments. They obtain the carbon required for the synthesis of new cell material by fixation of carbon dioxide from the atmosphere and energy from oxidation and reduction reactions.

21.3.3 Oxidation of elemental sulfur

Corrosive activities and rates can be dramatic where elemental sulfur is present. Railroad tracks through the elemental sulfur-producing areas of the Gulf Coast of Texas have high corrosion rates due to the activities of thiobacilli (Harris 1962). Steel troughs and drainage pipes failed after 15–30 days and 1.5–3 months, respectively, when used in Russian coal mines with acid-producing bacteria (see Beloglazov and Charoshavin 1995). The predicted lifetime of the materials when operated with waters at neutral pH is 1–2 years. Chromium–nickel and chromium–molybdenum steels are sensitive to attack by thiobacilli-contaminated environments containing elemental sulfur.

Formation of elemental sulfur during microbial oxidation of metal sulfides has been reported (Beyer *et al.* 1987) owing to incomplete oxidation, or oxidation by ferric iron (Equation 21.9):

$$FeS_2 + 2Fe^{3+} \rightarrow 3Fe^{2+} + S \qquad (21.9)$$

It has been demonstrated that *T. ferrooxidans* can excrete elemental sulfur during the oxidation of reduced sulfur compounds under special conditions (Hazeu *et al.* 1988). In acid solutions elemental sulfur can be produced from chalcopyrite (Equation 21.10):

$$CuFeS_2 + 4Fe^{3+} \rightarrow Cu^{2+} + 5Fe^{2+} + 2S \qquad (21.10)$$

Case histories of corrosion in the presence of elemental sulfur can be attributed to either (1) direct oxidation to H_2SO_4 or (2) electron transport from the metal through a metal sulfide to elemental sulfur.

The disproportionation of elemental sulfur in water occurs at a considerable rate at temperatures above 150 °C, but can occur at ambient temperature. The reaction yields H_2S and sulfuric acid, which inhibit repassivation promoting crevice corrosion and stress corrosion cracking. Elemental sulfur is a powerful oxidant for organic and inorganic materials. Metals and metal oxides react with sulfur, either dissolved or undissolved. This reaction proceeds at ambient or higher temperatures and is independent of the presence of oxygen to form sulfides, which act as catalysts for the cathodic reduction of elemental sulfur. The availability of sulfur in soil limits the activities of the sulfur-oxidizing bacteria.

Schmitt (1991) reviewed the effect of elemental sulfur on corrosion of construction materials, including carbon steels, ferric steels, austenitic steels, ferritic-austenitic steels (duplex steels), nickel and cobalt-based alloys and titanium. Wet elemental sulfur in contact with iron is aggressive and can result in the formation of iron sulfides or in stress corrosion cracking. Iron sulfides containing elemental sulfur initiate corrosion only when the elemental sulfur is in direct contact with the sulfide-covered metal. Iron sulfides are highly electron conductive and transport electrons from the metal to the elemental sulfur. The coexistence of hydrogen sulfide and elemental sulfur in aqueous systems, i.e., sour gases and oils, causes stress-oriented hydrogen-induced cracking of iron-containing alloys. The corrosion rate is enhanced in the presence of chlorides. Corrosion-resistant alloys experience lower corrosion rates than carbon steels, with the exception of the duplex steels that are corroded at the same rate as carbon steels. The resistance to sulfur increases with the content of nickel, chromium and molybdenum in the alloy. High-alloy stainless steels are susceptible to stress corrosion cracking at temperatures above 150 °C. Grade 2 titanium suffers sulfur-induced crevice corrosion above 130 °C; however, the Beta-C alloy retains passivity in sulfur-containing brines up to 300 °C.

Dowling (1992) proposed a mechanism for which standing moisture and steel/sulfur contact are requisites for corrosion of structural steel by solid elemental sulfur. The principal form of attack in S^o/H_2O media is not due to secondary acid generation resulting from hydrolysis of the sulfur. Instead Dowling (1992) demonstrated steel oxidation coupled to sulfur reduction through an electron conductive iron sulfide layer.

21.4 ACKNOWLEDGEMENTS

This work was performed under Program Element 0601153N, NRL Contribution Number NRL/BA/7303--99-0002. Robert K. Pope was supported by a CORE/NRL Postdoctoral Fellowship.

21.5 REFERENCES

Alanis, I., Berardo, L., De Cristofaro, N., Moina, C., and Valentini, C. (1986) A case of localized corrosion in underground brass pipes. In *Biologically Induced Corrosion* (ed. S.C. Dexter), pp. 102-108, NACE, Houston, TX.

Baas-Becking, G.M. and Moore, D. (1961) Biogenic sulfides. *Econ. Geol.* **56**, 259–272.

Beloglazov, S.M. and Charoshavin, A.N. (1995) Corrosion behavior of steel in coal mining waters in the presence of *thiobacillus thiooxidans* and *thiobacillus ferroxidans*. In *Proc. 3rd International EFC Workshop*, pp. 398–404, published by The European Federation of Corrosion by the Institute of Materials.

Berner, R.A. (1969) The synthesis of framboidal pyrite. *Econ. Geol.* **64**, 383–384.

Beyer, M., Eber, H.G., Assenmacher, H., and Frigge, J. (1987) Elemental sulfur in microbiologically desulfurized coals. *Fuel* **66**, 551–555.

Bibb, M. (1986) Bacterial corrosion in the South Africa power industry. In *Biologically Induced Corrosion 86*, NACE, Houston, TX.

Biestek, T. and Drys, M. (1987) Corrosion products forming on silver in various corrosive environments. *Powlocki Ochronne* (Warsaw) **9**, 2–5.

Birss, V.I. and Wright, G.A. (1981) The potentiodynamic formation and reduction of a silver sulfide monolayer on a silver electrode in aqueous sulfide solutions. *Electrochem. Acta* **27**, 1–7.

Blunn, G. (1986) Biological fouling of copper and copper alloys. In *Biodeterioration VI*, pp. 567–575, CAB International, Slough, UK.

Boivin, J., Laishley, E.J., Bryant, R.D., and Costerton, J.W. (1990) The influence of enzyme systems on MIC. In *Proc. Corros. 90*, No. 128, NACE, Houston, TX.

Bos, P. and Kuenen, G.J. (1990) Microbial treatment of coal. In *Microbial Metal Recovery* (eds. H.L. Ehrlich and C.L. Brierly), pp. 343–377, McGraw-Hill, New York, NY.

Bos, P., Boogerd, F.C., and Kuenen, G.J. (1992) Microbial desulfurization of coal. In *Environmental Microbiology*, pp. 375–403, Wiley-Liss, Inc., New York, NY.

Boyer, H.E. and Gall, T.L. (eds.) (1988) *Metals Handbook,* American Society for Metals, pp. 4-91–92, Desk Metals Park, OH.

Brozal, V.S., Hall, A., and deBruin, R. (1997) Effect of microbial flora in low grade coal on the corrosion of 3Cu12 steel. In *Biodeterioration and Biodegradation 9th International Symposium*, pp. 494–499.

Campbell, G.D., Lincoln, F.J., Power, G.P., and Rithchie, I.M. (1982) The anodic oxidation of silver in sulfide solutions. *Aust. J. Chem.* **35**, 1079–1085.

Chambers, L.A. and Trudinger, P.A. (1979) Microbiological fractionation of stable sulfur isotopes: a review and critique. *Geomicrobiol. J.* **1**, 249–293.

Cleland, J.H. (1995) Corrosion risks in ships' ballast tanks and the IMO pathogen guidelines. *Engineering Failure Analysis* **2**(1), 79–84.

Craig, J.R. and Scott, S.D. (1976) Sulfide phase equilibria. *Rev. Mineral.* **1**, CS1–CS110.

Crolet, J.-L. and Magot, M.F. (1996) Observations of non SRB sulfidogenic bacteria from oilfield production facilities. *Mat. Perf.* 60–64.

Cuthbert, M. (1962) Formation of bornite at atmospheric temperature and pressure. *Econ. Geol.* **57**, 38–41.

Daubree, G.A. (1862) Contemporary formation of copper pyrite by the action of hot springs at Bagnes-de-Bigorre. *Bull. Soc. Geol. France* **19**, 529–532.

de Gouvernain, M. (1875) Sulfiding of copper and iron by a prolonged stay in the thermal spring at Bourbon I'Archambault. *Compt. Rend.* **80**, 1297–1300.

Dexter, S.C. (ed.) (1986) *Biologically Induced Corrosion,* 1986 NACE International, Houston, TX, 363 pp.

Dowling, N.I. (1992) Corrosion of materials used in storage and handling of solid elemental sulphur. In *Materials Performance Sulphur and Energy*, pp. 104–115, published by the Canadian Institute of Mining, Quebec, Canada.

Duncan, S.J. and Ganiaris, H. (1987) Some sulphide corrosion products on copper alloys and lead alloys from London waterfront sites. In *Recent Advances in the Conservation and Analysis of Artifacts* (ed. E.J. Black), pp. 109–118, Summer Schools Press, University of London.

Eashwar, M., Subramanian, G., ChandraseKaran, P., and BalaKrishman, K. (1993) Microbiologically influenced corrosion of steel during putrefication of seawater: evidence for a new mechanism. *Corrosion* **49**(2), 108–113.

Evangelou, V.P. (1995) *Pyrite Oxidation and Its Control.* CRC Press, Boca Raton, FL.

Gettens, R.J. (1963) The Corrosion Products of Metal Antiquities, Annual Report to the Trustees of the Smithsonian Institution for 1963, pp. 547–568 (1969), The Freer Chinese Bronzes, Smithsonian Institution, Washington, D.C.

Goldschmidt, J. (1953) A simplexity principle. *J. Geol.* **61**, 539–551.

Gudas, J.P. and Hack, H.P. (1979) Sulfide-induced corrosion of copper-nickel alloys. *Corrosion* **35**, 67–73.

Hardy, J.A. and Bown, J.L. (1984) The corrosion of mild steel by biogenic sulfide films exposed to air. *Corrosion* **40**, 650–654.

Harrison, A.P., Jr. (1978) Microbial succession and mineral leaching in an artificial coil spoil. *Appl. Environ. Microbiol.* **36**, 861–869.

Harris, J.O. (1962) The role of soil microorganisms in corrosion. In *Proc. 7th Annual Appalachian Underground Corrosion Short Course* (ed. R.E. Hannah, Jr.), published by West Virginia University.

Hazeu, W., Batenburg-van der Vegte, W.H., Bos, P., van der Plas, R.K., and Kuenen, J.G. (1988) The production and utilization of intermediary elemental sulfur during the oxidation of reduced sulfur compounds by *T. ferrooxidans. Arch. Microbiol.* **150**, 574–579.

Heimann, R.B. (1989) Assessing the technology of ancient pottery: the use of ceramic phase diagrams. *Archeomater* **31**, 123–148.

Honneysett, D.G., Van Den Burgh, W.D., and O'Brien, P.F. (1985) Microbiological corrosion control in a cooling water system. *Mat. Perf.* **24**(10), 34.

Islander, R.I., Devinny, J.S., Mansfeld, F., Postyn, A., and Shih, H. (1991) Microbial ecology of crown corrosion in sewers. *J. Env. Eng.* **117**(6), 751–770.

Jack, T.R., Rogoz, E., Bramhill, B., and Roberge, P.R. (1994) The characterization of sulfate-reducing bacteria in heavy oil waterflood operations. In *Microbiologically Influenced Corrosion Testing* (eds. J.R. Kearns and B.J. Little), p. 108, ASTM, Philadelphia, PA.

Keevil, C.W. (1989) Detection of biofilms associated with pitting corrosion of copper pipework in Scottish hospitals. In *Biocorrosion* (eds. C.C. Gaylord and H.G. Morton), pp. 48–62, Biodeterioration Society, Kew.

King, R.A., Skerry, B.S., Moore, D.C.A., Stott, J.F.D., and Dawson, J.L. (1986) Corrosion behavior of ductile and grey iron pipes in environments containing sulphate-reducing bacteria. In *Biologically Induced Corrosion*, (ed. S.C. Dexter), No. 83, pp. 83–95, NACE, Houston, TX.

Kobrin, G. (ed.) (1993) *A Practical Manual on Microbiologically Influenced Corrosion,* 232 pp., 1993 NACE International, Houston, TX.

Kostov, I. and Minceva-Stefanova, J. (1981) *Sulphide Minerals.* Publishing House, Bulgarian Academy of Sciences, Sofia, Bulgaria.

Lee, W. and Characklis, W.G. (1993) Corrosion of mild steel under anaerobic biofilm. *Corrosion* 186–198.

Lee, W., Lewandwoski, Z., Okabe, S., Characklis, W.G., and Auci, R. (1993a) Corrosion of mild steel underneath aerobic biofilms containing sulfate-reducing bacteria. part I: at low dissolved oxygen concentration. *Biofouling* **7**, 197–206.

Lee, W., Lewandowski, Z., Nielsen, P.H., Morrison, M., and Characklis, W.G. (1993b) Corrosion of mild steel underneath aerobic biofilms containing sulfate reducing bacteria. part II: at high dissolved oxygen concentration. *Biofouling* **7**, 217–239.

Little, B.J., Wagner, P.A., and Jacobus, O.J. (1988) The impact of sulfate-reducing bacteria on welded copper-nickel seawater piping systems. *Mat. Perf.* **27**(8), 51–56.

Little, B.J., Wagner, P.A., Jacobus, O.J., and Janus, L. (1989) Evaluation of microbiologically induced corrosion in an estuary. *Estuaries* **12**(3), 138–141.

Little, B.J., Wagner, P.A., Ray, R.I., and McNeil, M.B. (1990) Microbiologically influenced corrosion in copper and nickel seawater piping systems. *J. Marine Tech. Soc.* **243**(3), 10–17.

Little, B., Wagner, P., and Mansfeld, F. (1991) Microbiologically influenced corrosion of metals and alloys. *Int. Mat. Rev.* **36**(6), 253–272.

Little, B.J. and Wagner, P.A. (1992) Standard practices in the United States for quantifying and qualifying sulphate reducing bacteria in microbiologically influenced corrosion. In *Proc., Redefining Int. Standards and Practices for the Oil and Gas Industry*. London, UK.

Little, B., Wagner, P., and Jones-Meehan, J. (1993) Sulfur isotope fractionation by sulfate-reducing bacteria in corrosion products. *Biofouling* **6**, 279–288.

Lowson, R.T. (1982) Aqueous oxidation of pyrite by molecular oxygen. *Chem. Rev.* **82**, 461–479.

MacDonald, D.D., Syrett, B.C., and Wing, S.S. (1979) Corrosion of Cu-Ni alloys 706 and 715 in flowing sea water-2. Effect of dissolved sulfide. *Corrosion* **35**(8), 367–378.

McNeil, M.B. and Little, B.J. (1990) Mackinawite formation during microbial corrosion. *Corrosion* **46**, 599–600.

McNeil, M.B., Jones, J., and Little, B.J. (1991) Mineralogical fingerprints for corrosion processes induced by sulfate-reducing bacteria. In *Proc. Corrosion 91*, No. 580, NACE, Houston, TX.

McNeil, M.B. and Little, B.J. (1992) Corrosion mechanisms for copper and silver objects in near-surface environments. *J. Amer. Inst. for Conserv.* **31**(3), 355–366.

McNeil, M.B. and Mohr, D.W. (1993) Formation of copper-iron sulfide minerals during corrosion of artifacts and implications for pseudogilding. *Geoarchaeol.* **8**(1), 23–33.

McNeil, M.B. and Odom, A.L. (1994) Thermodynamic prediction of microbiologically influenced corrosion by sulfate reducing bacteria. In *Microbiologically Influenced Corrosion Testing* (eds. J.R. Kearns and B.J. Little), pp. 173–179, STP 1232, published by ASTM, Philadelphia, PA.

Mittleman, N.W. and Danko, J.C. (1995) Corrosion of a concrete dam structure: evidence of microbially influenced corrosion activity. In *Proc. 1995 International Conference on Microbially Influenced Corrosion,* pp. 15–1 to 15–7, Welding Society and NACE, Houston, TX.

Mohagheghi, A., Updegraff, D.M., and Goldhaber, M.B. (1985) The role of sulfate-reducing bacteria in the deposition of sedimentary uranium ores. *Geomicrobiol. J.* **4**, 153–173.

Mor, E.D. and Beccaria, A.M. (1975) Behavior of copper in artificial seawater containing sulphides. *Br. Corr. J.* **10**, 33–38.

Nordstrom, D.K. (1982) Aqueous pyrite oxidation and the consequent formation of secondary iron minerals. In *Acid Sulfate Weathering* (eds. J.A. Kittrick, D.S. Fanning, and L.R. Hossner), pp. 37–56, *Soil Sci. Soc. Am. Spec. Pub.* **10**.

Nordstrom, D.K. and Southam, G. (1997) Geomicrobiology of sulfide mineral oxidation. In *Reviews in Mineralogy Geomicrobiology: Interactions Between Microbes and Minerals Volume 35* (eds. J.F. Banfield and K.H. Nealson), pp. 361–390, Mineralogical Society of America, Washington, D.C.

North, R.F. and Pryor, M.J. (1970) The influence of corrosion product structure on the corrosion rate of Cu-Ni alloys. *Corr. Sci.* **10**, 297–311.

North, N.A. and Macleod, I.D. (1986) Corrosion of metals. In *Conservation of Archaeological Objects* (ed. C. Pearson), pp. 69–98, Butterworths, London.

Parsons, H.W. and Ingraham, T.R. (1970) The hydrogen sulphide route to sulfur recovery from base metal sulphides, part I: The generation of H_2S from these metal sulphides. Circular No. 242, Ottawa Department of Natural Resources, Mines Branch Information.

Pollard, A.M., Thomas, R.G., and Williams, P.A. (1989) Synthesis and stabilities of basic copper (II) chlorides atacamite, paratacamite, and botallackite. *Mineral Mag.* **53**, 557–563.

Pope, D.H., Duquette, D.J., Wayner, P.C., and Johannes, A.H. (1984) *Microbiologically Influenced Corrosion: A State of the Art Review.* Materials

Technology Institute of the Chemical Process Industries, Columbus, OH.
Pope, D.H. (1986) A study of microbiologically influenced corrosion in nuclear power plants and a practical guide for countermeasures. EPRRINP-4592 Project 1166-6, final report.
Pope, D.H., Zintel, T.P., Kuruvilla, A.K., and Siebert, O.W. (1988) Organic acid corrosion of carbon steel: A mechanism of microbiologically influenced corrosion. In *Proc. Corrosion 88*, No. 79, NACE, Houston, TX.
Pourbaix, M. (1966) Atlas of electrochemical equilibria in aqueous solutions. NACE International, Houston, TX.
Putnis, A. (1977) Electron diffraction study of phase transformations in copper sulfides. *Am. Mineral.* **62**, 107–114.
Rawlings, D.E. and Woods, D.R. (1995) Development of improved biomining bacteria. In *Bioextraction and Biodeterioration of Metals*, (eds. C.C. Gaylarde and H.A. Videla), pp. 63–84, The Biology of World Resources Series 1, Cambridge Press, New York, NY.
Ribbe, P.H. (ed.) (1976) Sulfide mineralogy. Mineralogical Society of America, *Rev. Mineral.* **1**, p. 284, Washington, D.C.
Roseboom, E.H. (1966) An investigation of the system Cu-S and some natural copper sulfides between 25 degrees and 700 degrees. *Econ. Geol.* **61**, 641–672.
Rowlands, J.C. (1965) Corrosion of tube and pipe alloys due to polluted seawater. *J. Appl. Chem.* **15**, 57–63.
Sanders, P.F. and Hamilton, W.A. (1986) Biological and corrosion activities of sulphate-reducing bacteria in industrial process plant. In *Biologically Induced Corrosion*, No. 47, NACE, Houston, TX.
Schmitt, G. (1991) Effect of elemental sulfur on corrosion in sour gas systems. *Corrosion* **47**(4), 285–308.
Sinclair, J.D. (1982) The tarnishing of silver by organic sulfur vapors. *J. Electrochem. Soc.* **129**, 33–40.
Syrett, B.C. (1977) Accelerated corrosion of copper in flowing pure water contaminated with oxygen and sulfide. *Corrosion* **33**, 257–262.
Syrett, B.C. (1980) The mechanism of accelerated corrosion of copper-nickel alloys in sulfide polluted seawater. In *Proc. Corrosion 80,* No. 33, NACE, Houston, TX.
Syrett, B.C. (1981) The mechanism of accelerated corrosion of copper-nickel alloys in sulphide-polluted seawater. *Corr. Sci.* **21**, 187–209.
Tatnall, R. (1981) Case histories: bacteria induced corrosion. *Mat. Perf.* **19**, 41–48.
Volpe, L. and Peterson, P.J. (1989) The atmospheric sulfidation of copper in a tubular corrosion reactor. *Corr. Sci.* **29**, 1179–1186.
von Wolzogen Kuhr, C.A.H. and van der Vlugt, L.S. (1934) Graphication of cast iron as an electrobiochemical process in anaerobic soils. *Water (The Hague)* **18**, 147–165.
Wagman, D.D., Evans, W.H., Parker, V.B., Schumm, R.H., Halow, I.I., Bailey, S.M., Churney, K.L., and Nuttall, R.L. (1982) The NBS tables of chemical thermodynamic properties: selected values for inorganic and C_1 and C_2 organic substances in SI units. *J. Phys. Chem.,* ref. data 11, supplement 2.
Wikjord, A.G., Rummery, T.E., Doern, F.E., and Owen, D.G. (1980) Corrosion and deposition during the exposure of carbon steel to hydrogen sulfide water solutions. *Corr. Sci.* **20**, 651–671.

22

Recent developments in research on biogenic sulfuric acid attack of concrete

Elke Vincke, Joke Monteny, Anne Beeldens, Nele De Belie, Luc Taerwe, Dionys Van Gemert and Willy H. Verstraete

22.1 INTRODUCTION

The prediction of the life-time of concrete structures under environmental conditions implies the examination of the material and structure at micro and macro levels by experts of different fields (Sagrov 1998). Up to now, concrete has been studied mainly by two different groups of scientists. Chemists and physicists have considered the mechanisms of cement hydration, development of microstructures, chemical resistance, etc. Mechanical researchers have modelled the behaviour of concrete structures under certain conditions. However, a third group of increasing importance in the research on the influence of environmental impacts on concrete structures are the biologists. Surfaces of concrete structures, when exposed to non-sterile environments, become the site of an array of aerobic and anaerobic biological activities, many of which are deleterious to the surface. Since the discovery in 1945 of a type of rapid corrosion of a concrete surface by *Thiobacillus* populations (Parker 1945), much effort has focused on the understanding of this corrosive process (Pomeroy and Parkhurst 1977; Sand and Bock 1984; Nielsen and Hvitved-Jacobsen 1988; Mori *et al.* 1992; Coleman and Gaudet 1993; Nielsen *et al.* 1998). However, there are still many unknown factors in the corrosion process caused by microorganisms. This paper presents an overview on recent developments in research on corrosion by sulfuric acid produced by sulfur oxidizing bacteria; in other words the biogenic sulfuric acid corrosion of concrete.

22.2 MECHANISM OF BIOGENIC SULFURIC ACID ATTACK

Considering the different ways of microbiologically influenced corrosion (MIC) of concrete, the most important mechanism is biogenic sulfuric acid corrosion (Vancalbergh 1996). The term biogenic sulfuric acid attack (BSA) is used for a corrosion process caused by biologically produced sulfuric acid. The BSA corrosion of concrete in sewer systems is a worldwide phenomenon and is considered to have a great economic impact. In Los Angeles County approximately 10% of the sewer pipes are subject to significant sulfide corrosion and the costs for the rehabilitation of these pipelines are roughly estimated at €400 million (Sydney *et al.* 1996). MIC also gives rise to indirect destruction of pavement and roads and also to environmental problems. The restoration of the overall damaged sewerage network in Germany, indispensable for ecological as well as economical reasons, is estimated to cost about €1.10^5 million (Kaempfer and Berndt 1998). Estimates for sewer repair in Flanders (Belgium) suggest that biogenic sulfuric acid corrosion of sewers can amount to €5 million per year. An example of biogenic sulfuric acid corrosion in a sewerage system in Flanders is given in Figure 22.1. There is no clear view on the extend of damage but a cost estimate of a few € per inhabitant per year seems appropriate; the latter has to be compared with the value of treating the sewage which is of the order of €20 per inhabitant per year.

Fundamental in corrosion control, specifically in a sewerage system, is the understanding of the main factors that cause the corrosion (Figure 22.2).

The corrosion mechanism was described, among others, by Sand *et al.*

Figure 22.1. Crown corrosion in a sewerage system in Flanders.

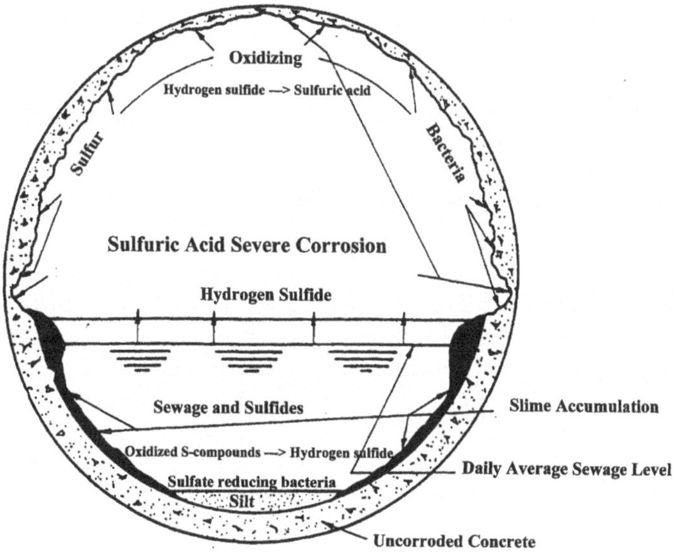

Figure 22.2. Schematic representation of the sulfur-cycle occurring in sewer pipes (After: De Ceukelaire 1989).

(1987), Sand *et al.* (1992), Coleman and Gaudet (1993) and Heuer and Kaskens (1993). The main source of biogenic sulfuric acid is the sulfur compound hydrogen sulfide, H_2S. It is produced by sulfate reducing bacteria (SRB, e.g. *Desulfovibrio* sp.). The latter are active under anaerobic conditions and reduce oxidized sulfur compounds to H_2S. These microorganisms live in the sewage, in the mud at the bottom of the pipelines and in the slime layer, the biofilm, coating the surfaces of pipelines above and below the water level. The produced sulfide is mainly a result of microbial reduction of sulfate, sulfite and thiosulfate. However, the presence of several organic sulfur compounds, such as S-containing amino acids (methionine, cysteine) in wastewater is also important because they might contribute to the overall sulfide production (Nielsen *et al.* 1992).

In the sulfur cycle and in the microbial corrosion, H_2S production often seems to be the limiting factor, as H_2S is the substrate for conversion to sulfuric acid by sulfur oxidizing bacteria (thiobacilli). Sulfuric acid is a very aggressive acid that is able to rapidly destroy the concrete.

The release of H_2S from the water phase depends on different factors: like pH and temperature in the pipeline and turbulence in the sewage flow. The latter can be caused by high sewage flow levels, changes in velocity, high debris build-up, blockages in air flow and direct connection of local lines into large interceptors (DeHollander 1998). However, Monteith *et al.* (1997) found water temperature to have less effect on the emission of H_2S from the sewage. The latter authors found that wind velocity over the opening in the reach and water flow rate have the greatest effect on the wastewater ventilation rate.

Once the H_2S has reached the atmosphere, it may react with oxygen to elemental sulfur, which is deposited on the slime layer coating the walls. Sulfur is a substrate for many thiobacilli, such as *Thiobacillus thiooxidans*, *Thiobacillus neapolitanus* and *Thiobacillus intermedius*. They will metabolize the sulfur to sulfuric acid. The energy obtained from the oxidation is used for CO_2 fixation for cell mass production. The sulfuric acid formation and the ensuing degree of corrosion is largely dependent on the number of *Thiobacillus thiooxidans* on the concrete surface (Sand *et al.* 1987; Derangère and Cochet 1991). The sulfuric acid produced may attack the inner surface of the concrete pipe and other parts of the treatment and transportation facilities, such as pumping stations, manholes and reservoirs.

Mori *et al.* (1991) determined corrosion rates of 4.3-4.7 mm per year in the sewer pipes of Ohmuta station (Japan). Sewers are presupposed to have working life-spans of 80-100 years. At such corrosion rates however, an 88 mm thick pipe hardly lasts for 20 years.

From the above-described process, it can be concluded that both anaerobic and aerobic conditions in the sewer pipeline are responsible for the formation of sulfuric acid and the corrosion problems. This duality presents opportunities for the prevention of deterioration of concrete structures. Understanding the complicated microbial processes and the influence of environmental conditions is a corner stone for the remediation of biogenic sulfuric acid corrosion.

22.3 SULFIDE FORMATION IN SEWERS

22.3.1 Sulfide formation

Microbial processes in wastewater proceed during transportation and the dissolved oxygen (DO) concentration determines if processes take place under aerobic, anaerobic or intermediary conditions. Variable conditions occur in pressure mains when injecting air or pure oxygen and in gravity sewers with varying slope.

Sulfide generation is the most serious problem resulting from anoxic and anaerobic conditions and long residence times in pipelines (Hvitved-Jacobsen *et al.* 1988; Boon 1995). It is a very annoying problem, especially in regions with a relatively warm climate (Elmaleh *et al.* 1998b; Delgado *et al.* 1999). This is due to some characteristics of H_2S : its unpleasant odour (odour thresholds for H_2S in the range of 1-4 ppb by volume), the high toxicity (fatal at gas concentrations within the range of 300-500 ppm by volume in a few minutes; toxicity has been ranked with hydrogen cyanide (U.S. Environmental Protection Agency 1985)) and its strong corrosive properties on all kinds of metal and concrete materials (Matos and de Sousa 1992; Boon *et al.* 1998; Smet *et al.* 1998).

Under anaerobic conditions, oxidized sulfur compounds are reduced by sulfate reducing bacteria resulting in the formation of sulfide and the maintenance or the growth of the sulfate reducing biomass. The change of free energy is of the same order of magnitude as that for methane production:

$$SO_4^{2-} + 8\,H^+ + 8e^- \rightarrow S^{2-} + 4\,H_2O \quad \Delta G^{o\prime} = 170.1\text{ kJ/mol S} \qquad (22.1)$$
$$CO_2 + 8\,H^+ + 8e^- \rightarrow CH_4 + 2\,H_2O \quad \Delta G^{o\prime} = 192.8\text{ kJ/mol C} \qquad (22.2)$$

The sulfate reducing bacteria use readily biodegradable organic matter, like lactate, pyruvate and some aromatic substrates (benzoate), but in general they do not use higher carbohydrates such as maltose, cellobiose,

Table 22.1. Sulfate reducing bacteria and their substrates

	Electron acceptor	Electron donor	Example (form, width/μm, length/μm)
Group 1 Incomplete oxidizers: Excretion of acetate	Oxidized S-compounds	Lactate, formate propionate, butyrate, pyruvate, molecular hydrogen	*Desulfovibrio* (vibrio, 0.5-0.8, 1.5-2) *Desulfotomaculum* (rod, 0.7-1.2, 2-4)
		Short and long-chain fatty acids, alcohols, aromatic compounds, molecular hydrogen	*Desulfomonas pigra* (rod, 0.8-1.3, 1.2-5) *Desulfovibrio thermophilus* (rod, 0.5, 1.2-2.5) *Desulfolobus* (rod or oval, 0.6-1.3, 1.5-2.5)
Group 2 Complete oxidizers: Excretion of CO_2	Oxidized S-compounds	Branched-chain fatty acids, acetone, phenolic compounds, indole, substrates mentioned above	*Desulfococcus* (sphere, 1.5-3) *Desulfobacterium* (oval rod, 0.7-2, 1.5-2.8) *Desulfonema* (filament, 2.5-8, 50-1000, length one cell 2.5-3)

glucose and fructose (Grusenmeyer *et al.* 1985). The electron donors oxidized by sulfate reducing bacteria are always low molecular mass compounds. Almost all of these compounds are fermentation products from the anaerobic bacterial degradation of carbohydrates, proteins and other components of biomass (Widdel and Hansen 1991).

The group of sulfate reducing bacteria is characterized by the ability to transfer substrate hydrogen to sulfate as the terminal electron acceptor which results in the reduction of sulfate to sulfide. The group of sulfate reducing bacteria is very heterogeneous with regard to substrate utilization and metabolic pathways. No species can use all substrates and no substrate can be utilized by all the sulfate reducing bacteria. For example, *Desulfovibrio* can use lactate and pyruvate as substrate, but no acetate (Table 22.1).

Within the nutritionally diverse sulfate reducing bacteria, two major metabolic groups may be distinguished (Widdel 1988). The first group comprises species oxidizing their substrates incompletely to acetate. The second group embraces the sulfate reducers, which are principally able to oxidize their organic substrates, including acetate, completely to carbon dioxide.

In general the incompletely oxidizing sulfate reducers are nutritionally less versatile than the completely oxidizing species and the incomplete oxidizers may grow significantly faster than the complete oxidizers. For example under optimum conditions, *Desulfovibrio* species using hydrogen,

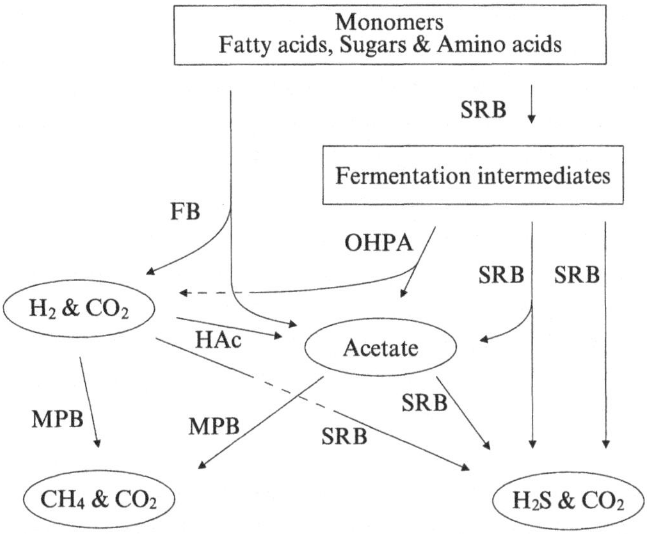

Figure 22.3. Pathways of organic compound degradation under methanogenic and sulfidogenic conditions. FB-Fermentative bacteria, OHPA-Obligate hydrogen producing acetogens, HAc-Homoacetogenic bacteria, MPB-Methane producing bacteria, SRB-Sulfate reducing bacteria (After: Colleran *et al.* 1995).

lactate or pyruvate may reach a doubling time of 3-4 h, whereas the doubling time for the complete oxidizers under optimal conditions is of the order of 20 h or more (Widdel 1988).

In sulfate containing environments, three broad groups of organisms are important for the sulfate reduction: some heterotrophs (aerobic, anaerobic or facultative), hydrogen producing acetogenic bacteria and sulfate reducing bacteria. The heterotrophs not only give rise to fermentation products that serve as the substrates for the other groups, but they also generate the conditions of anaerobiosis and low-redox potential (< -150 mV) necessary for the growth of these organisms (Smet *et al.* 1998). Although the sulfate reducing bacteria as a group have the capacity to utilize a wide range of the products of heterotrophic metabolism, it is considered that such compounds as lactate and higher fatty acids are at least partly converted to hydrogen and acetate by the so-called obligate hydrogen producing acetogens (OHPA) and the fermentative bacteria before serving as substrates for sulfate reducing bacteria (Hamilton 1985 and Figure 22.3).

For common substrates such as hydrogen, formate and acetate, a competition exists between methanogens and sulfate reducing bacteria and the competitive advantage of the SRB relative to the methanogens is markedly increased by decreasing the substrate concentration in the presence of sulfate (Isa *et al.* 1986). Another important factor is temperature. In some instances, increasing temperature causes a significant change from methane production to sulfate reduction in acetate conversion (Rintala and Lettinga 1992).

Additional factors that may be of importance in the competition are adherence properties, mixed substrate utilization, affinity for sulfate of sulfate reducers, relative numbers of bacteria, and conditions such as pH and sulfide concentration (Oude Elferink *et al.* 1994). In general, sewers in which water rich in carbohydrates, together with relative high levels of sulfates enters can be considered to be conducive to intensive H_2S production.

Literature on the sensitivity of sulfate reducing bacteria to sulfide toxicity is contradictory (see the review in Chapter 20). Isa *et al.* (1986) concluded that sulfate reducing bacteria were hardly affected by high levels (500 mg L^{-1}) of hydrogen sulfide. However, Widdel (1988) reported inhibition of *Desulfotomaculum acetooxidans* at hydrogen sulfide concentrations of 85 mg L^{-1} S^{2-}. Hilton and Oleszkiewicz (1988) concluded that sulfate reducing bacteria are more sensitive to elevated levels of dissolved total sulfide than methane producing bacteria. Values for the free H_2S concentration at which methanogenesis was inhibited for 50% vary between 50 mg L^{-1} and 270 mg L^{-1}.

Sulfate reducers are in general localized in a biofilm, present on the inner wall of a pipeline. The combined use of microsensors and molecular techniques is a fairly new and promising tool in the study of, for instance, these sulfate reducing bacteria and the relation between these bacteria and other microorganisms. Such a combination was first used by Ramsing *et al.* (1993) to study sulfate reducers in a trickling filter biofilm. Recently, Schramm *et al.* (1996) and Ferris *et al.* (1997) also used the technique. Molecular techniques can be used to identify bacteria responsible for certain processes in the biofilm (like sulfide generation). Moreover, the distribution of the different microorganisms in the microzonations of the biofilm can also be studied when applying, for example, the fluorescent *in situ* hybridisation (FISH) technique. On the other hand, the processes that occur in these microzonations can be studied with microsensors with a high resolution (to the micrometer level). Santegoeds *et al.* (1998) used a molecular technique in combination with microsensors to study the development of a biofilm in a wastewater treatment plant. With denaturing gel electrophoresis (DGGE) of PCR-amplified 16S rRNA gene fragments (specific for each species), they showed an increasing complexity of the biofilm community with time. Lens *et al.* (1995) investigated the microzonation of microbial activity in UASB aggregates by using microelectrodes. Okabe *et al.* (1998) showed with fluorescently labelled 16S rRNA-targeted oligonucleotide probes for SRB and with microelectrodes, that the abundance of the fluorescently labelled SRB was the highest in a restricted zone located in the middle anaerobic part of the biofilm. The overall message is that SRB are very diverse and very abundant throughout nature, even in quite aerobic environments such as trickling filters, activated sludge flocs and well-drained soils (Manz *et al.* 1998). However, they are rarely dominant but rather active in situations where their end products can easily diffuse out or be further converted.

If a DO concentration of 0.2-0.5 mg L^{-1} or another more thermodynamically favoured electron acceptor such as nitrate is present in water in a substantial amount, sulfide production does not occur (Hvitved-Jacobsen *et al.* 1998; Elmaleh *et al.* 1998a). Elmaleh *et al.* (1998a) have shown that sulfide generation started up from the moment DO has disappeared and the ORP value was below -100 mV. Pomeroy (1974) has shown that sulfide generation depends on the sulfate concentration when the latter is below 3.33 mg L^{-1} SO_4^{2-}-S. At higher concentrations, sulfate does not influence the sulfide generation. Ingvorsen *et al.* (1984) stated that no sulfide generation occurs at a sulfate concentration below 1 mg L^{-1} SO_4^{2-}-S. Furthermore, sulfide generation increases with COD (total and soluble COD) concentration. An equation that relates the sulfide generation rate with the organic matter content and temperature is in general as follows (Elmaleh *et al.* 1998b):

$$r = a \times 1.07^{T-20} \times \mathrm{COD}^{b} \tag{22.3}$$

r: Sulfide generation rate in g m^{-2} h^{-1} S^{2-}
COD: Chemical oxygen demand in g m^{-3}
T: Temperature in °C

The temperature expression used for this equation has been previously proposed by different authors (see 22.3.2). The exponent b and the constant a can be obtained from the representation of $\ln(r/(1.07^{T-20}))$ versus ln(COD).

The temperature of the sewage is an important factor for the solubility of H_2S. At a temperature of 20 °C the solubility of H_2S is 3.85 g L^{-1} water and with an increase of the temperature with 1 °C, the solubility decreases with 2.5%.

Kitagawa *et al.* (1998) showed that there is a little effect of the organic matter concentration on the sulfide generation rate when the fluctuation of soluble organic matter concentration is slight. The investigations of Nielsen *et al.* (1998) showed that no or only a very low sulfide production occurs at dissolved COD concentrations below 50 mg L^{-1}. These data suggest a low competition of the SRB for organics.

The generation of hydrogen sulfide is an important step in the corrosion process. The corrosive attack can occur starting from a hydrogen sulfide concentration in the sewage of 0.5 mg L^{-1} S^{2-} water, depending among others on turbulence and temperature. A hydrogen sulfide concentration of 5 mg L^{-1} S^{2-} at the discharge of a pressure line is often observed in practice under moderate climate conditions.

22.3.2 Modelling

There is no simple and economical method of avoiding septicity of sewage within the sewerage system or of removing sulfide from air under all conditions of operation. Each sewerage system is particular. Therefore, it is necessary to examine in detail the causes of septicity and the generation of sulfide. Also, for the effective and low capital use of the methods (described in 22.3.3) to inhibit the SRB and to decrease the H_2S concentration, it is important to follow up and evaluate the different conditions in the sewerage system. Several empirical equations are available to predict the sulfide production in sewer pipes (Table 22.2). The symbols and units used are the same as those in the original publications. For models describing sulfide production in pressure mains, the important parameters to quantify the sulfide production rate are the concentration of sulfate, the concentration of organic matter and the temperature.

To further improve the model predicting the sulfide generation rate in pressure lines, several studies have introduced a biofilm model, considering substrate transfer rate (Fick's first law of diffusion) and substrate uptake rate (combination of Michaelis-Menton kinetics and Fick's second law of diffusion) in biofilms (Nielsen and Hvitved-Jacobsen 1988; Holder *et al.* 1985, 1989).

Most of the empirical equations proposed for forecasting the generation of hydrogen sulfide in pipes make use of BOD_5 or COD as electron donor, although sulfate reducing bacteria only consume soluble organic matter, more specifically substances as outlined in Table 22.1. However, Hvitved-Jacobsen *et al.* (1988) proposed an equation as a function of the soluble COD (Table 22.2).

Table 22.2. Different equations for predicting the hydrogen sulfide generation rate in sewer pipes

Authors	Equation
Boon and Lister (1975)	$r = 0.228 \times 10^{-3} \times \text{COD} \times 1.07^{T-20}$
Pomeroy and Parkhurst (1977)	$r = 1 \times 10^{-3} \times \text{BOD}_5 \times 1.07^{T-20}$
Hvitved-Jacobsen *et al.* (1988)*	$r = 1.5 \times 10^{-3}\sqrt{\text{COD}_s - 50} \times 1.07^{T-20}$
Boon (1995)**	$r = 1.52 \times 10^{-3} \times \text{COD}(1 + 0.004D)/D$

r: Hydrogen sulfide generation rate in g m^{-2} h^{-1} S^{2-}
*COD: Chemical oxygen demand < 500 mg L^{-1}
***r*: Hydrogen sulfide generation rate in g m^{-3} min^{-1} S^{2-}
D: Diameter in cm
T: Temperature in°C
BOD_5: Biological oxygen demand five days in mg L^{-1}

Nielsen *et al.* (1992) showed that not only the sulfate concentration was important, but also the presence of other oxidized sulfur compounds like sulfite and thiosulfate. The hydrogen sulfide formation from organic sulfur compounds was however, insignificant compared with sulfide formation from sulfate reduction.

Further details concerning modelling of sulfide transformations are described in Chapter 6.

22.3.3 Measures for control

The aim of the following measures is to prevent corrosive microbial attack by intervening in the microbial process of sulfuric acid formation. Instead of protection against the effect of the microbial attack, these measures interfere with the cause of the corrosion, i.e. the hydrogen sulfide concentration. Recently, several methods have been proposed to control build-up of hydrogen sulfide concentration in sewers.

22.3.3.1 Decrease of sulfate concentration

At low sulfate concentrations (< 1 mg L^{-1} SO_4^{2-}-S) no sulfate reduction occurs. Different systems for sulfate removal have been described, e.g. membrane and ion-exchange processes (electrodialysis, ion-exchanger, reverse osmosis). However, most of the systems are not useful for industrial sites and wastewater with a high COD concentration. Another possibility is the removal of sulfate with addition of chemical compounds, such as $Ca(OH)_2$, $Mg(OH)_2$ and $Al_2(OH)_5Cl$. However, high costs for the addition of chemicals and processing of chemical sludge make these sulfate removal methods less useful.

22.3.3.2 *Elimination of H_2S*

Hydrogen sulfide can be removed from air in two different ways: chemically and biochemically (Boon *et al.* 1998). Alkaline and oxidative scrubbers (Drust and Deacon 1994; Smet *et al.* 1998), incorporating sodium hydroxide and oxidants, i.e. hypochlorite or ozone, are used for removing hydrogen sulfide. However, these scrubbers have high capital and operating costs because of the addition of chemical and/or oxidant. Recently, Boon *et al.* (1998) have developed a catalytic-iron filter placed in a mild steel pipework for chemical sulfide oxidation, based on a filter system of the 19th century. The initial results of this low operating cost filter (lower chemical cost) are very encouraging. At an air-flow rate of 220 m^3 h^{-1} loaded to two of these filters, an overall removal efficiency of H_2S of 92% was achieved, almost irrespective of the inlet H_2S concentration (which varied between 48 and 714 mg m^{-3} H_2S-S).

Bioscrubbers, biotrickling filters and biofilters (Pomeroy 1982; Brennan *et al.* 1996; Smet *et al.* 1998) are used to biochemically oxidize sulfides. These systems are based on the sorption of volatile contaminants in an aqueous phase or biofilm, followed by the biodegradation of the sorbed pollutants. The biological scrubbers have the advantage of low operating cost, because they do not require continuous addition of chemicals, but they may be adversely affected by high concentrations of hydrogen sulfide. Besides *Thiobacillus* sp., numerous H_2S converting microorganisms can be applied, such as the colourless sulfur bacteria *Thiotrix* and *Beggiatoa*, methylotrophs such as *Hyphomicrobium* sp., cyanobacteria and fungi.

For example, Haskoning (Dutch Consulting Engineers and Architects) constructed a type of stripper at the point of discharge of a pressurised pipeline (Heuer and Kaskens 1993). A free-fall weir creates high turbulence so that most of the hydrogen sulfide formed in the pressurized pipelines escapes from the sewage. To prevent concrete corrosion, the air with the stripped hydrogen sulfide is sucked away and led through a biofilter to eliminate the hydrogen sulfide. Heuer and Kaskens (1993) showed that in the outgoing air of the biofilter the hydrogen sulfide concentration was reduced to less than 0.01 mg m^{-3} H_2S-S, so normally no further corrosive action should occur.

22.3.3.3 *Chemical oxidation or precipitation of sulfide*

One of the conventional treatment methods involves the addition of large quantities of various chemicals in the flowing wastewater to control the sulfide concentration in the liquid phase. The chemicals used are, among others, ferric or ferrous chloride, ferric nitrate, hydrogen peroxide and sodium nitrate (Sydney *et al.* 1996). However, if chlorine or lead salts are considered, the legal requirements concerning the production of carcinogens and the addition of heavy metals to wastewater should be checked.

22.3.3.4 Microbial treatment and ozonation

Ozone is a relatively cheap and powerful oxidant (Laplanche *et al.* 1994). In the water phase, sulfur containing compounds are efficiently oxidized with ozone (Anderson 1984). The removal of H_2S is also possible from the gas phase, but for other sulfur compounds this oxidation is too slow. In the City of Tampa, Florida, an innovative combination of microbial treatment and diluted ozone has been successfully used for reducing hydrogen sulfide emissions from a large pumping station (Richman 1997).

22.3.3.5 Air injection: controlling sulfide generation

Air injection into the pressure mains is another useful and effective method to control the hydrogen sulfide concentration in the sewer atmosphere. By keeping the sewage under aerobic conditions, it is possible to prevent hydrogen sulfide generation. Several studies of the air injection method have been reported (Grusenmeyer *et al.* 1985; Ochi *et al.* 1998). These researchers found that the reaeration from the gaseous phase to the wastewater was affected by sewage flow velocity and oxygen concentration in the gaseous phase. For optimal use of this method, an oxygen balance in air-injected pressure mains was made and the minimum DO concentration to keep the sewage in aerobic conditions was investigated. In the study of Ochi *et al.* (1998), air injection completely eliminated sulfide presence at the pipe outlet when the DO at the pipe end was 0.2 mg L^{-1} or higher. In these experiments, the oxygen consumption rate in the bulk water (R_r) and the oxygen consumption rate in biofilm (R_e) were set at 13.5 mg L^{-1} h^{-1} and 1.6 g m^{-2} h^{-1}, respectively. An air quantity of 1.5 times that consumed by R_r and R_e was injected (0.045 m^3 per min). They found a DO of 0.5 mg L^{-1} at the outlet. The injection of oxygen represents in a first approximation a cost of €1.2 per inhabitant per year. Hence, these aeration costs are of the same order of magnitude as those resulting from the biogenic sulfuric acid corrosion.

22.3.3.6 New methods

The above mentioned methods are in general very effective for controlling the hydrogen sulfide concentration, but they are also expensive. Therefore, researchers are looking for new or adapted methods, i.e. effective but less expensive compared with the above-mentioned methods. A rather new method is the addition of a specially manufactured nitrate solution (Nutriox™, Bentzen *et al.* 1995) or another highly soluble biological oxidant to the pressure line to prevent septicity and to remove already formed hydrogen sulfide. By following up the exact conditions (DO, ORP) in the sewerage system, optimal dosing is possible. Elmaleh *et al.* (1998a) also stated that relatively low concentrations of oxidized nitrogen compounds ($(NO_2^- + NO_3^-)$-N < 4 mg L^{-1}) prevent anaerobic conditions, which are important for the H_2S production and undesirable for the good operation of the sewer system.

22.4 SULFURIC ACID PRODUCTION IN SEWERS

22.4.1 Sulfuric acid production

The sulfide in the liquid phase coexists in several forms, such as dissolved H_2S, HS^- and S^{2-}. The dissolved H_2S gas is volatilized from the liquid to the gaseous phase (or the sewer headspace). Insufficient ventilation of these areas leads to the second step, an enrichment of the H_2S in the atmosphere at the pipe walls where it is dissolved in the moisture or biofilm present on the crown of the sewer or is sorbed directly into the pores of the concrete. It diffuses into the concrete in both the air and water phases and the dissolved H_2S is then oxidized to elemental sulfur by oxygen. At the crown or the walls above the water line of the sewer, sulfur or sulfide oxidizing bacteria (SOB) oxidize the solubilized sulfur compounds to sulfuric acid (Kelly and Harrison 1989). New concrete has a rather low permeability and only a small fraction of its pores (5%) are large enough for the penetration of microorganisms (> 1 μm). However, small interconnected voids allow diffuse transfer of dissolved compounds and chemical gradients are quickly established. The neutralization of the concrete alkalinity and the dissolution of calcium hydroxide by acid enlarges the porosity. Consequently, the corrosion process is associated with slowly increasing penetration of the concrete by microorganisms (Islander *et al.* 1991). The corrosion product formed by the biogenic sulfuric acid attack is a layer of white precipitate that accumulates on the concrete surface. It consists of a combination of sand particles, unreacted cement particles and various sulfates, produced by re-precipitation of ions dissolved from the cement. Bock and Sand (1986) identified the main part of the material as gypsum ($CaSO_4.2H_2O$) with no structural strength. This corrosion layer can change the microenvironment;

Start corrosion:
Lime carbonation with decrease of the pH

$Ca(OH)_2 + CO_2 \rightarrow CaCO_3 + H_2O$

$2\ CO_2 + Ca(OH)_2 \rightarrow Ca(HCO_3)_2$

Corrosion reactions:
$Ca(OH)_2 + H_2SO_4 \rightarrow CaSO_4.2H_2O$
(gypsum)

$3\ CaO.Al_2O_3.12H_2O + 3\ (CaSO_4.2H_2O) + 14\ H_2O \rightarrow 3\ CaO.Al_2O_3.3CaSO_4.32H_2O$
(ettringite)

$3\ CaO.2SiO_2.nH_2O + 3\ H_2SO_4 \rightarrow 3\ CaSO_4.2H_2O + 2\ SiO_2 + nH_2O$

Figure 22.4. Overview on the corrosion reactions of concrete structures.

it holds water and reduces the effects of dry periods on the microorganisms. It can, however, also act as a barrier and slow down the transport of chemical compounds, such as sulfuric acid. Moreover, $CaSO_4$ can react with Ca-aluminate, present in the cement mixture and ettringite can be formed (Figure 22.4). Ettringite ($CaO.Al_2O_3.3CaSO_4.32H_2O$) is an expansive product and can lead to increased internal pressure resulting in small cracks. Gypsum is formed at pH levels less than 3 on the surface of the concrete sewer pipe. Ettringite is produced in the concrete at pH levels higher than 3 (Mori *et al.* 1992). Moreover, the calcium loss diminishes the resistance of the concrete structure.

Aerobic oxidation of hydrogen sulfide to sulfuric acid is complex and may occur in many steps. The oxidation of sulfide can occur abiotically with the production of thiosulfate or tetrathionate on new concrete surfaces. However, on corroded concrete, presumably at lower pH, elemental sulfur is produced (Parker 1945). Mori *et al.* (1992) showed that the surface of a corroded concrete pipe had a very low pH of 1.9 and a high moisture content of 20.8%. Colourless *Thiobacillus*-like bacteria are capable of oxidation of elemental sulfur, as reported by numerous authors (Kelly 1982; Trüper 1984; Hazeu *et al.* 1988). The pH of uncorroded concrete is approximately 12, and sulfate producing bacteria cannot grow when directly exposed to such a high pH. However, when the concrete is exposed to the air, the pH of the surface decreases owing to carbonation. In addition, H_2S is an acidic gas and it contributes to neutralization of the concrete surface. After neutralization of the concrete, sulfate producing bacteria are able to grow on the surface. *Thiobacillus thiooxidans* especially dominates at very low pH.

Hutchinson *et al.* (1969) found that elemental sulfur allowed growth of

T. thiooxidans at lower pH (down to pH 1) than other substrates. Sand *et al.* (1987) found that if the main nutrition source is known, e.g. H_2S, thiosulfate, methylmercaptan, it is possible to predict the composition of the thiobacilli microbiota. With the nutrient source H_2S, which is chemically oxidized to sulfur and subsequently microbiologically oxidized to sulfate, *T. thiooxidans* dominates and a severe corrosion occurs. This bacterium occupies a constrained ecological niche that is characterized by extremely low pH (pH 1-5) and low-organic conditions. *T. thiooxidans* assimilates carbon dioxide and obtains energy exclusively from the aerobic oxidation of reduced sulfur forms such as H_2S (Kelly 1982). A new finding by Cho and Mori (1995) is the coexistence with *T. thiooxidans* of an acid-resistant and H_2S-oxidizing fungus on severely corroded sewer pipes. This fungus oxidizes H_2S to thiosulfate, probably as a sulfide detoxification process. Thiosulfate can then be used by *T. thiooxidans* as energy source. Moreover, the existence of an association is reasonable to assume because the heterotroph can degrade the organics excreted by *T. thiooxidans*. These organics (pyruvate, oxalate) work inhibitory at certain concentrations (2.10^{-5}, 7.10^{-5} M) on *T. thiooxidans*. A less strong corrosion occurred with thiosulfate as nutrient and *T. neapolitanus*, *T. intermedius* and *T.novellus* as the dominant thiobacilli (Table 22.3). However, methylmercaptan and other organic sulfides such as dimethylsulfide can, according to Sand *et al.* (1987), be excluded as nutrients for thiobacilli and hence are not a source of sulfuric acid. This is in contrast to the results of other researchers (Sivelä and Sundman 1972). This discrepancy can possibly be explained by the aerobic degradation of methylmercaptan and other organic sulfides by heterotrophic bacteria and fungi.

The establishment of *T. thiooxidans* populations in corroded concrete pipes is presumed to be preceded by other *Thiobacillus* species (Milde *et al.* 1983). The surface pH of an uncorroded concrete pipe after neutralization by carbonation and H_2S is approximately 2 pH-units too high for the growth of *T. thiooxidans* (optimal pH 5 and lower). However, at neutral pH, colonization by mid-pH thiobacilli, such as *T. intermedius*, *T. neapolitanus* and *T. thioparus* is feasible. These bacteria, which also generate acid by oxidizing H_2S, offer a mechanism by which the local pH is decreased and make the growth of *T. thiooxidans* possible. Subsequent colonization of *T. thiooxidans* further decreases the pH.

It is interesting to make a rough calculation of relative rates at which H_2S-production by sewage can corrode the concrete in the ceiling of the sewer. Consider a H_2S-flux of 0.2 g m^{-2} h^{-1} S^{2-} from the sewer atmosphere to the biofilm on the concrete surface layer in a sewer half filled with sewage.

Table 22.3. Important Thiobacilli species for biogenic sulfuric acid corrosion and their characteristics

Species	Characteristics	Description
T. thioparus	Strictly autotrophic, aerobic, energy derived from oxidation of thiosulfate, sulfur, hydrogen sulfide and tetrathionate	Thin, short rods, 0.5 by 1.0-3.0 μm, motile with polar flagellum
T. novellus	Facultatively autotrophic, aerobic, utilisable organic compounds repress the oxidation of thiosulfate	Short rods, coccoidal or ellipsoidal cells, 0.4-1.0 by 0.6-4.0 μm, non-motile
T. intermedius	In media containing both sulfur and organic both are oxidized simultaneously in contrast to *T. novellus*	Thin, short rods, 0.5 by 1.0-2.0 μm
T. neapolitanus	Strictly autotrophic, but will assimilate organic compounds in the presence of an oxidizable sulfur compound, derives energy by the oxidation of thiosulfate, sulfur, and hydrogen sulfide	Short rods, 0.5 by 1.0-1.5 μm, non-motile
T. thiooxidans	Strictly autotrophic, aerobic, rapid oxidation of sulfur and oxidation of other reduced sulfur compounds	Short rods, 0.5 by 1.0-2.0 μm, occurring singly, in pairs or in short chains

A rough calculation learns that in concrete with 350 kg cement, containing 63% CaO, 3.7 mol CaO is present in the outer 1 mm surface of the sewer pipe. On a concrete surface of 1 m^2, maximum 54.8 mol H_2SO_4 can be formed per year. So, the base equivalents present in the upper 1 mm layer of the concrete are estimated as about one tenth the acid equivalents formed. Hence, two things are possible. On the one hand it can be that only a part (< 10%) of the formed H_2SO_4 reacts with the concrete. On the other hand, H_2SO_4 can be formed and penetrate into deeper pores to react with concrete compounds in the deeper concrete layers. The latter warrants further elucidation considering that corrosion rates of 4.3-4.7 mm per year were measured (Mori *et al.* 1991).

22.4.2 Measures for control

Since 20% of the damage of concrete structures in sewerage systems seems to be caused by sulfuric acid or sulfate attack (Kaempfer and Berndt 1998), it is necessary to take adequate measures against this type of corrosion. Most of the conventional solutions have a protective character. The object of these measures is to enhance or improve the resistance of the entire concrete structure or the concrete surface. There are several conventional measures, appearing to be suitable for enhancing the chemical resistance of concrete. In addition, some new approaches need to be explored.

22.4.2.1 Special cement additions and treatments for enhanced resistance to acids

The resistance of conventional types of concrete prepared by using standard types of cement is not sufficient to prevent corrosion in those areas of wastewater treatment plants in which corrosion by biogenic sulfuric acid cannot be ruled out because of operating conditions, long retention times, high sulfate content of the wastewater, etc. Investigations of Schmidt *et al.* (1997) have shown that concrete correctly produced using a high-alumina cement has a higher resistance to sulfuric acid and a higher abrasion resistance. Whereas high alumina cement concrete does not give good results when it is submerged in an H_2SO_4 bath, it gives better results than other concrete types when tested in a microbial simulation chamber. This is an indication that the substrate-bacterium interaction plays an important, perhaps crucial, role in the biogenic sulfuric acid attack. A possible explanation is that the hydrate of calcium aluminate is much more stable than the calcium hydroxide. In addition to this, aluminium hydroxide forms gels of high volume which subsequently block up the pores in the concrete. In this way, penetration of aqueous solutions is to a large extent inhibited. The addition of 10% silica fume to portland cement also gives a better sulfuric acid resistance, although it is less important compared with the other cement improvements. Another solution that can reduce corrosion is the use of sulfate-resisting cement. One can also add fibre mesh to the concrete and treat the concrete with linseed oil (Schmidt *et al.* 1997). Most of these measures, however, work very well with a pure chemical attack of H_2SO_4, but give less protection against microbial attack. The latter clearly reveals that the bacteria form H_2SO_4 in a different way or at different sites in the concrete. These aspects need to be examined in more detail, also because these special cement mixtures can cost up to five times more than the usual cement mixture.

Another possible measure is reducing the water/cement factor to no more than 0.45 and decreasing the depth of penetration by water to less than 2.0 cm (Schmidt *et al.* 1997). This method is more a postponement-most of the concrete pipes have a water/cement ratio of 0.5 or lower-than a protection against the corrosive attack of sulfuric acid.

22.4.2.2 Polymer cement concrete

The use of polymers (for example styrolacrylester) often leads to a denser microstructure, smaller discontinuous pores, a better bond between the aggregate and the cement matrix, and bridged microcracks relative to conventional concrete. Shaker *et al.* (1997) have investigated the durability of reinforced concrete structures with addition of a polymer, more specifically a styrenebutadiene latex modified concrete. They found a denser microstructure and a strongly improved resistance to a sulfate solution. Generally, the porosity or pore volume of the polymer-modified concrete decreases from 20% to 5% in layer radii of 0.2 μm or more, and increases greatly in smaller radii of 75 nm or less. However, more fundamental research on the microbial corrosion should be done in relation to pore size distribution and the use of polymers in general. The price of concrete pipes with addition of polymers is about 20-50% higher than for normal concrete pipes.

22.4.2.3 Acid-resistant coatings and linings

Kaempfer and Berndt (1998) have investigated the behaviour of different polymer-modified mortars towards attack of biogenic sulfuric acid. It was determined that polymer modified mortars can be used as linings for renovation and restoration in sewer systems with high acid loads. It was estimated that the resistance to sulfuric acid attack of polymer modified mortars is up to 10 times higher than for normal cement mortars. Some other linings used against biogenic sulfuric acid are those with thermoplastics, duroplastics, and synthetic-resin concrete. On condition that the surface of the concrete wall is cleaned and the coefficient of adhesion is sufficient, linings of 5-15 mm can be applied. Most of the linings comprise two separate components: a support fabric and a coating of an appropriate polymer. For example, for the repair of sewer pipes, a liner made up of a felt material of selected thickness and coated with a linear low-density polyethylene coating can be used.

Different coatings for concrete pipes are also available: coatings containing bituminous materials, coatings based on polyurethane, epoxy-resin, unsaturated polyester resin, and the combination of synthetic resin and bituminous raw materials (Kaempfer and Berndt 1998).

The cost for concrete pipes with these linings or coatings is dependent on the type and the thickness of the lining or coating and can be, roughly estimated, two to three times more expensive than the concrete pipe without linings or coatings.

However, a substantial safety factor should be ensured because once the coating or lining fails, the corrosion can act unlimited on the concrete surface. For example, calcareous aggregates are recommended as a second line of defense for concrete sewers where some form of protection of the concrete is contemplated.

22.4.2.4 New methods

More and more researchers are looking for alternatives to reduce the costs of the technical implications related to concrete corrosion. Some of the alternative measures are listed below.

Addition of biocides

When using biocides, it is important to take into consideration that a quantity of the applied biocides into the liquid stream could affect the downstream biological processes. Biocides are used in treatment processes on the pipe surface, mostly in combination with other treatments, for example the crown spray process (see further). However, the use of biocides in the crown spray process and other processes may not be practical at the present time, because the biocides currently used are not specific to sulfide and sulfur oxidizing bacteria (Sydney *et al.* 1996). Research by Sydney *et al.* (1996) showed that quaternary ammonium compounds clearly possess biocidal properties against sulfur oxidizing bacteria, but the exact mechanism for deactivation is not known. Another problem is the alkalinity of concrete, as most of the organic biocides are unstable in such high pH-values.

Many compounds (for example phenol derivatives and quaternary ammonium salts) have been considered and tried for the combination of the necessary biocidal activity and persistence with lack of undesirable side effects. Sanderson and Stewart (1997) did some experiments with monochloramine. They found that it was more effective to deliver monochloramine in a short concentrated dose (2-4 mg L^{-1}) than in a longer dose of lower concentration (< 0.5 mg L^{-1}), owing to the production of monochloramine-neutralizing biomass constituents. They also expected the same for any biocide to which microorganisms can adapt.

Metals

Other investigators (Maeda *et al.* 1996) are investigating the addition of metals to inhibit sulfur oxidizing bacteria. Some research was done on the addition of selenium, which resulted in a concrete structure more resistant against thiobacilli. Recently, an antibacterial concrete based on Portland cement supplemented with 0.1% Ni by mass was developed by Maeda *et al.* (1996). They found that with *T. thiooxidans* both cellular activities of elemental sulfur oxidation and CO_2 incorporation were strongly inhibited by a concentration of the nickel inhibitor of 0.1% by mass of cement in concrete. This indicates that Ni binds to cells and inhibits enzymes involved in sulfur oxidation, resulting in an inhibition of cell growth.

Table 22.4. Treatment costs associated with various corrosion control methods evaluated by the County Sanitation Districts of Los Angeles County (After: Padival 1995; Sydney et al. 1996)

Chemical treatment method	Estimated total annual costs, €10^6
Crown spray process	0.5[a]
Addition of chemicals in the sewage	
Ferrous chloride	3.7[b]
Ferric nitrate	14[b]
Hydrogen peroxide	21[b]
Oxygen injection in the pipe	9[b]

[a] Based on full-scale program with 50% magnesium hydroxide to treat 136 km of corroded sewers. The costs include the amortized capital investment (operations and management), and chemicals.
[b] Based on previous full-scale evaluations for treating the average outfall system flow of $1.2.10^6$ m^3/d. The costs include the amortized capital investment, operations and management, and chemicals.

Crown spray process

The crown spray process is a new and innovative process to control sewer crown corrosion (Sydney *et al.* 1996). This new technology controls the crown corrosion directly by deactivating the sulfur or sulfide oxidizing bacteria present and neutralizing the acid generated on the crown. The basic spraying system consists of a spray head mounted on a float. The float is pulled through the sewer at a controlled rate to spray the crown at a predetermined application rate. Experiments show that magnesium hydroxide is the most effective deactivating chemical for the crown spray process. A single application of 50% magnesium hydroxide maintained the sprayed sewer crown at pH 9 for 9 months and the number of sulfur or sulfide oxidizing bacteria remained at the treated surface at 1 to 2 log orders of magnitude below untreated levels. Reapplication of the spray may control corrosion of the treated surface indefinitely. The crown spray process should be used to treat corroded sewers whose structural integrity has not been compromised, because the application of a chemical coating will not restore the structural integrity of a compromised sewer. Economic comparisons of the crown spray process with the conventional sewer treatment methods, i.e. the addition of large quantities of various chemicals in the flowing wastewater, heavily favour the crown process (Table 22.4).

Competition between microorganisms

The competitive success of an organism in a particular environment depends upon its ability to distinguish itself in some important way from potential competitors. Thiobacilli are characterised by slow growth. This is a disadvantage compared with faster growing species when the growth rates of both are controlled by the same limiting factor. However, in the sewer environment, attachment to the pipe wall and even penetration into the crown concrete, which becomes increasingly likely as corrosion proceeds, may protect thiobacilli, making them more difficult to displace via competitive methods. The low pH environment that is established by the acidophilic thiobacilli, even at depth in rapidly corroding concrete, may provide a formidable barrier to encroachment by species that cannot tolerate these extreme conditions. Padival *et al.* (1995) examined the competitive nature of heterothrophs, more specific yeasts isolated from the sewer crown, against thiobacilli by changing nutrient conditions (N and C limited). They found that it was possible to displace established populations of both *T. thiooxidans* and *T. neapolitanus* in continously stirred tank reactors. However, experiments more linked to wastewater collection systems and sewer pipes should be done to take specific environmental conditions, like surface attachment, into consideration.

22.5 ACKNOWLEDGEMENTS

This manuscript was funded by a research grant of the Fund for Scientific Research Flanders (FWO-Vlaanderen) and the Flemish Institute for the Improvement of Scientific Technological Research in the Industry (IWT). The authors gratefully acknowledge Eva Top, Jan Kielemoes, Sarah Philips and Laurent Verschuere for critically reading the manuscript.

22.6 REFERENCES

Anderson, R. (1984) Sewage treatment processes: the solution to the odour problem. In *Characterization and Control of Odoriferous Pollutants in Process Industries* (eds. Vigneron, S., Hermia, J. and Chaouki, J.), pp. 471-496, Amsterdam, Nederland Elsevier Science B.V.

Bentzen, G., Smith, A.T., Bennett, D., Webster, N.J., Reinholt, F., Sletholt, E. and Hobson, J. (1995) Controlled dosing of nitrate for prevention of H_2S in a sewer network and the effects on the subsequent treatment processes. *Wat. Sci. Technol.* **31**(7), 293-302.

Bock, E. and Sand, W. (1986) Applied electron microscopy on the biogenic destruction of concrete and blocks-use of the transmission electron microscope for identification of mineral acid producing bacteria. In *Proc. 8th Int. Conf. on Cement Microscopy*, (eds. Bayles, J., Gouda, G.R. and Nisperos, A.), International Cement Microscopy Association, Duncanville, Texas.

Boon, A.G. and Lister, A.R. (1975) Formation of sulfide in rising main sewers and its prevention by injection of oxygen, *Prog. Wat. Tech.* **7**(2), 289-300.

Boon, A.G. (1995) Septicity in sewers: causes, consequences and containment. *Wat. Sci. Technol.* **31**(7), 237-253.

Boon, A.G., Vincent, A.J. and Boon, K.G. (1998) Avoiding the problems of septic sewage. *Wat. Sci. Technol.* **37**(1), 223-231.

Brennan, B.M., Donlon, M. and Bolton, E. (1996) Peat biofiltration as an odour control technology for sulphur-based odours. *J. CIWEM* **10**(3), 180-198.

Cho, K.-S. and Mori, T. (1995) A newly isolated fungus participates in the corrosion of concrete sewer pipes. *Wat. Sci. Technol.* **31**(7), 263-271.

Coleman, R.N. and Gaudet, I.D. (1993) *Thiobacillus neopolitanus* implicated in the degradation of concrete tanks used for potable water storage. *Water Res.* **27**, 413-418.

Colleran, E., Finnegan, S. and Lens, P. (1995) Anaerobic treatment of sulphate-containing waste streams. *Ant. Leeuwenhoek* **67**, 29-46.

De Ceukelaire, L. (1989) Mineralogie van beton in verband met verweringsverschijnselen. Ph.D. thesis volume I, University of Gent, Belgium.

DeHollander, G.R. (1998) Gaseous emissions from wastewater facilities. *Water Environ. Res.* **70**(4), 580-585.

Delgado, S., Alvarez, M., Rodríguez-Gómez, L.E. and Aguiar, E. (1999) H_2S generation in a reclaimed urban wastewater Pipe. Case study: Tenerife (Spain). *Water Res.* **33**(2), 539-547.

Derangère, D. and Cochet, C. (1991) Concrete corrosion in individual septic tank systems. *Eur. Water Pollut. Contr.* **1**(6), 24-28.

Drust, J. and Deacon, R. (1994) A case history of odour control using chemical scrubbers. *Proc. Symp. on Odour Control and Prevention in the Water Industry.* CIWEM, Newcastle upon Tune, United-Kingdom, 14 April 1994.

Elmaleh, S., Delgado, S., Alvarez, M., Rodríguez-Gómez, L.E. and Aguiar, E. (1998a) Septicity and H_2S build-up in a reclaimed wastewater pipe. AWT98-Advanced Wastewater Treatment, Recycling and Reuse, pp. 1019-1022, Milano, Italy, 14-16 September 1998.

Elmaleh, S., Delgado, S., Álvarez, M., Rodríguez-Gómez, L.E. and Aguiar, E. (1998b) Forecasting of H_2S build-up in a reclaimed wastewater pipe. *Wat. Sci. Technol.* **38**(10), 241-248.

Ferris, M.J., Nold, S.C., Revsbech, N.P. and Ward, D.M. (1997) Population structure and physiological changes within a hot Spring microbial community following disturbances. *Appl. Environ. Microbiol.* **63**(4), 1367-1374.

Grusenmeyer, S., Verstraete, W. and Defrancq, J. (1985) Waterstofsulfide (H_2S) in het milieu : vorming, gevolgen en mogelijke remediëring. *Water* **22**, 106-112.

Hamilton, W.A. (1985) Sulfate-reducing bacteria and anaerobic corrosion. *Ann. Rev. Microbiol.* **39**, 195-217.

Hazeu, W., Battenburg, W.H., Bos, P., van der Pas, R. K. and Kuenen, J.G. (1988) The production and utilization of intermediary elemental sulphur during the oxidation of reduced sulphur compounds by *Thiobacillus ferrooxidans*. *Arch. Microbiol.* **150**, 574-579.

Heuer, J.J.M.B. and Kaskens, H.J. (1993) Prevention of concrete corrosion and odour annoyance with biofiltration. *Wat. Sci. Technol.* **27**(5, 6), 207-218.

Hilton, B.L. and Oleszkiewicz, J.A. (1988) Sulfide-induced inhibition of anaerobic digestion. *J. Environ. Eng.* **114**, 1277-1282.

Holder, G.A., Vaughan, G. and Drew, W. (1985) Kinetic studies of the microbiological conversion of sulfate to hydrogen sulfide and their relevance to sulfide generation within sewers. *Wat. Sci. Technol.* **17**(2-3), 183-188.

Holder, G.A., van Oorschot, R. and Hauser, J. (1989) Experimental and theoretical studies of sulfide generation in sewerage systems. *Wat. Sci. Technol.* **21**(8-9), 757-768.

Holder, G.A. (1994) Application of biofilm theory to the prediction of sulfide production from bioslimes. *Wat. Sci. Technol.* **29**(10-11), 537-543.

Hutchinson, M., Johnstone, K.I. and White, D. (1969) Taxonomy of the genus *Thiobacillus*: The outcome of numerical taxonomy applied to the group as a whole. *J. Gen. Microbiol.* **57**, 397-410.

Hvitved-Jacobsen, T., Jütte, B., Nielsen, P.H, and Jensen, N. A. (1988) Hydrogen sulfide control in municipal sewers. In *Pretreat. in Chem. Wat. and Wast. Treat., Proc. of the 3rd International Gothenburg Symposium*, (eds. Hahn H.H. and Klute R.), pp.239-247, Gothenburg, Sweden, June 1-3 1988.

Hvitved-Jacobsen, T., Vollertsen, J. and Tanaka, N. (1998) Wastewater quality changes during transport in sewers - an integrated aerobic and anaerobic model concept for carbon and sulfur microbial transformations. *Water Qual. Int. '98* **8**, 252-259.

Ingvorsen, K., Zehnder, A.J.B. and Jorgensen, B.B. (1984) Kinetics of sulfate and acetate uptake by *Desulfobacter postgatei*. *Appl. Environ. Microbiol.* **47**, 403-587.

Isa, Z., Grusenmeyer, S. and Verstraete, W. (1986) Sulfate reduction relative to methane production in high-rate anaerobic digestion : Microbiological aspects. *Appl. Environ. Microbiol.* **51**(3), 580-587.

Islander, R.L., Devinny, J.S., Mansfeld, F., Postyn, A. and Shih, H. (1991) Microbial ecology of crown corrosion in sewers. *J. Environ. Eng.* **117**(6), 751-770.

Kaempfer, W. and Berndt, M. (1998) Polymer modified mortar with high resistance to acid and to corrosion by biogenous sulfuric acid. *Proc. of the IXth ICPIC Congress*, Bologna, Italy, 681-687.

Kelly, D.P. (1982) Biochemistry of the chemolithotrophic oxidation of inorganic sulfur. *Phil. Trans. R. Soc. Lond. Ser. B.* **298**, 499-528.

Kelly, D.P. and Harrison, A.P. (1989) The Genus *Thiobacillus*. In *Bergey's Manual of Systematic Bacteriology*, (eds. Staley, M.P., Bryant, M.P., Pfenning, N. and Holt, J.G.), pp. 1842-1858, Waverly Press, Baltimore, USA.

Kitagawa, M., Ochi, T. and Tanaka, S. (1998) Study on hydrogen sulfide generation rate in pressure mains. *Wat. Sci. Technol.* **37**(1), 77-85.

Laplanche, A., Bonnin, C. and Darmon, D. (1994) Comparative study of odors removal in a wastewater treatment plant by wet scrubbing and oxidation by chlorine or ozone. In *Characterization and Control of Odoriferous Pollutants in Process Industries* (eds Vigneron, S., Hermia, J. and Chaouki, J.), pp. 277-294, Amsterdam, Nederland, Elsevier Science B.V.

Lens, P., de Beer, D., Cronenberg, C., Ottengraf, S. and Verstraete, W. (1995) The use of microsensors to determine population distributions in UASB aggregates. *Wat. Sci. Technol.* **31**(1), 273-280.

Maeda, T., Negishi, A., Nogami, Y. and Sugio, T. (1996) Nickel inhibition of the growth of a sulfur-oxidizing bacterium isolated from corroded concrete. *Bioscience, Biotechnology, Biochemistry* **60**(4), 626-629.

Manz, W., Eisenbrecher, M., Neu, T.R. and Szewzyk, U. (1998) Abundance and spatial organization of gram-negative sulfate-reducing bacteria in activated sludge investigated by in situ probing with specific 16S rRNA targeted oligonucleotides. *FEMS Microbiol. Ecol.* **25**(1), 43-61.

Matos, J.S. and de Sousa, E.R. (1992) The forecasting of hydrogen sulfide gas build-up in sewerage collection systems. *Wat. Sci. Technol.* **26**(3-4), 915-922.

Milde, K.W., Sand, W.W. and Bock, E. (1983). Thiobacilli of the corroded concrete walls of the Hamburg sewer system. *J. Gen. Microbiol.* **129**, 1327-1333.

Monteith, H., Bell, J. and Harvey, T. (1997) Assessment of factors controlling sewer ventilation rates. *Proc. Control Odors Volatile Organic Chem. Emissions Conf.*, Houston, Texas, U.S.A.

Mori, T., Koga, M., Hikosaka, Y., Nonaka, T., Mishina, F., Sakai, Y. and Koizumi, J. (1991) Microbial corrosion of concrete sewer pipes, H_2S production form sediments and determination of corrosion rate. *Wat. Sci. Technol.* **23**, 1275-1282.

Mori, T., Nonaka, T., Tazaki, K., Koga, M., Hikosaka, Y. and Noda, S. (1992) Interactions of nutrients, moisture and pH on microbial corrosion of concrete sewer pipes. *Water Res.* **26**(1), 29-37.

Nielsen, P.H. and Hvitved-Jacobsen, T.H. (1988) Effect of sulfate and organic matter on the hydrogen sulfide formation in biofilms of filled sanitary sewers. *J. Wat. Pollut. Control Fed.* **60**, 627-632.

Nielsen, P.H., Raunkjaer, K., Norsker, N.H., Jesen, N.A. and Hvitved-Jacobsen, T. (1992) Transformation of wastewater in sewer systems-a Review. *Wat. Sci. Technol.* **25**(6), 17-31.

Nielsen, P.H., Raunkjaer, K. and Hvitved-Jacobsen, T.H. (1998) Sulfide production and wastewater quality in pressure mains. *Wat. Sci. Technol.* **37**(1), 97-104.

Ochi, T. Kitagawa, M. and Tanaka, S. (1998) Controlling sulfide generation in force mains by air injection. *Wat. Sci. Technol.* **37**(1), 87-95.

Okabe, S., Matsuda, T., Satoh, H., Itoh, T., Watanabe, Y. (1998) Sulfate reduction and sulfide oxidation in aerobic mixed population biofilms. *Wat. Sci. Technol.* **37**(4-5), 131-138.

Oude Elferink, S.J.W.H., Visser, A., Hulshoff, P.L.W. and Stams, A.J.M. (1994) Sulfate reduction in methanogenic bioreactors. *FEMS Microbiol. Rev.* **15**, 119-136.

Padival, N.A.,Weiss, J.S. and Arnold, R.G. (1995) Control of *Thiobacillus* by means of microbial competition : Implications for corrosion of concrete sewers. *Water Environ. Res.* **67**(2), 201-205.

Parker, C.D. (1945) The Corrosion of Concrete 1. The Isolation of a species of bacterium associated with the corrosion of concrete exposed to atmosphere containing hydrogen sulfide. *Aust. J. Exp. Biol. Med. Sci.* **23**, 81-90.

Pomeroy, R.D. (1974) *Process Design Manual for Sulphide Control in Sanitary Sewerage Systems.* U.S. Environmental Protection Agency-Technology Transfer.

Pomeroy, R.D. and Parkhurst, J.D. (1977) The forecasting of sulfide build-up rates in sewers. *Prog. Wat. Technol.* **9**, 621-628.

Pomeroy, R.D. (1982) Biological treatment of odorous air. *J. Wat. Pollut. Control Fed.* **54**(12), 1541-1545.

Ramsing, N.B., Kühl, M. and Jorgensen, B.B. (1993) Distribution of sulfate-reducing bacteria, O_2, and H_2S in photosynthetic biofilms determined by oligonucleotide probes and microelectrodes. *Appl. Environ. Microbiol.* **59**(11), 3840-3849.

Richman, M. (1997) Microbial system controls odors in Florida collection system. *Water Environ. Technol.* **9**(10), 20-29.

Rintala, J. and Lettinga, G. (1992) Effects of temperature elevation from 37 to 55 °C on anaerobic treatment of sulfate-rich acidified wastewaters. *Environ. Technol.* **13**, 801-810.

Sagrov, S. (1998) Durability control of concrete sewer pipes - a system for durability documentation. *Water Qual. Int. '98* **8**, 260-267.

Sand, W. and Bock, E. (1984) Concrete corrosion in the Hamburg sewer systems. *Environ. Technol. Lett.* **5**, 517-528.

Sand, W., Bock, E. and White, D.C. (1987) Biotest system for rapid evaluation of concrete resistance to sulfur-oxidizing bacteria. *Mater. Perform.* **26**, 14-17.

Sand, W., Dumas, T. and Marcdargent, S. (1992) Tests for biogenic sulfuric acid corrosion in a simulation chamber confirm the on site performance of calcium aluminate based concretes in sewage applications. *Proc. of the International Symposium on Microbiologically Influenced Corrosion Testing.*, Miami, Florida, USA, 16-17 November, 1992.

Sanderson, S.S. and Stewart, P.S. (1997) Evidence of bacterial adaptation to monochloramine in *Pseudomonas aeruginosa* biofilms and evaluation of biocide action model. *Biotechnol. Bioeng.* **56**(2), 201-209.

Santegoeds, C.M., Muyzer, G. and de Beer, D. (1998) Biofilm dynamics studied with microsensors and molecular techniques. *Wat. Sci. Technol.* **37**(4-5), 125-129.

Shaker, F.A., El-Dieb, A.S. and Reda, M.M. (1997) Durability of styrene-butadiene latex modified concrete. *Cem. Concrete Res.* **27**(5), 711-720.

Schmidt, M., Hormann, K., Hofmann, F.-J. and Wagner, E. (1997) Beton mit erhöhtem Widerstand gegen Säure und Biogene Schwefelsäurekorrosion. *Concrete Precasting Plant Technol.* **4**, 64-70.

Schramm, A., Larsen, L.H., Revsbech, N.P., Ramsing, N.B., Amann, R. and Schleifer, K.-H. (1996) Structure and function of a nitrifying biofilm as determined by in situ hybridization and the use of microelectrodes. *Appl. Environ. Microbiol.* **62**(12), 4641-4647.

Sivelä, S.K. and Sundman, V. (1972) Demonstration of *Thiobacillus* type bacteria which utilize methyl sulfides. *Arch. Microbiol.* **103**, 303-304.

Smet, E., Lens, P. and Van Langenhove, H. (1998) Treatment of waste gases contaminated with odorous sulfur compounds. *Environ. Sci. Technol.* **28**(1), 89-117.

Sydney, R., Esfandi, E. and Surapaneni, S. (1996) Control concrete sewer corrosion via the crown spray process. *Water Environ. Res.* **68**(3), 338-347.

Trüper, H.G. (1984) Microorganisms and the sulphur cycle. In *Sulphur, its Significance for Chemistry, for the Geo-, Bio-, and Cosmosphere and Technology. Studies in inorganic chemistry,* (eds. Muller, A. and Krebs, B.), pp. 351-365, Elsevier Science Publishers, Amsterdam.

U.S. Environmental Protection Agency (1985) Control of volatile organic compound emissions from industrial wastewater. Research Triangle Park, N.C.

Vancalbergh, L. (1996) Corrosieverschijnselen in rioleringen-Voorspelbaar! *Water* **88**, 129-136.

Widdel, F. (1988) Microbiology and ecology of sulfate- and sulfur-reducing bacteria. In *Biology of Anaerobic Microorganisms*. (ed. Zehnder, A.J.B.), pp. 469-585, John Wiley, New York.

Widdel, F. and Hansen, T.A. (1991) The dissimilatory sulfate- and sulfur-reducing bacteria. In *The Prokaryotes*, (eds. Balows, A; Truper, H.G.; Dworkin, M.; Hardner, W. and Schleifer, K.), Springer-Verlag, New York.

Index

A

Acid mine drainage (AMD), 176, 335, 338, 346, 401
Acid sulfate soils, 339, 345, 368
Acidianus
  sp., 52
  A. ambivalens, 67
Acidification, 36, 179, 291, 335
Acidiphilium sp., 201
Acidithiobacillus sp., 54; *see also Thiobacillus* sp.
Activated sludge
  bulking, 450
  process, 449
Aerosols, 37
Air–water partition coefficient, 282
Alliins, 356
Allium sp., 360
Alkanolamines, 248, 267
Aluminium, 180, 334, 397
Ammonium
  formation, 416
  toxicity, 161, 242, 298
  sulfate, 362
Ammonification, 440
APS, 24, 48, 356
  kinase, 75
  reductase, 59, 70, 423
Aromatic sulfonates, 377
Atmospheric pollution, 36, 265, 281, 363
ATP sulfurylase, 59, 69, 356
Azo-dyes, 378

B

Barites, 34
Beggiatoa sp., 52, 168, 425, 453
Benzothiophene, 88
Biofilter, 286
Biohydrometallurgy, 106
Bioleaching, 105, 341
Biomineralogy, 107, 500
Biomining, 106, 177
Bioscrubber, 284
Biosorption, 461
Biotrickling filter, 284
Brassica sp., 359
Bulking sludge, 450

C

Cadmium, 187, 337, 368, 395
Carbon disulfide, CS_2, 241, 282
Carbonyl sulfide, COS, 35, 241, 282
Cathodic depolarisation, 493
Cat clays, 339
Cement, 148, 516
Chalcopyrite, 34, 109, 180, 506
Chlorobiaceae, 52
Chromatiaceae, 52
Claus
  reaction, 245
  process, 244, 267
  tail gas treatment, 244
Climate interactions, 37
Clostridium sp., 73, 382
Cobalt, 96, 159, 397
Comamonas sp., 384
Competition
  aerobic, 345, 460, 529
  anaerobic, 154, 521
Compost, 186, 286
Concrete, 147, 515, 528
Control
  adaptive, 220
  disturbance monitoring, 218
  fault detection and isolation (FDI), 224,
  fuzzy, 221,
  model predictive, 219,
  robust, 221
Control of
  sludge bulking, 458
  sulfide formation, 159, 485, 525
  corrosion, 149, 525, 531
Copper, 188, 331, 398, 494, 498
Corrosion,
  crown, 517
  pit, 503
  processes, 493, 516, 528

D

Desulfurisation, 87, 237
Desulfonation, 380
Desulfobacterium sp., 520
Desulfobulbus sp., 68, 480
Desulfococcus sp., 68, 520
Desulforomonas sp., 71, 520
Desulfosporosinus orientis, 200

Desulfotomaculum sp., 68, 199, 479, 520
Desulfovibrio sp., 68, 200, 479, 520
Denitrification
  autotrophic, 427
  drinking water, 431
  wastewater, 168, 427
Diesel oil, 88
Dimethyl disulfide (Me_2S_2), 282
Dimethyl sulfide (DMS, Me_2S), 35, 282, 361
Dimethylsulphoniopropionate (DMSP), 37, 357
Dissimilatory
  sulfate reduction, 48, 197, 393
  metal reduction, 406
Disulfide, 12
Disulfites, 17
Dithionates, 19
Dithionites, 19

E
Electronic nose, 310
Ettringite, 529

F
Fertiliser, 362
Filamentous microorganisms, 452
Flavocytochrome, 58
Floc
  activated sludge, 449
  anaerobic, 482
  formation, 450

G
Gas
  Claus tail, 245
  flue gas, 240, 276
  impurities, 240
  liquified natural (LNG), 241
  liquified petroleum (LPG), 241
  natural, 238
  refinery, 238, 269
  Syngas, 239
  waste gas, 281
Gibbs free energy, 39, 156, 394, 519
Glucosinolate, 356
Granulation, 155, 457, 482
Gypsum, 34, 240, 333, 362, 528
Gypsiferous soil, 367

H
Heavy metals, 167, 188, 335, 393
Hot potassium carbonate process, 252
Hydrogen sulfide, H_2S, 282
  abatement, 159, 485, 525
  corrosion, 147, 518
  dissociation, 10, 145, 468
  effects on treatment systems, 155, 457, 467
  free H_2S, 10, 145
  health aspects, 146, 309
  oxidation, 40, 51, 107, 267, 420, 505
  removal from gases, 246, 266, 290
  removal from wastewater, 159, 485
  prediction, 137, 524
  toxicity, 468, 476
Hyphomicrobium sp., 294
Hydrodesulfurisation (HDS), 87
  catalyst, 95
  deep HDS, 97
  processes, 89
Hybrid reactor, 484

I
Inhibition
  calcium, 471
  heavy metals, 397
  IC_{50} values, 473
  sodium, 471
  sulfide, 468, 476, 522
  sulfite, 470
  tests, 472
Inoculation, 291, 296, 386
Iron, 242, 398, 495
  minerals, 41
  oxidation, 107, 179, 332, 493, 507
  reduction, 184, 256, 493

J
Jarozite, 116, 334

L
Lead, 34, 395, 495
Liming, 184, 344
Limestone, 297, 430
Linear alkylbenzene sulfonates (LAS), 378
Leptospirillum sp., 107, 177
LO-Cat process, 244, 256, 267

M
Maërl, 291, 427
Marcasite, 506
Membrane
  bioreactor, 287
  extraction, 159
  separation, 259
Mercaptans, 282, 294, 307
Metal enriched soil, 337, 368
Metal
  hydroxide, 395

removal, 183, 342, 401
sulfide, 10, 332, 395, 495
Methanethiol (MeSH), 282
Methanogenic bacteria
competition, 154, 164, 483, 521
organisms, 480, 521
stoichiometry, 156
specific activity, 472
Microaerobic treatment systems, 167
Microbiologically influenced corrosion (MIC), 491, 503, 516
Microthrix sp., 450
Mixed waste gases, 298
Modelling
biofilm, 212
odour, 325
odour dispersion, 319
reactor, 208
sulfide formation, 139, 524
sulfide toxicity, 482
Moisture content, 287
Molybdate, 96, 159, 394, 406

N
Neural network, 223
Nickel, 96, 159, 242, 397, 495
Nitrate, 168, 417
Nitrobacter sp., 416
Nitrogen cycle, 417
Nitrosomonas sp., 416

O
Ochre, 179, 338
Odour, 146, 282, 305
concentration, 308
emission, 144, 314
emission capacity, 317
map, 318
measurement, 306
potential, 313
Oil, 87
Olfactometry, 308

P
Packed bed reactor, 89, 427, 478
PAPS, 24, 48, 75, 356
Paracoccus versutus, 60
Phytoremediation, 182, 366
Polysulfanes
organic, 25
inorganic, 13
Polysulfides, 12, 51
Polysulfido complexes, 14
Polythionates, 20
Pyrite
formation, 10, 39
degree of pyritization (DOP), 42
solubilisation, 108, 176, 506

R
Rectisol process, 254
Redox, 2, 34, 168, 180, 395, 493
Reed beds, 186
Rhodanese, 55

S
Scrubbing
wet, 260, 276, 284
spray dry, 262
Sebkhas, 38
Sediment, 38, 337
Selector, 460
Selenium, 394, 407
Selexol process, 254, 267
Sensors, 213, 310
Sewage, 131, 305, 519
Sewage sludge, 339
Sewer
corrosion, 131, 515
gravity, 132, 137
pressure mains, 132, 135
Shewanella putrefaciens, 407
Silver, 494, 502
Steel
carbon, 494
stainless, 494
Stretford process, 244, 267
Solonchaks, 367
Solvent, 246, 288
Sphalerite, 34, 108, 506
Stability diagrams
sulfur, 3, 34
iron–sulfur, 497
copper–sulfur, 499
Sulfate, 23
calcium sulfate, *see* Gypsum
inorganic sulfate rich wastewater, 182, 277,
lead sulfate, 34, 333
organic sulfate rich wastewater, 153
removal, 161, 192, 276
Sulfate reduction
acidophilic, 196
assimilatory, 49, 74, 356
competition, 154, 164, 521
detection, 502, 522
electron donor, 156, 163, 196, 277
organisms, 68, 520

Sulfate reduction (*cont.*)
  stoichiometry, 156, 330, 519
  pathways, 51
  psychrophilic, 199
SulFerox process, 244, 256, 267
Sulfide, *see* Hydrogen sulfide
Sulfide minerals
  biological oxidation, 115, 176, 331, 507
  chemical oxidation, 15, 112, 506
  direct oxidation, 109, 177
  indirect oxidation, 110, 177
Sulfite, 16, 51
  disproportionation, 23, 68
  reductase, 58, 70, 423
Sulfolobus sp., 52, 107
Sulfonated xenobiotics, 378
Sulfonic acids
  aminobenzene, 383
  aminonaphthalene, 385
Sulfonil process, 255
Sulfur (atom, general)
  allotropes, 4
  catenation, 3
  assimilation pathways, 74, 357
  deficiency, 362
  metabolism, 47, 357
  oxidation state, 2, 33
  reuse, 349
  speciation, 3, 34
Sulfur (elemental), 50, 268, 507
  corrosion, 491
  disproportionation, 23, 508
  gaseous, 7
  liquid, 6
Sulfur bacteria
  colourless, 52, 268, 293
  green, 52
  purple, 52
Sulfur cycle
  atmospheric, 35
  biofilm, 168, 409, 493
  sedimentary, 38
  total, 35, 48
Sulfur dioxide
  emission, 16, 36, 265, 363
  effects, 36, 265, 363
  removal, 260, 273
Sulfur isotopes
  characteristics, 43
  fractionation, 43
Sulfur oxidation pathways
  branched thiosulfate pathway, 54
  multi-enzyme-complex pathway, 60
  tetrathionate pathway, 62
Sulfur sol
  LaMer, 21
  Raffo, 21
  Selmi, 21
  Weimarn, 9
Sulfuretum, 48, 168
Sulfuric acid, 23
  biogenic attack, 516
  production, 185, 331, 528
Sulfurous acid, 16
Surfactants, 344, 377
Syntrophic bacteria, 155, 477, 521

T
Tetrathionate, 20, 51
  hydrolase, 63
  pathway, 62
  reductase, 73
Thiobacillus sp., 52, 291, 531
  Tb. concretivorus, 505
  Tb. denitrificans, 420
  Tb. ferrooxidans, 107, 177
  Tb. versutus, 60, 424
Thiophene, 88, 176
Thiomargarita namibiensis, 424
Thiomicrospira sp., 52, 424
Thionic fluvisol, 368
Thioploca sp., 52
Thiosphaera sp., 52, 424
Thiospira sp., 52
Thiosulfate, 50, 429, 492
  reductase, 56, 71
  disproportionation, 24, 68
Thiothrix sp., 52, 453
Thiovulum sp., 52
Threshold odor number (TON), 308
TOMES, 60
Toxicity tests, 472
Two-phase bioreactor, 288
Type 021N filaments, 455

U
Uranium, 394, 407

V
Volatile organic compounds (VOC), 310
Volatile organic sulfur compounds (VOSC)
  emission, 35, 355
  properties, 282
  removal, 294
Volatile sulfur compounds (VSC)
  emission, 35, 355
  properties, 282
  removal, 290

W
Wellman–Lord process, 263
Wetlands, 185, 347
Wollinella succinogenes, 73

X
Xenobiotics, 167, 377

Z
Zeolite, 257
Zinc, 159, 194, 243, 397, 503

IWA Publishing's authorised EU representative for General Product Safety Regulations is Diane D'Arras, 15 rue Duret, 75116 Paris, France, e-mail: safety@iwap.co.uk.

Printed and bound by CPI Group (UK) Ltd, Croydon, CR0 4YY

12/05/2026

02108863-0006